Secretion and Its Control

Society of General Physiologists Series • Volume 44

Secretion and Its Control

Society of General Physiologists • 42nd Annual Symposium

Edited by
Gerry S. Oxford
University of North Carolina
at Chapel Hill
and
Clay M. Armstrong
University of Pennsylvania

Marine Biological Laboratory
Woods Hole, Massachusetts

7–10 September 1988

New York

Library of Congress Catalog Card Number 89-61702
ISBN 0-87470-045-0
Printed in the United States of America

Contents

Cell Biology of Secretion

Calcium and Calcium Channels in Secretion

Exocytosis

Preface

Secretion is one of the mechanisms used by cells for the control of their composition, and to introduce materials into their environment. It is a part of the larger function of selective transport of materials across membranes. Nutrients and essential amino acids must be admitted to the cell, and the breakdown products must be excreted. Also, salt content of the cell is carefully regulated through secretion, in part for purposes of cell volume control.

In many of these cases, the transported material is dealt with molecule-by-molecule, e.g., by the Na-dependent glucose transporter that selectively binds glucose molecules and transports them inside cells one by one. To be contrasted with this is vesicular transport, in which a vesicle buds off or fuses with the surface membrane, thus mediating the simultaneous transport of thousands of molecules.

Endocytosis is the process of inward vesicular transport. In this process, vesicles bud off the inner surface of the cell membrane, carrying extracellular material to organelles in the cell interior. The transported species are selected to some degree by adsorption to specific receptors that are elaborated by the cell and placed in the membrane at the sites of endocytosis.

The reverse process, exocytosis, involves the fusion of vesicles elaborated in the cell interior with the surface membrane, and the expulsion of the vesicle contents to the outside. An original use of this mechanism may have been simply for metabolic house-cleaning, e.g., the expulsion of excess water by way of vacuoles in some plant cells. On the other hand, molds secrete enzymes into their environment to perform digestion externally, and then they incorporate the digested material.

In multicellular organisms the uses of exocytosis are highly varied, as are the materials transported. Some of our cells secrete digestive enzymes, although, being less fastidious than molds, our digestion is done internally. Other cells secrete, for example, immunoglobins, mucus, or tears. Of particular fascination to the physiologist are the cells that secrete hormones and neurohumors in order to modify the behavior of other cells and thus to coordinate the parts of a complex organism.

At the cellular level the role of exocytosis in altering the composition of the surface membrane is now clearly recognized, and it is addressed by several papers in this volume. The papers deal with the elaboration of proteins in the cell interior and the routing to specific parts of the cell surface. Alteration of membrane composition is an important mechanism for adapting a cell to changes in its environment.

To speculate further on its possible roles, we find it surprising that exocytosis is seldom discussed in theories of memory. The potential for a cell to modify the receptor content of its membrane, and thus its behavior and responsiveness, as a consequence of transmitters secreted by other cells, is perhaps too obvious to mention. When it is remembered that transmitters increase the calcium permeability of postsynaptic cells, and that calcium is a trigger par excellence for exocytosis, the case seems irresistible.

This volume contains many approaches to the subject of exocytosis, and its near-neighbor, endocytosis. Our intent was to be eclectic with regard to method, but uniform in selecting contributors of high merit. We hope that we have succeeded, and that you will enjoy this volume as much as we enjoyed organizing and attending the symposium.

Our thanks go to all the participants, and in particular to those who contributed manuscripts in a timely fashion. We also thank the Society of General Physiologists, which is certainly the most interesting society addressing cellular function that we have encountered, for sponsoring the symposium. We would like also to express our appreciation to the National Institutes of Health, the National Science Foundation, the Office of Naval Research, Glaxo Research Laboratories, Pfizer, Inc., Wyeth-Ayerst Laboratories, Eli Lilly and Company, and Carl Zeiss, Inc. for their generous financial support of this symposium, without which it would certainly not have reached our goals and expectations. Finally, the patience and good humor of Ms. Jane Leighton in organizing the symposium, and the patience and diligence of Ms. Susan Lupack in copyediting this volume deserve special thanks.

Gerry S. Oxford
Clay M. Armstrong

Cell Surface Receptors and Second Messengers

Chapter 1

Role of Inositol Lipid Second Messengers in Regulation of Secretion: Studies of Thyrotropin-releasing Hormone Action in Pituitary Cells

Marvin C. Gershengorn

Division of Endocrinology and Metabolism, Department of Medicine, Cornell University Medical College and The New York Hospital, New York, New York

Introduction

Diverse types of extracellular signal molecules are stimulators of secretion. Many of these secretagogues initiate their action by interacting with receptors in the plasma membrane of their target cells. Hence, mechanisms to transduce the signal at the cell surface into enhanced secretion must be present. Although a number of biochemical pathways appear capable of mediating stimulated secretion, a central role for Ca^{2+} in coupling stimulus to secretion was recognized many years ago (for review, see Douglas, 1978). More recently, an important regulatory link between phosphatidylinositol (PI) turnover and cellular Ca^{2+} homeostasis was appreciated, and stimulation of PI metabolism was proposed as a mechanism of signal transduction (Michell, 1975). Since 1981, two products of the metabolism of the PI family of lipids (or inositol lipids or phosphoinositides), inositol 1,4,5-trisphosphate (I-1,4,5-P_3) (for review, see Berridge, 1987) and 1,2-diacylglycerol (1,2-DG) (for review, see Nishizuka, 1986), have been recognized as second messenger molecules that interact with Ca^{2+} and play important roles in the mechanism of stimulated secretion in many cell types.

My laboratory has been engaged in studies to define the molecular mechanism that serves as a signal transducing pathway for thyrotropin-releasing hormone (TRH) stimulation of prolactin (PRL) secretion from pituitary cells (for review, see Gershengorn, 1986). We have used cloned rat mammotropic tumor (GH_3) cells as our model system. This culture system was chosen because it represents a homogeneous population of cells in which biochemical changes induced during TRH stimulation can be presumed to be occurring in all the cells. Herein, I present the findings, primarily from my own laboratory, that form the basis for a detailed working hypothesis to explain signal transduction by TRH.

TRH and PRL Secretion

TRH was first shown to stimulate PRL secretion from GH_3 cells in 1971 (Tashjian et al., 1971) and then, two years later, from cells isolated from normal pituitary glands (Vale et al., 1973). Hence, although it was well established that TRH stimulates PRL secretion, the kinetics of the response were not appreciated. It was shown (Kolesnick and Gershengorn, 1985*b*, 1986) that the effect of TRH is biphasic with a rapid first phase ("burst") of secretion at a higher rate that lasts for ~2 min followed by a second ("sustained") phase at a lower rate. Fig. 1 (top, left) shows the effect of TRH on the accumulation of PRL in the medium perfusing GH_3 cells. In this figure, the basal rate of secretion, calculated from the slope of the line drawn by linear regression analysis from replicate perfusions in three experiments, has been normalized to 1.0 "ng-equiv" PRL/min. Without an apparent latency, TRH caused a burst of PRL secretion at a rate 3.8-fold that of basal. This first phase was transient. After 3 min, the stimulated rate of secretion had decreased to 2.0-fold basal. Based upon biochemical evidence that TRH stimulates inositiol lipid metabolism in GH_3 cells (see below), two pathways, one initiated by an elevation of the concentration of free (or ionized) Ca^{2+} in the cytoplasm ($[Ca^{2+}]_i$) and the other by 1,2-DG activation of protein kinase C (PKC), seem likely to act in concert to mediate stimulated secretion. Pharmacological agents were, therefore, used to stimulate specifically one or the other of these pathways. Fig. 1 (top, right) illustrates that depolarization of GH_3 cells with high extracellular K^+, which rapidly elevates $[Ca^{2+}]_i$ (Gershengorn et al., 1984; Gershengorn and Thaw, 1985) causes a burst of PRL secretion, at a rate 4.3-fold higher than basal, but does not cause sustained secretion. The rate of PRL secretion after 3 min of perfusion with medium containing 25 mM KCl was not different from basal. We have obtained similar results when $[Ca^{2+}]_i$ was elevated with calcium ionophores and with the Ca^{2+} channel agonist, Bay K8644. Phorbol 12-myristate 13-acetate (PMA, 12-*O*-tetradecanoylphorbol 13-acetate), which activates PKC in GH_3 cells (Sobel and Tashjian, 1983; Drust and Martin, 1984), causes a sustained secretory response but does not stimulate a burst of secretion (Fig. 1, bottom left). Similar stimulation of sustained secretion, but at a lower rate, in the absence of a "burst" was found with synethetic 1,2-diglycerides. Most importantly, when cells were stimulated by high K^+ and PMA simultaneously (Fig. 1, bottom right) a pattern of stimulated PRL secretion very similar to that caused by TRH was observed. K^+ plus PMA stimulated a rapid increase in secretion to

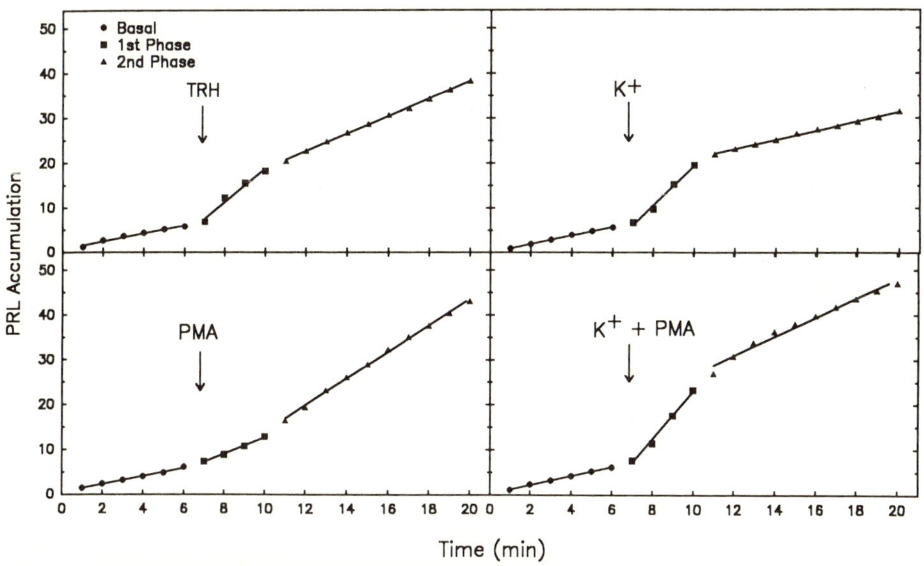

Figure 1. Effects of TRH, K^+ depolarization, PMA, or K^+ depolarization plus PMA on PRL secretion. GH_3 cells were adsorbed onto polyacrylamide beads, placed in a column, and perfused with basal medium. Perfusate was collected at 1-min intervals and PRL was measured by radioimmunoassay. After 7 min, 1 μM TRH (*top left*), 25 mM KCl (*top right*), 0.1 μM PMA (*bottom left*), or 25 mM KCl plus 0.1 μM PMA was added to the perfusion medium for 2 min. The data are presented as PRL accumulation.

5.2-fold the basal rate that was followed after 3 min by a second phase of secretion at a rate 2.2-fold basal. From these data, we suggested that the burst phase of secretion stimulated by TRH may be mediated by an elevation of $[Ca^{2+}]_i$ and the sustained phase may be mediated by 1,2-DG activation of PKC.

We (Gershengorn and Thaw, 1985) showed that TRH stimulates a biphasic elevation of $[Ca^{2+}]_i$ in GH_3 cells (see below). More direct evidence that the rapid elevation of $[Ca^{2+}]_i$ is the mediator of the burst phase of PRL secretion caused by TRH has been obtained. We predicted that if $[Ca^{2+}]_i$ was prevented from increasing during TRH action then the first phase of secretion would be abolished. We took advantage of our previous finding that pretreatment of GH_3 cells with arachidonic acid prevented the elevation of $[Ca^{2+}]_i$ caused by TRH (Kolesnick et al., 1984; Kolesnick and

Gershengorn, 1985*a*). Fig. 2 shows that in GH_3 cells pretreated with arachidonic acid the burst phase of PRL secretion is abolished (Kolesnick and Gershengorn, 1985*b*). In parallel experiments we showed that abolition of the burst phase of secretion in arachidonic acid–pretreated cells was not caused by depletion of preformed PRL and that K^+ depolarization, which caused an elevation of $[Ca^{2+}]_i$ in arachidonic acid–treated cells, still caused a burst secretion. In these experiments, TRH stimulation of burst PRL secretion was 7.6 ± 0.8 ng/10^6 cells per min from control cells and was 1.4 ± 0.3 ng/10^6 cells per min from arachidonic acid–pretreated cells but K^+ depolarization still caused a burst increase in secretion to 8.7 ± 1.1 ng/10^6 cells per min. Based on these findings, we propose that an elevation of $[Ca^{2+}]_i$ is necessary and sufficient to cause burst phase secretion.

In contrast to the effects of inhibiting an increase in $[Ca^{2+}]_i$ on burst secretion, the sustained phase of secretion stimulated by TRH was not inhibited even though the secondary elevation of $[Ca^{2+}]_i$ usually caused by enhanced influx of extracellular Ca^{2+}

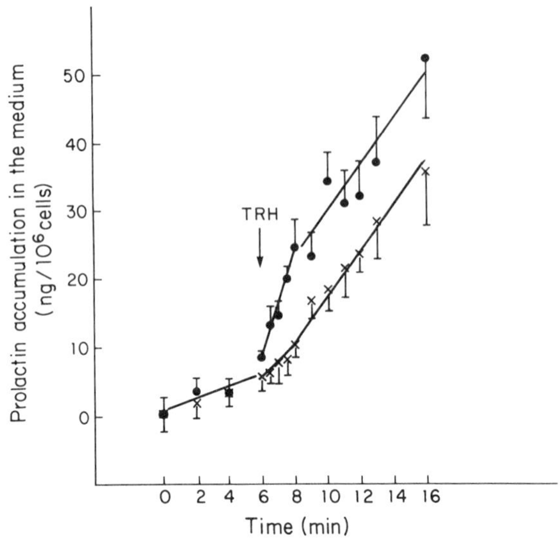

Figure 2. Effect of pretreatment with arachidonic acid on TRH-stimulated PRL secretion. GH_3 cells were incubated in suspension in basal medium. Arachidonic acid (30 μM) was added to one set of incubates. After 15 min, cells were washed and resuspended in fresh medium and 1 μM TRH was added after 6 min. PRL accumulation was assessed by measuring PRL by radioimmunoassay at the times indicated.

was abolished (Fig. 2). In the experiments described in the preceding paragraph, even though the burst phase of secretion was inhibited by 80% there was no difference in PRL secretion rates during the sustained phases stimulated by TRH in control (3.3 ± 0.2 ng/10^6 cells per min) and arachidonic acid–pretreated cells (3.2 ± 0.2 ng/10^6 cells per min). Hence, if PKC is the mediator of the sustained phase of secretion, the level of $[Ca^{2+}]_i$ usually found in resting cells appears to be sufficient for it be activated.

We used another approach to implicate PKC as the mediator of the sustained phase of secretion. We took advantage of the observation that prolonged pretreatment of GH_3 cells with PMA would down-regulate PKC (Ballester and Rosen, 1985). We pretreated GH_3 cells with PMA for 16 h and then measured PMA- and TRH-stimulated PRL secretion (Rodriguez et al., 1987). Prolonged treatment with PMA had only a small effect on basal PRL secretion. As expected, secretion stimulated by PMA was inhibited by 78% in cells pretreated with PMA. Sustained secretion stimulated by TRH was inhibited by 61% in PMA-pretreated cells. These data show that sustained secretion stimulated by TRH is inhibited in cells in which PKC activity is decreased, and they are consistent with the idea that sustained secretion caused by TRH is mediated, at least in part, by activation of PKC.

In summary, it appears that the burst phase of PRL secretion is caused by a rapid elevation of $[Ca^{2+}]_i$ and that the sustained phase of secretion may be mediated, at least in part, by activation of PKC.

General Model of Stimulus-Response Coupling

Fig. 3 illustrates a model for signal transduction that appears to adequately delineate the general mechanism of action of a number of extracellular signal molecules. This has been termed the inositol lipid signalling pathway. In this mechanism, stimulus binds to its receptor on the surface of its target cell and, with the apparent involvement of a guanine nucleotide–binding regulatory (G) protein that couples the receptor-stimulus complex to the effector enzyme, rapidly activates a phospholipase C. This phospholipase C specifically hydrolyzes phosphatidylinositol 4,5-bisphosphate (PIP_2) to form the two second messengers, I-1,4,5-P$_3$ and 1,2-DG. I-1,4,5-P$_3$ acts to release

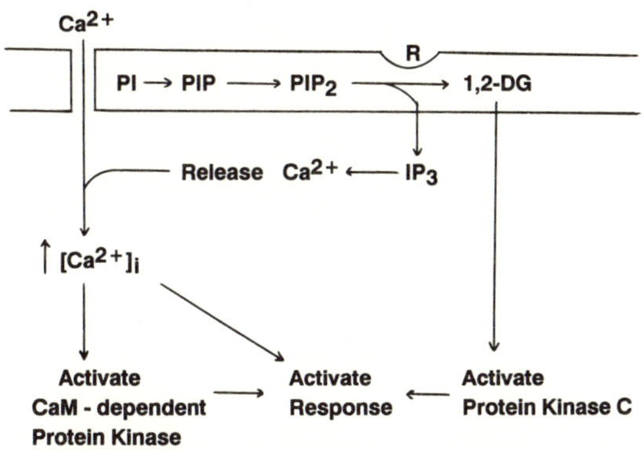

Figure 3. General model of stimulus-response coupling. See text. *R*, receptor.

Ca^{2+} previously sequestered within intracellular stores, perhaps within the endoplasmic reticulum or other specialized structures (but not within mitochondria), and thereby elevates $[Ca^{2+}]_i$. The elevation of $[Ca^{2+}]_i$ is prolonged by the enhancement of the influx of extracellular Ca^{2+}. The molecular mechanism by which Ca^{2+} influx is increased has not been defined in most cells but appears to involve activation of voltage-sensitive Ca^{2+} channels in "excitable" cells and of voltage-insensitive channels (perhaps those regulated by coupling proteins, other second messengers, or directly by receptors) in "nonexcitable" cells. The elevation of $[Ca^{2+}]_i$ leads to activation of Ca^{2+} sensitive proteins, such as calmodulin-dependent protein kinase, which via the phosphorylation of other proteins causes stimulation of the cellular response, or of regulatory proteins that directly stimulate the response. In parallel with the I-1,4,5-P$_3$-Ca^{2+} limb of this pathway, 1,2-DG causes activation of PKC that phosphorylates regulatory proteins and stimulates the cellular response. The two bifurcating limbs of this signal transduction pathway can act in an additive, synergistic or coordinate manner to stimulate the final cellular response. In the following sections, I will describe the special features of this signal transduction pathway as they apply to TRH stimulation of PRL section.

Mediators of the First Phase of Secretion

TRH binding to its receptor within the plasma membrane of GH_3 cells triggers the hydrolysis of PIP_2 to I-1,4,5-P_3 and 1,2-DG. This idea is based primarily on data derived from experiments performed in cells prelabeled with *myo*-[^{3}H]inositol to isotopic equilibrium in order to measure changes in the cellular content of even minor substances, such as PIP_2 and I-1,4,5-P_3, more readily. It is important to note that PIP_2 is a minor member of the inositol lipid family in GH_3 cells (comprising only 2.5%, whereas PI makes up 88%, lysoPI 6.2%, and PI 4-monophosphate (PIP) 2.8%) and that the inositol lipids comprise only ~10% total lipids within these cells. Fig. 4 shows that TRH causes a rapid (within seconds) decrease in the content of PIP_2 and a concomitant increase in I-1,4,5-P_3. Because I-1,4,5-P_3 can only be formed in mamma-

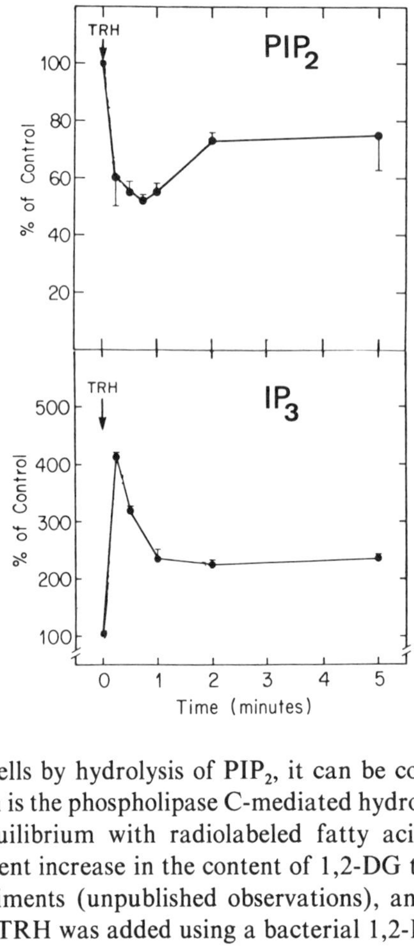

Figure 4. Effects of TRH on PIP_2 and I-1,4,5,-P_3. GH_3 cells, which had been labeled to isotopic steady-state with *myo*-[^{3}H]inositol, were suspended in basal medium and stimulated with 1 μM TRH at the time indicated. PIP_2 and I-1,4,5,-P_3 were measured by the methods of Rebecchi and Gershengorn (1983).

lian cells by hydrolysis of PIP_2, it can be concluded that a proximate event in TRH action is the phospholipase C-mediated hydrolysis of PIP_2. As expected, in cells labeled to equilibrium with radiolabeled fatty acids, we showed that TRH stimulates a transient increase in the content of 1,2-DG that peaks between 15 and 30 s; in recent experiments (unpublished observations), an increase in 1,2-DG content was shown after TRH was added using a bacterial 1,2-DG kinase to measure unlabeled 1,2-DG. Hence, the decrease in PIP_2, and the increments in I-1,4,5-P_3 and 1,2-DG may be accounted for, at least in part (see below), by this mechanism.

The molecular details of the mechanism by which the TRH-receptor complex may be coupled to the phospholipase C enzyme are not known. There is evidence, however, that a guanine nucleotide–binding regulatory protein is involved (Straub and Gershengorn, 1986). In membranes isolated from GH_3 cells, GTP (and its nonhydrolyzable analogues) and TRH stimulate phospholipase C-mediated hydrolysis of PIP_2. More importantly, Fig. 5 illustrates that when TRH and GTP are added together there is a synergistic activation of the phospholipase C.

It was proposed that I-1,4,5-P_3 could function as an intracellular messenger to mobilize Ca^{2+} from a pool(s) within cells. To determine whether I-1,4,5-P_3 could serve as a mediator to release previously sequestered Ca^{2+} from GH_3 cell stores, we developed a preparation of saponin-permeabilized cells to allow I-1,4,5-P_3 access to intracellular Ca^{2+} pools (Gershengorn and Thaw, 1983). Permeabilized cells sequestered Ca^{2+} in an ATP-dependent manner into two functionally distinct stores. These were tentatively identified as mitochondrial and nonmitochondrial pools based on their sensitivity to agents that release Ca^{2+} from mitochondria. The nonmitochondrial pool had a high affinity for Ca^{2+} and could be saturated when permeabilized cells were incubated in buffer with as little as 200 nM $[Ca^{2+}]_{free}$. By contrast, the mitochondrial pool exhibited a lower affinity for Ca^{2+} but a greater capacity for accumulation that

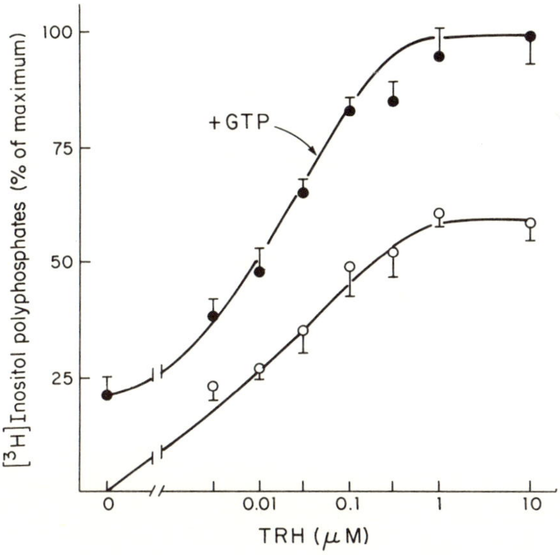

Figure 5. The effect of TRH and GTP in activating phospholipase C in membranes. Membranes prepared from GH_3 cells that had been prelabeled with *myo*-[^{3}H]inositol were suspended in fresh "cytosol-like" buffer in the presence of the indicated concentrations of TRH without (○) or with 100 μM GTP (●). After 5 min at 37°C, inositol polyphosphates were quantified by anion-exchange chromatography. The data are presented as the percentage of maximal activation. Reproduced with permission from Straub and Gershengorn (1986).

was not fully saturated even at $[Ca^{2+}]_{free}$ up to 1,000 nM. When permeabilized cells were incubated in buffer with a $[Ca^{2+}]_{free}$ initially between 200 and 1,000 nM, $[Ca^{2+}]_{free}$ was buffered to 130 nM, a level identical with that maintained in intact cells (see below). Fig. 6 shows that in permeabilized cells, which had accumulated $^{45}Ca^{2+}$ to a steady-state level, I-1,4,5-P_3 caused a very rapid loss of $^{45}Ca^{2+}$ followed by a slow reuptake. This effect was observed even when mitochondrial Ca^{2+} accumulation was blocked. I-1,4,5-P_3 induced release of Ca^{2+} from a nonmitochondrial pool within permeabilized cells was concentration-dependent; the half-maximal effect was at 1 μM. The effect of I-1,4,5-P_3 was specific as there was no effect of inositol bisphosphate (IP_2), inositol monophosphate (IP_1), and inositol at similar or even higher concentrations. These data are compatible with the hypothesis that I-1,4,5-P_3 could mobilize Ca^{2+} from an intracellular pool to rapidly elevate $[Ca^{2+}]_i$ during TRH action.

Fig. 7 illustrates the effects of TRH on $[Ca^{2+}]_i$ in GH_3 cells (Gershengorn and Thaw, 1985). TRH stimulates an immediate severalfold elevation of $[Ca^{2+}]_i$, followed

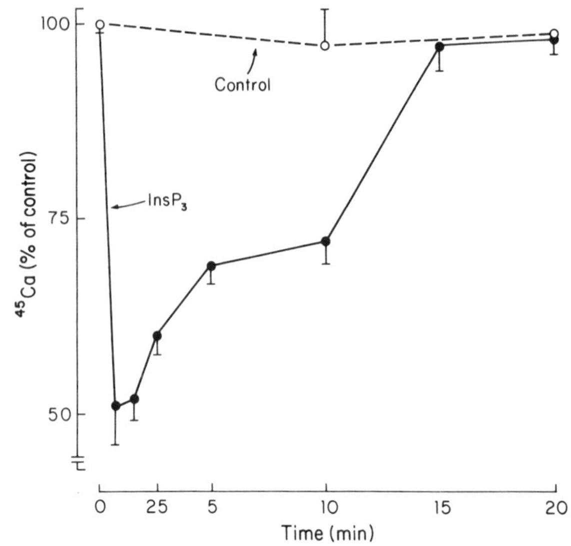

Figure 6. Effect of I-1,4,5,-P$_3$ on ^{45}Ca accumulated by saponin-permeabilized GH$_3$ cells. Permeabilized cells were loaded with ^{45}Ca in the presence of ATP. After 15 min, 10 μM I-1,4,5,-P$_3$ or vehicle (control) was added and the permeabilized cell-associated ^{45}Ca was measured.

by a more prolonged secondary increase, i.e., a biphasic elevation of $[Ca^{2+}]_i$. Basal $[Ca^{2+}]_i$ was 120 ± 18 nM and TRH stimulated a rapid increase in $[Ca^{2+}]_i$ that reached a peak of 520 ± 29 nM at <10 s, followed by a decline over 1.5 min. This was followed by a sustained elevation of $[Ca^{2+}]_i$ to 260 ± 14 nM that lasted for at least 12 min ("second phase") (see below). The first phase elevation of $[Ca^{2+}]_i$ appears to be caused in large part, if not completely, by the mobilization of cellular Ca^{2+}. This idea is based primarily on two observations. We showed that in cells incubated in medium depleted of Ca^{2+} or in the presence of EGTA, the first phase elevation was not affected (Gershengorn and Thaw, 1985). Other investigators have shown that the first phase elevation was abolished by pretreating cells with a low dose of a Ca^{2+} ionophore to deplete intracellular Ca^{2+} stores (Albert and Tashjian, 1984). Furthermore, because the first phase elevation of $[Ca^{2+}]_i$ was not affected in cells in which the mitochondrial pool was depleted, it appears that the TRH-responsive pool was nonmitochondrial, perhaps within the endoplasmic reticulum or other specialized structures (Gershengorn and Thaw, 1983).

In summary, these data are consistent with the idea that TRH rapidly activates a phospholipase C enzyme that causes the hydrolysis of PIP$_2$ which leads to the

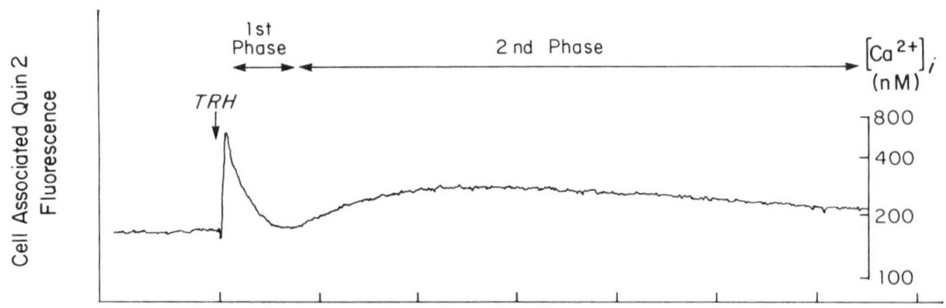

Figure 7. Effect of TRH on $[Ca^{2+}]_i$. GH$_3$ cells were loaded with quin 2, a fluorescent Ca^{2+} probe, and the $[Ca^{2+}]_i$ was measured and calculated as described by Gershengorn and Thaw (1983).

generation of I-1,4,5-P_3. I-1,4,5-P_3 causes the release of Ca^{2+} from an intracellular pool and thereby elevates $[Ca^{2+}]_i$. Ca^{2+}, directly or indirectly (e.g., by activating a calmodulin-dependent protein kinase) then activates exocytosis which leads to a burst of PRL secretion.

Mediators of the Second Phase of Secretion

In the general model of signalling via inositol lipids depicted in Fig. 3, the only substrate for the phospholipase C enzyme is PIP_2. It was proposed, therefore, that concomitant with the hydrolysis of PIP_2 there is enhanced flux of PI to PIP and then to PIP_2 that provides additional precursor for the continued generation of I-1,4,5-P_3 and 1,2-DG. It was suggested that this enhanced conversion of PI to PIP_2 could account for the decrease in PI observed during stimulation (Rebecchi et al., 1983). Because PIP_2 is formed from PI by stepwise phosphorylation catalyzed by distinct kinases first at the 4-position to form PIP, and then at the 5-position, it was suggested that these kinases were activated during cell stimulation. In GH_3 cells (Rebecchi and Gershengorn, 1983), as in most cell types, however, I-1,4,5-P_3 is increased only transiently during stimulation, while 1,2-DG appears to be increased more persistently. The transient nature of the increase in I-1,4,5-P_3 has been ascribed to its very rapid metabolism. However, it was possible that the transient nature of the increase in I-1,4,5-P_3 may be secondary to its transient formation and that the prolonged hydrolysis is of PI rather than PIP_2.

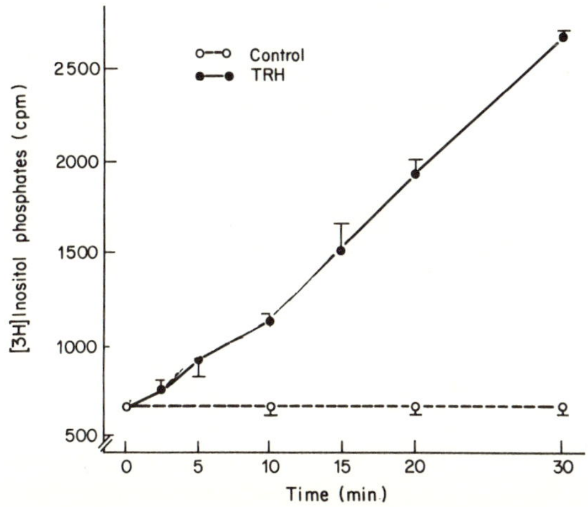

Figure 8. Effect of TRH on the formation of IP_1. GH_3 cells that had been prelabeled with *myo*-[^{3}H]inositol were suspended in a basal medium containing 10 mM LiCl to prevent dephosphorylation of IP_1; 1 μM TRH was added at the time indicated. IP_1 was measured by anion-exchange chromatography at the indicated times. The amount of IP_1 produced after 30 min was equal to approximately 20 times the PIP_2 level at the start of the incubation.

Fig. 8 illustrates that during TRH stimulation of GH_3 cells there is continuous generation of IP_1. IP_1 could have been formed from the sequential dephosphorylation of I-1,4,5-P_3 to IP_2 and then to IP_1 if PIP_2 were continuously hydrolyzed during TRH action, or it could have been formed directly from hydrolysis of PI if after the initial hydrolysis of PIP_2 the substrate for phospholipase C-mediated hydrolysis changed to PI. With either mechanism, there would be a continued generation of 1,2-DG. Although it has not been possible to show a sustained production of 1,2-DG because of its rapid metabolism, it can be inferred from the continuous formation of phosphatidic acid (PA), the phosphorylation product of 1,2-DG (Rebecchi et al., 1981). If PIP_2 turnover were continuously enhanced during activation of the inositol lipid signal transduction mechanism, it should be possible to show that PIP_2 was persistently resynthesized during cell stimulation. This is so because the content of PIP_2 is insufficient to account for the IP_1 formed during prolonged stimulation. We attempted to demonstrate this in GH_3 cells by measuring the incorporation of inorganic [^{32}P]phosphate ($^{32}P_i$) into inositol lipids and PA under conditions in which the specific radioactivity in the lipids was continually increasing (Imai and Gershengorn, 1986). Under these conditions, small increases in the rate of PIP_2 resynthesis would be magnified. In cells incubated with $^{32}P_i$ for 1 min, TRH rapidly and persistently (for at least 30 min) enhanced the rate of ^{32}P labeling of PA. After a lag of 1 min, TRH markedly and persistently increased the ^{32}P labeling of PI also. In contrast, TRH caused only a transient increase in ^{32}P labeling of PIP_2 that lasted <2 min. There was no rapid effect of TRH on the ^{32}P labeling of PIP. The temporal dissociation of PIP_2 and PI turnover was also shown in experiments in which cells were first stimulated by

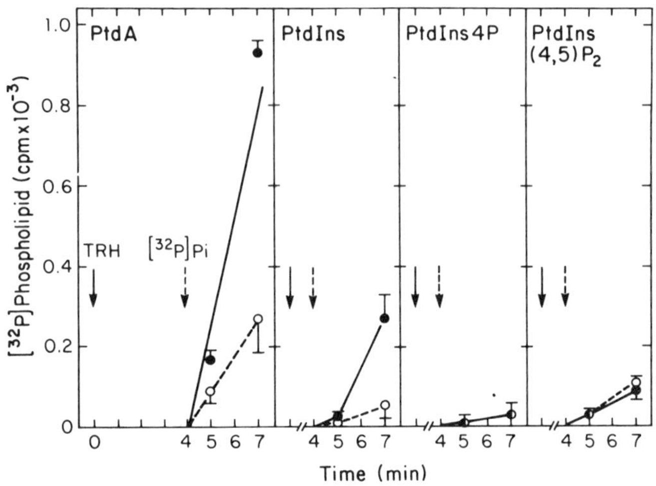

Figure 9. Effect of prestimulating cells with TRH and then adding $^{32}P_i$. GH_3 cells were preincubated for 4 min with 1 μM TRH, and then $^{32}P_i$ was added. The cells were collected at the times indicated and the levels of PA (PtdOH), PI (PtdIns), PIP (PtdIns4P), and PIP_2 (PtdIns(4,5)P_2O) were measured by thin-layer chromatography. Reproduced with permission from Imai and Gershengorn (1986).

TRH and then incubated with $^{32}P_i$. Fig. 9 shows that in cells exposed to TRH for 4 min and then to $^{32}P_i$, the rate of ^{32}P labeling of PA and PI was increased, but that of PIP_2 was not affected. Lastly, persistent turnover of PA and PI did not require an initial hydrolysis of PIP_2 because turnover of PA and PI could be terminated by the displacement of TRH from its receptor, and restarted by reoccupying the receptors with TRH when PIP_2 turnover had ceased (Gershengorn and Paul, 1986; Imai and Gershengorn, 1986). These data demonstrated that turnover of PIP_2 was stimulated only transiently whereas turnover of PI and PA was stimulated persistently by TRH in GH_3 cells. Hence, it may be concluded that inositol lipid turnover in GH_3 cells does not occur via continued hydrolysis of PIP_2 accompanied by enhanced flux of PI to PIP to PIP_2, but that there is direct and persistent hydrolysis of PI. Direct hydrolysis of PI may be supported by the sustained elevation of $[Ca^{2+}]_i$ (Vallar et al., 1988).

The sustained elevation of $[Ca^{2+}]_i$ is caused by a different mechanism than is the rapid elevation. This was likely considering the prolonged duration of this elevation, as intracellular pools might have become depleted with continued release of Ca^{2+}, and that I-1,4,5-P_3 was not being generated persistently to release intracellular Ca^{2+}. We showed that the second phase elevation was caused by enhanced influx of extracellular Ca^{2+}. This was based primarily on two observations. First, the second phase elevation

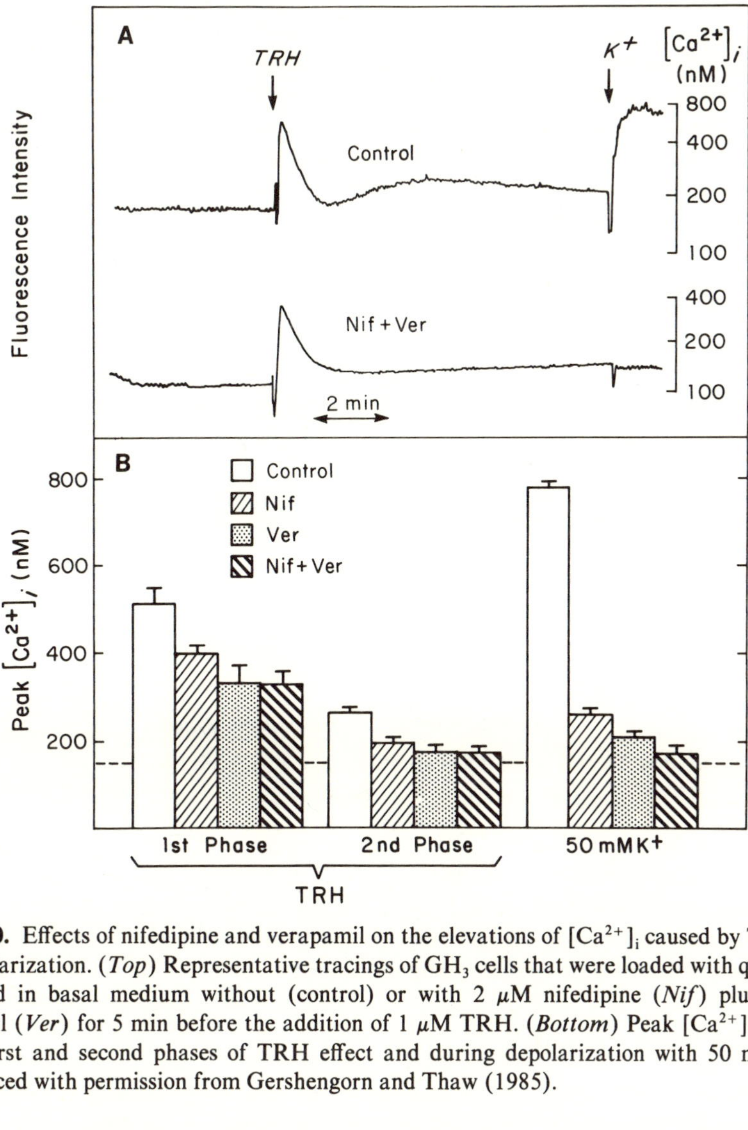

Figure 10. Effects of nifedipine and verapamil on the elevations of $[Ca^{2+}]_i$ caused by TRH and K^+ depolarization. (*Top*) Representative tracings of GH_3 cells that were loaded with quin 2 and incubated in basal medium without (control) or with 2 μM nifedipine (*Nif*) plus 75 μM verapamil (*Ver*) for 5 min before the addition of 1 μM TRH. (*Bottom*) Peak $[Ca^{2+}]_i$ attained during first and second phases of TRH effect and during depolarization with 50 mM KCl. Reproduced with permission from Gershengorn and Thaw (1985).

of $[Ca^{2+}]_i$ was completely abolished in cells incubated in medium containing no added Ca^{2+} or with EGTA (Gershengorn and Thaw, 1985). And second, that in cells exposed to organic Ca^{2+} channel–blocking agents, such as verapamil or nifedipine, the second phase elevation was markedly inhibited (Fig. 10) (Geras et al., 1982; Gershengorn et al., 1984). Studies that complement the findings of these direct measurements of

$[Ca^{2+}]_i$ have been performed. Electrophysiological experiments have shown that TRH stimulates an increase in the frequency of Ca^{2+}-dependent action potentials (compiled in Gershengorn, 1982) that correlates with the second phase elevation of $[Ca^{2+}]_i$. Moreover, recent observations suggest that 1,2-DG activation of PKC may be necessary if not sufficient to activate these voltage-sensitive Ca^{2+} channels (Albert et al., 1987; Dufy et al., 1987). This may be an additional effect of sustained activation of PKC. Hence, the enhanced influx of extracellular Ca^{2+} that maintains the sustained elevation of $[Ca^{2+}]_i$ appears to occur through voltage-sensitive Ca^{2+} channels in GH_3 cells.

We have recently obtained evidence that shows the importance of PI in the sustained phase of elevation of $[Ca^{2+}]_i$ and of PRL secretion stimulated by TRH. By pretreating GH_3 cells with TRH in the presence of LiCl, which inhibits the conversion of IP_1 to inositol and thereby prevents its reutilization, and then incubating them in an inositol-free medium, we were able to deplete the cells of 47% of their PI without affecting the content of PIP_2 (Rodriguez et al., 1987). In PI-depleted cells, we found

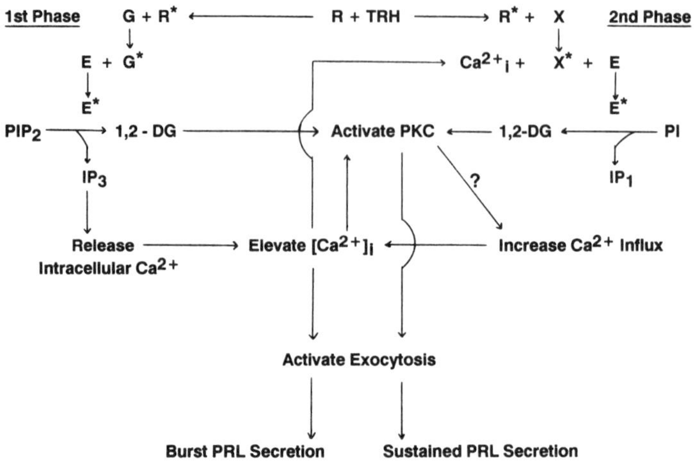

Figure 11. Model of TRH stimulation of PRL secretion. See text.

that although the rapid elevation of $[Ca^{2+}]_i$ and the burst phase of PRL secretion were unaffected, the sustained elevation of $[Ca^{2+}]_i$ was inhibited by 50% and the stimulation of sustained secretion was inhibited by 40%. Hence, these data show that PI is involved in the sustained effects of TRH. However, they do not show that it is the direct hydrolysis of PI that these effects are dependent on.

Model of TRH Stimulation of PRL Secretion

Based on the observations reviewed herein, the following working model can be proposed as the sequence of intracellular events involved in the mechanism of TRH stimulation of the two phases of PRL secretion from GH_3 cells (Fig. 11). The binding of TRH to its plasma membrane receptor initiates both phases of TRH action. In the first phase, the receptor-TRH complex, via a G protein, activates a phospholipase C that specifically causes the hydrolysis of PIP_2 to yield I-1,4,5-P_3 and 1,2-DG. I-1,4,5-P_3, the water-soluble product, diffuses from the plasma membrane to a nonmitochondrial pool of Ca^{2+} (which is, perhaps, the endoplasmic reticulum) and causes the release of Ca^{2+}. The movement of Ca^{2+} from a sequestered pool(s) into the cytoplasm results in the rapid elevation of $[Ca^{2+}]_i$, which couples stimulus to burst secretion. The elevation of $[Ca^{2+}]_i$ may activate exocytosis directly, or through the phosphorylation of proteins involved in the exocytotic process via activation of a Ca^{2+}- and calmodulin-dependent protein kinase(s), or it may be due to both. The role of 1,2-DG, and its activation of PKC during the burst phase of PRL secretion, is unclear. During the second phase, the elevation of $[Ca^{2+}]_i$ is extended by a delayed, but prolonged, TRH-induced enhancement of the influx of extracellular Ca^{2+}. Enhanced Ca^{2+} influx appears to occur via voltage-sensitive Ca^{2+} channels that may be "opened" by activation of PKC in combination with changes in plasma membrane polarization. The sustained elevation of $[Ca^{2+}]_i$ may support the change in the substrate of hydrolysis from PIP_2 to PI. Direct hydrolysis of PI then leads to the sustained generation of 1,2-DG and to the activation of PKC without formation of I-1,4,5-P_3, and, therefore, no signal for mobilization of intracellular Ca^{2+} results. The details of how the receptor-TRH complex activates a phospholipase C that hydrolyzes PI are not known. There is no evidence, however, that this involves a G protein and is, therefore, denoted by an *X* in Fig. 11. 1,2-DG activation of PKC leads to the sustained phase of secretion most likely by phosphorylating proteins that have regulatory functions in the exocytotic process.

References

Albert, P. R., and A. H. Tashjian, Jr. 1984. Relationship of thyrotropin-releasing hormone-induced spike and plateau phases in cytosolic free Ca^{2+} concentrations to hormone secretion. *Journal of Biological Chemistry*. 259:15350–15363.

Albert, P. R., G. Wolfson, and A. H. Tashjian, Jr. 1987. Diacylglycerol increases cytosolic free Ca^{2+} concentration in rat pituitary cells. Relationship to thyrotropin-releasing hormone action. *Journal of Biological Chemistry*. 262:6577–6581.

Ballester, R., and O. M. Rosen. 1985. Fate of immunoprecipitable protein kinase C in GH_3 cells treated with phorbol 12-myristate 13-acetate. *Journal of Biological Chemistry*. 260:15194–15199.

Berridge, M. J. 1987. Inositol trisphosphate and diacylglycerol: two interacting second messengers. *Annual Review of Biochemistry*. 56:159–193.

Douglas, W. W. 1978. Stimulus-secretion coupling: variations on the theme of calcium-activated exocytosis involving cellular and extracellular sources of calcium. *Ciba Foundation Symposia*. 54:61–90.

Drust, D. S., and T. F. J. Martin. 1984. Thyrotropin-releasing hormone rapidly activates protein phosphorylation in GH_3 pituitary cells by a lipid-linked, protein kinase C-mediated pathway. *Journal of Biological Chemistry*. 259:14520–14530.

Dufy, B., S. Jaken, and J. L. Barker. 1987. Intracellular Ca^{2+}-dependent protein kinase C activation mimics delayed effects of thyrotropin-releasing hormone on clonal pituitary cell excitability. *Endocrinology*. 121:793–802.

Geras, E., M. J. Rebecchi, and M. C. Gershengorn. 1982. Evidence that stimulation of thyrotropin and prolactin secretion by thyrotropin-releasing hormone occur via different calcium-mediated mechanisms: studies with verapamil. *Endocrinology*. 110:901–906.

Gershengorn, M. C. 1982. Thyrotropin releasing hormone. A review of the mechanisms of acute stimulation of pituitary hormone release. *Molecular and Cellular Biochemistry*. 45:163–179.

Gershengorn, M. C. 1986. Mechanism of thyrotropin releasing hormone stimulation of pituitary hormone secretion. *Annual Review of Physiology*. 48:515–526.

Gershengorn, M. C., E. Geras, V. S. Purrello, and M. J. Rebecchi. 1984. Inositol trisphosphate mediates thyrotropin-releasing hormone mobilization of nonmitochondrial calcium in rat mammotropic pituitary cells. *Journal of Biological Chemistry*. 259:10675–10681.

Gershengorn, M. C., and M. E. Paul. 1986. Evidence for tight coupling of receptor occupancy by thyrotropin-releasing hormone to phospholipase C-mediated phosphoinosoitide hydrolysis in rat pituitary cells: use of chlordiazepoxide as a competitive antagonist. *Endocrinology*. 119:833–839.

Gershengorn, M. C., and C. Thaw. 1983. Calcium influx is not required for TRH to elevate free cytoplasmic calcium in GH_3 cells. *Endocrinology*. 113:1522–1524.

Gershengorn, M. C., and C. Thaw. 1985. Thyrotropin-releasing hormone (TRH) stimulates biphasic elevation of cytoplasmic free calcium in GH_3 cells. Further evidence that TRH mobilizes cellular and extracellular Ca^{2+}. *Endocrinology*. 116:591–596.

Imai, A., and M. C. Gershengorn. 1986. Phosphatidylinositol 4,5-bisphosphate turnover is transient while phosphatidylinositol turnover is persistent in thyrotropin-releasing hormone-stimulated rat pituitary cells. *Proceedings of the National Academy of Science*. 83:8540–8544.

Kolesnick, R. N., and M. C. Gershengorn. 1985*a*. Arachidonic acid inhibits thyrotropin-releasing hormone-induced elevation of cytoplasmic free calcium in GH_3 pituitary cells. *Journal of Biological Chemistry*. 260:707–713.

Kolesnick, R. N., and M. C. Gershengorn. 1985*b*. Direct evidence that burst but not sustained secretion of prolactin stimulated by thyrotropin-releasing hormone is dependent on elevation of cytoplasmic calcium. *Journal of Biological Chemistry*. 260:5217–5220.

Kolesnick, R. N., and M. C. Gershengorn. 1986. Thyrotropin-releasing hormone stimulation of prolactin secretion is coordinately but not synergistically regulated by an elevation of cytoplasmic calcium and 1,2-diacylglycerol. *Endocrinology*. 119:2461–2466.

Kolesnick, R. N., I. Musacchio, C. Thaw, and M. C. Gershengorn. 1984. Arachidonic acid mobilizes calcium and stimulates prolactin secretion from GH_3 cells. *American Journal of Physiology*. 246 (*Endocrinology and Metabolism 9*):E458–E462.

Michell, R. H. 1975. Inositol phospholipids and cell surface receptor function. *Biochimica et Biophysica Acta*. 415:81–147.

Nishizuka, Y. 1986. Studies and perspectives of protein kinase C. *Science*. 233:305–312.

Rebecchi, M. J., and M. C. Gershengorn. 1983. Thyroliberin stimulates rapid hydrolysis of phosphatidylinositol 4,5-bisphosphate by a phosphodiesterase in rat mammotropic pituitary cells. *Biochemical Journal*. 216:287–294.

Rebecchi, M. J., R. N. Kolesnick, and M. C. Gershengorn. 1983. Thyrotropin-releasing hormone stimulates rapid loss of phosphatidylinositol and its conversion to 1,2-diacylglycerol and phosphatidic acid in rat mammotropic pituitary cells. *Journal of Biological Chemistry*. 258:227–234.

Rebecchi, M. J., M. E. Monaco, and M. C. Gershengorn. 1981. Thyrotropin releasing hormone rapidly enhances [^{32}P]orthophosphate incorporation into phosphatidic acid in cloned GH_3 cells. *Biochemical and Biophysical Research Communications*. 101:124–130.

Rodriguez, R., A. Imai, and M. C. Gershengorn. 1987. Phosphatidylinositol depletion in GH_3

rat pituitary cells inhibits sustained responses to thyrotropin-releasing hormone. Reversal with *myo*-inositol. *Molecular Endocrinology.* 1:802–807.

Sobel, A., and A. H. Tashjian, Jr. 1983. Distinct patterns of cytoplasmic protein phosphorylation related to regulation of synthesis and release of prolactin by GH cells. *Journal of Biological Chemistry.* 258:10312–10324.

Straub, R. E., and M. C. Gershengorn. 1986. Thyrotropin-releasing hormone and GTP activate inositol trisphosphate formation in membranes isolated from rat pituitary cells. *Journal of Biological Chemistry.* 261:2712–2717.

Tashjian, A. H., Jr., N. J. Barowsky, and D. K. Jensen. 1971. Thyrotropin releasing hormone: direct evidence for stimulation of prolactin production by pituitary cells in culture. *Biochemical and Biophysical Research Communications.* 43:516–523.

Vale, W., R. Blackwell, G. Grant, and R. Guillemin. 1973. TRF and thyroid hormones on prolactin secretion by rat anterior pituitary cells in vitro. *Endocrinology.* 93:26–33.

Vallar, L., L. M. Vicentini, and J. Meldolesi. 1988. Inhibition of inositol phosphate production is a late Ca^{2+}-dependent effect of D_2 dopaminergic receptor activation in rat lactotroph cells. *Journal of Biological Chemistry.* 263:10127–10134.

Chapter 2

Identification of G Protein-gated and G Protein-modulated Ionic Channels. Molecular Basis for G Protein Action

Lutz Birnbaumer, Anthony M. J. VanDongen, Juan Codina, Atsuko Yatani, Rafael Mattera, Rolf Graf, and Arthur M. Brown

Departments of Cell Biology, Molecular Physiology and Biophysics, and Medicine, Baylor College of Medicine, Houston, Texas 77030

Introduction

The description of a receptor-sensitive signal-transducing adenylyl cyclase system by Sutherland and co-workers in 1956 (Rall et al., 1956) was expanded in 1970 by the discovery that the system is regulated not only by hormones but also by GTP (Rodbell et al., 1971). Involvement of a GTP regulatory step in light perception was discovered in 1975 (Wheeler and Bitensky, 1977). The S49 cell mutant with a reduced capacity to synthesize cAMP was described in 1975. This mutant, now designated cyc^-, was found to be affected in those aspects of cAMP formation that are GTP dependent. By 1980, the GTP-binding component involved in light perception (phototransduction) termed transducin and the GTP-binding regulatory component of adenylyl cyclase, originally termed G/F and N and now designated as G_s, had been purified (Northup et al., 1980).

When compared, the GTP-binding proteins involved in phototransduction and in hormone receptor action were found to be quite similar: both were activated by GTP under the influence of receptor, hormone receptor for G_s and rhodopsin for transducin, and both were $\alpha\beta\gamma$ heterotrimers formed of homologous but distinct α subunits, interchangeable β subunits (Manning and Gilman, 1983), and small γ subunits (Fung, 1983; Hildebrandt et al., 1984). The α subunits of both bound and hydrolyzed GTP and the purified trimeric holoproteins dissociate on interaction with nonhydrolyzable GTP analogues to give two products: a free α-G nucleotide complex and a $\beta\gamma$ dimer. Of these, it is the former that regulates the effector functions: adenylyl cyclase in the case of G_s and a cGMP-specific phosphodiesterase in the case of transducin, (Fung et al., 1981), while the latter plays a key regulatory role in what appears to be a shuttling of the α subunit between effector and receptor.

Parallel to those in depth studies on the biochemical and molecular basis of phototransduction and activation of adenylyl cyclase, other studies led to the discovery first that hormonal stimulation as well as hormonal inhibition of adenylyl cyclase proceeded through intervening GTP-dependent steps (Hildebrandt et al., 1983), and also that actions of hormones and neurotransmitters affecting cell metabolism through means other than regulation of cAMP formation, are also dependent on a GTP-binding process. It is now recognized that hormonal regulation of phosphoinositide hydrolysis by a phospholipase of the C type, which releases inositol trisphosphate (IP_3) plus diacylglycerol (DAG) (Cockroft and Gomperts, 1985; Litosch et al., 1985), activation of a phospholipase A_2 with consequential release of arachidonic acid (Burch et al., 1986), and, as is expanded on below, the stimulation of several types of K^+ channels (Yatani et al., 1987*a*, *b*; VanDongen et al., 1988) and two dihydropyridine-sensitive voltage-gated Ca^{2+} channels (Yatani et al., 1987*c*, 1988*a*) all occur with the obligatory participation of G proteins. G proteins that regulate phospholipases and the K^+ channel are termed G_p (Crockroft and Gomperts, 1985) and G_k (Breitweiser and Szabo, 1985), respectively. Very recently it was discovered that smell receptors in cilia of the olfactory neuroepithelium activate adenylyl cyclase through a unique "olfactory" G protein, G_{olf}, the α subunit of which is highly homologous to G_s but encoded in a separate gene expressed exclusively in these cells (Jones and Reed, 1988).

Signal Transduction by G Proteins: Receptors, G Proteins, and Effectors

About 80% of all known hormones and neurotransmitters, as well as many neuromodulators and auto- and paracrine factors that regulate cellular interactions, termed primary messengers, elicit cellular responses by combining to specific receptors that are coupled to effector functions by G proteins (Table I) according to a general scheme shown in Fig. 1 (Birnbaumer and Brown, 1989).

Even though the primary messengers are many, the number of distinct receptors that mediate their action is even larger. In contrast to receptors, the number of final effector functions regulated by these receptors and the number of G proteins that provide for receptor-effector coupling is much lower, probably not much more than 15 each.

At this time we know of 12 G proteins. Several of them have been purified to better than 95% purity, which led to the disclosure of a complex heterotrimeric $\alpha\beta\gamma$ subunit structure with the general properties shown in Fig. 2 *A*.

Based on the amino acid sequence of their α subunits, these G proteins can be grouped into four homology groups that include (Table II, Fig. 2 *B*): (*a*) G_s represented by four isoforms (splice variants) derived from a single gene, (*b*) three G_i and one G_o, all substrates for pertussis toxin (see below) and each encoded in a separate

	Input	Transduction			Output vs. Output
Location	Extracellular milieu		Plasma membrane		Intracellular milieu or plasma membrane
		G protein-coupled receptor		G protein-regulated effector	
Functional elements	Primary messenger →	R →	G Protein →	E →	Secondary messenger; Membrane potential
Molecular diversity					
No. known	~40	80	12	6	cAMP, IPs, DAG, Ca^{2+} cGMP, AA
No. estimated	?	~100?	Up to 15?	~12?	

Figure 1. Flow of information through G protein-dependent signal transduction systems as found in vertebrates.

gene, (*c*) one $G_{Z/X}$, for which the gene is known but the protein is not, and (*d*) two classes of sensory G proteins: of which two tranducins, T-r, expressed in rod cells, and T-c, expressed in cone cells, constitute a structural group of their own, and one, G_{olf}, belongs structurally to the adenylyl cyclase-stimulating G_s group.

There are about as many effector functions as there are G proteins. Effectors include adenylyl cyclase, the cGMP-specific phosphodiesterase of photoreceptor cells, phospholipases of the C and A_2 type, and various classes of ionic channels. These latter types include one specific for K^+ that conducts preferentially in the inward direction, another specific for Ca^{2+} that is strongly dependent on membrane potential for its activity and sensitive to dihydropyridines, and a group of neuronal monovalent cation channels either K^+ or both K^+ and Na^+ selective. As a consequence of ion channel modulation, occupancy of G protein-coupled receptors by primary messenger leads not only to changes of intracellular second messengers, but also of the cell's membrane

TABLE I
Examples of Receptors Acting on Cells via G Proteins

Type of receptor	Membrane function system affected	Effect	Coupling Examples of Target: Protein involved	Coupling Examples of Target: Cell(s)/organs
Neurotransmitters				
Adrenergic				
β-1	AC	S*	G_s	Heart, fat, sympathetic synapase
	Ca channel	S	G_s	Heart, skeletal muscle
β-2	AC	S	G_s	Liver, lung
α-1	PhL C	S	G_{plc}	Smooth muscle, liver
	PhL A_2	S	G_{pla}	FRTL-1 cells
α-2a, -2b	AC	I	G_i	Platelet, fat (human)
	Ca channel	C	G_o (G_p?)	NG-108, sympathetic presynapse
Dopamine				
D-1	AC	S	G_s	Caudate nucleus
D-2	AC	I	G_i	Pituitary lactotrophs
Acetylcholine				
Muscarinic M_1	PhL C	S	G_{plc}	Pancreatic acinar cell
	K channel (M)	C	?	CNS, sympathetic ganglia
Muscarinic M_2	AC	I	G_s	Heart
	K channel	O	G_k (G_i?)	Heart, CNS
	PhL C	S	G_p	Heart, transfected cells
$GABA_B$	Ca channel	C	G_o (G_p?)	Neuroblastoma N1E
	K channel	O	G_i (G_k?)	Sympathetic ganglia
Purinergic P1				
Adenosine A-1 or Ri	AC	I	G_i	Pituitary, CNS, heart
	K channel	O	G_k (G_i?)	Heart
Adenosine A-2 or Ra	AC	S	G_s	Fat, kidney, CNS
Purinergic P_{2X} and P_{2Y}	PhL C (PIP_2)	S	G_{plc}	Turkey erythrocytes
	PhL C (PC)	S	G (?)	Liver
Serotonin (5HT)				
S-la (5HT-la)	AC	I	G_i	Pyramidal cells
	K channel	O	G_k (G_i?)	Pyramidal cells
S-lc (5HT-1c)	PhL C	S	G_{plc}	Aplysia
S-2 (5HT-2)	AC	S	G_s	Skeletal muscle
Histamine				
H-1	PhL C	S	G_{plc}	Smooth muscle, macrophages
	PhL A_2 (?)			
H-2	AC	S	G_s	Heart
H-3	AC	I	G_i	Presynaptic CNS, lung, mast
Peptide hormones				
Pituitary				
Adrenocorticotropin	AC	S	G_s	Fasciculata, glomerulosa
Opioid (μ, κ, δ)	AC	I	G_i	NG-108
	Ca channel	C	G_o (G_p?)	NG-108

TABLE I (continued)

Type of receptor	Membrane function system affected	Effect	Coupling Examples of Target	
			Protein involved	Cell(s)/organs
Luteinizing hormone	AC	S	G_s	Granulosa, luteal, Leydig
Follicle-stimulating hormone	AC	S	G_s	Granulosa
Thyrotropin	AC	S	G_s	Thyroid, FRTL-5
	Phospholipase?	S	G_p(?)	Thyroid
Melanocyte-stimulating hormone	AC	S	G_s	Melanocytes
Hypothalamic				
Corticotropin-releasing hormone	AC	S	G_s	Corticotroph, hypothalamus
Growth hormone-releasing hormone	AC	S	G_s	Somatotroph
Gonadotropin-releasing hormone	PhL A_2	S	G_{pla}	Gonadotroph
	PhL C	S	G_{plc}	Gonadotroph
	Ca channel	O	G_i-type	GH_3
Thyrotropin-releasing hormone	PhLC	S	G_{plc}	Lactotroph, thyrotroph
	AC	I	G_i	GH_4C_1
Somatostatin	AC	I	G_i	Pituitary cells, endocrine, pancreatic
	K channel	O	G (G_i?)	Pituitary cells, endocrine, pancreatic
	Ca channel	C	?	Pituitary cells
Other hormones				
Chorionic gonadotropin	AC	S	G_s	Granulosa, luteal, Leydig
Glucagon	AC	S	G_s	Liver, fat, heart
	Ca pump	I	G_s(?)	Liver, heart (?)
	PhL C	S	?	Liver
Cholecystokinin	PhL C	S	G_{plc}	Pancreatic acini
Secretin	AC	S	G_s	Pancreatic duct, fat
Vasoactive intestinal peptide	AC	S	G_s	Pancreatic duct, CNS
	PhL C	S	G_{plc}	Sensory ganglia, CNS
Vasopressin				
V-1a (vasopressor, glycogenolytic)	PhL C	S	G_{plc}	Smooth muscle, liver, CNS
	AC	I	G_i	Liver
V-1b (pituitary)	PhL C	S	G_{plc}	Pitutary
V-2 (antidiuretic)	AC	S	G_s	Distal and collecting tubule
Oxytocin	PhL C	S	G_{plc}	Uterus, CNS
Angiotensin II	PhL C	S	G_{plc}	Liver, glomerulosa cells
	AC	I	G_i	Liver, glomerulosa cells
	CA channel	S	G_i-type	Y1 adrenal cells

TABLE I (continued)

Type of receptor	Membrane function system affected	Effect	Coupling Protein involved	Examples of Target Cell(s)/organs
Other regulatory factors				
Chemoattractant (fMet-Leu-Phe or fMLP)	PhL C	S	G_{plc}	Neutrophils
Thrombin	PhL C	S	G_{plc}	Platelets, fibroblasts
	AC	I	G_i	Platelets
Bombesin	PhL C	S	G_{plc}	Fibroblasts
IgE	PhL C	S	G_{plc}	Mast cells
Bradykinin	PhL C	S	G_{plc}	Lung, fibroblasts, NG-108
	PhL A_2	S	G_{pla}	Fibroblasts, endothelial cells
	K channel	S	G_k (G_i?)	NG-108
	AC	I	G_i	NG-108
Neurokinin/tachykinin				
NK1 (substance P)	PhL C	S	G_{plc}	CNS, salivary gland, endothelium
NK2 (neurokinin A or substance K)	PhL C	S	G_{plc}	CNS, sympath, smooth muscle
NK3 (neurokinin B)	PhL C (?)	S	G_{plc}	CNS, smooth muscle
Neuropeptide Y	K channels	S	G	Heart
Tumor necrosis factor	?	?	?	Monocytes
Colony-stimulating factor	?	?	?	Monocytes
Prostanoids				
Prostaglandin E_1, E_2	AC	I	G_i	Fat, kidney
Prostacyclin (PGI_2, PGE_1, PGE_2)	AC	S	G_s	Luteal cells, endothelium, kidney
Thromboxanes	PhL C	S	G_{plc}	Platelets
Platelet-activating factor	PhL C	S	G_{plc}	Platelets
Leukotriene D_4, C_4	PhL A_2	S	G_{pla}	Endothelial cells
Sensory				
Light (rhodopsins)	cGMP-PDE	S	Tr(G_{t-r})	Retinal rod cells (night)
	cGMP-PDE	S	Tc(G_{t-c})	Retinal cone cells (color)
Olfactory signals	AC	S	G_{olf}	Olfactory cilia
	Phospholipases?		G_p?	
Taste signals	AC	S	G_s	Taste epithelium
	Phospholipases	S	G_p?	

AC, adenyl cyclase; PhL C, unless denoted otherwise, phospholipase C with specificity for phosphatidylinositol bisphosphate; PhL A_2, phospholipase A_2 (substrate specificity unknown); PIP_2, phosphatidylinositol bisphosphate; PC, phosphatidylcholine.
*C, closing; I, inhibition, O, opening; S, stimulation.

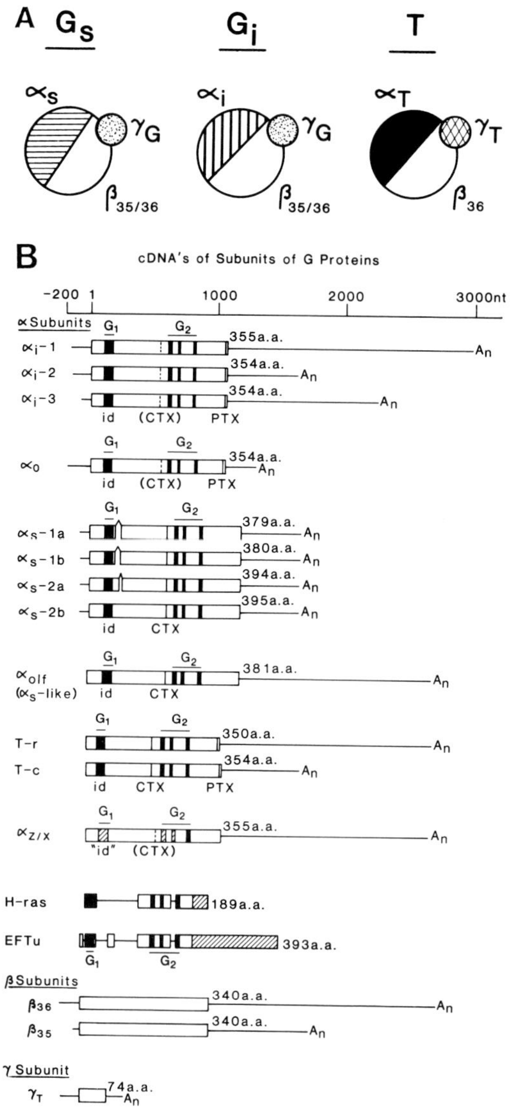

Figure 2. Schematic representation of subunit composition of typical vertebrate G proteins (*A*) and of G protein α subunit mRNA molecules as deduced from cDNA cloning (*B*). (*A*) G proteins have molecular weights (M_r) of 80,000–95,000. Except for transducin, which is easily washed off membranes with low salt solutions, all other known G proteins are tightly bound to membranes and can only be extracted with detergents. All G proteins involved in receptor-effector coupling, including transducin, are formed of distinct Gα subunits (M_r, 39,000–50,000), one of two types of Gβ subunits (Gβ_{36}, M_r = 36,000 or Gβ_{35}, M_r 35,000) and two types of Gγ subunits with M_r of 7–10,000. One Gγ (type t) is water soluble with a high proportion of charged amino acids. It is expressed only in retina and hence found in transducin (G$_t\gamma$). The other type, of which there appear to be two: Gγ_a and Gγ_b, are very hydrophobic and are found in all other G proteins studied to date. They are thought to play a role in anchoring G proteins to the plasma membrane. Gβ and Gγ subunits have not been purified separate from each other, only as dimers or complexed to Gα subunits as holo-G proteins. Relative abundance of Gβ36, Gβ_{35} and the various Gγ subunits varies significantly from cell to cell. Extreme cases are retinal rod cells, which have exclusively Gβ_{36} and G$_t\gamma$, and placenta, which expresses mostly Gβ_{35} and Gγ subunits other than G$_t\gamma$. Within any given tissue, the different Gα subunits interact with a common pool of G$\beta\gamma$ dimers and G$\beta\gamma$ heterogeneity is the same for each holo-G protein. (*B*) Open boxes represent the open reading frames or coding sequences and lines represent 5′ and 3′ untranslated sequences which may be incomplete. Black boxes within the open reading frames of α subunits denote sequences highly homologous to those known in bacterial elongation factor (EF) TU to be involved in GTP binding and hydrolysis. Sequences homologous to these are present also in the *ras* molecules. Hatched areas in EF TU and *ras* are nonhomologous to Gα. The mRNA molecules encoding the β_{36}, β_{35}, and γ_T are shown for comparison. The position of amino acids ADP-ribosylated by CTX and PTX are indicated. *i.d.*, identity box: a stretch of 18 invariant amino acids in all (filled) except $\alpha_{Z/X}$, which differs not only within the i.d. box but also in two of the other three regions involved in GTP binding (hatched). The scale is in nucleotides.

TABLE II
List and Properties of G Proteins as Known through Analysis of Their α Subunits

Subunit	No. of genes	Substrate for toxin	Trimeric form(s) purified	Length (amino acids)	Molecular weight		Regulated effector systems
					Calculated*	SDS-PAGE‡	
G_s	1	CTX	Yes				AC, S§
Four splice variants							DHP-sensitive Ca^{2+} channels, S
$G_s\alpha$-short (Ser^-)				379	44,093		
$G_s\alpha$-short (Ser^+)				380	44,180	42–45 kD	
$G_s\alpha$-long (Ser^-)				394	45,578		
$G_s\alpha$–long (Ser^+)				395	45,665	50–52 kD	
G_{olf}	1	CTX	No	381		42 kD	AC, S
T	2	CTX, PTX					
T-rod			Yes	350	39,964	39 kD	cGMP-PDE, S
T-cone			No	354	40,100	?	cGMP-PDE, S
G_i	3	PTX					
$G_i\alpha$-1			Yes	355	40,359	41 kD	K^+ channel, S; AC, I
$G_i\alpha$-2			Yes	354	40,456	40 kD	K^+ channel, S
$G_i\alpha$-3			Yes	354	40,538	40–41 kD	K^+ channel, S
G_o	1	PTX	Yes	354		39 kD	K^+ and K^+/Na^+ channels, S Ca^{2+} channel, I
$G_{Z/X}$	1	(CTX?)	No	355	40,920	?	?

*Molecular weights are for human G_s and G_i proteins, for rat G_{olf}, and bovine transducins and G_o. Species differences are minor, 3–5 amino acids at most.

‡Ranges are given where laboratories differ in their assessment.

§I, inhibition; S, stimulation.

potential, which itself is a potent regulator of cell function. Fig. 1 presents a general diagram of the elements involved in G protein-dependent signal transductions. The diversity of G proteins and their involvement in receptor-effector coupling has been reviewed quite extensively by Stryer and Bourne, 1986; Gilman, 1987; Lockrie and Simon, 1988; and Birnbaumer and Brown, 1989. The reader is referred to these articles for more detailed information on structure-function relations.

Mechanism of Action of a G Protein: Molecular Basis for Receptor-Effector Coupling

The mechanism by which binding of ligand to receptor is thought to increase or decrease the activity of an effector, with the intermediary participation of a G protein,

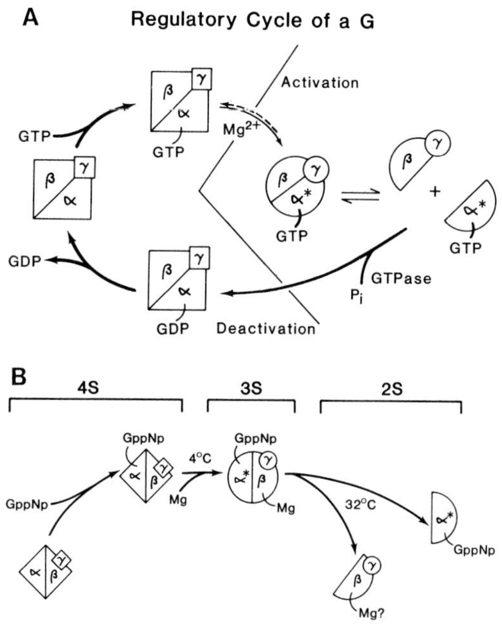

Figure 3. Regulatory cycle of a G protein. (*A*) As it is thought to occur in the presence of GTP. (*B*) As it occurs with solubilized G protein in the presence of nonhydrolyzable GTP analogues (GMP-P(NH)P and GTPγS). Squares and semi-squares represent inactive conformations as they relate to modulation of effector functions. Circular and semi-circular shapes represent activated forms of the G protein. Activation is both GTP- and Mg^{2+}-dependent and stabilized by subunit dissociation to give an activated Gα*GTP complex plus the G$\beta\gamma$ dimer. Gα subunit–mediated GTP hydrolysis leads to α subunit deactivation, increased affinity for G$\beta\gamma$ and reassociation to give an inactive holo-G protein with GDP bound to it. Reinitiation of the activation cycle requires release of GDP and renewed binding of GTP. Specificity of action is encoded in Gα. Different Gα subunits associate with a common pool of G$\beta\gamma$ dimers.

has received less attention in recent times and deserves some discussion. At the heart of receptor action lie two characteristic reactions of G proteins: (*a*) the binding of GTP followed by dissociation of the G protein to give a GTP-ligand α subunit (Gα*GTP) plus G$\beta\gamma$, and (*b*) in situ hydrolysis of GTP, which leads to reassociation of G protein subunits (Fig. 3 *A*). In the test tube, with isolated G proteins, this cycle proceeds only very slowly, if at all, with GTP. But upon substituting GTP analogues (GMP-P(NH)P or GTPγS) for GTP, the subunit dissociation reaction proceeds readily provided the reaction is supplemented with Mg^{2+} (Fig. 3 *B*), which leads to mixtures of $\beta\gamma$ dimers plus activated α-G nucleotide analogue complexes from which the activated α subunits can be isolated biochemically.

It is now accepted that the role of the receptor (Fig. 4) is to catalyze the activation reaction, i.e., binding of GTP followed by subunit dissociation, and that effector stimulation (or inhibition) is due to its interaction with the Gα*GTP complex. In so doing, effectors act as monitors of the membrane levels of Gα*GTP complexes. On hydrolysis of GTP, the Gα changes its conformation and loses its affinity for effector. As a consequence, Gα dissociates from the effector. Because of this, the GTPase reaction is equivalent to an inactivation reaction. This is followed by reassociation with Gβγ to give a stable GDP-liganded trimeric G protein (Gαβγ-GDP), which requires the aid of receptor to reinitiate the cycle. The receptor reinitiates the cycle because in the presence of GTP, it promotes the formation of Gα*GTP·Gβγ through a series of equilibrium reactions involving GDP dissociation, GTP binding, and change in protein

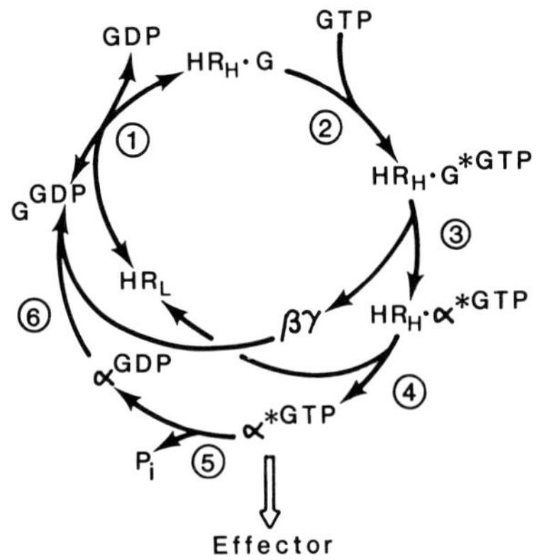

Figure 4. Catalysis of G protein activation by hormone-receptor complex. The scheme incorporates three overlapping and mutually dependent regulatory cycles: *1,* the G protein undergoes the cyclical dissociation-reassociation reaction, with receptor interacting only with the trimeric form of the G protein; *2,* the G protein oscillates between GDP, nucleotide-free, and GTP states, driven energetically forward by its capacity to hydrolyze GTP; and (*c*) receptor intervenes as a catalyst by stabilizing an "activated" GTP-liganded form of the G protein, that however is no longer stable as a trimer and decomposes into free Gβγ dimer plus receptor-Gα*GTP, which in turn, because of the absence of the Gβγ portion of the G protein, loses its ability to stay associated with receptor and decomposes further into activated Gα*GTP plus the free form of the receptor. The receptor is shown in two forms or states. One has low (*L*) affinity for agonist and the other has high (*H*) affinity for agonist. These are forms that it adopts when it is free and associated with Gαβγ, respectively. Although the assumption is made that regulation of effector is possible only after Gβγ and receptor have dissociated, it is in fact not yet known at what point the Gα subunit becomes competent to associate with effector. Likewise, it is also not known at what point the Gα subunit acquires the capacity to hydrolyze GTP. In spite of these uncertainties, note that without subunit dissociation, receptors could not act catalytically. Note also that since the Gβγ dimers are necessary for interaction of the G protein with the receptor, their overall effect is to stimulate Gα activation. All G proteins involved in coupling receptors to effectors conform to this scheme.

conformation. The cycle would come to a rest here if it were not that the "starred" conformation of Gα has a low affinity for Gβγ, which therefore dissociates. This in turn leads to the release of the hormone-receptor complex. Thus protein-protein interactions between receptor and G protein and between G protein and effector are not only based on the conformation of the individual subunits, but also on the multimeric/polymeric state transitions of the G protein.

The receptor-stimulated cycle (Fig. 4) allows the activated Gα*GTP to regulate the effector and, at the same time, the receptor that led to its formation is free to mediate the activation of another G protein by GTP. Two important consequences of this mechanism are (*a*) that the receptor signal is amplified, and (*b*) that several receptors may be engaged simultaneously in activating the same G protein pool. Amplification was well demonstrated for phototransduction where a single photon can lead to activation of up to 10 rhodopsin molecules (Fung et al., 1981), and for β-adrenergic stimulation of adenylyl cyclase (Tolkovsky and Levitzki, 1978). The second consequence leads to synergism between hormones at low concentrations, but lack of additivity at high concentrations, such as may happen in tissues with multiple receptors of different ligand specificity but the same cellular action. For example in fat, ACTH, β-adrenergic, secretin, and glucagon receptors all potentiate each other to induce cAMP-mediated lipolysis by catalyzing the activation of G_s that stimulates adenylyl cyclase (Birnbaumer and Rodbell, 1969). In heart atria, muscarinic M2, adenosine A1, and in some species neuropeptide Y receptors have a bradycardic effect as a result of activation of K^+ channels stimulated by G_k (see below).

Modification of G Proteins by Bacterial Toxins

Although G proteins are not targets for any known drugs, many G proteins are the targets of two relevant bacterial toxins: cholera toxin (CTX) and pertussis toxin (PTX). These toxins are enzymes that covalently modify specific $G\alpha$ subunits by transferring the ADP-ribose moiety from NAD onto arginine (CTX), or cysteine (PTX) of the $G\alpha$ protein (ADP-ribosylation reaction). G proteins are therefore often referred to as toxin substrates. The G protein specificity of CTX and PTX differs as does the effect that these toxins have on G protein function. Because of their G protein specificity, and because of the effect they have on G protein function, CTX and PTX are powerful tools that can be used to investigate possible involvement of a G protein in a cellular response (Scherer et al., 1987).

CTX ADP-ribosylates G_s, G_{olf}, and transducin (T) (Table II). The effect is both to eliminate the requirement of receptor participation in the activation of the G protein by GTP, and to inhibit the GTPase activity of $G\alpha$*GTP. A persistently active G protein is obtained. With CTX it is therefore possible to activate one of these G proteins and bypass the need for receptor. Given that G_{olf} and transducins are only in sensory epithelia, effects of CTX are quite reliably ascribed to G_s activation and consequential cAMP formation. Effects of CTX have been observed that could not be mimicked by cAMP (Askamit et al., 1985; Imboden et al., 1986), indicating that G_s may have effects in addition to those of stimulating adenylyl cyclase, such as shown below for stimulation of Ca^{2+} channels, or that CTX may affect other unknown (G?) protein(s).

In contrast to CTX, PTX ADP-ribosylates a much wider spectrum of G proteins, including G_i (the mediator of adenylyl cyclase inhibition), at least one form of G_p (the activator(s) of membrane phospholipase activity of C and A_2 type), G_k (the stimulator of a class of K^+ channels), and possibly other G proteins as defined by their function. Also in contrast to CTX, PTX does not facilitate activation of its substrates, rather it blocks the ability of the G protein to interact with receptors and hence causes functional inactivation of all signal transduction pathways in which a PTX substrate is involved.

Biochemical purification and molecular cloning has led to the molecular identification of six PTX-sensitive $G\alpha$ subunits, each the product of a separate gene (Table II, Fig. 2 *B*). Of these, the function of T-r and T-c in retinal rod and cone cells as stimulators of cGMP-phosphodiesterase is well established, but the function or functions of all the other PTX substrates, three G_i and one G_o, are still being unravelled.

In addition to G proteins that are substrates for either or both CTX and PTX, there are G proteins with α subunits that are not affected by these toxins. The G protein(s) mediating activation of phospholipase C in liver in response to epinephrine and vasopressin, and in the pituitary in response to TRH is (are) of this kind. They have not been purified and are known only through their functional expression. Recently a new G protein α subunit was cloned independently in the United States and in Japan (Fong et al., 1988; Matsuoka et al., 1988). We refer to it as $\alpha_{Z/X}$ and its heterotrimeric $\alpha\beta\gamma$ form as $G_{Z/X}$. Although $G_{Z/X}$ may be one of the toxin-insensitive G_p proteins, this is mere speculation at this time. $\alpha_{Z/X}$ is expressed primarily if not exclusively in brain and thymus, so that even if it has G_p activity there must be homologues to fulfill this function in tissues where there is a toxin-insensitive G_p and $\alpha_{Z/X}$ is not expressed.

Direct Regulation of Ionic Channels by G Proteins

The Inwardly Rectifying G Protein-gated K^+ Channel

The possibility of direct regulation of an ionic channel by a G protein (Yatani et al., 1987*a*) emerged from the study of the mechanism by which muscarinic acetylcholine receptors (mAChR) activate the cardiac K^+ channel that mediates vagal regulation of chronotropy. This channel is commonly referred to as the "muscarinic" K^+ channel, the K^+ current generated by its opening is referred to as I_{ACh}, and the G protein that is thought to mediate the effect is referred to as G_k (Breitweiser and Szabo, 1985; Pfaffinger et al., 1985). After discovery of its direct regulation by a G protein of the G_i type, and of the presence of similar hormone-regulated K^+ channels in other cells (Yatani et al., 1987*b*; Brown and Birnbaumer, 1988), this type of channel is referred to as a G_k-sensitive K^+ channel. As presented below, all three types of G_i proteins activate this K^+ channel, and there are other K^+ channels gated selectively by the G_o G protein. Because of these developments, we shall distinguish between the various G protein-gated K^+ channels by referring to them as either G_i-gated or G_o-gated.

G protein-gated K^+ channels are modulators of cellular function. Activation of these K^+ channels causes cells to hyperpolarize and become less excitable. As a consequence, secretion is attenuated in endocrine and nerve cells (Hartzell et al., 1977, Hill-Smith and Purves, 1978; Nakajima et al., 1986; Gregerson et al., 1989) whereas in heart, they cause a decrease in chronotropy (reviewed by Hartzell, 1981).

Acetylcholine inhibits adenylyl cyclase in heart membranes (Murad et al., 1962; Kurose and Ui, 1983; Mattera et al., 1985). Yet involvement of soluble second messengers (i.e., decrease in cAMP levels) in the action of muscarinic receptors on atrial cells is unlikely on grounds of the rapidity with which acetylcholine modifies K^+ channel activity (Trautwein et al., 1982; Nargeot et al., 1983), and was demonstrated so in electrophysiological experiments using the giga-seal patch-clamp approach. Information came from experiments using the cell-attached configuration (Soejima and Noma, 1984) and from the whole-cell recording configuration (Breitweiser and Szabo, 1985; Pfaffinger et al., 1985). Direct regulation of the channel by a G_i G protein was shown in experiments with inside-out membrane patches (Codina et al., 1987*a*, *b*; Yatani et al., 1987*a*, *b*; Kirsch et al., 1988; Mattera et al., 1988).

Fig. 5 summarizes experiments that demonstrated direct (or membrane delimited) regulation of muscarinic K^+ channels by a G protein. The cell-attached patch-clamp experiments showed that atrial muscarinic K^+ channels in the membrane patch are insensitive to muscarinic stimulation when acetylcholine is added to the bath, but become activated when the agonist is superfused onto the outer surface of the isolated membrane patch (Fig. 5 *A*; Soejima and Noma, 1984). This indicated that the muscarinic receptors had to be in close proximity of the channel regulated by them and eliminated a soluble cytoplasmic second messenger as mediator of receptor activation. The actual coupling mechanism was not addressed by this experiment, however possibilities include direct receptor regulation of the channel (receptor-channel coupling), activation of a G protein by the receptor followed by direct G protein-channel coupling, or receptor-mediated formation of a diffusible yet membrane-delimited second messenger (e.g., diacylglycerol and protein kinase C–mediated phosphorylation).

The involvement of a G protein in the coupling of heart muscarinic receptors to K^+ channels was established in whole-cell voltage-clamp experiments in which atrial cells were "perfused" with pipette solutions containing GTP or the GTP analogue GMP-P(NH)P (Fig. 5 *B*; Breitweiser and Szabo, 1985; Pfaffinger et al., 1985). One set of experiments established that acetylcholine is able to increase the K^+ currents

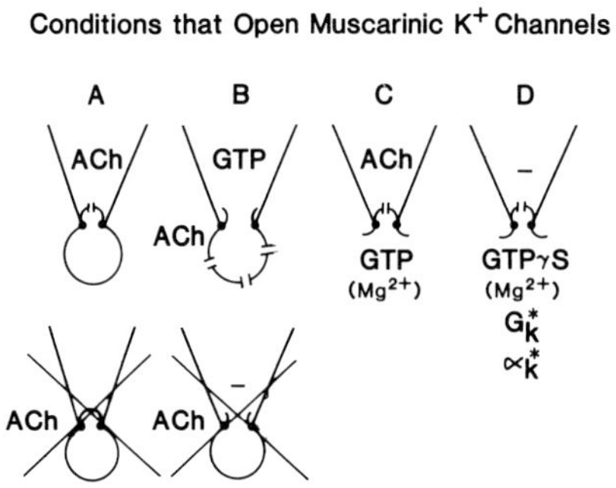

Figure 5. Summary of experiments leading to the demonstration of direct (membrane delimited and independent of soluble second messengers and protein phosphorylation) G protein gating of K^+ channels. *A*, Results from Soejima and Noma, 1984; *B*, experiments of Pfaffinger et al., 1985; *C* and *D*, experiments shown in Fig. 6.

only when the pipette solution contains GTP and that PTX treatment of cells abolishes the muscarinic response (Pfaffinger et al., 1985). The other set of experiments used amphibian atrial cells and measured not only K^+ current but also Ca^{2+} currents (Breitweiser and Szabo, 1985). In these cells acetylcholine induces a K^+ current (inwardly rectifying I_{ACh}) and attenuates β-adrenergic receptor-induced increases in a specific Ca^{2+} current (the so-called slow inward I_{ca}). The minimal pipette/cytoplasm exchange allowed hormonal regulation of ionic currents even without addition of GTP to the pipette solution. Addition of GMP-P(NH)P to the pipette had no effect unless hormones were added to the extracellular bathing fluid, presumably because of the slow action of the nucleotide analogue in the absence of hormonal stimulation, accentuated by the prevailing low (submillimolar) intracellular Mg^{2+} concentrations. ACh induced a persistent I_{ACh} current and isoproterenol induced a persistent slow $I_{Ca^{2+}}$. These results indicated the involvement of a guanine nucleotide binding protein in the action of both types of receptors. Furthermore, the persistent nature of the acetylcholine response indicated that the G protein intervening between the muscarinic receptor and the K^+ channel, termed generically G_k (Breitweiser and Szabo, 1985), resembles the stimulatory G protein of adenylyl cyclase in its activation kinetics in that activation by GTP analogue is slow in the absence of hormonal stimulation and fast in its presence. In agreement with the postulate that muscarinic regulation of K^+ currents is independent of cAMP levels, stimulation of cells held with a GMP-P(NH)P–containing pipette in the presence of isoproterenol had no effect on the induction of K^+ currents by acetylcholine even though persistent effects of isoproterenol, and therefore activation of G_s, were obtained (Breitweiser and Szabo, 1985).

Direct action of a G protein on the heart muscarinic receptor-sensitive K^+ channel was demonstrated in a cell-free system using inside-out membrane patches (Fig. 5, *C* and *D*; Yatani et al., 1987*a*; Cerbai et al., 1988). In these experiments addition to the bath, i.e., to the cytoplasmic face of the K^+ channel-containing membrane, of GTPγS or an activated PTX-sensitive G protein purified from human erythrocyte membranes, results in K^+ channel activation. The effect of GTPγS required Mg^{2+} and proceeded with a lag. The effect of the PTX-sensitive G protein preactivated with GTPγS developed much faster. Since these studies were carried out in the absence of ATP (Yatani et al., 1987*a*), they ruled out the involvement of a phosphorylation reaction in the action of muscarinic receptors. Similar studies with pituitary GH_3 cells showed the existence in these cells of a similar K^+ channel stimulated by a PTX-sensitive G protein that is under regulatory control of somatostatin and muscarinic receptors (Yatani et al., 1987*b*). G protein-regulated K^+ channels are highly selective with respect to the G protein with which they interact. Highly purified G_s, even at 100-fold higher concentrations than needed to obtain half-maximal effects with G_i, had no effect on the K^+ channel (Yatani et al., 1987*a*, *b*).

Fig. 6 summarizes many of the properties of receptor mediated regulation of G_i-sensitive K^+ channels as seen in atrial and pituitary membrane patches. The experiments show that G protein regulation of K^+ channels by receptors is critically dependent on GTP, that activation of G_k by GTP requires the participation of agonist-occupied receptor, that the endogenous G_k is uncoupled from regulation by receptor by PTX, and that a PTX-uncoupled system is readily reconstituted by the addition of exogenous unactivated G_k, provided GTP is present in the bath. Further, resolved GTPγS-activated α subunits of hRBC G_i (α_i^*) mimic the actions of GTPγS-activated Gi (G_i^*) with comparable potency, while resolved $\beta\gamma$ dimers do not (Codina et al., 1987*a*, *b;* Kirsch et al., 1988). Thus, the experiments also indicate that the most plausible mechanism used by muscarinic receptors to stimulate G protein-gated K^+ channels is the catalysis of the activation of membrane G_k by GTP and formation of activated free α_k^* which in turn acts as mediator to stimulate the K^+ channel.

Properties of G protein-gated K^+ channels. The microscopic kinetic properties of the G_i-sensitive K^+ channels have been measured and are in many ways typical of other ionic channels (Codina et al., 1987*a*, *b;* Yatani et al., 1987*a*, *b;* Cerbai et al., 1988; Kirsch et al., 1988). The effect of activated G_i is clearly one of increasing the opening probability of the channel. Once stimulated by G_i the channel opens in bursts and clusters of bursts as illustrated by the NP_o diaries of activity shown in Fig. 7. Fig. 7 also expresses channel activation data as cumulative NP values, allowing for comparison of activities on a quantitative basis, as may be required when comparing effects of increasing concentrations of holo-G or resolved α subunits. Frequency histograms of open times are well fit by first-order decay functions with time constants of ~1 ms. Open amplitude histograms are fit well by Gaussian frequency distributions. Slope conductances were calculated from data obtained with 130–140 mM K^+ on both sides of the membrane range from ~40 pS in guinea pig atria to ~50 pS in GH_3 cells. In both systems the channels conduct very poorly in the outward direction, and the microscopic properties of the channels stimulated by receptor plus GTP are indistinguishable from those seen on stimulation by GTPγS or exogenous addition of preactivated G_i or its activated α subunit.

As summarized in Table I, among the receptors that catalyze activation of G_k-type G protein by GTP are the heart atrial muscarinic, adenosine and neuropeptide Y receptors (Yatani, A., L. Birnbaumer, and A. M. Brown, unpublished observations), hippocampal 5HT1a and $GABA_B$ receptors (Newberry and Nicoll, 1984; Andrade et al., 1986; Sasaki and Sato, 1987; Thalman, 1988), and endocrine cell somatostatin and muscarinic receptors (Yatani et al. 1987*b*). In rat atria the muscarinic K^+ channel is also stimulated by neuropeptide Y receptors (Yatani, A., L. Birnbaumer, and A. M. Brown, unpublished observations). While it is clear at this time that in atrial and pituitary cells receptors cause activation of G_i-gated K^+ channels, studies in other cell types may reveal the involvement of G_o- rather than G_i-gated channels.

The Dihydropyridine-sensitive, Voltage-gated Ca^{2+} Channel

Stimulation by G proteins. Two lines of evidence, one obtained with atrial membrane patches in the inside-out configuration (Yatani et al; 1987*c*), the other with cardiac sarcolemmal and skeletal muscle T tubule membranes incorporated into planar lipid bilayers (Yatani et al., 1987*c*, 1988*a;* Birnbaumer et al., 1988; Imoto et al., 1988; Mattera et al., 1989), indicate that dihydropyridine-sensitive voltage-gated Ca^{2+} channels of these tissues are positively modulated by G_s, the stimulator of adenylyl cyclases.

In contrast to K^+ channels, which survive patch excision very well, voltage-gated Ca^{2+} channels of the kind putatively regulated in intact cells by G protein-coupled receptors are extremely unstable after patch excision. The left panels of Fig. 8 illustrate this point for cardiac guinea pig ventricle Ca^{2+} channels. The channels are readily seen in the cell-attached patch throughout repeated cycles of activation by the depolarizing test potential. However, after patch excision, activity is observed only during the initial tests and rapidly extinguishes thereafter. The right panels of Fig. 8 show that when GTPγS is added to the bath into which the membrane patch is excised, a significant increase resulted in both average activity during the depolarizing test periods and the number of tests during which depolarization evoked channel activation. As summarized in Fig. 9, this stabilizing effect could not be mimicked by protocols that increased adenylyl cyclase activity, increased cAMP levels, or provided for phosphorylation by the catalytic unit of cAMP-dependent protein kinase (protein kinase A). Also, inhibition of protein kinase A by addition of excess protein kinase inhibitor (PKI) did not inhibit the effect of isoproterenol plus GTPγS that stabilizes these Ca^{2+} channels (Yatani et al., 1987*c*).

The fact that isoproterenol is required in the pipette to obtain the effect of GTPγS suggested that the responsible G protein might be G_s, with β-adrenergic receptors in the membrane patch acting locally to accelerate the activation of G_s before irreversible inactivation of the channel molecules would occur. Indeed, addition of GTPγS-activated G_s, but not GTPγS-activated G_k, substituted for the combination of isoproterenol in the pipette and GTPγS or GTP in the bath.

It has been well established that cardiac ventricle dihydropyridine-sensitive Ca^{2+} currents are increased by about two to threefold by adrenergic stimuli via a protein

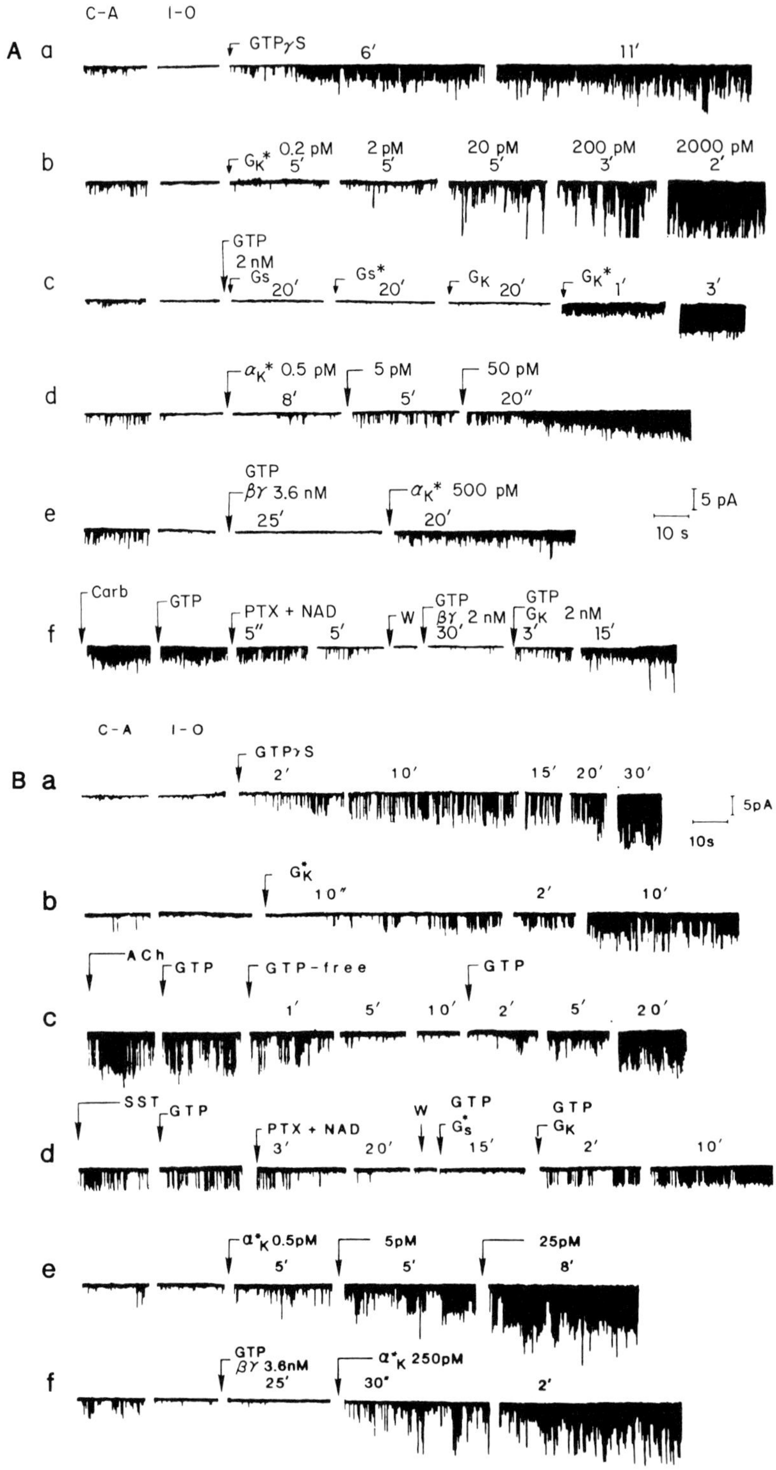

C-A
I-O
A
a
GTPγS
6′
11′
b
G_K* 0.2 pM 5′
2 pM 5′
20 pM 5′
200 pM 3′
2000 pM 2′
c
GTP 2 nM
Gs 20′
Gs* 20′
G_K 20′
G_K* 1′
3′
d
α_K* 0.5 pM 8′
5 pM 5′
50 pM 20″
e
GTP βγ 3.6 nM 25′
α_K* 500 pM 20′
5 pA
10 s
f
Carb
GTP
PTX + NAD 5″
5′
W
GTP βγ 2 nM 30′
GTP G_K 2 nM 3′
15′
C-A
I-O
B
a
GTPγS 2′
10′
15′
20′
30′
5pA
10s
b
G*_K 10″
2′
10′
c
ACh
GTP
GTP-free 1′
5′
10′
GTP 2′
5′
20′
d
SST
GTP
PTX + NAD 3′
20′
W
GTP G*_s 15′
GTP G_K 2′
10′
e
α*_K 0.5pM 5′
5pM 5′
25pM 8′
f
GTP βγ 3.6nM 25′
α*_K 250pM 30″
2′

Figure 6. Properties of G_k-mediated regulation of G protein-sensitive K^+ channels as seen in inside-out membrane patches of guinea pig atrial cells and GH_3 rat pituitary tumor cells. Each line represents a separate experiment in which single-channel K^+ currents were recorded before (cell-attached or *C-A*) and after membrane patch excision to the inside-out configuration (*I-O*). Records were obtained at holding potentials that varied between −80 and −100 mV and using symmetrical 130–140 mM KC or K-methanesulfonate solutions in 5 mM HEPES, pH 7.5, containing in addition either 1.8 mM $CaCl_2$ in the pipette, or 5 mM EGTA and 2 mM $MgCl_2$ in the 100-μl bathing chamber. Other additions are shown or described for each experiment. Numbers above the records denote time elapsed between the indicated additions and the beginning of the segment of record shown. Routinely, the first addition was made between 5 and 10 min after patch excision and subsequent additions were at 5–25-min intervals, depending on the purpose of the experiments. 5-min intervals were used when dose-response relationships were studied; 25-min intervals were used when substances added had no apparent effect. In some instances the bathing solutions were exchanged by perfusion at 1–2 ml/min. (*A*) Experiments with atrial membrane patches from adult guinea pigs. Cells were obtained by collagenase digestion and used without further culturing. *Experiment a:* stimulation of single-channel K^+ currents by activation of the atrial G_k protein with GTPγS (100 μM in the bath). *Experiment b:* stimulation of the atrial K^+ channels in a dose-dependent manner by exogenously added human erythrocyte pertussis toxin substrate referred to as G_k, (formerly referred to as N_i or G_i), preactivated by incubation with GTPγS and Mg^{2+}, and dialyzed extensively to remove free GTPγS to ineffective levels (G_k^*). Threshold effects of G_k^* were obtained with between 0.2 and 1 pM in separate membrane patches. *Experiment c:* the bathing solution contained 100 μM GTP throughout, proteins were added at 2 nM each. Lack of stimulatory effects of 1 nM of either nonactivated or GTPγS-activated G_s or of nonactivated G_k. *Experiment d:* mimicry of the effect of GTPγS-activated holo-G_k by equally low concentrations of the resolved GTPγS-activated α subunit of the protein (α_k^*). *Experiment e:* lack of effect of resolved human erythrocyte $\beta\gamma$ exposed to a G_k^*-sensitive membrane patch. *Experiment f:* stimulation of G_k-sensitive K^+ channels by the muscarinic receptor agonist carbachol (*carb*) present in the pipette throughout, maintenance of receptor–G protein effector coupling after patch excision into bathing solution with 100 μM GTP, and uncoupling of atrial G_k by treatment with PTX and NAD added to the bathing solution. Lack of recoupling effect of resolved human erythrocyte $\beta\gamma$ and reconstitution of acetylcholine receptor–K^+ channel stimulation by addition of native unactivated human erythrocyte G_k in the presence of GTP. Note that this differs from the result in *experiment c* and demonstrates that the exogenously added G_k requires receptor participation for activation by GTP. (*B*) Experiments with inside-out membrane patches from rat GH_3 pituitary tumor cells. Cells were grown as monolayers on cover slips and membrane patches excised from their upper membrane surface. *Experiment a:* effect of GTPγS added to the bathing medium. *Experiment b:* representative time course of activation of GH_3 cell K^+ channel by a saturating concentration (2 nM) of GTPγS-activated human erythrocyte G_k (G_k^*). *Experiment c:* dependence on GTP and reversibility of receptor-mediated stimulation of G_k-sensitive K^+ channel. Acetylcholine (ACh) was present in the pipette solution throughout. 100 μM GTP was present in the bathing medium, then removed and readded as shown. *Experiment d:* stimulation of GH_3H cell G_k-sensitive K^+ channels by somatostatin (*SST*): demonstration of PTX sensitivity of the GH_3 G_k protein and reconstitution of the signal transduction pathway by addition of unactivated native human erythrocyte G_k in the presence of GTP. *Experiments e and f:* mimicry of the effects of GTPγS-activated G_k by resolved GTPγS-activated α subunit of human erythrocyte G_k and lack of effect of resolved, α subunit-free $\beta\gamma$ dimers. (Adapted from Yatani et al., 1987*a*, *b*; Codina et al., 1987*a*,*b*.)

kinase A–mediated phosphorylation reaction (Kameyama et al., 1986). It is mimicked quantitatively by injection of pure active catalytic subunit of protein kinase A and it is inhibited by injection of PKI. However, the experiments presented above indicated that the same cardiac dihydropyridine-sensitive Ca^{2+} channels, can also be positively modulated by G_s, provided the channels are prestimulated either by physiological

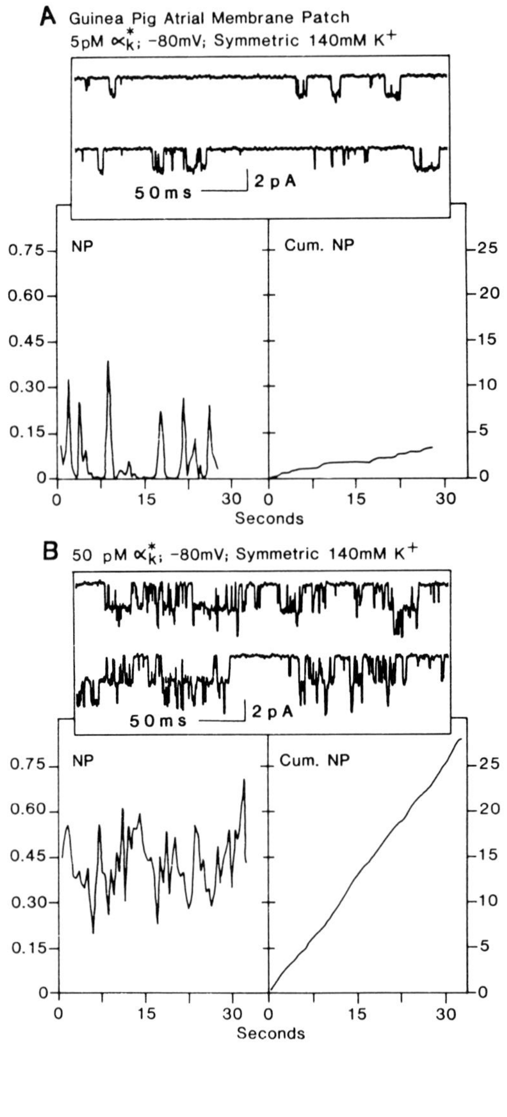

Figure 7. Single-channel K^+ currents in guinea pig atrial membranes after patch excision and addition of either 5 pM (*A*) or 500 pM (*B*) of GTPγS-activated resolved α subunit of G_k (α_k^*), quantification of activity, *NP* and cumulative *NP* values. Records of single-channel currents, such as shown in the top insets, were subdivided into consecutive 200-ms segments to quantify the proportion of time during which single-channel currents are observed in each segment. The variation of these values per 200 ms, representing the product of the opening probability (P_o) of each of the channels present in the membrane patch times *N*, the number of channels in the patch, as a function of time are the *NP* diaries shown in the lower left panels. Accumulation of *NP* values as a function of time gives cumulative *NP* values over the same time period (Cum *NP*). These are equivalent to a time course of channel activity and are shown in the lower right panels of *A* and *B*. The slopes of the Cum *NP* curves (Cum *NP*/min) refer to the maximum cumulative *NP*/min values that can be elicited from the same membrane patch and provide individual points that can serve to construct dose-response curves for the action of a G protein. Concentrations of G_k^* and α_k^* giving half maximal stimulations of guinea pig atrial K^+ channels vary between 5 and 60 pM, with a mean of ~20 pM for either G_k^* or α_k^*.

means as obtained with isoproterenol or by pharmacological means as obtained with the dihydropyridine agonist, Bay K 8644. On rare occasions it was possible to restimulate single-channel Ca^{2+} currents with GTPγS-activated G_s after they had ceased to respond to the test potential. The question arose therefore whether the G protein actually stimulated, or simply stabilized channel molecules. To obtain an answer to this question it was necessary to obtain a cell-free system in which

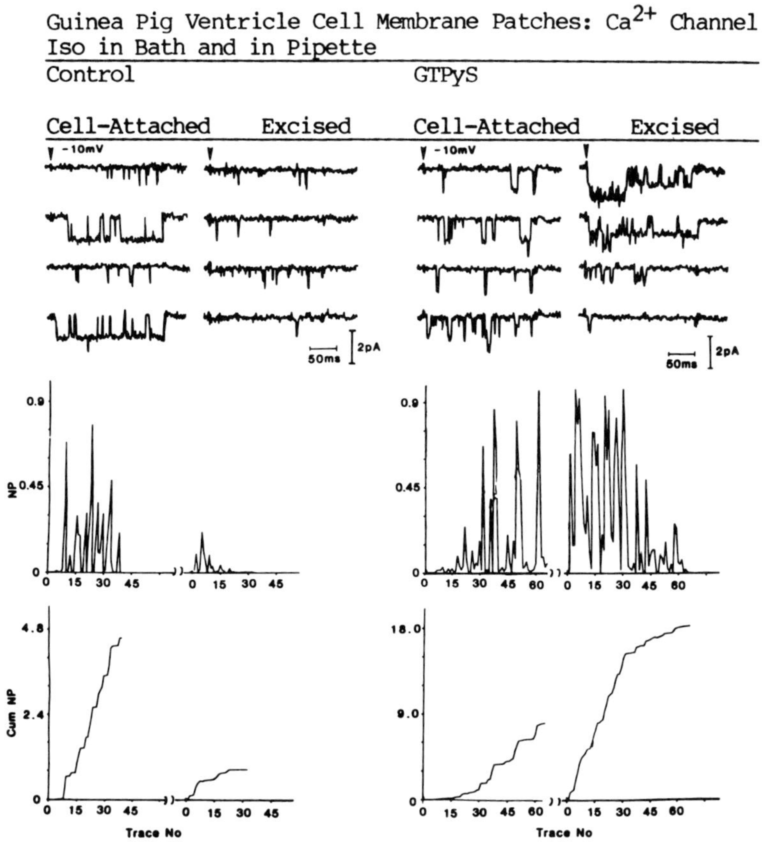

Figure 8. Transient survival of heart ventricle voltage-gated Ca^{2+} channels after membrane patch excision and stabilization through isoproterenol-dependent activation of a G protein with GTPγS. Guinea pig cardiac ventricle cells were patch clamped in the cell-attached mode with isoproterenol in both bath and pipette. Single Ca^{2+} channel currents were recorded during 200-ms test depolarizations from holding potentials of -90 to -20 mV applied at a frequency of 2 Hz. (*Left panels*) Bath solution without GTPγS. (*Top*) Representative single-channel current activities recorded during test potentials before and after excision of a membrane patch into bathing solution without GTPγS. (*Bottom*) *NP* diaries of the Ca^{2+} channel activities of the same membrane patch during 45 successive test periods immediately before and after excision. (*Right panels*) Bath solution with 100 μM GTPγS. (*Top*) Representative single-channel Ca^{2+} currents as obtained in test cycles before and after excision of a representative membrane patch. (*Bottom*) *NP* diaries of Ca^{2+} channel activities in the same membrane patch before and after its excision into a solution with GTPγS. On the average, the presence of GTPγS had two effects: (*a*) it increased the number of test cycles during which single Ca^{2+} channel openings were observable, and (*b*) it increased the cumulative *NP* values as quantified during the first 30 test cycles. (Adapted from Yatani et al., 1987*c*.)

dihydropyridine-sensitive Ca^{2+} channels were more stable. Incorporation of Ca^{2+} channels into planar phospholipid bilayers (Coronado and Affolter, 1986; schematized in Fig. 10) provided such a system. Initial experiments were conducted with Ca^{2+} channels of bovine cardiac sarcolemmal vesicles. Conclusive data were obtained with Ca^{2+} channels from skeletal muscle T tubules.

Skeletal muscle T tubule membranes are inside-out vesicles with respect to the orientation of their cytoplasmic surface and incorporate into bilayers exposing their extracellular surface to the *trans* chamber (Affolter and Coronado, 1986). After incorporation in the presence of Bay K 8644 and using Ba^{2+} as the charge carrier, the dihydropyridine-sensitive Ca^{2+} channels exhibit an activity at 0 mV that is stable over long periods of up to 30 min or until bilayer breakdown. As summarized in Fig. 10, addition of GTPγS-activated G_s, or GTPγS, which activates endogenous coincorporated skeletal muscle T tubule G protein, results in clear increases in NP_o, i.e., in the product of the opening probability of incorporated channels (P_o) times the number of channels incorporated (N). This can be mimicked by cholera toxin–treated G_s in the

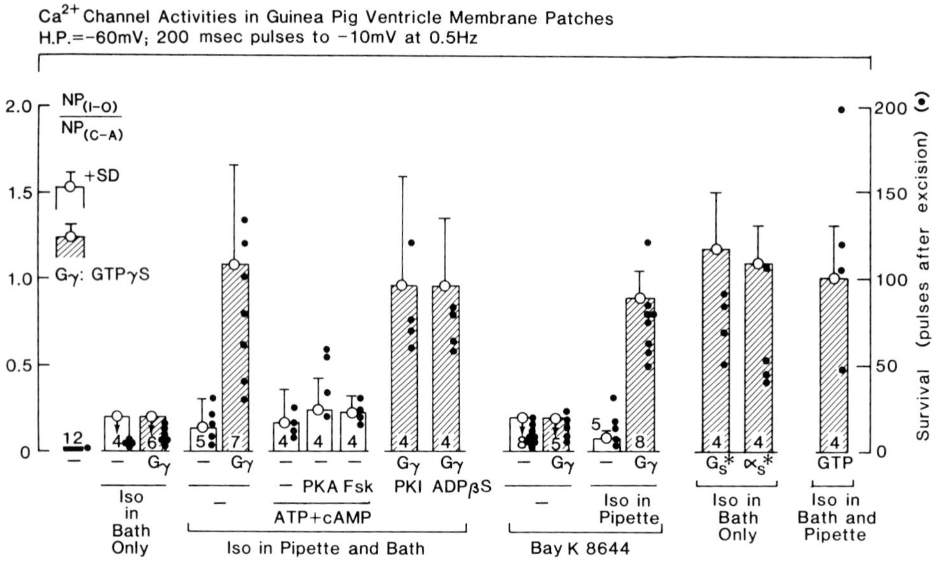

Figure 9. Summary of experiments with guinea pig cardiac ventricle voltage-gated, dihydropyridine-sensitive Ca^{2+} channels, such as shown in Fig. 8, in which additions to pipette and bath solutions were varied as shown. Bars represent the ratio between cumulative *NP* values during the first 30 test cycles after patch excision and those prior to patch excision (left ordinate). Closed circles denote last test cycle during which, after patch excision, at least one single-channel opening was observed (right ordinate). *Iso,* isoproterenol; *PKA,* catalytic unit of cAMP-dependent protein kinase, *PKI,* inhibitor of PKA; *lpt,* leupeptin. When added, GTPγS was 100 μM; GTP, 1 mM; ATP, 1 mM; cAMP, 500 μM, Bay K 8644, 3 μM; iso, 10^{-5} M; GTPγS-activated G_s ($^*_s$) or GTPγS-activated α_s (α^*_s), 100–200 pM. (Adapted from Yatani et al., 1988*c*.)

presence of GTP and by the resolved GTPγS-activated α subunit of G_s. Stimulations by the variously activated G_s preparations averaged two to threefold as seen in over 30 independent experiments. Importantly, stimulation of incorporated Ca^{2+} channels is not dependent on presence of Bay K 8644 (Fig. 11). Indeed, in one series of 12 experiments without Bay K 8644, the relative increases in NP_o values averaged over 2–5 min were 18-fold.

The above results were all obtained in the absence of ATP and hence they were not mediated by channel phosphorylation. This indicated that these Ca^{2+} channels, like the muscarinic K^+ channels and adenylyl cyclase, are under direct control of a G protein, which, as will be shown below, is undoubtedly G_s. In agreement with whole-cell recordings (Kameyama et al., 1986) from ventricle cells, as well as single-channel current records from inside-out membrane patches of GH_3 cells (Armstrong and Eckert, 1987), addition to the *cis* chamber of catalytic unit of protein kinase A and ATP or ATPγS also stimulated single Ca^{2+} channel currents by

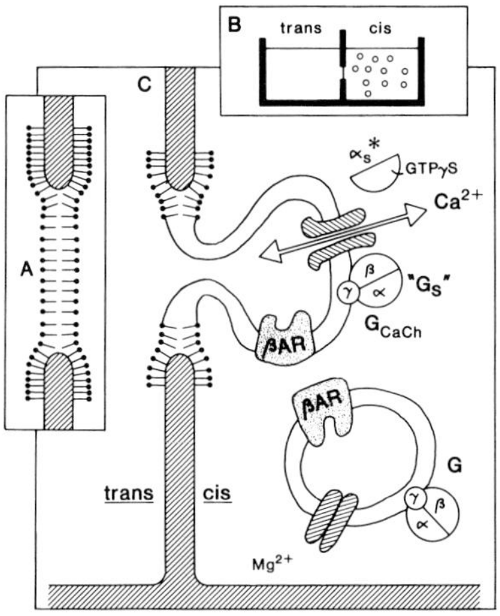

Additions: cis	Additions: trans	Stimulation of Ca^{2+} Channel
GTP & Iso	–	–
–	GTP & Iso	–
GTP	Iso	+
GTPγS	–	±(lag)
GTPγS	Iso	+(fast)
–	G_s^*	–
G_s^*	–	+
α_s^*	–	+
GTP & G_s	–	–
GTP & G_s^{CTX}	–	+

Figure 10. Strategy of incorporation of dihydropyridine-sensitive Ca^{2+} channels into planar phospholipid bilayers to study their direct regulation by a G protein (*left*); and summary of results obtained upon addiiton of guanine nucleotides, agonist (isoproterenol) and/or GTPγS-activated G_s (*right*). Phosphatidylcholine and phosphatidylethanolamine in decane are painted onto a hole of 0.1–0.3 mm diam in a partition separating two chambers of 0.15–0.5 ml referred to as *cis* and *trans,* with respect to the side from which membranes are forced to be fused with the lipid bilayer. Before membrane addition the *trans* chamber is filled with buffer A (50 mM NaCl, 2 mM $MgCl_2$, and 5 mM HEPES, pH 7.0), and the *cis* chamber with buffer A containing 100 mM $BaCl_2$. Incorporation of ionic channels into the lipid bilayer is obtained after addition of membrane vesicles in 10–20 μl aliquots to the *cis* chamber. After incorporation has occurred, and to prevent further incorportations from occurring during the course of the experiments, one of two protocols is followed. In one, the liquid in the *cis* chamber is extensively exchanged with the same solution, buffer A plus 100 mM $BaCl_2$, free of membranes. In the other, the liquids of both chambers are exchanged, the *trans* chamber receiving the original *cis* solution (buffer A plus 100 mM $BaCl_2$), and the *cis* chamber receiving buffer A. In both cases Ba^{2+} is used as the charge carrier, the difference being that single-channel currents are in the *cis* to *trans* direction in the first and in the *trans* to *cis* direction in the second of these protocols. The figure depicts a membrane vesicle that is inside-out with respect to the orientation of its cytoplasmic side, as is the case with skeletal muscle T tubules, before and after its fusion to the lipid bilayer, and illustrates the sidedness that results (Adapted from Yatani et al., 1988*a*.)

increasing NP_o values, which in a series of four experiments in the presence of Bay K 8644 averaged twofold. The top portion of Fig. 12 depicts our current concept as to how G_s is involved in a dual manner in positively modulating the activity of voltage-gated Ca^{2+} channels.

Hormonal stimulation of Ca^{2+} channels mediated by PTX-sensitive G protein(s). Electrophysiological studies on the effects of angiotensin II on Ca^{2+} currents in Y1 adrenal cells and gonadotropin releasing hormone on Ca^{2+} currents in GH_3 pituitary cells, carried out in the whole-cell configuration, have shown that the hormones added to the bath increase Ca^{2+} currents. In neither case was the effect mimicked by inclusion of cAMP in the pipette solution or addition of forskolin, and in both cases the hormonal effects were abolished by pretreatment of the cells with PTX (Hescheler et al., 1988; Hirsch et al., 1988). While it might be tempting to suggest that these G protein-dependent effects of the hormones might be direct, it is equally possible that the effects be mediated via intermediary small molecular weight second messengers with or without subsequent activation of a protein kinase.

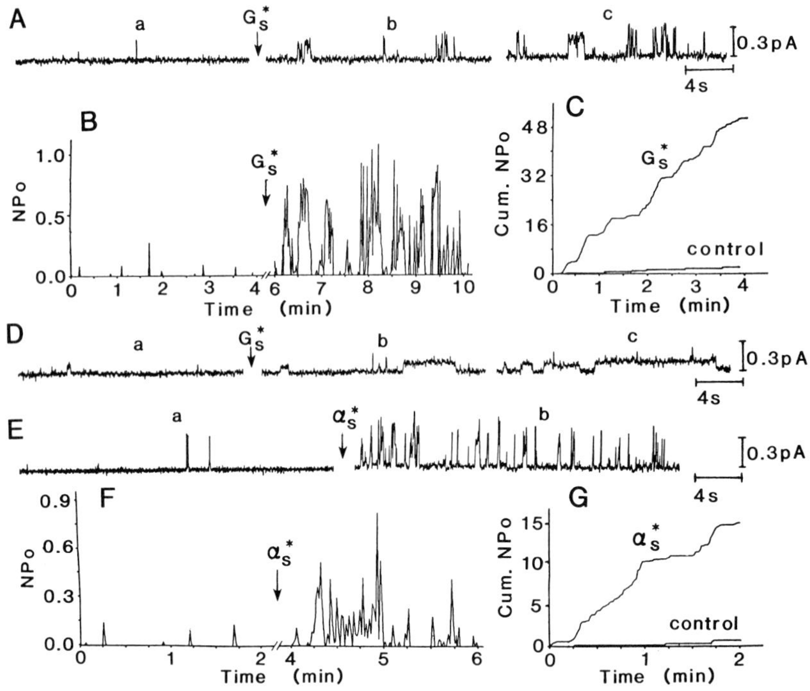

Figure 11. Experiments showing direct regulation of the skeletal muscle T tubule dihydropyridine-sensitive Ca^{2+} channel by purified human erythrocyte G_s and α_s, as seen in the absence of dihydropyridine agonist (Bay K 8644). Shown are both single-channel currents and the corresponding cumulative NP_o curves before and after addition of GTPγS-activated G_s or GTPγS-activated α_s. The holding potentials were either 0 or +20 mV, and experiments were done with $BaCl_2$ in the *cis* chamber. Average stimulations of activity were between 10- and 20-fold. (Adapted from Yatani et al., 1988*a*.)

Inhibitory regulation of Ca^{2+} channels by G proteins. Several experiments point to involvement of a G protein in the action of certain receptors that inhibit Ca^{2+} channels and that this G protein is G_o. First, whole-cell voltage-clamp measurements of Ca^{2+} currents in chick dorsal root ganglia showed that both norepinephrine and GABA, acting through α-adrenergic and $GABA_B$ receptors, respectively, inhibit voltage-gated Ca^{2+} currents (Holz et al., 1986; Rane and Dunlap, 1986). This effect is sensitive to PTX and blocked by injection of GDPβS into the cells (Holz et al., 1986). In other studies, GTPγS, included in the pipette solution, was found to inhibit whole-cell Ca^{2+}

currents (Scott and Dolphin, 1987). Second, whole-cell voltage-clamp experiments in which Ca^{2+} currents with similar properties to those studied in dorsal root ganglia were measured in neuroblastoma x glioma cells showed them to be inhibited also in a PTX-sensitive manner by opioid receptors (Hescheler et al., 1986). Dialysis of such PTX treated cells with pipette solutions containing G_o or G_i protein reconstituted the system, requiring about 10 times lower concentrations of G_o than G_i (Hescheler et al., 1986). Third, injection of a G_o antibody into snail neurons blocks the PTX-sensitive inhibition of Ca^{2+} currents by dopamine (Harris-Warrick et al., 1988).

While clearly implicating G_o as the mediator of the actions of these receptors in

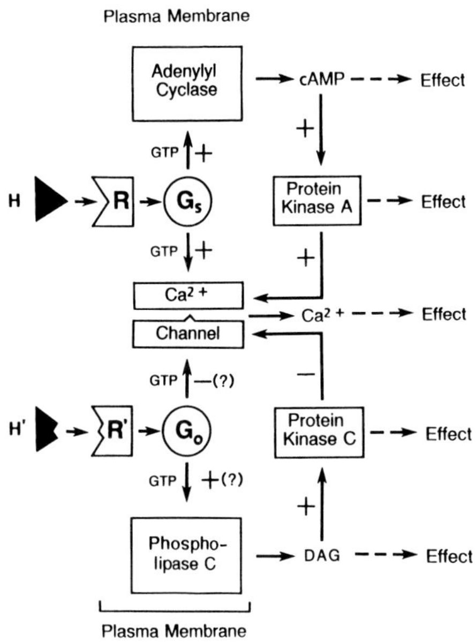

Figure 12. Diagram of dual effects of G_s to stimulate adenylyl cyclase and the voltage-gated dihydropyridine-sensitive Ca^{2+} channel and the dual regulation of this channel by G_s and phosphorylation by cAMP-dependent protein kinase, and of possible direct or indirect inhibitory regulation of Ca^{2+} channels by Ca^{2+} channel inhibitory receptors acting via G_o with putative G_p activity. G_s is shown to stimulate both adenylyl cyclase and the dihydropyridine-sensitive Ca^{2+} channel. G_o is assumed to mediate the inhibitory effect of some R_i-type receptors as experimental evidence dictates (Hescheler et al., 1986; Ewald et al., 1988). G_o is also assumed to be a G_p, i.e., a stimulator of phospholipase C, as suggested from studies in which G_o reconstituted receptor-mediated activation of polyphosphoinositide-specific phospholipase C in PTX-treated membranes (Kikuchi et al., 1986). In turn, the channel is shown to be under stimulatory influence of cAMP-dependent protein kinase (Kameyama et al., 1986; Armstrong and Eckert, 1987) and G_s, and under inhibitory influence of protein kinase C (Rane and Dunlap, 1986) and G_o. The R_s receptor thought to act in a dual way to stimulate Ca^{2+} channel activity is the beta-adrenergic receptor. R_i-type receptors acting to inhibit Ca^{2+} currents in this way are: $GABA_B$ in dorsal root ganglia (Holz et al., 1986), opioid receptors in NG 108-15 cells (Hescheler et al., 1986), somatostatin receptors in pituitary AtT-20 cells (Lewis et al., 1986), dopamine receptors in snail neurons (Harris-Warrick et al., 1988), neuropeptide Y in sensory neurons (Ewald et al., 1988), and possibly, muscarinic receptors in cardiac ventricle cells.

these systems, these experiments did not however address the actual mechanism by which G_o or a closely related G protein, cause inhibition of Ca^{2+} currents. Specifically they could not address whether the effect of the G protein might be direct, without involvement of a second messenger system, or indirect, through perhaps activation of a phospholipase and eventual channel phosphorylation. The latter is a distinct possibility since in dorsal root ganglia the inhibitory effect of neurotransmitters is mimicked in PTX-inactivated cells by protein kinase C activators such as oleylacetyl-glycerol (OAG) (Rane and Dunlap, 1986). However, in view of the synergistic positive regulation of cardiac ventricle and skeletal muscle voltage-gated Ca^{2+} channels by both G_s and protein kinase A-mediated phosphorylation, a direct inhibitory regulation of voltage-gated Ca^{2+} channels by G_o, independent of protein kinase C-mediated phosphorylation, should not be ruled out. The lower portion of Fig. 12 depicts in a schematic way the possibility of dual inhibitory regulation of Ca^{2+} channels by G_o: directly and via protein kinase C phosphorylation due to its phospholipase C stimulatory activity.

Assignment of Function to Purified and Cloned G Proteins

As should be evident from the previous paragraphs there are two nomenclatures for referring to G proteins. One has a functional basis. It defines G_i as the mediator of hormonal inhibition of adenylyl cyclase, G_p as stimulators of phospholipases of the C and A_2 types, and G_k as the stimulator of K^+ channels. The other has a structural basis, and although it began with the naming of purified proteins with real or perceived functional identity, it has now changed and refers to cloned α subunits of which there are currently 12 vertebrate forms known: four α_s, three α_i, one α_o, one $\alpha_{Z/X}$; and three sensory: two α_t and one α_{olf}. Except for α_s, α_{olf}, and $\alpha_{Z/X}$, all others are substrates for PTX.

A large part of current efforts centers in several laboratories on defining how many G proteins there are, which effector function is regulated by which G proteins and which G protein regulates more than one effector function. It is likely that not all G proteins have been found as yet. Additionally, postranslational modifications may create as yet unsuspected, functionally important G protein diversity.

Superimposed on the problem of finding the complete set of signal transducing G proteins, is the problem of how correct functional assignments may be when they are based on activities of purified proteins. This problem of course relates directly to the purity of preparations used, leading to the question: is the activity elicited with a given protein fraction due to its majority component or to one of the contaminants? G proteins are so similar, that even though the purified preparations appear to have no significant contaminants, comigrations of α subunits cannot be excluded. This applies particularly, but not necessarily only, to PTX-sensitive G proteins.

The same problem also applies to the assignment of structure to a given function present in a purified protein. For example, α subunits of human erythrocyte G_i protein were recently isolated by us by 10% SDS-PAGE followed by electroelution. Their amino termini were blocked. To obtain a partial amino acid sequence, they were proteolyzed and several of the proteolytic fragments sequenced (Codina et al., 1988). Two such fragments, isolated from independent proteolytic reactions, each had a different sequence composition that identified it as having originated from an α subunit encoded by the α_i-3 gene (Codina et al., 1988). Based on the assumption that in each case the sequence had been derived from the majority α subunit present in the starting protein, and on the basis that the starting protein is a potent activator of the G protein-sensitive K^+ channels, it was concluded that α_i-3 is the α subunit responsible for G_k activity of the purified protein (Codina et al., 1988). However, the possibility existed, and could not be eliminated, that by chance the sequenced peptides had been derived from the contaminant, in which case the conclusion might not have been justified.

Finally, just as there are functionally identified G proteins that need to find their structural counterparts, and purified and/or partially purified G proteins of which the function is either unknown or doubtful, there are G protein α subunits known through cloning in need of functional identification. This applies not only to the new $\alpha_{Z/X}$, but also to each of the three G_i molecules and the various splice variants of G_s.

Expression of Cloned α Subunits as a Means to Assign Functions to Individual G Proteins

The advent of recombinant DNA techniques, bacterial expression, as well as in vitro synthesis of cloned molecules has opened the possibility to test and/or confirm the assignment of function to purified polypeptides and/or to cloned molecules for which no function is known.

A search for an adequate bacterial expression vector to synthesize α subunits encoded in their respective cDNA molecules led to the use of pT7-plasmids developed by Tabor and Richardson at Harvard University (Tabor and Richardson, 1985; Tabor et al., 1987). Expression is carried out in *Escherichia coli* K38 cells carrying the pT7 plasmid with the cDNA insert of interest α_s, α_i-1 through -3 and α_o) for the present studies. Because of the nature of the recombinant construction, 27 nucleotides are added to the 5′ end of the region that codes for the α subunit cDNAs, which leads to the addition of nine amino acids to the amino terminus of the expressed α subunits. As shown in Fig. 13, the analysis of total bacterial cell proteins after induction of expression, reveals the accumulation of polypeptides that depend on the composition and orientation of the cDNA inserted into the pT7-7 vector, and which can account for up to 5–8% of total cell protein. α subunits made in this way were then tested for function: α_s for reconstitution of the *cyc*$^-$ adenylyl cyclase system and stimulation of skeletal muscle Ca^{2+} channels in lipid bilayers, and one of the α_i molecules, suspected of being an α_k, for stimulation of single-channel K^+ currents in inside-out atrial membrane patches.

There are advantages and drawbacks in synthesizing α subunits of G proteins in bacteria. The principal advantage is that the product obtained is absolutely free of any other α subunit. A second advantage is that, potentially, this approach could inform us about the function of a G protein that, although cloned, has not yet been purified and about which there is no functional information available. In this respect, a test for specificity of action of *recombinant* α subunits obtained from bacteria expressing the sense strand of the cDNA inserts shown in Fig. 13, was conducted using activation of *cyc*$^-$ adenylyl cyclase as an assay. Only *recombinant* α_s, but neither *recombinant* α_o nor *recombinant* α_i-1, -2, or -3, stimulated the adenylyl cyclase system. Thus, as far as this assay is concerned, *recombinant* α subunits have the effector specificity expected from studies with native proteins.

The drawbacks are of two kinds. One is that even though the levels of α subunit protein synthesized by bacteria can be very high, making purification a relatively easy task, only a fraction, 5–10%, can be recovered as a soluble protein after bacterial lysis. However, purification is possible (Graziano et al., 1987, 1989), so that this drawback is more of a technological than conceptual kind.

The second drawback is more serious, for it relates to the "intrinsic" activity of the bacterially made proteins. On assaying the soluble portion of *recombinant* α_s molecules for *cyc*$^-$ reconstituting activity, it was found that these molecules are active, but only at high concentrations. In fact, equal stimulation of adenylyl cyclase activity requires about 100-fold higher concentrations of *recombinant* α_s than of native α_s. The same was found with *recombinant* α_s made by using a different expression plasmid that does not add nine amino acids to the amino terminal of the molecule (Graziano et al., 1987), indicating that reduced "intrinsic" activity lies in the mode of biosynthesis and not in the primary composition of the polypeptide chain. The reasons for this low intrinsic activity were studied in detail and attributed to a lowered affinity of the α_s molecule for the catalytic unit of adenylyl cyclase (Graziano et al., 1989), while other properties of bacterially made α_s subunits are normal. These include: rates of GTP hydrolysis, GTP binding, and GDP release and interaction with $\beta\gamma$ dimers and receptors. It is likely that the lowered affinity of *recombinant* α_s is due to omission by bacteria of one or more posttranslational modifications that eukaryotic cells can, but bacteria cannot, perform.

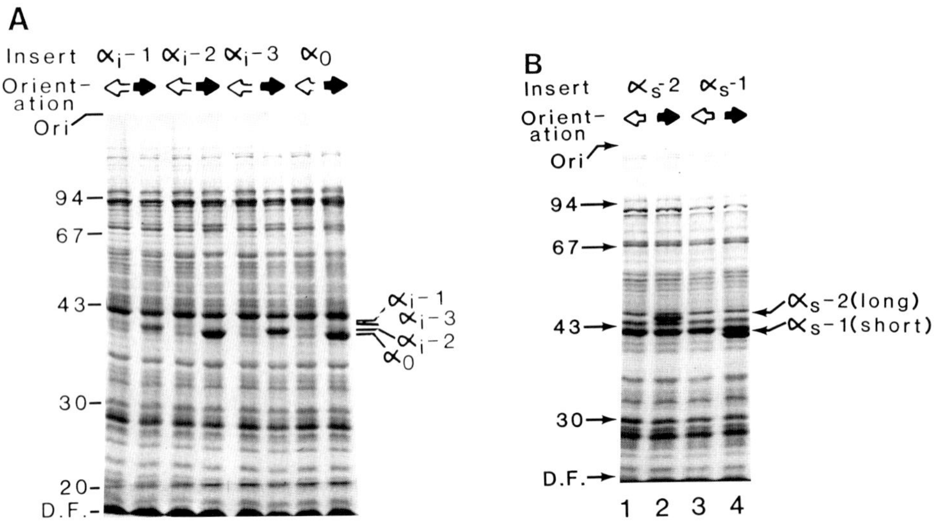

Figure 13. Bacterial expression of α subunit fusion proteins. cDNAs were inserted by blunt-end ligation into the Bam HI site of pT7-7 plasmid. *Recombinant* pT7-α_x plasmids with cDNAs inserted in either the antisense (3′ to 5′, *open arrows*) or sense (5′ to 3′, *filled arrows*) direction were transfected into *E. coli* K38 cells and expressed after infection of the cells with phage mGPl-2. Cells were harvested, dissolved in Laemmli's sample buffer and analyzed by SDS-PAGE in 10% polyacrylamide gel slabs. The slabs were stained with Coomassie blue and photographed. Note the appearance of insert-specific and orientation-dependent α subunit bands that constitute between 5 and 8% of the total cell protein.

Effects of *recombinant* α subunits on activity of ionic channels. As mentioned earlier it has been well established that the cardiac and GH_3 cell's voltage-gated dihydropyridine-sensitive Ca^{2+} channel is positively regulated by cAMP-dependent phosphorylation (Kameyama et al., 1986; Armstrong and Eckert, 1987). Our finding that G_s preparations stimulate the Ca^{2+} channel directly, suggests that these channels are apparently under dual and redundant regulation by G_s. The question that needed therefore to be addressed was whether it was G_s that caused the stimulation of Ca^{2+} channel activity, or whether it may have been the action of a closely related but nevertheless distinct G protein. The ability to express *recombinant* α_s molecules in bacteria that are active as stimulators of adenylyl cyclase activity, albeit with low potency, offered the opportunity to test whether the biological activity seen with *cyc*⁻

membranes had a correlate in stimulating the activity of Ca^{2+} channels incorporated into lipid bilayers. Fig. 14 illustrates that indeed *recombinant* α_s, but not recombinant α_i-3, stimulates the activity of the skeletal muscle T tubule Ca^{2+} channel. The *recombinant* α_s tested in the experiment of Fig. 14 is the 379 amino acid (short, Ser^-) splice variant. The same results were obtained with *recombinant* α_s 380 (short, Ser^+) and 394 (long, Ser^-) amino acid long as subunits (Mattera, R., M.P. Graziano, A.

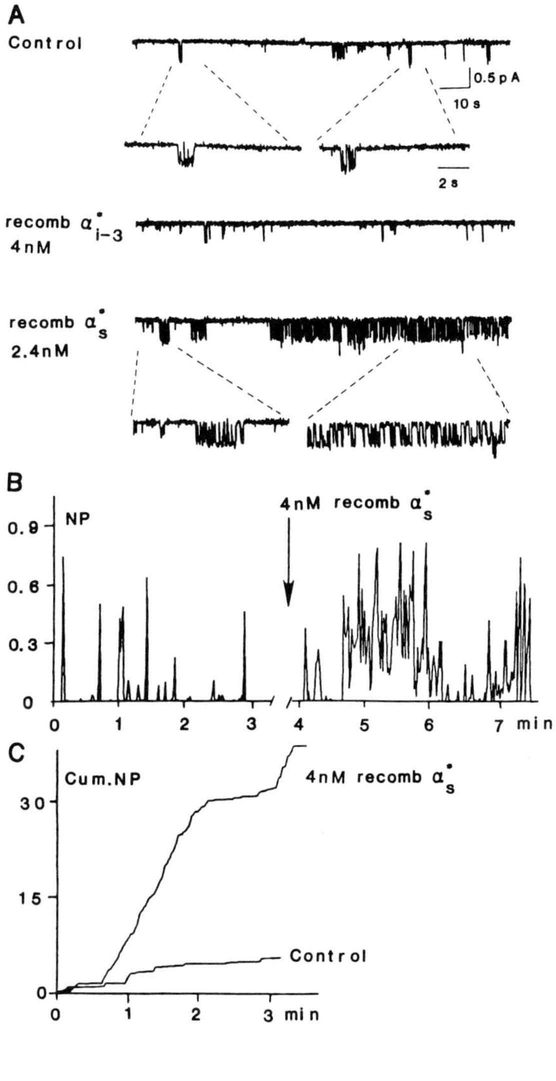

Figure 14. Stimulation of skeletal muscle T tubule dihydropyridine-sensitive Ca^{2+} channel incorporated into lipid bilayers by the short form of *recombinant* a_s. *Recombinant* α_s (Asp^{71}-Glu^{72} version) and *recombinant* α_i-3 were prepared in their GTPγS-activated form and partially purified by diethylaminoethyl (DEAE)-Sephacel chromatography. (*A*) Single-channel Ba^{2+} currents (*trans* to *cis*) before and after addition first of GTPγS-activated *recombinant* α_i-3 and then of GTPγS-activated *recombinant* α_s. (*B* and *C*) NP_o diaries of Ca^{2+} channel activities before and after addition of GTPγS-activated recombinant α_s to the *cis* chamber (*B*) and cumulative NP_o values as a function of time of the same experiment (*C*).

Yatani, J. Codina, L. Birnbaumer, A.G. Gilman, and A.M. Brown, unpublished observations).

Expression in bacteria and testing for K^+ channel stimulatory activity allowed for a direct testing of the hypothesis that α_i-3 has α_k activity. As shown in Fig. 15, recombinant α_i-3, activated either by GTPγS or AlF_4^-, does indeed stimulate the atrial "muscarinic" K^+ channel.

All three *recombinant* α_i proteins have α_k activity. When *recombinant* α_i-1 and *recombinant* α_i-2 were tested, it was found that they also stimulated the atrial muscarinic K^+ channel, in a manner indistinguishable in terms of potency and channel kinetic properties from that seen with *recombinant* α_i-3 (Yatani et al., 1988*b*). Stimulation of the muscarinic K^+ channel required about 30–50 times higher concentrations of any of the *recombinant* α_i than native hRBC α_i-3.

The following tests proved that the effects observed with each of the *recombinant* $\alpha_i s$ are intrinsic properties of the expressed molecules. (*a*) Boiling prior to addition to the bath obliterated the activity of each *recombinant* α_i. (*b*) Threshold concentrations of GTPγS for activation of atrial muscarinic K^+ channels in inside-out membrane patches when 100 nM carbachol was in the pipette were between 10 and 100 nM, which is at least 10 times the maximum GTPγS added with saturating concentrations (1,000 pM) of any of the GTPγS-activated partially purified and ultrafiltrated *recombinant* α_i preparations. (*c*) Prior addition of 100 μM GDPβS, which blocks

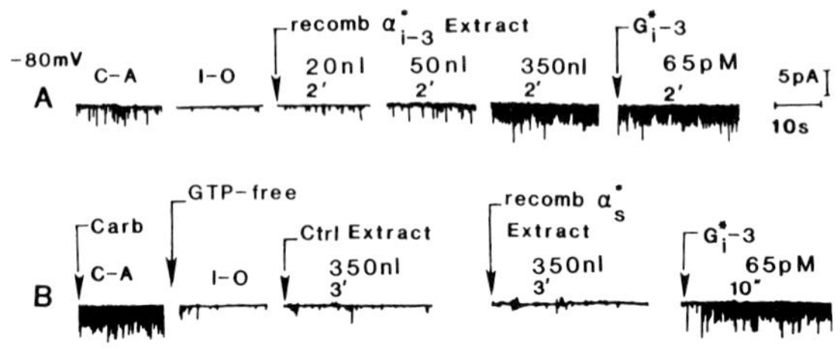

Figure 15. Stimulation of atrial G_k-gated channels by GTPγS-treated lysate supernatants (*Extracts*) obtained from bacteria that synthesized *recombinant* α_i-3, but not from control bacteria (*Ctrl*) or from bacteria that synthesized *recombinant* α_s. Two representative experiments are shown in which a gigaseal was established between an atrial cell and a patch pipette without (*A*) or with (*B*) 100 nM carbachol. (*A*) Effect of increasing volumes of an extract containing 120 nM GTPγS-activated *recombinant* α_i-3 (*recomb* α_i^**-3 Extract*). (*B*) Failure of either a GTPγS-treated extract obtained from cells transformed with pT7-α_i-3 with the cDNA inserted in the antisense orientation or of 350 nl of 90 nM (final concentration 315 pM) GTPγS-activated *recombinant* α_s (*recomb* α_s^* *Extract*), to stimulate single K^+ channel currents when added to an atrial patch that responded to 65 pM human erythrocyte G_i^*-3. The numbers above the records indicate elapsed minutes (′) or seconds (″) between addition of test substances and the initiation of the record shown.

carbachol-mediated effects of 10 μM GTP, did not interfere with the effects of GTPγS-activated *recombinant* α_i-1, -2, or -3.

The results with *recombinant* α_i-2 and -1, were surprising and required independent confirmation so as to eliminate the possibility that their channel-stimulating activity might be artifactual, such as could occur because of incomplete or abnormal posttranslational processing by bacteria. To this effect, both native G_i-2 and G_i-1 were purified to ~90% purity from human erythrocytes and bovine brain respectively (Birnbaumer et al., 1988; Yatani et al., 1988*b*). The purified proteins were activated with GTPγS and Mg^{2+}, and the activated α_i^* subunits were separated from the $\beta\gamma$ dimers by DEAE-Toyopearl chromatography. Both human erythrocyte α_i^*-2 and bovine brain α_i^*-1 exhibited potent α_k activity, which on a quantitative basis is essentially the same as that of the human erythrocyte α_i^*-3 studied previously.

Inhibition of G_k Action by Anti-α Subunit Monoclonal Antibody (MAb) 4A

Further support for the contention that α subunits are mediators of receptor stimulation, came from studies with a MAb 4A. This antibody was made against the α subunit of transducin (Hamm and Bownds, 1984; Hamm et al., 1987). The epitopes of α_t recognized by MAb 4A have been defined (Hamm et al., 1988) and are also present on the α subunit of G_is. When added to inside-out guinea pig atrial membrane patches stimulated by carbachol and GTP, MAb 4A blocks muscarinic activation of atrial K^+ channels in an irreversible manner (Fig. 16). MAb 4A also immunoneutralizes K^+

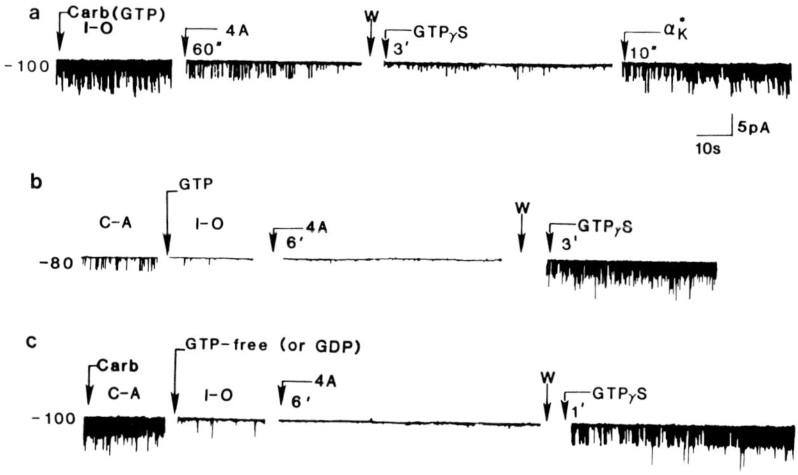

Figure 16. Monoclonal anti-transducin antibody 4A (MAb 4A) interacts with α_k and blocks irreversibly carbachol-stimulated K^+ currents, but does not interact with unactivated G_k: evidence against a stimulatory effect of naturally formed $\beta\gamma$ dimers. *Experiment a:* exposure of the cytoplasmic aspect of a membrane patch stimulated by carbachol in the pipette and GTP in the bath to MAb 4A results in a persistent block of the membrane G_k to activation by exogenously added GTPγS. *Experiments b and c:* exposure of the membrane patch to MAb 4A under conditions that do not lead to G_k activation, also do not lead to a persistent block to activation by GTPγS after MAb 4A washout. Other experiments showed that the block is specific for MAb 4A, neither nonspecific IgG nor another anti-α-transducin monoclonal antibody with a different epitope specificity blocked the effect of carbachol. When GDP (100 μM) was used instead of GTP in experiments such as shown in the middle, MAb 4A had likewise no persistent effect. (Adapted from Yatani et al., 1988*c*.)

channel activation by GTPγS-preactivated hRBC G_i-3 or GTPγS-preactivated hRBC α_i-3 (Yatani et al., 1988*c*).

Conflicting Data from Other Investigators

Clapham, Neer, and collaborators (Logothetis et al., 1987) initially reported that neither bovine brain GTPγS-activated α_i (α^{*}_{41}) nor bovine brain GTPγS-activated α_o (α^{*}_{39}) stimulated atrial muscarinic K^+ channels. More recently they have reported that the α_o-type protein does indeed have a stimulatory effect while the α_i-type molecule continues consistently to fail in stimulating the muscarinic channel (Logothetis et al., 1988). Our findings differed (Yatani et al., 1987*a*), as our purified bovine brain G_o has only ~1% the G_k activity of purified hRBC G_i, now known to be G_i-3. We have been unable to determine whether this activity is due to a weak activity intrinsic to G_o or to the presence in the G_o preparation of contaminating G_i-type protein(s). Since we find that all three types of G_i have G_k activity, contamination with any G_i (or α_i) would suffice.

One explanation for the discrepancies between the two laboratories with respect to activities of G_i proteins is that the bovine brain G_i prepared in Dr. Neer's laboratory contains not only G_i—positive results would be expected regardless of the exact identity (α_i-1, α_i-2, or α_i-3) of their α^*_{41} PTX substrate—but also an inhibitor of its action on K^+ channels. As yet no attempts have been made to test the effects of mixtures of active α^*_{39} plus "inactive" α^*_{41}, which may shed light on the possible presence of such an inhibitor.

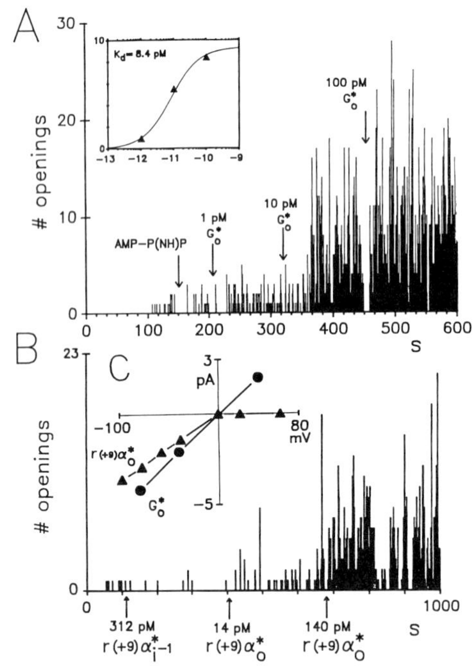

Figure 17. Bovine brain G_o (*A*) and *recombinant* α_o, but not *recombinant* α_i-1 (*B*). gate single K^+ channel currents in hippocampal pyramidal cells of neonatal rats. (*A*) The effect of increasing concentrations of G^*_o on a 55 pS K^+ channel is shown by plotting the number of openings per 0.8 s as a function of time of continuous recording. (*Inset*) Openings per 0.8 s averaged for 1 min as a function of G^*_o concentration. (*B*) Lack of effects of 312 pM *recombinant* α^*_i-1, and stimulatory effect of increasing concentrations of *recombinant* α^*_o on a 40 pS K^+ channel, plotted as number of openings per 0.2 s as a function of the time of continuous recording shown. Single K^+ channel currents were recorded from excised inside-out membrane patches as in Fig. 6 in the presence of 0.2 mM AMP-P(NH)P. Holding potential was −80 mV and both pipette and bath solutions contained 140 mM K-methanesulfonate, 1 mM EGTA, 1 mM $MgCl_2$, and 10 mM HEPES adjusted to pH 7.4 with Tris-base (Adapted from VanDongen et al., 1988.)

Results obtained by Clapham, Neer, and collaborators with $\beta\gamma$ dimers also differ from results obtained by us. As shown in Fig. 7, addition of hRBC $\beta\gamma$ dimers has no stimulatory effects on either atrial or GH_3 G_k-gated K^+ channels, and, when added at high (2–4 nM) concentrations to atrial membrane patches, stimulated by carbachol in the pipette and GTP in the bath, it caused inhibition of signal transduction. In contrast, Clapham and collaborators found that bovine brain (Logothetis et al., 1987) and human placental (Logothetis et al., 1988) $\beta\gamma$ dimers activate the muscarinic K^+

channel when used at similar concentrations, which are much higher than those required by us to obtain stimulation with α_i subunits.

It is not possible at this time to reconcile the opposing results obtained by the two laboratories. As pointed out elsewhere (Kirsch et al., 1988), there are differences in the buffer systems, specifically the detergents, used to keep $\beta\gamma$ dimers in solution. We use Lubrol PX, which according to Clapham and collaborators inhibits G protein-gated K^+ channel activity, but which at the concentrations used by us does not; and Neer's proteins are in CHAPS (3-[(3-cholamido-propyl)dimethyl-ammonio] – 1-propanesulfonate), which we find is a potent activator of these channels (Kirsch et al., 1988), an effect that Clapham and collaborators do not find (Logothetis et al., 1988).

Assuming that $\beta\gamma$ dimers under special conditions, or that special $\beta\gamma$ dimers may stimulate atrial muscarinic K^+ channels, one may question whether such observed effects are of physiological importance. Two lines of evidence seem to deny this possibility. One is that effects of $\beta\gamma$ dimers when obtained, require about 100-fold higher concentrations than α subunits (Kirsch et al., 1988). Since receptor-mediated activation of a G protein generates equimolar concentrations of α subunits and $\beta\gamma$ dimers, it is difficult to see how, even if both are stimulatory in nature, $\beta\gamma$ dimers could be mediating an effect when α subunits saturate the system at concentrations that are below those required for minimal effects with $\beta\gamma$ dimers.

The other line of evidence emerges from the experiments with MAb 4A (Fig. 16). If, as it is commonly assumed, G protein activation is accompanied by subunit dissociation, then the MAb 4A block of muscarinic stimulation proves that physiologically formed $\beta\gamma$ dimers do not stimulate the K^+ channel. If, on the other hand, G proteins do not dissociate under physiologic conditions, positive effects of in vitro added $\beta\gamma$ dimers are likely to be a pharmacologic curiosity of little physiologic consequence.

G Protein Gating as a Tool to Discover Novel Ionic Channels: Neuronal G_o-gated K^+ Channels

One of the properties of the "muscarinic" K^+ channels is that they are essentially silent in the absence of stimulation by an activated G protein (G_k). That is, in the absence of activated G protein their P_o is close to zero. The possibility existed that not only G_i proteins regulate K^+ channels but also the structurally closely related G_o. Since nervous tissue is rich in G_o, central nervous system neurons, specifically hippocampal pyramidal cells, were placed into culture and studied for the potential presence of both G_i- and G_o-gated K^+ channels. Application of purified bovine brain G_o^* to the cytoplasmic aspect of inside-out membrane patches of cultured hippocampal pyramidal cells resulted in the appearance of three new types of single-channel K^+ currents, which is consistent with the existence of the three nonrectifying of K^+ channels having sizes of 13, 40, and 55 pS, respectively, plus an inwardly rectifying K^+ channel with a slope conductance of 40 pS. No such channel activities were observed with hRBC G_i^*-3 or hRBC α_i^*-3. The effect of increasing concentrations of bovine brain G_o^* on the 55-pS–type K^+ channel is shown in Fig. 17 *A*. Note that in contrast to earlier observations with the same preparation of G_o^* added to guinea pig atrial membrane patches (see Fig. 6 *A*) the hippocampal K^+ channel is highly sensitive to G_o^*. Significant activation was obtained at 1 pM and half-maximal effects were obtained at ~10 pM.

To confirm that the channels observed on addition of G_o^* are indeed gated by G_o and not by a contaminant, the effects of partially purified *recombinant* α^* molecules, obtained by the pT7 expression method discussed above were examined. All of the above-mentioned types of K^+ channel were stimulated by *recombinant* α_o^*, under conditions where prior addition of one of the *recombinant* α_i^* preparations, active on guinea pig atrial muscarinic K^+ channels, had no effect (Fig. 17 *B*). The G_o-gated channels were stimulated in the absence of Ca^{2+} or ATP, presence of AMP-P(NH)P, added routinely to inhibit ATP-sensitive 70-pS K^+ channels, and EGTA did not interfere with the actions of G_o or *recombinant* α_o. Thus, in hippocampal pyramidal cells of the rat, G_o is a G_k, and the K^+ channels gated by it are several and differ from those present in atrial cells in various aspects including G protein specificity.

Conclusions

Signal transduction by G proteins is a fundamental and widespread mechanism used by a wide variety of hormones, neurotransmitters, and auto- and paracrine factors to regulate cellular functions. G proteins modulate not only cAMP formation, but also intracellular Ca^{2+} mobilization, arachidonic acid release and, very importantly, membrane potential. The latter is a trigger for neurotransmitter release. In tissues such as secretory cells, membrane potential is the main regulator of Ca^{2+} entry. In heart, action potentials play the dual role of determining the frequency of contraction, and through modulation of their duration, Ca^{2+} entry and the force of contraction is determined. More subtle changes in resting membrane potential alter the cell's predisposition to be stimulated by other factors and hormones.

The mechanism by which G proteins are activated provides for amplification, reversal of action, and continued monitoring of hormone. For example, an amplification by few receptor molecules may act catalytically to activate many G protein molecules and reversal of action results from a turnoff mechanism inherent in $G\alpha$ subunits via hydrolysis of GTP to GDP. Continued monitoring of the primary messenger level can be accomplished since each activation cycle requires not only GTP but also occupied receptor.

Work is in progress to unravel a complicated network of interactions between receptors, G proteins, and effector systems, that not only affects regulation of metabolic activities of organs such as liver, heart, and fat, but also of the integrative functions of the central nervous system.

The present results show that three distinct $G_i\alpha$ subunits, all PTX substrates, are equipotent activators of one effector, an atrial K^+ channel. This raises issues concerning specificity and important questions relating to the identity of the G proteins that mediate the other PTX-sensitive effects described earlier (Ohta et al., 1985; Murayama and Ui, 1985; Burch et al., 1986; Kikuchi et al., 1986; Lewis et al., 1986; Hescheler et al., 1988; Hirsch et al., 1988). Our results would imply that one or more of the $G_i\alpha$ subunits are multifunctional and involved, at least partially, in at least one of these responses. That a $G_i\alpha$ may be multifunctional may be concluded on analyzing data available on G_i-1. Thus, GTPγS-activated $G_i\alpha$-1 chains derived from bovine brain preparations are active in stimulating atrial K^+ currents (Szabo et al., 1985), and what was probably a mixture of $G_i\alpha$-1 and $G_i\alpha$-2 (Mumby et al., 1988) inhibited *cyc*$^-$ adenylyl cyclase (Roof et al., 1985). Also, injection of nonactivated brain G_i, possibly a mixture of G_{40} (G_i-2) and G_{41} (G_i-1), into PTX-treated adrenal Y-1 (Hescheler et al.,

1988) or pituitary GH_3 cells (Hirsch et al., 1988) reconstitutes stimulation of Ca^{2+} currents under voltage-clamp conditions, an effect that cannot possibly be due to activation of K^+ channels. It is of course not known whether isoforms with respect to K^+ channel regulation may not also be isoforms with respect to any other effector systems that they regulate. Use of individual recombinant $G_i\alpha$ subunits, which exhibit fidelity in the quality of the effects they elicit and which are unequivocally free of any other $G_i\alpha$, should aid in defining which G_i protein has which other actions.

The finding reported here that all three forms of G_i are isoforms with respect to K^+ channel regulation, leads to the conclusion that one or more of these G proteins must be multifunctional. The same can be said for G_o, which clearly stimulates more than one neuronal G protein-gated K^+ channel, and may in addition affect (inhibit) certain voltage gated Ca^{2+} channels. A summary scheme of G protein actions on K^+ channels is illustrated in Fig. 18. In the world of signal transduction, a G protein may not simply regulate an effector, but it executes a signal transduction program that coordinates responses to single receptor inputs.

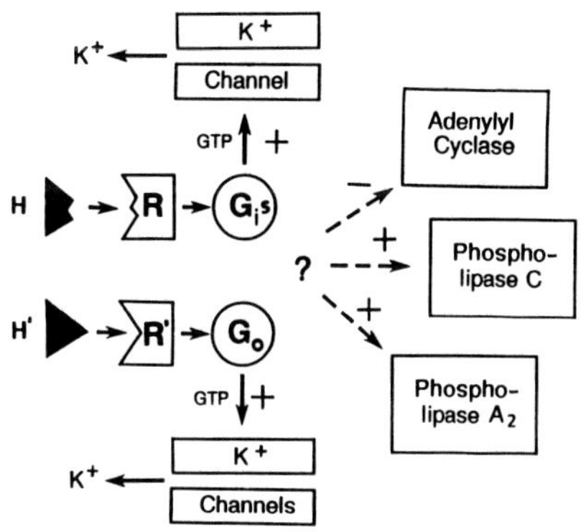

Figure 18. The scheme depicts the existence of G_i-gated and G_o-gated K^+ channels and the fact that each of the G proteins is pleotypic in its effects, i.e., able to regulate more than one effector function.

References

Affolter, H., and R. Coronado. 1986. Sidedness of reconstituted calcium channels from muscle transverse tubules as determined by D600 and D890 blockade. *Biophysical Journal.* 49:767–771.

Aksamit, R. R., P. S. Backlund, Jr., and G. L. Cantoni. 1985. Cholera toxin inhibits chemotaxis by a cAMP-independent mechanism. *Proceedings of the National Academy of Sciences.* 82:7475–7479.

Andrade, R., R. C. Malenka, and R. A. Nicoll. 1986. A G protein couples serotonin and $GABA_B$ receptors to the same channels in hippocampus. *Science.* 234:1261–1265.

Armstrong, D., and R. Eckert. 1987. Voltage-activated calcium channels that must be phosphorylated to respond to membrane depolarization. *Proceedings of the National Academy of Sciences.* 84:2518–2522.

Birnbaumer, L., and A. M. Brown. 1989. Signal transduction by G proteins. *Biochimica et Biophysica Acta.* In press.

Birnbaumer, L., J. Codina, R. Mattera, A. Yatani, R. Graf, J. Olate, J. Sanford, and A. M. Brown. 1988. Receptor-effector coupling by G proteins. Purification of human erythrocyte G_i-2 and G_i-3 and analysis of effector regulation using recombinant α subunits synthesized in *E. coli. Cold Spring Harbor Symposium on Quantitative Biology.* 53:229–239.

Birnbaumer, L., and M. Rodbell. 1969. Adenyl cyclase in fat cells. II. Hormone receptors. *Journal of Biological Chemistry*. 244:3477–3482.

Breitwieser, G. E., and G. Szabo. 1985. Uncoupling of cardiac muscarinic and beta-adrenergic receptors from ion channels by a guanine nucleotide analogue. *Nature*. 317:538–540.

Brown, A. M., and L. Birnbaumer. 1988. Direct G protein gating of ion channels. *American Journal of Physiology*. 254(*Heart and Circulation Physiology* 23):H401–H410.

Burch, R. M., A. Luini, and J. Axelrod. 1986. Phospholipase A_2 and phospholipase C are activated by distinct GTP-binding proteins in response to alpha$_1$-adrenergic stimulation in FRTL5 cells. *Proceedings of the National Academy of Sciences*. 83:7201–7205.

Cerbai, E., U. Kloeckner, and G. Isenberg. 1988. The α subunit of the GTP binding protein activates muscarinic potassium channels of the atrium. *Science*. 240:1782–1784.

Cockroft, S., and B. D. Gomperts. 1985. Role of guanine nucleotide binding protein in the activation of polyphosphoinositide phosphodiesterase. *Nature*. 314:534–536.

Codina, J., A. Yatani, D. Grenet, A. M. Brown, and L. Birnbaumer. 1987*a*. The α subunit of the GTP-binding protein G_k opens atrial potassium channels. *Science*. 236:442–445.

Codina, J., D. Grenet, A. Yatani, L. Birnbaumer, and A. M. Brown. 1987*b*. Hormonal regulation of pituitary GH_3 cell K^+ is mediated by its alpha subunit." *FEBS Letters*. 216:104–106.

Codina, J., J. Olate, J. Abramowitz, R. Mattera, R. G. Cook, and L. Birnbaumer. 1988. α_i-3 cDNA encodes the α subunit of G_k, the stimulatory G protein of receptor-regulated K^+ channels. *Journal of Biological Chemistry*. 263:6746–6750.

Coronado, R., and H. Affolter. 1986. Insulation of the conduction pathway of muscle transverse tubule calcium channels from the surface charge of bilayer phospholipid. *Journal of General Physiology*. 87:933–953.

Ewald, D. A., P. C. Sternweis, and R. J. Miller. 1988. Guanine nucleotide binding protein G_o induced coupling of neuropeptide Y receptors to Ca^{2+} channels in sensory neurons. *Proceedings of the National Academy of Sciences*. 85:3633–3637.

Fong, H. K., K. K. Yoshimoto, P. Eversole-Cire, and M. I. Simon. 1988. Identification of a GTP-binding protein α subunit that lacks an apparent ADP-ribosylation site for pertussis toxin. *Proceedings of the National Academy of Sciences*. 85:3066–3070.

Fung, B. K.-K. 1983. Characterization of transducin from bovine retinal rod outer segments. I. Separation and reconstitution of the subunits. *Journal of Biological Chemistry*. 256:10495–10502.

Fung, B. K.-K., J. B. Hurley, and L. Stryer. 1981. Flow of information in the light-triggered cyclic nucleotide cascade of vision. *Proceedings of the National Academy of Sciences*. 78:152–156.

Gilman, A. G. 1987. G proteins: transducers of receptor-generated signals. *Annual Review of Biochemistry*. 56:615–649.

Graziano, M. P., P. J. Casey, and A. G. Gilman. 1987. Expression of cDNAs for G proteins in *Escherichia coli*. Two forms of $G_{s\alpha}$ stimulate adenylate cyclase. *Journal of Biological Chemistry*. 262:11375–11381.

Graziano, M. P., M. Frissmuth, and A. G. Gilman. 1989. Expression of G_s alpha in *Escherichia coli*. Purification and properties of two forms of the protein. *Journal of Biological Chemistry*. 264:409–418.

Gregerson, K. A., L. Einhorn, M. M. Smith, and G. S. Oxford. 1989. Modulation of potassium channels by dopamine in rat pituitary lactotrophs: a role in the regulation of prolactin secretion? *In* Secretion and Its Control. G. S. Oxford and C. M. Armstrong, editors. Rockefeller University Press, New York.

Hamm, H. E., and M. D. Bownds. 1984. A monoclonal antibody to guanine nucleotides binding protein inhibits the light-activated cyclic GMP pathway in frog rod outer segments. *Journal of General Physiology*. 84:265–280.

Hamm, H. E., D. Deretic, A. Arendt, P. A. Hargrave, B. Koenig, and K. P. Hofmann. 1988. Site of G protein binding to rhodopsin mapped with synthetic peptides from the alpha subunit. *Science*. 241:832–835.

Hamm, H. E., D. Deretic, K. P. Hofmann, A. Schleicher, and B. Kohl. 1987. Mechanism of action of monoclonal antibodies that block the light activation of the guanyl nucleotide-binding protein, transducin. *Journal of Biological Chemistry*. 262:10831–10838.

Harris-Warrick, R. M., C. Hammond, D. Paupardin-Tritsch, V. Homburger, B. Rouot, J. Bockaert, and H. M. Gershenfeld. 1988. An α_{40} subunit of a GTP-binding protein immunologically related to mammalian G_o mediates a dopamine-induced decrease of calcium current in snail neurons. *Neuron*. 1:27–32.

Hartzell, H. C. 1981. Mechanisms of slow postsynaptic potentials. *Nature*. 291:539–544.

Hartzell, H. C., S. W. Kuffler, R. Stickgold, and D. Yoshikami. 1977. Synaptic excitation and inhibition resulting from direct action of acetylcholine on two types of chemoreceptors on individual amphibian parasympathetic neurones. *Journal of Physiology*. 271:817–846.

Hescheler, J., W. Rosenthal, K.-D. Hirsch. M. Wulfern, W. Trautwein, and G. Schultz. 1988. Angiotensin II-induced stimulation of voltage-dependent Ca^{2+} currents in an adrenal cortical cell line. *EMBO Journal*. 7:619–624.

Hescheler, J., W. Rosenthal, W. Trautwein, and G. Schultz, 1986. The GTP-binding protein, N_o, regulated neuronal calcium channels. *Nature*. 325:445–447.

Hildebrandt, J. D., J. Codina, R. Risinger, and L. Birnbaumer. 1984. Identification of a gamma subunit associated with the adenylyl cyclase regulatory proteins N_s and N_i. *Journal of Biological Chemistry*. 259:2039–2042.

Hildebrandt, J. D., R. D. Sekura, J. Codina, R. Iyengar, C. R. Manclark, and L. Birnbaumer. 1983. Stimulation and inhibition of adenylyl cyclases is mediated by distinct proteins. *Nature* 302:706–709.

Hill-Smith, I., and R. D. Purves. 1978. Synaptic delay in the heart: an ionophoretic study. *Journal of Physiology*. 270:31–54.

Hirsch, K.-D., K. Spicher, W. Trautwein, W. Rosenthal, J. Hescheler, and G. Schultz. 1988. Cyclic AMP-independent, dual regulation of voltage-dependent Ca^{2+} currents by LHRH and somatostatin in a pituitary cell line. *EMBO Journal*. 7:1627–1633.

Holz, G. G., S. G. Rane, and K. Dunlap. 1986. GTP-binding proteins mediate transmitter inhibition of voltage-dependent calcium channels. *Nature*. 319:670–672.

Imboden, J. B., D. M. Shoback, G. Pattison, and J. D. Stobo. 1986. Cholera toxin inhibits the T-cell antigen receptor-mediated increases in inositol trisphosphate and cytoplasmic free calcium. *Proceedings of the National Academy of Sciences*. 83:5673–5677.

Imoto, Y., A. Yatani, J. P. Reeves, J. Codina, L. Birnbaumer, and A. M. Brown. 1988. The α subunit of G_s directly activates cardiac calcium channels in lipid bilayers. *American Journal of Physiology*. 255:H722–H728.

Jones, D. T., and R. R. Reed. 1988. Olfactory signal transduction utilizes a novel GTP-binding protein. *Cold Spring Harbor Symposia on Quantitative Biology*. 53:In press.

Kameyama, M., J. Hescheler, F. Hofmann, and W. Trautwein. 1986. Modulation of Ca current during the phosphorylation cycle in the guinea pig heart. *Pflügers Archiv*. 407:121–128.

Kikuchi, A., O. Kozawa, K. Kaibuchi, T. Katada, M. Ui, and Y. Takai. 1986. Direct evidence for involvement of a guanine nucleotide-binding protein in chemotactic peptide-stimulated formation of inositol bisphosphate and trisphosphate in differentiated human leukemic (HL-60) cells. Reconstitution with G_i or G_o of the plasma membranes ADP-ribosylated by pertussis toxin. *Journal of Biological Chemistry*. 261:11558–11562.

Kirsch, G., A. Yatani, J. Codina, L. Birnbaumer, and A. M. Brown. 1988. The alpha subunit of G_k activates atrial K^+ channels of embryonic chick, neonatal rat and adult guinea pig. *American Journal of Physiology*. 254:H1200–H1205.

Kurose, H., and M. Ui. 1983. Functional uncoupling of muscarinic receptors from adenylate cyclase in rat cardiac membranes by the active component of islet-activating protein, pertussis toxin. *Journal of Cyclic Nucleotide and Protein Phosphorylation Research*. 9:305–318.

Lewis, D. L., F. F. Weight, and A. Luini. 1986. A guanine nucleotide-binding protein mediates the inhibition of voltage-dependent calcium current by somatostatin in a pituitary cell line. *Proceedings of the National Academy of Science*. 83:9035–9039.

Litosch, I., C. Wallis, and J. N. Fain. 1985. 5-Hydroxytryptamine stimulates inositol phosphate production in a cell-free system from blowfly salivary glands. Evidence for a role of GTP in coupling receptor activation to phosphoinositide breakdown. *Journal of Biological Chemistry*. 260:5464–5471.

Lochrie, M. A., and M. I. Simon. 1988. G protein multiplicity in eukaryotic signal transduction systems. *Biochemistry*. 27:4957–4965.

Logothetis, D. E., D. Kim, J. K. Northup, E. J. Neer, and D. E. Clapham. 1988. Specificity of action of guanine nucleotide-binding regulatory protein subunits on the cardiac muscarinic K^+ channel. *Proceedings of the National Academy of Sciences*. 85:5814–5818.

Logothetis, D. E., Y. Kurachi, J. Galper, E. J. Neer, and D. E. Clapham. 1987. The $\beta\gamma$ subunits of GTP-binding proteins activate the muscarinic K^+ channel in heart. *Nature*. 325:321–326.

Manning, D. R., and A. G. Gilman. 1983. The regulatory components of adenylate cyclase and transducin. A family of structurally homologous guanine nucleotide-binding proteins. *Journal of Biological Chemistry*. 258:7059–7063.

Matsuoka, M., H. Itoh, T. Kozasa, and Y. Kaziro. 1988. Sequence analysis of cDNA and genomic DNA for a putative pertussis toxin insensitive guanine nucleotide-binding regulatory protein alpha subunit. *Proceedings of the National Academy of Sciences*. 85:5384–5388.

Mattera, R., B. J. R. Pitts, M. S. Entman, and L. Birnbaumer. 1985. Guanine nucleotide regulation of a mammalian myocardial receptor system. Evidence for homo- and heterotropic cooperativity in ligand binding analyzed by computer assisted curve fitting. *Journal of Biological Chemistry*. 260:7410–7421.

Mattera, R., A. Yatani, G. E. Kirsch, R. Graf, J. Olate, J. Codina, A. M. Brown, and L. Birnbaumer. 1988. Recombinant α_i-3 subunit of G protein activates G_k-gated K^+ channels. *Journal of Biological Chemistry*. 264:465–471.

Mattera, R., M. P. Graziano, A. Yatani, Z. Zhou, R. Graf, J. Codina, L. Birnbaumer, A. G. Gilman, and A. M. Brown. 1989. Single bacterially synthesized splice variants of the α subunit of the G protein G_s activate both adenylyl cyclase and dihydropyridine-sensitive Ca^{2+} channels. *Science*. 243:804–807.

Mumby, S, I.-H. Pang, A. G. Gilman, and P. C. Sternweis. 1988. Chromatographic resolution and immunologic identification of the α_{40} and α_{41} subunits of guanine nucleotide-binding regulatory proteins from bovine brain. *Journal of Biological Chemistry*. 263:2020–2026.

Murad, F., Y.-M. Chi, T. W. Rall, and E. W. Sutherland. 1962. Adenyl cyclase. III. The effect of catecholamines and choline esters on the formation of adenosine 3′,5′-phosphate by preparations from cardiac muscle and liver. *Journal of Biological Chemistry*. 237:1233–1238.

Murayama, T., and M. Ui. 1985. Receptor-mediated inhibition of adenylate cyclase and stimulation of arachidonic acid release in 3T3 fibroblasts. Selective susceptibility to islet-activating protein, pertussis toxin. *Journal of Biological Chemistry*. 260:7226–7233.

Nakajima, Y., S. Nakajima, R. J. Leonard, and K. Yamagucchi. 1986. Acetylcholine raises excitability of inhibiting the fast transient potassium current in cultured hippocampal neurons. *Proceedings of the National Academy of Sciences*. 83:3022–3026.

Nargeot, J., J. M. Nerbonne, J. Engels, and H. A. Lester. 1983. Time course of the increase in the myocardial slow inward current after a photochemically generated concentration jump of intracellular cAMP. *Proceedings of the National Academy of Sciences*. 80:2395–2399.

Newberry, N. R., and R. A. Nicoll. 1984. Direct hyperpolarizing action of baclofen on hippocampal pyramidal cells. *Nature*. 308:450–452.

Northup, J. K., P. C. Sternweis, M. D. Smigel, L. S. Schleifer, E. M. Ross, and A. G. Gilman. 1980. Purification of the regulatory component of adenylate cyclase. *Proceedings of the National Academy of Sciences*. 77:6516–6520.

Ohta, H., F. Okajima, and M. Ui. 1985. Inhibition by islet-activating protein of a chemotactic peptide-induced early breakdown of inositol phospholipids and Ca^{2+} mobilization in guinea pig neutrophils. *Journal of Biological Chemistry*. 260:15771–15780.

Pfaffinger, P. J., J. M. Martin, D. D. Hunter, N. M. Nathanson, and B. Hille. 1985. GTP-binding proteins couple cardiac muscarinic receptors to a K channel. *Nature*. 317:536–538.

Rall, T. W., E. W. Sutherland, and W. D. Woçilait. 1956. The relationship of epinephrine and glucagon to liver phosphorylase. *Journal of Biological Chemistry*. 218:483–495.

Rane, S. O., and K. Dunlap. 1986. Kinase C activator 1,2-oleylacetyl-glycerol attenuates voltage-dependent calcium current in sensory neurons. *Proceedings of the National Academy of Sciences*. 83:184–188.

Rodbell, M., L. Birnbaumer, S. L. Pohl, and H. M. J. Krans. 1971. The glucagon-sensitive adenyl cyclase system in plasma membranes of rat liver. V. An obligatory role of guanyl nucleotides in glucagon action. *Journal of Biological Chemistry*. 246:1877–1992.

Roof, D. J., M. L. Applebury, and P. C. Sternweis. 1985. Relationships within the family of GTP-binding proteins isolated from bovine central nervous system. *Journal of Biological Chemistry*. 260:16242–16249.

Sasaki, K., and M. Sato. 1987. A single GTP-binding protein regulates K^+-channels coupled with dopamine, histamine and acetylcholine receptors. *Nature*. 325:259–262.

Scherer, N. M., M.-J. Toro, M. L. Entman, and L. Birnbaumer. 1987. G. protein distribution in canine cardiac sarcoplasmic reticulum and sarcolemma. Comparison to rabbit skeletal membranes and brain and erythrocyte G proteins. *Archives of Biochemistry and Biophysics*. 259:431–440.

Scott, R. H., and A. Dolphin. 1987. Activation of a G protein promotes agonist responses to calcium channel ligands. *Nature*. 330:760–762.

Soejima, M., and A. Noma. 1984. Mode of regulation of the ACh-sensitive K-channel by the muscarinic receptor in rabbit atrial cells. Pflügers Archiv. 400:424–431.

Stryer, L., and H. R. Bourne. 1986. G Proteins: a family of signal transducers. *Annual Reviews of Cell Biology*. 2:391–419.

Szabo, G., I. H. Pang, and P. C. Sternweis. 1985. *Biophysical Journal*. 53:424*a*. (Abstr.)

Tabor, S., H. E. Huber, and C. C. Richardson. 1987. *Escherichia coli* thioredoxin confers processivity on the DNA polymerase activity of the gene 5 protein of bacteriophage T7. *Journal of Biological Chemistry*. 262:16212–16223.

Tabor, S., and C. C. Richardson. 1985. A bacteriophage T7RNA polymerase/promoter system for controlled exclusive expression of specific genes. *Proceedings of the National Academy of Sciences*. 82:1074–1078.

Thalmann, R. H. 1988. Evidence that guanosine triphosphate (GTP)-binding proteins control a synaptic response in brain: effect of pertussis toxin and GTP-γ-S on the late inhibitory postsynaptic potential of hippocampal CA_3 neurons. *Journal of Neuroscience*. 8:4589–4602.

Tolkovsky, A. M., and A. Levitzki. 1978. Mode of coupling between the β-adrenergic receptor and adenylate cyclase in turkey erythrocytes. *Biochemistry*. 17:3795–3810.

Trautwein, W., J. Taniguchi, and A. Noma. 1982. The effect of intracellular cyclic nucleotides and calcium on the action potential and acetylcholine response of isolated cardiac cells. *Pflügers Archiv*. 392:307–314.

VanDongen, A., J. Codina, J. Olate, R. Mattera, R. Joho, L. Birnbaumer, and A. M. Brown. 1988. Newly identified brain K^+ channels gated by the G protein G_o. *Science*. 242:1433–1437.

Wheeler, G. L., and M. W. Bitensky. 1977. A light-activated GTPase in vertebrate photoreceptors: regulation of light-activated cyclic GMP phosphodiesterase. *Proceedings of the National Academy of Sciences*. 74:4238–4242.

Yatani, A., J. Codina, A. M. Brown, and L. Birnbaumer. 1987*a*. Direct activation of mammalian atrial muscarinic K channels by a human erythrocyte pertussis toxin-sensitive G protein, G_k. *Science*. 235:207–211.

Yatani, A., J. Codina, R. D. Sekura, L. Birnbaumer, and A. M. Brown. 1987*b*. Reconstitution of somatostatin and muscarinic receptor mediated stimulation of K^+ channels by isolated G_k protein in clonal rat anterior pituitary cell membranes. *Molecular Endocrinology*. 1:283–289.

Yatani, A., J. Codina, Y. Imoto, J. P. Reeves, L. Birnbaumer, and A. M. Brown. 1987*c*. A G protein directly regulates mammalian cardiac calcium channels. *Science*. 238:1288–1292.

Yatani, A., Y. Imoto, J. Codina, S. L. Hamilton, A. M. Brown, and L. Birnbaumer. 1988*a*. The stimulatory G protein of adenylyl cyclase, G_s, also stimulates dihydropyridine-sensitive Ca^{2+} channels. Evidence for direct regulation independent of phosphorylation by cAMP-dependent protein kinase or stimulation by a dihydrophridine agonist. *Journal of Biological Chemistry*. 263:9887–9895.

Yatani, A., R. Mattera, J. Codina, R. Graf, K. Okabe, E. Padrell, R. Iyengar, A. M. Brown, and L. Birnbaumer. 1988*b*. The G protein-gated K^+ channel is stimulated by three distinct $G_i\alpha$ subunits. *Nature*. 336:680–682.

Yatani, A., H. Hamm, J. Codina, M. R. Mazzoni, L. Birnbaumer, and A. M. Brown. 1988*c*. A monoclonal anti-G-alpha antibody blocks muscarinic atrial K^+ channels. *Science*. 241:828–831.

Chapter 3

G Protein Activation Mechanisms of the Cardiac K^+ Channel, $i_{K.ACh}$

David E. Clapham and Donghee Kim

Department of Pharmacology, Mayo Foundation, Rochester, Minnesota 55905

Introduction

Pfaffinger et al. (1985), Breitwieser and Szabo (1985), and Kurachi et al. (1986) found almost simultaneously that the muscarinic-gated potassium channel ($i_{K.ACh}$) in cardiac atrial cells was gated by a guanine nucleotide regulatory protein (G protein; GTP-binding protein). $i_{K.ACh}$ is an inwardly rectifying, K^+-selective current activated by acetylcholine (ACh). ACh is released from the vagus nerve to hyperpolarize and thus slow the rate of atrial and pacemaker cardiac cells. Pfaffinger et al. (1985) and Kurachi et al. (1986) found that the G protein was pertussis toxin sensitive. Pretreating cells with pertussis toxin abolished the whole-cell current induced by ACh (Pfaffinger et al., 1985). Similarly, application of the active A protomer of pertussis toxin to inside-out patches of cell membranes blocked the opening of single inwardly rectifying K^+ channels gated by GTP (Kurachi, 1986). Subsequently, reports showed that application of the purified $\beta\gamma$ subunit from bovine brain (Logothetis et al., 1987) or the α subunit purified from human erythrocytes (Codina et al., 1987) activated $i_{K.ACh}$. The fact that both G protein subunits activated the channel, with the α subunit activating the channel at lower concentrations, led us to investigate more extensively the mechanism of $\beta\gamma$ activation. Brown, Birnbaumer, and colleagues (1987) contended that the $\beta\gamma$ effect was due to contamination of $\beta\gamma$ by activated α subunits, or that the detergent used to suspend the hydrophobic $\beta\gamma$, alone induced channel activity. In this paper we summarize experiments which show that the effect of the $\beta\gamma$ preparation on the K^+ channel is specifically due to the $\beta\gamma$ protein. We also propose a new pathway involving phospholipase A_2 for $\beta\gamma$-dependent activation of the channel.

Fig. 1 *A* summarizes the dissociation model of G protein action (for review, see Neer and Clapham, 1988). Fig. 1 *B* is a schematic representation of the various possibilities for G protein localization in the membrane.

Methods

Atria from 14-d-old embryonic chicks and 1-d-old rats were dissociated as single cells and used within 1 d of tissue culture (Logothetis et al., 1987). Standard patch-clamp methods were used to record single-channel currents from cell-attached and inside-out membrane patches. The solutions in the pipette and bath contained 118.5 mM KCl, 21.5 mM KOH, 2 mM $MgCl_2$, 5 mM EGTA, and 10 mM HEPES (pH 7.2, titrated with HCl). Exceptions are noted. The pipette tip was moved into the mouth of one of a series of polyethylene tubes where it was perfused by the test solution (Logothetis et al., 1987). All experiments were carried out at room temperature (20–22°C). Currents were recorded with a List model EPC7 (Darmstadt, FRG) patch-clamp amplifier and digitized on a VCR recorder. Original records were played back from digital tape onto a thermal array recorder (Gould Inc., Oxnard, CA)(f_o = 2 kHz). Records analyzed quantitatively for activity were obtained as described (Logothetis et al., 1987) with an INDEC 11/73 computer system. The current integral was divided by single-channel amplitude (i) to obtain Np_o, the product of the number of channels (N) and probability of opening (p_o) (Logothetis et al., 1987). Averaged Np_o ($\overline{Np_o}$) is given as a percent Np_o over the time noted.

A problem with the inside-out mode of patch recording is the spontaneous sealing over of patches into vesicles; a vesicle no longer has an accessible inner membrane and gives false negative results. Persistence of unattenuated single-channel openings indicates that the patch has not resealed. Chicken and rat membrane patches have

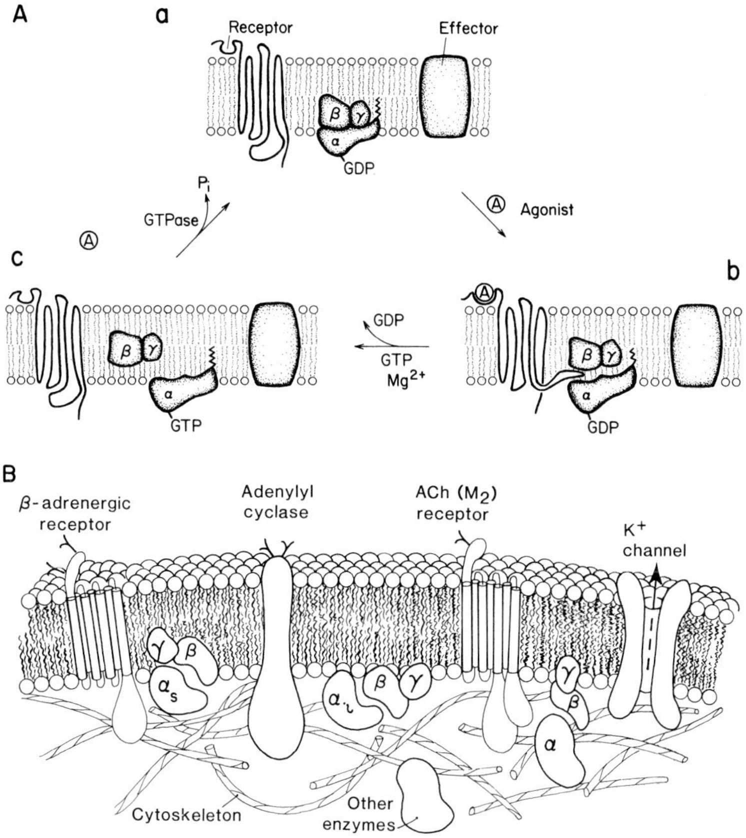

Figure 1. (*A*) Simple scheme for G protein signal transduction. From the unliganded state (*a*), the receptor binds agonist A (for example, epinephrine or acetylcholine), which produces a change (*b*) in receptor-G protein interaction, allowing GTP, in the presence of Mg^{2+}, to replace GDP on the α subunit. The activated α-GTP subunit and the $\beta\gamma$ subunits dissociate and one or both interact(s) with effectors (for example, adenylyl cyclase, K^+ channel). Alternatively, free $\beta\gamma$ may bind other α subunits. The intrinsic GTPase activity of the α subunit hydrolyzes GTP to GDP, releasing inorganic phosphate (P_i), and α-GDP which recombines with $\beta\gamma$ (*c*), ending the activation cycle. Nonhydrolyzable analogues of GTP such as Gpp(NH)p or GTPγS produce persistent activation of α subunits and persistent dissociation of α from $\beta\gamma$ (from Lewis et al., 1989). (*B*) Schematic view of the organization of receptors, G proteins and effectors in the plasma membrane. Receptors coupled to the G proteins, such as rhodopsin, and β-adrenergic and muscarinic receptors have a similar structure that appears to include seven transmembrane helices and regions of homology in the cytoplasmic loops. The relative location of α and $\beta\gamma$ subunits in the membrane is unknown but possibilities include loose attachment of α to the membrane, α association with the bilayer via a fatty acid, and α association with the cytoskeleton. $\beta\gamma$ is lipophilic mainly due to the γ subunit (from Neer and Clapham, 1988).

basal activity (usually > 1 opening per 10 s), thus providing a continuous monitor of the state of the inside-out patch. Vesicle formation is seen by rounding and attenuation of the current steps. GTP (when ACh was in the pipette), GTP[γ-S], or $\beta\gamma$ were often applied to patches at the end of negative experiments to ensure that the patch had not resealed into a vesicle. Data from patches that had sealed over were rejected.

Purification of $\beta\gamma$ and α subunits from brain, placenta, and retina has been previously described (Logothetis et al., 1988).

Results

$\beta\gamma$ Activates $i_{K.ACh}$

Using inside-out patches, the intracellular surface of chick atrial cell membranes was exposed to 100 μM GTP in the presence and absence of bound receptor (ACh in the pipette). Only when ACh was present in the pipette were $i_{K.ACh}$ channels activated by GTP. Net current increased ~100–200-fold when GTP was applied to the intracellular surface (ACh receptor bound). Fig. 2 shows a representative experiment in which ACh

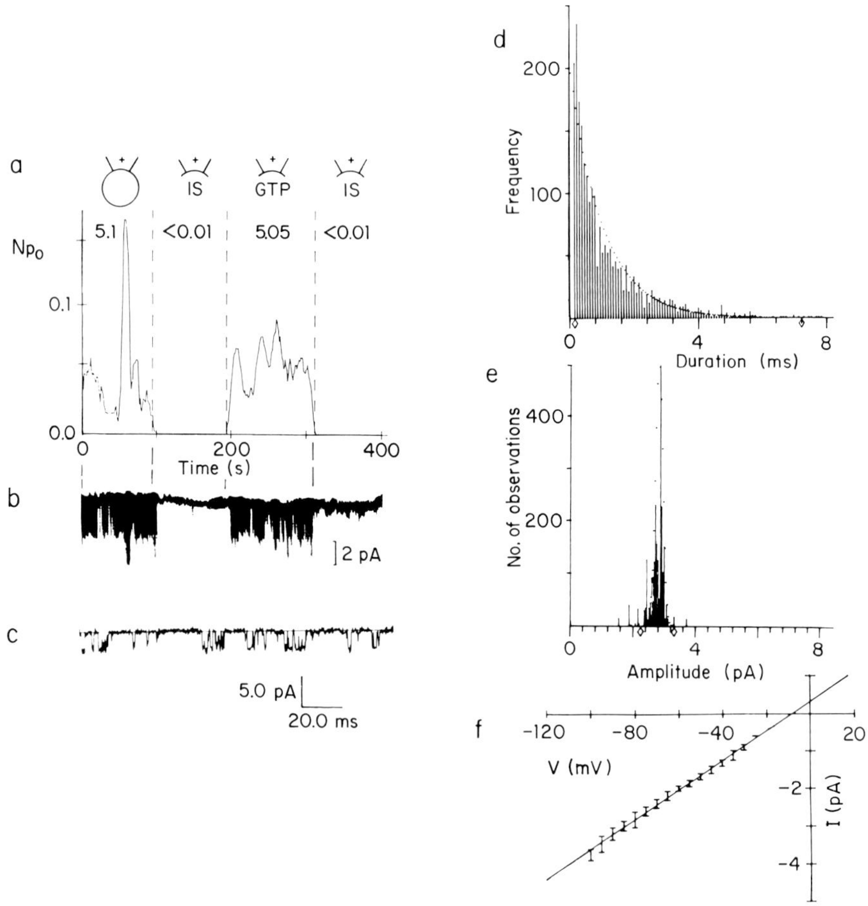

Figure 2. Muscarinic potassium channel activation shown in cell-attached and inside-out patch recordings from a chick embryonic atrial cell. (*a–c*) Representative experiment showing GTP-dependent channel activity with agonist in the pipette and with or without GTP at the intracellular surface. An overall average (Np_o) is shown at the top of the plot for each condition ($\times 10^{-2}$). (*d*) Distribution of single-channel open time durations from the inside-out patch recording in GTP shown in *a* and *b*, fitted by a single exponential; decay time constant, $\tau = 1.03$ ms. (*e*) Distribution of current amplitudes. Mean amplitude was 2.83 ± 0.12 pA (mean ± SD). (*f*) *I-V* relation of ACh-induced single-channel current (from Logothetis et al., 1987).

was present in the pipette. Channel activity is represented as the integral of the single-channel current divided by the single-channel amplitude (Np_o at -80 mV, the holding potential in these experiments). As is shown in this figure, channel activity declined dramatically after the patch was detached from the cell and GTP diffused away. As soon as GTP was reapplied to the patch, channel activity returned in a

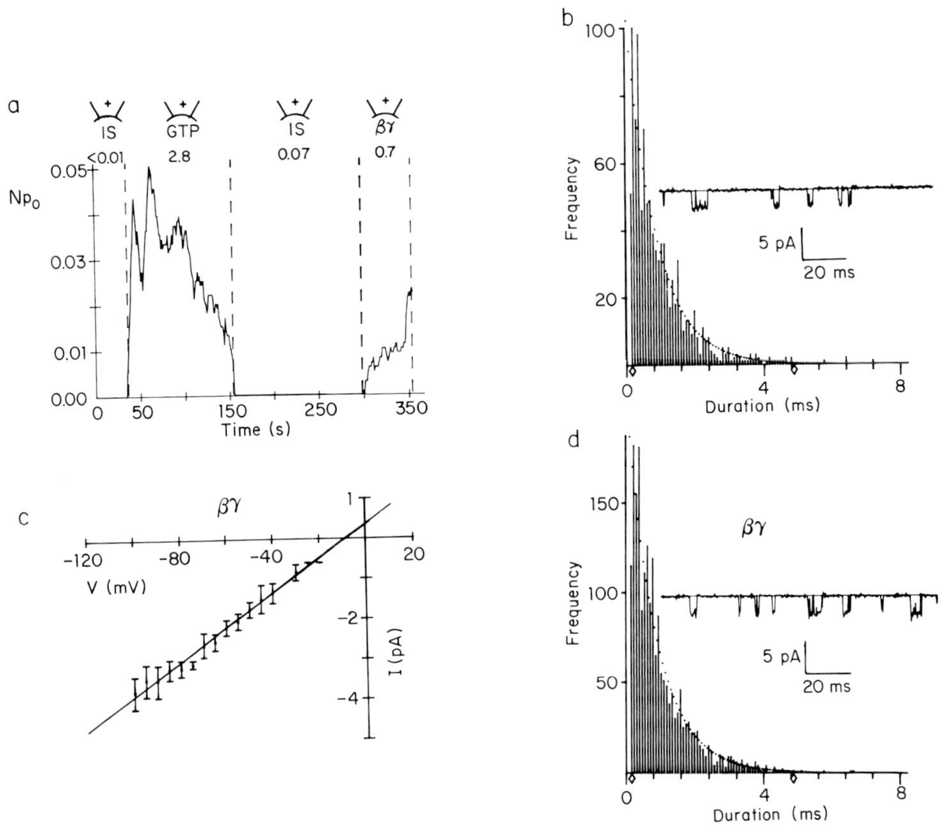

Figure 3. Properties of $\beta\gamma$-induced single-channel currents (*a*, *c*, and *d*) and currents induced by the native G proteins with no pertussis toxin pretreatment (*a* and *b*). (*a*) Channel activity where either the endogenous or exogenous subunits of the G proteins activated $i_{K.ACh}$. (*b*) Distribution of single-channel open time durations from the inside-out patch recording exposed to GTP as in *a*, decay time constant 0.88 ms. (*c*) *I-V* relation of the $\beta\gamma$-induced single-channel current between -100 and -20 mV membrane potential from inside-out patches. The single-channel conductance was 42 pS. (*d*) Distribution of single-channel open time durations from the inside-out patch recording exposed to $\beta\gamma$; decay time constant, 0.86 ms (from Logothetis et al., 1987).

reproducible fashion. The channels induced by GTP had a mean amplitude of 2.9 pA at -80 mV, a conductance of 35–40 pS, and a mean open time of ~1 ms. The channels were K^+ selective and rectified inwardly in the presence of intracellular Mg^{2+}. Rectification ceased when Mg^{2+} was removed from the intracellular surface of the patch (Logothetis et al., 1987), which is similar to the Mg^{2+}-dependent rectification of other inwardly rectifying K^+ channels (Matsuda et al., 1987).

Fig. 3 shows that channels activated by extracellular ACh in the presence of intracellular GTP were the same as those activated by the $\beta\gamma$ subunit. Channels activated by either GTP or $\beta\gamma$ had a conductance of 40 pS and a mean open time of ~1 ms. A 25-pS substate was often present under both conditions but these substates have not been analyzed in any detail. We also found that $\beta\gamma$-dependent activation could be

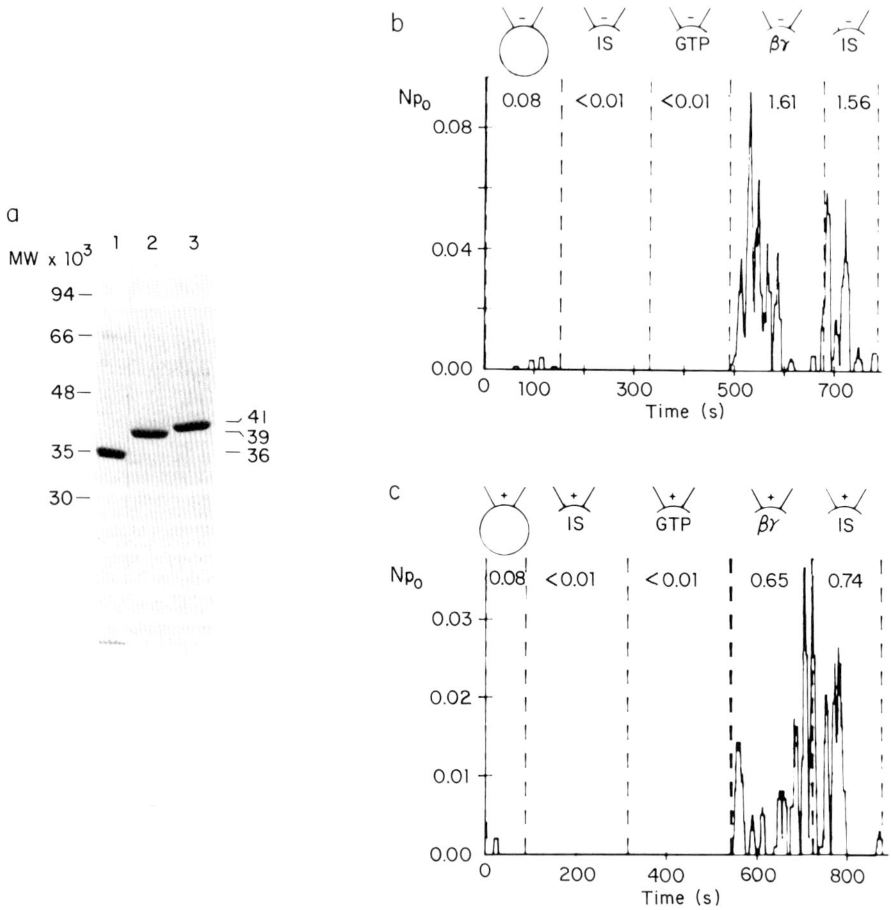

Figure 4. (*a*) One-dimensional sodium dodecyl sulfate–polyacrylamide gel stained with Coomassie blue showing purified $\beta\gamma$, α_{39}, and α_{41} components and (*b* and *c*) time plots of channel activity (Np_o) from cell-attached patches and inside-out patches exposed to the purified $\beta\gamma$ subunits of *a*. (*a*) The β subunit (lane *1*) was a doublet of proteins of molecular weights 35,000 and 36,000. α_{39} and α_{41} are in lanes *2* and *3*, respectively. (*b*) Effect of $\beta\gamma$ complex applied at 23 nM ($\beta\gamma$ M_r 44,000). (*c*) Application of $\beta\gamma$ complex (23 nM) to a patch from a pertussis toxin–pretreated cell (from Logothetis et al, 1987).

prevented by preincubation of the subunits with excess α_{41}-GDP or α_{39}-GDP. Presumably, α subunits combined with $\beta\gamma$ subunits to form an inactive heterotrimer. Similarly, $\beta\gamma$ was found to activate $i_{K.ACh}$ in Mg^{2+}-free solution. Since Mg^{2+} is required for dissociation of α and $\beta\gamma$ to form active α-GTP, it is doubtful that contaminating inactive α subunits in the $\beta\gamma$ preparation could bind GTP and stimulate $i_{K.ACh}$ in the absence of Mg^{2+} (Logothetis et al., 1987).

$\beta\gamma$ Activation Is Not Due to α Impurities

Codina et al. (1987) have shown that in its activated state α_{40} (α_{40}-GTPγS), purified from human erythrocytes, activated the K^+ channel at concentrations of ~1 pM. The fact that α activated $i_{K.ACh}$ at such low concentrations has raised the issue of the purity of the $\beta\gamma$ preparation that we used. $\beta\gamma$ was initially used in the nanomolar range and produced activity corresponding to ~10 pM α_{40}-GTPγS, a 100-fold difference in potency. We tested the purity of $\beta\gamma$ and found that α (of all types) made up 0.01% of the $\beta\gamma$ preparation in the purest of preparations.

Fig. 4 shows an ADP-ribosylation gel of the $\beta\gamma$ subunit. The total ADP-ribosylation substrates comprised ~0.01% of the total protein. Other preparations had up to 1% contaminating α subunit (but showed no major difference in their effective threshold dose from the protein containing 0.01%). We then tested the lowest threshold dose for $\beta\gamma$-induced activation. We found that $\beta\gamma$ began to activate the channel at ~200 pM concentrations using the $\beta\gamma$ preparation with 0.01% α subunit (Fig. 6 *B*). Thus, for 1 nM $\beta\gamma$, α would have to activate the channels at 0.1 pM, a concentration we have never observed opening the channels even with pure, preactivated α subunits. It should be remembered that 0.01% α represents total ribosylation substrate of all kinds and that only activated α subunits can stimulate the channel. However, the contaminating α subunit should be inactive since the purification of $\beta\gamma$ does not involve GTPγS and all $AlFl^{4-}$ has been removed from the protein by a sequence of one (and in some cases, two) purification steps (Logothetis et al., 1988). Even if fluoride were present, activation of the G protein by $AlFl^{4-}$ is rapidly reversible (Higashijima et al., 1987). As we shall show below, $\beta\gamma$ purified without an activating regime ($\beta_{35}\gamma$ from human placenta) also stimulated the channel (Logothetis et al., 1988). Perhaps the most compelling data to eliminate the α contamination issue is the finding of $i_{K.ACh}$ stimulation by 10 pM rat brain $\beta\gamma$ applied to guinea pig atria (Kurachi et al., 1989*a*). Kurachi and co-workers also found that rat brain $\beta\gamma$ preincubated for hours in Mg^{2+}-free solution still stimulated channel activity. Prolonged preincubation in Mg^{2+}-free solution irreversibly inactivates α subunits.

Specificity of $\beta\gamma$ Activation

The specificity of $\beta\gamma$-induced channel activation has been tested with $\beta\gamma$ derived from bovine retina (transducin $\beta\gamma$, $\beta\gamma_T$), human placenta ($\beta_{35}\gamma$), and bovine brain $\beta\gamma$ (a mixture of $\beta_{35}\gamma$ and $\beta_{36}\gamma$). As shown in Fig. 5, transducin $\beta\gamma$, tested up to 400 nM, did not activate $i_{K.ACh}$ while 20 nM placental $\beta\gamma$ did. CHAPS (3-[(3-cholamidopropyl)dimethylammonio]-1-propanesulfonate) detergent, used to suspend the hydrophobic $\beta\gamma$, did not activate the channels by itself (n = 32) or in the presence of transducin $\beta\gamma$ (Fig. 5). Since $\beta\gamma_T$ differs from β_{35} or $\beta_{36}\gamma$ mostly in the γ subunit, one possibility is that it is the γ subunit that activates the K^+ channel. Alternatively, γ may change the structure of β in such a way that it is not effective in activating $i_{K.ACh}$.

The failure of Codina et al. (1987) to activate $i_{K.ACh}$ with $\beta\gamma$ may have been due to the presence of the detergent Lubrol in their preparation. Lubrol is used in most G protein purification schemes. Lubrol, even at 0.001%, blocked $i_{K.ACh}$. For this reason, we always substituted CHAPS for Lubrol. Even in the presence of 184 μM CHAPS, the maximum amount of detergent we used, $\beta\gamma$ was ~80% aggregated (Logothetis et al., 1988). Thus, the true concentrations of $\beta\gamma$ at the patch is probably less than that calculated from amino acid analysis of protein. Furthermore, in preparations in which

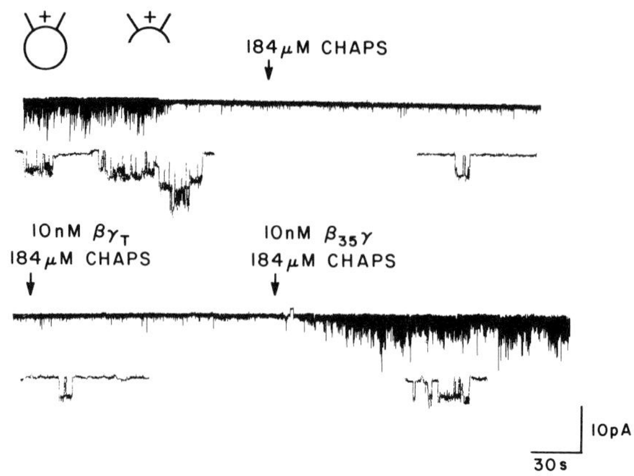

Figure 5. Detergent (CHAPS) does not activate $i_{K.ACh}$ alone or in the presence of transducin. Transducin $\beta\gamma$ ($\beta\gamma_T$) does not activate $i_{K.ACh}$ even in the presence of CHAPS although $\beta_{35}\gamma$ (placental) does activate the channel (from Logothetis et al., 1988).

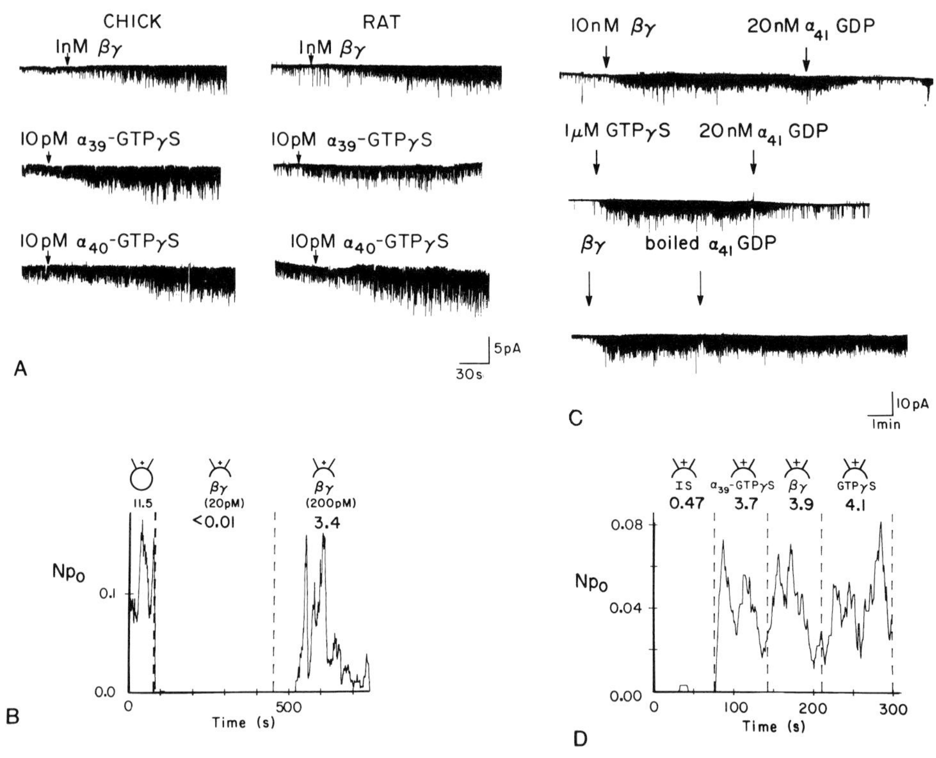

Figure 6. (*A*) $\beta\gamma$,α_{39}-GTP[γ-S], and α_{40}-GTP[γ-S] activate $i_{K.ACh}$ in embryonic chick and neonatal rat atria. Holding potential was −80 mV. The conductance (35–40 pS) and kinetics (mean open time, 1.0 ms at −80 mV) of rat and chick single-channel $i_{K.ACh}$ currents were indistinguishable. (*B*) Picomolar concentration of the $\beta\gamma$ subunits activate the muscarinic-gated K^+ channel. (*C*) Reversal of $\beta\gamma$ or GTP[γ-S]-activated $i_{K.ACh}$ by α_{41}-GDP in neonatal rat atria. (*D*) α_{39}-GTP[γ-S]-activated channel activity was not increased further by $\beta\gamma$ (10 nM) or GTP[γ-S] (100 μM) (from Logothetis et al., 1988).

detergents were removed ($\beta\gamma$ largely aggregated), $\beta\gamma$ still activated $i_{K.ACh}$ in the nanomolar range. Thus, it seems unlikely that detergents are the cause of $\beta\gamma$-dependent activation.

If $\beta\gamma$ activates the channel in a specific manner, it should be possible to turn off channel activity with the inactive form of α, α-GDP. Fig. 6 *C* shows that α_{41}-GDP, applied to the inside surface of patches previously activated by $\beta\gamma$, returned channel activity to baseline. Similarly, GTPγS-induced channel activation was reversed by α_{41}-GDP. Previously boiled α_{41}-GDP did not reverse activity. One interpretation of the reversal of GTPγS-induced activation is that only $\beta\gamma$ activates the channel. A more conservative interpretation is that α_{41}-GDP competes with $\beta\gamma$ or other α's for the binding site on the K^+ channel. We conclude from this experiment that $\beta\gamma$ alone can activate the K^+ channel.

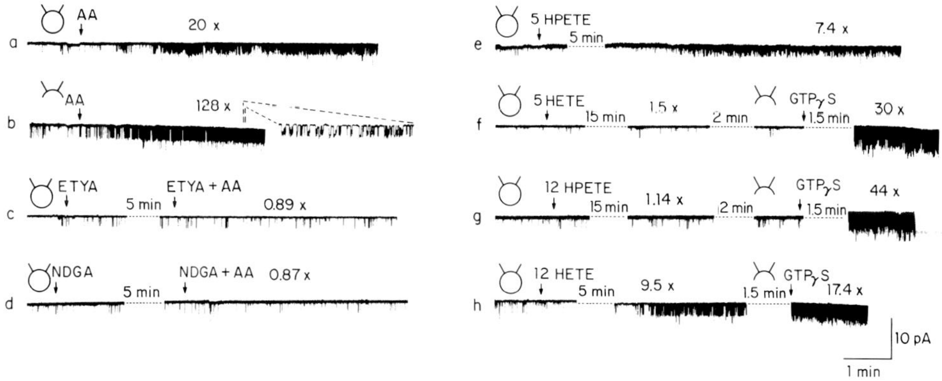

Figure 7. Arachidonic acid and its lipoxygenase derivatives activate $i_{K.ACh}$. Arachidonic acid activated the channel when applied to (*a*) cell-attached patches (50 μM) or (*b*) inside-out patches (10 μM). (*c* and *d*) ETYA (10 μM), which inhibits arachidonic acid metabolism to cyclooxygenase and lipoxygenase derivatives, and NDGA (20 μM), which inhibits metabolism of arachidonic acid to lipoxygenase derivatives; both block arachidonic acid (50 μM)-induced activation of the K^+ channel. (*e–h*) Arachidonic-acid derived metabolites of the lipoxygenase pathway, 5-HPETE and 12-HETE (*e* and *h*), activated the channel (100 μM). 5-HETE, 12-HPETE (*f* and *g*), and 15-HETE (not shown) did not activate $i_{K.ACh}$ in cell-attached patches (from Kim et al., 1989).

Both α and $\beta\gamma$ Activate $i_{K.ACh}$

To summarize to this point, both α-GTPγS and $\beta\gamma$ subunits activate $i_{K.ACh}$ (Fig. 6 *A*). As Codina et al. (1988) have shown, human erythrocyte α, now identified as $\alpha_{i\text{-}3}$ clearly activates $i_{K.ACh}$. We have repeated the experiments with their α_{40}-GTPγS with essentially the same result. We have also found that α_{39}-GTPγS (α_o; bovine brain) activates the channel and would not be surprised to find that several α subunits may be equally effective in this regard (Logothetis et al., 1988). Both α and $\beta\gamma$ appear to act on the same population of channels since addition of $\beta\gamma$ after α activation (Fig. 6 *D*) or α after $\beta\gamma$ activation does not increase the net channel Np_o. Since both subunits are effective, the next question concerns the mechanism of action of G proteins in vivo.

$\beta\gamma$ Activates $i_{K.ACh}$ via PLA_2

Kurachi and co-workers made an important discovery that led us to a new set of experiments on the mechanism of $\beta\gamma$'s action. They found that arachidonic acid applied to cells activated the $i_{K.ACh}$ channel (Kim et al., 1989; Kurachi et al., 1989*b*). Fig. 7 shows cell-attached and inside-out patches in which 10–100 μM arachidonic acid activated $i_{K.ACh}$. Arachidonic acid and its derivatives have been shown to act as

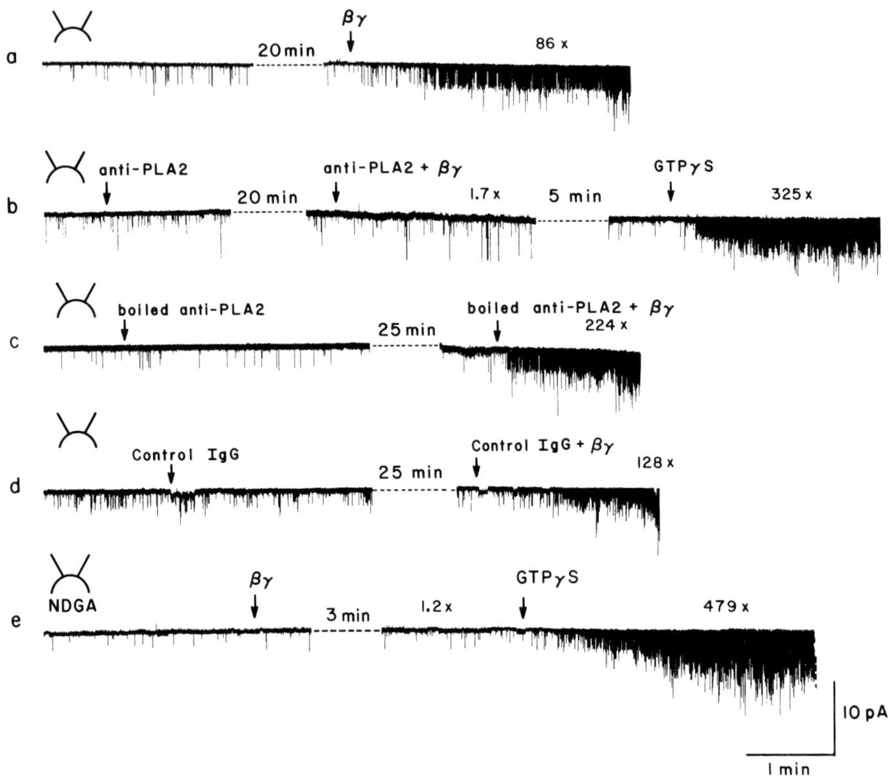

Figure 8. Anti-PLA_2 antibody inhibits $\beta\gamma$ activation of $i_{K.ACh}$. (*a*) 2 nM $\beta\gamma$ activates a 40-pS K^+ channel ($i_{K.ACh}$) in an inside-out patch from neonatal rat atria. Membrane potential held at -80 mV. (*b*) An inside-out patch was held in bath solution containing 100 nM anti-PLA_2 antibody for 20 min. When 2 nM $\beta\gamma$ was subsequently applied to the patch, channels were not activated. GTPγS applied 5 min later activated $i_{K.ACh}$. (*c*) Boiled anti-PLA_2 antibody (100 nM) applied 25 min to an inside-out patch does not prevent activation of the channel by 2 nM $\beta\gamma$. (*d*) Preincubation of the patch with rabbit anti-rat IgG (200 nM) did not block $\beta\gamma$ activation. (*e*) Preincubation of the patch with 1 μM NDGA blocked $\beta\gamma$ (2 nM)-dependent activation but not GTPγS-dependent stimulation (from Kim et al., 1989).

second messengers for ion channels in *Aplysia* (Piomelli et al., 1987). Arachidonic acid is lipophilic and rapidly metabolized to a multitude of products. Three main types of enzymes in this pathway are the lipoxygenases, epoxygenases, and cyclooxygenases. Since $\beta\gamma$ has been shown to stimulate phospholipase A_2 (PLA_2) in rod outer segment (Jelsema and Axelrod, 1987), we tested the idea that $\beta\gamma$ generated an increase in arachidonic acid via stimulation of membrane-bound PLA_2.

Fig. 8 shows the key result which suggests that $\beta\gamma$ activates $i_{K.ACh}$ via stimulation of PLA_2. Affinity-purified antibody to PLA_2 blocked activation of $i_{K.ACh}$ by 2 nM $\beta\gamma$ ($n = 5$). This antibody, raised in rabbit against porcine pancreatic PLA_2, had previously been shown to inhibit the generation of lysophospholipids and thus to block the function of PLA_2 (Bar-Sagi et al., 1988). Inside-out patches were incubated in 200

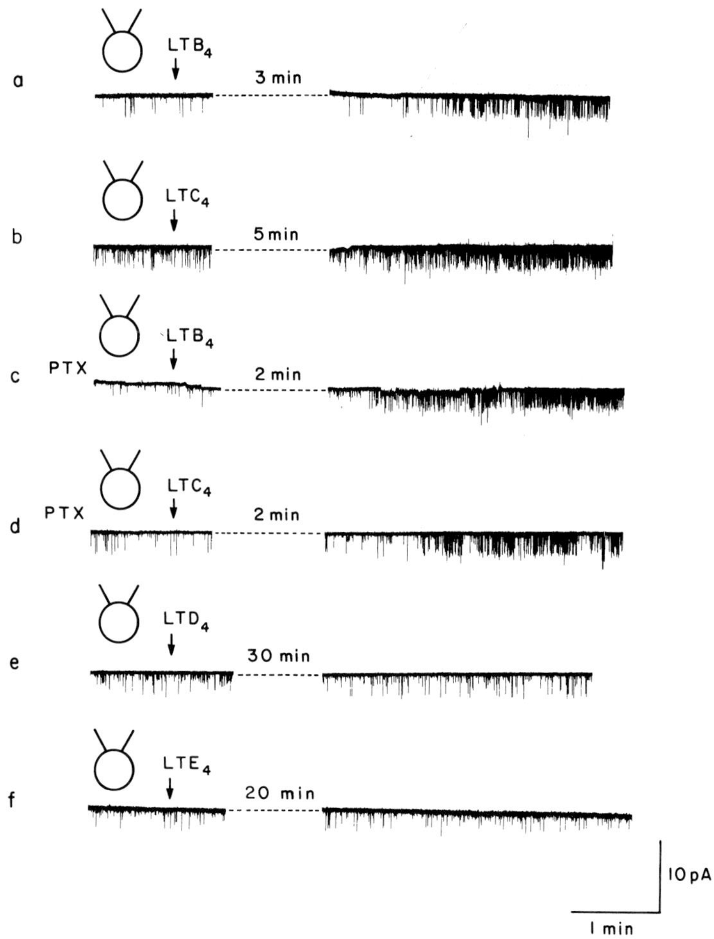

Figure 9. (*a* and *b*) Leukotrienes B_4 and C_4 (10 μM) activated $i_{K.ACh}$ in cell-attached patches of atrial membrane. (*c* and *d*) Pertussis toxin pretreatment (100 ng/ml for 12 h) of cells did not affect activation by either leukotriene B_4 or C_4. Previous work has shown that pertussis toxin blocks muscarinic activation of the channel. (*e* and *f*) Leukotrienes D_4 and E_4 (10 μM) did not activate the channel (modified from Kim et al., 1989).

nM anti-PLA_2 antibody for ~20 min. Channels were shown to be present in each patch by applying GTP before the application of antibody (ACh in pipette). All five patches incubated with antibody to PLA_2 failed to respond to 2 nM $\beta\gamma$. The same concentration of $\beta\gamma$ activated 8/8 patches in prior controls. 1–10 nM $\beta\gamma$ activated 97% of patches ($n = 123$) in rat and chick atria (Logothetis et al., 1987; Kim et al., 1989). Antibody inactivated by boiling, and control preimmune IgG antibody did not block $\beta\gamma$-induced activation (Fig. 8; Kim et al., 1989).

We next tested arachidonic acid metabolites and blockers of specific enzymes to delineate second messengers involved in this pathway. ETYA (eicosatetraenoic acid) (20 μM), which inhibits both the lipoxygenase and cyclooxygenase pathways, blocked activation by arachidonic acid. The cyclooxygenase inhibitor indomethacin (10 μM) did not affect channel activation while the lipoxygenase inhibitor, NDGA (nordihydroguaiaretic acid), blocked activation (Figs. 7 and 8). To rule out generation of arachidonic acid by the sequential activation of phospholipase C and diacylglycerol lipase, we inhibited phosphatidylinositol turnover by incubating some cells in 200 μM neomycin. No inhibition of activation was noted. Also, phorbol esters, which activate protein kinase C, failed to activate the channel (Logothetis and Clapham, unpublished).

The lipoxygenase-derived metabolites of arachidonic acid, 5-HPETE (hydroperoxyeicosatraenoic acid) and 12-HETE (hydroxyeicosatetraenoic acid), activated $i_{K.ACh}$ while 5-HETE, 12 HPETE, and 15-HETE did not (Fig. 7). Distal metabolites of the 5-lipoxygenase pathway, the leukotrienes, were also effective in that leukotriene B_4 and C_4 but not D_4 and E_4 activated the K^+ channel (Fig. 9). Activation of the channel via leukotrienes suggests a receptor mechanism, since specific receptors for leukotrienes have been identified in some cells. If so, leukotriene-dependent activation was not mediated via a pertussis toxin-sensitive G protein (Kim et al., 1989).

Discussion

Both α and $\beta\gamma$ subunits of GTP-binding proteins activate $i_{K.ACh}$ when applied to the inside surface of detached patches of embryonic chick or neonatal rat cardiac atria. More than one type of α subunit may activate the channel. We do not know if both $\beta_{35}\gamma$ and $\beta_{36}\gamma$ are equally effective in channel stimulation but $\beta_{35}\gamma$ from placenta alone activated the channel, while transducin $\beta\gamma$ did not. We do not know if α stimulation of $i_{K.ACh}$ is direct or indirect but $\beta\gamma$ stimulation appears to be via PLA_2. Fig. 10 summarizes the results of this study.

There are three main possibilities for activation by arachidonic acid and its metabolites. First, these lipophilic compounds may insert into the membrane and simply destabilize the $\alpha\beta\gamma$ heterotrimer, resulting in the release of α subunit and α-dependent activation. If so, the α must be made available for binding by GTP. This would explain why several compounds and particularly 12-HETE and 5-HPETE might activate the channel. (There is no known physiological pathway for conversion of 12-HETE to 5-HPETE). However, other eicosanoids of similar structure, 15-HETE, 12-HPETE, and 5-HETE, did not also stimulate activity, suggesting that such destabilization would have to be specific for certain compounds. Second, no free GTP was available in the bath for released α to bind. A second possibility is that one or more of the 5-lipoxygenase products was effective in activating the channel directly. A third possibility is that a distal metabolite, such as a leukotriene, may be released upon $\beta\gamma$ activation and bind to a specific G protein-linked receptor, again releasing an active α subunit. If so, the G protein would have to be pertussis toxin insensitive. Since there are many unstable metabolites of arachidonic acid, the resolution of this issue will depend on specific and selective antagonists.

Which subunit activates the channel when the muscarinic receptor is bound?

Experiments with an antibody specific for α subunits (but not α_{i3} alone) suggest that the α subunit is sufficient for activation (Yatani et al., 1988). Our experiments, which show GTPγS activation despite binding of $\beta\gamma$ by anti-PLA$_2$ antibody, support a role for α in muscarinic coupling (Kim et al., 1989). This role need not be exclusive of $\beta\gamma$, and resolution of this point may depend on generation of antibodies specific to the functional site of $\beta\gamma$. Another possible explanation for both α and $\beta\gamma$-induced channel activation is that $\beta\gamma$ may be released from a separate receptor. The above experiments were reconstitution studies and cannot replicate conditions in vivo. An intriguing possibility is that some receptors release $\beta\gamma$ to stimulate production of arachidonic acid and channel-activating second messengers while others release α subunits. Although these pathways may intersect at the channel, there is no reason to assume they use the same receptor subtype.

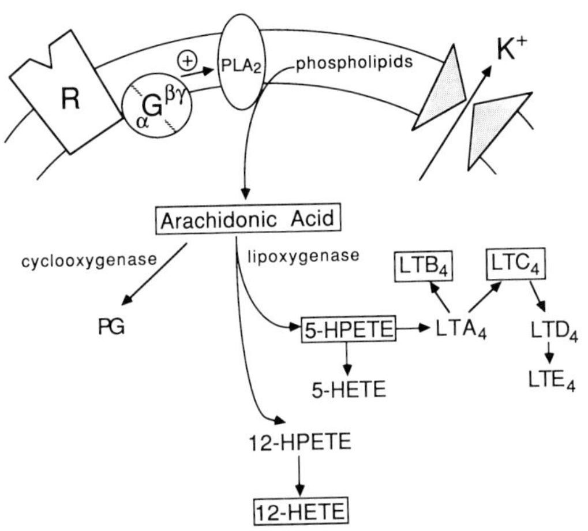

Figure 10. Proposed mechanism of the G protein $\beta\gamma$ subunit activation of the cardiac muscarinic K^+ channel. $\beta\gamma$, released from the G protein bound to the muscarinic receptor or, more probably, from another G protein-linked receptor, activates PLA$_2$ and lipoxygenase-derived compounds are released. Boxed metabolites/compounds activated the channel when applied to the patch. Since these compounds are unstable, we do not know if 5-HPETE, 12-HETE, and the leukotrienes B_4 and C_4 are all direct mediators or whether the leukotrienes and distal metabolites are activators. Activation may occur via destabilization and release of α from the G protein $\alpha\beta\gamma$ heterotrimer, direct activation of the channel, or leukotriene receptor binding (from Kim et al., 1989).

Acknowledgments

We thank Rose Goldsmith for culturing heart cells and Julie Snell for typing the manuscript.

D. E. Clapham and D. Kim were supported by grants from the National Institutes of Health.

References

Bar-Sagi, D., J. P. Suhan, F. McCormick, and J. Feramisco. 1988. Localization of phospholipase A_2 in normal and *ras*-transformal cells. *Journal of Cell Biology*. 106:1649–1658.

Birnbaumer, L., and A. M. Brown. 1987. G protein coupling of K^+ channels. *Nature*. 327:21–22.

Breitwieser, G., and G. Szabo. 1987. Uncoupling of cardiac muscarinic and β-adrenergic receptors from ion channels by a guanine nucleotide analogue. *Nature*. 317:538–540.

Codina, J., J. Olate, J. Abramowitz, R. Mattera, R. G. Cook, and L. Birnbaumer. 1988.α_{i3} cDNA encodes the α subunit of G_k, the stimulatory G protein of receptor-regulated K^+ channels. *Journal of Biological Chemistry*. 263:6746–6750.

Codina, J., A. Yatani, D. Grenet, A. M. Brown, and L. Birnbaumer. 1987. The α subunit of the GTP binding protein G_k opens atrial potassium channels. *Science*. 236:442–445.

Higashijima, T., K. M. Ferguson, P. C. Sternweis, E. M. Ross, M. D. Smigel, and A. G. Gilman. 1987. The effect of activating ligands on the intrinsic fluorescence of guanine nucleotide-binding regulatory proteins. *Journal of Biological Chemistry*. 262:752–756.

Jelsema, C. L., and J. Axelrod. 1987. Stimulation of phospholipase A_2 activity in bovine rod outer segments by the $\beta\gamma$ subunits of transducin and its inhibition by the α subunit. *Proceedings of the National Academy of Sciences*. 84:3623–3627.

Kim, D., D. L. Lewis, E. J. Neer, L. Graziadei, D. Bar-Sagi, and D. E. Clapham. 1989. G protein $\beta\gamma$ subunits activate the cardiac K^+ channel via phospholipase A_2. *Nature*. 337:557–560.

Kurachi, Y., H. Ito, T. Sugimoto, T. Katada, and M. Ui. 1989*a*. Activation of atrial muscarinic K channels by low concentrations of Br subunits of rat brain G protein. *Pflügers Archiv*. 413:325–327.

Kurachi, Y., H. Ito, T. Sugimoto, T. Shimizu, I. Miki, and M. Ui. 1989*b*. Arachidonic acid metabolites as intracellular modulators of the G protein-gated cardiac K^+ channel. *Nature*. 337:555–557.

Kurachi, Y., T. Nakajima, and T. Sugimoto. 1986. Acetylcholine activation of K^+ channels in cell-free membranes of atrial cells. *American Journal of Physiology*. 251:H681–H684.

Lewis, D. L., J. D. Lechleiter, D. Kim, C. Joshi, and D. E. Clapham. 1989. Intracellular regulation of ion channels and cell membranes. *Mayo Clinic Proceedings*. In press.

Logothetis, D. E., D. Kim, J. K. Northup, E. J. Neer, and D. E. Clapham. 1988. Specificity of action of guanine nucleotide-binding regulatory protein subunits on the cardiac muscarinic K^+ channel. *Proceedings of the National Academy of Sciences*. 85:5814–5818.

Logothetis, D. E., Y. Kurachi, J. Galper, E. J. Neer, and D. E. Clapham. 1987. The $\beta\gamma$ subunits of GTP-binding proteins activate the muscarinic K channel in heart. *Nature*.325:321–326.

Matsuda, H., A. Saigusa, and H. Irisawa. 1987. Ohmic conductance through the inwardly rectifying K channel and blocking by internal Mg. *Nature*. 325:156–159.

Neer, E. J., and D. E. Clapham. 1988. Roles of G protein subunits in transmembrane signalling. *Nature*. 333:129–134.

Pfaffinger, P. J., J. M. Martin, D. D. Hunter, N. M. Nathanson, and B. Hille. 1985. GTP-binding proteins couple cardiac muscarinic receptors to a K^+ channel. *Nature*. 317:536–638.

Piomelli, D., A. Volterra, N. Dale, S. A. Siegelbaum, E. R. Kandel, J. H. Schwartz, and F. Belardetti. 1987. Lipoxygenase metabolites of arachidonic acid as second messengers for presynaptic inhibition of *Aplysia* sensory cells. *Nature*. 328:38–43.

Yatani, A., H. Hamm, J. Codina, M. R. Mazzoni, L. Birnbaumer, A. M. Brown. 1988. A monoclonal antibody to the α subunit of G_k blocks muscarinic activation of atrial K^+ channels. *Science*. 241:828–831.

Chapter 4

Regulation of Ion Channels in *Aplysia* Neurons by Autoactive Peptides and Second Messengers

Karen J. Loechner and Leonard K. Kaczmarek

Departments of Pharmacology and Cellular and Molecular Physiology, Yale University School of Medicine, New Haven, Connecticut 06510

Introduction

In many types of neurons, synaptic stimulation of a brief train of action potentials causes subsequent changes in the properties of the neuron that last for a substantial time beyond the duration of the stimulus (Kaczmarek and Levitan, 1987). Properties that may be altered as a result of stimulation include the intrinsic electrical excitability of the neuron as well as its secretory properties. In this brief review, we shall describe the long-lasting changes in excitability that occur after brief stimulation of the bag cell neurons within the abdominal ganglion of *Aplysia*. In vivo, the bag cell neurons serve to generate a prolonged sequence of behaviors that constitutes the fixed action pattern of egg-laying. We shall first give an account of the modulation of voltage-dependent potassium and calcium currents by the second messengers cyclic AMP and protein kinase C. We will then discuss preliminary data on the relation of these changes in electrical excitability to the secretion of peptides by these neurons.

Changes in Excitability during an Afterdischarge

The peptidergic bag cell neurons of *Aplysia* provide a model system for the study of modulation of neuronal excitability. These cells are normally silent, exhibiting

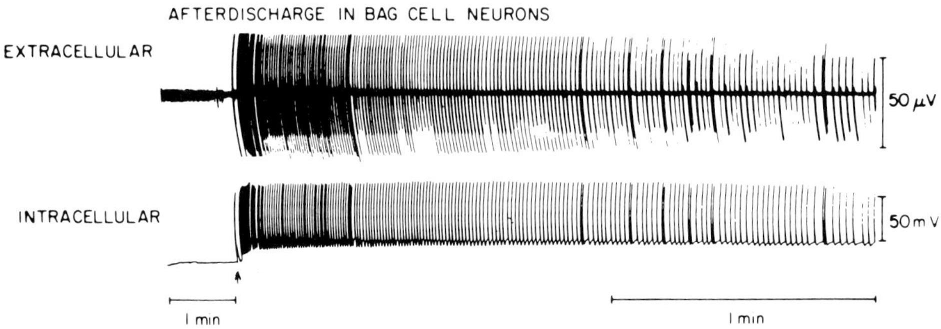

Figure 1. (*Top*) Extracellular recording of an afterdischarge from an intact cluster of bag cell neurons within the abdominal ganglion. (*Bottom*) Intracellular recording of an afterdischarge from a single bag cell neuron within the same cluster.

relatively negative resting potentials. However, upon brief electrical stimulation or application of peptides from the reproductive tract, these neurons fire repetitively for a period of 15–30 min that has been termed the afterdischarge (Kupferman and Kandel, 1970; Heller et al., 1980; Dudek and Blankenship, 1977). In vivo, the afterdischarge is followed by a sequence of behaviors that culminates in egg-laying (Dudek et al., 1979). Fig. 1 illustrates both extracellular (top) and intracellular (bottom) recording of an afterdischarge from bag cell neurons. The electrical activity can be divided into two phases according to action potential frequency. The initial fast phase lasts for ~1 min, during which time the cells fire at 4–5 Hz. This fast phase is followed by a second, slower phase which lasts the remainder of the afterdischarge, and during which the neurons fire at ~0.5 Hz (Kaczmarek et al., 1982).

During the afterdischarge several neuroactive peptides are released by the bag cell neurons (Arch, 1972; Stuart et al., 1980). These peptides include egg-laying hormone (ELH) and α-, β-, and γ-bag cell peptides (BCP). These peptides are

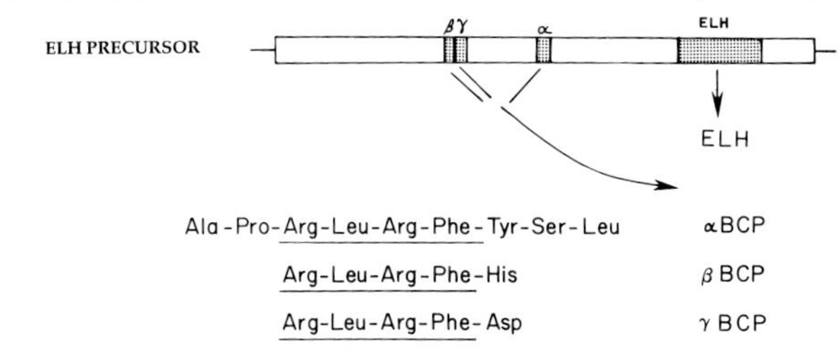

Figure 2. Schematic of the ELH precursor and several of its products, including α-, β-, and γ-BCP.

proteolytically cleaved from a common precursor protein, as illustrated in Fig. 2 (Berry, 1981; Scheller et al., 1983; Newcomb and Scheller, 1987). ELH is a 36 amino acid peptide which, upon injection in vivo, induces all the behaviors associated with the fixed action pattern of egg-laying (Kupfermann, 1970; Chiu et al., 1979). α-, β-, and γ-BCP are smaller peptides and share the amino acid sequence -Arg-Leu-Arg-Phe-. Evidence indicates that the bag cell neurons possess autoreceptors for these peptides, such that exposure of the cells to α-, β-, or γ-BCP produces changes in their excitability (Rothman et al., 1983; Brown and Mayeri, 1986; Kauer et al., 1987; Loechner and Kaczmarek, 1987). The modulation by autoreceptors may be a general feature for many peptide-secreting neurons. For example, electrophysiological and biochemical studies in mammalian brain have suggested that neuropeptides may act at autoreceptors either to increase or to decrease further release of the peptide and also that of other neurotransmitters that are colocalized with the peptide (Hokfelt, 1987).

The action potentials elicited by the bag cell neurons during the afterdischarge are not fixed in amplitude. Fig. 3 illustrates that an enhancement of action potentials, measured from a single bag cell neuron within the bag cell cluster, occurs during an afterdischarge. Within several minutes after the onset of the afterdischarge, there is a marked potentiation of both the height and width of the action potential (Kaczmarek et al., 1982). These changes would be expected to allow for greater calcium influx for each action potential elicited and presumably, an increase in release of the neuroactive peptides (Katz and Miledi, 1967). Evidence from our laboratory has shown that there exist several intracellular mechanisms underlying the modulation of action potentials.

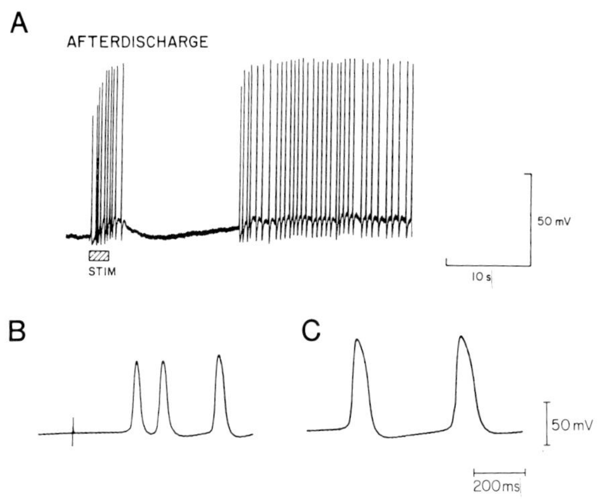

Figure 3. (*A*) Intracellular recording of an afterdischarge from a single bag cell neuron within the abdominal ganglion. (*B* and *C*) Action potentials recorded at the onset (*B*) and within 10 min (*C*) into the afterdischarge, demonstrating enhancement of action potential height and width.

The Role of Cyclic AMP

Cyclic AMP levels have been shown to increase within the first few minutes of the afterdischarge, and return to basal levels before its conclusion (Kaczmarek et al., 1978). As a result of the increased levels of cyclic AMP, one would expect activation of a cyclic AMP-dependent protein kinase, and changes in the phosphorylation state of proteins that are substrates for this enzyme can be detected during the afterdischarge (Jennings et al., 1982).

The fluctuations in cyclic AMP levels during the afterdischarge may be controlled, at least in part, by autoreceptors for the bag cell peptides. Application of α-, β-, and γ-BCP to clusters of bag cell neurons has been shown to produce changes in levels of cyclic AMP (Kauer et al., 1987; Loechner and Kaczmarek, 1987). As shown in Fig. 4, incubation of bag cell clusters within the abdominal ganglion with β-BCP leads to a significant increase in cyclic AMP levels compared with those in control animals. In contrast, the addition of γ-BCP results in a significant decrease in cyclic AMP levels. Application of α-BCP has also been shown to decrease cyclic AMP levels (Kauer et al., 1987). It is possible, therefore, that the elevation in cyclic AMP seen

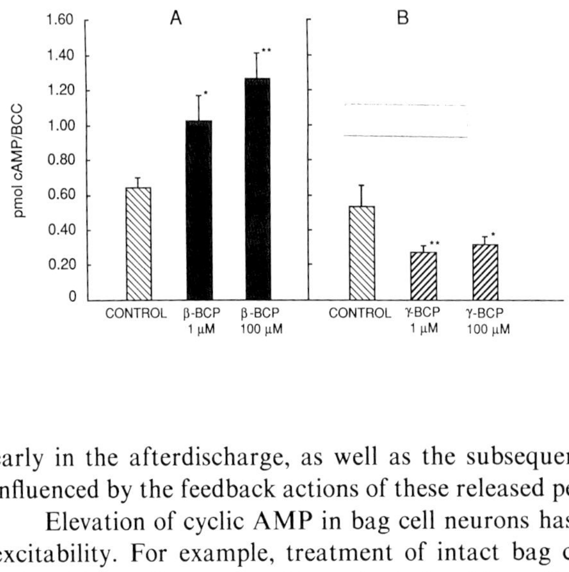

Figure 4. (*A*) Application of β-BCP (1 or 100 μM, in artificial seawater/protease inhibitors) increases cyclic AMP levels in intact clusters of bag cell neurons as measured using radioimmunoassay. Controls received artificial seawater/protease inhibitors. Data are expressed as picomole cyclic AMP/bag cell cluster. (*B*) Application of γ-BCP decreases cyclic AMP levels at both 1 and 100 μM. Protocol as in *A*. $*P < 0.01$, $**P < 0.005$.

early in the afterdischarge, as well as the subsequent decrease in cyclic AMP, are influenced by the feedback actions of these released peptides.

Elevation of cyclic AMP in bag cell neurons has been correlated with increased excitability. For example, treatment of intact bag cell clusters as well as cultured isolated bag cell neurons with nonhydrolyzable cyclic AMP analogues can trigger an afterdischarge (Kaczmarek et al., 1978; Kaczmarek and Strumwasser, 1981). Pharmacological elevation of cyclic AMP can also modulate action potential size and shape as measured with a single intracellular microelectrode in isolated cells. As shown in Fig. 5, injection of the cyclic AMP analogue 8-benzylthiocyclic AMP causes a marked increase in the height and width of action potentials (Kaczmarek and Strumwasser, 1981). Similar effects can be obtained after an injection of the catalytic subunit of the cyclic AMP-dependent protein kinase, or after the addition of forskolin, a direct activator of adenylate cyclase in bag cell neurons, plus theophylline, a phosphodiesterase inhibitor (Kaczmarek et al., 1980; Kauer and Kaczmarek, 1985). Moreover, the enhancement of action potentials by forskolin/theophylline can be blocked or reversed by microinjection of the protein inhibitor of the cyclic AMP-dependent protein kinase (Kaczmarek et al., 1984; Conn et al., 1988*a*).

At least part of the mechanism of enhancement of action potentials by elevation of cyclic AMP has been shown to involve modulation of voltage-dependent potassium currents, presumably via activation of a cyclic AMP-dependent protein kinase, which alters the properties of the potassium channels either by direct phosphorylation of the channels or by phosphorylation of a protein which can then modify the channels. The addition of 8-benzylthio-cyclic AMP to isolated bag cell neurons leads to a decrease in amplitude of outward potassium currents as measured with two-microelectrode voltage clamp (Kaczmarek and Strumwasser, 1984).

Changes in potassium currents in the bag cell neurons can be produced by the addition of α-, β-, and γ-BCP in a manner that is generally consistent with their effects

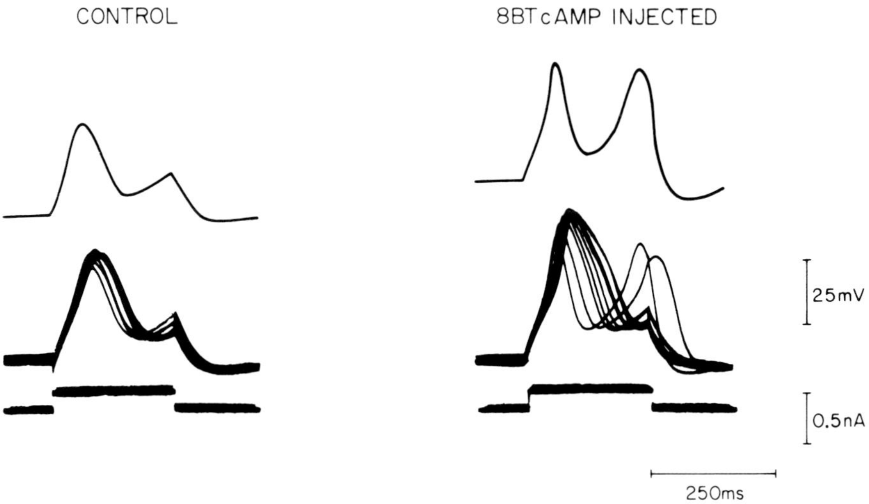

Figure 5. Enhancement of action potentials elicited by depolarizing current in isolated bag cell neurons by injection of 8-benzylthio-cyclic AMP. Upper traces show only a single action potential. Lower traces show the superimposed response to a series of 10 stimuli delivered at a frequency of 1/s.

on cyclic AMP levels. For example, application of β-BCP decreases the amplitude of a delayed voltage-dependent potassium current, as measured with whole-cell patch-clamp technique in isolated bag cell neurons (Fig. 6 *A*) (Loechner and Kaczmarek, 1987). This potassium current, which is activated upon depolarization positive to -50 mV, is likely to play a role in the repolarization of the bag cell neuron during an action potential. Fig. 6 *B* illustrates that β-BCP can also produce an increase in the rate of inactivation of the current, an effect that is seen in some but not all cells. These effects of β-BCP are consistent with the fact that β-BCP causes an increase in cyclic AMP in these cells. In contrast, application of either α- or γ-BCP causes an increase in the amplitude of this current, as is shown in Fig. 7. Again, these effects are consistent with the results that both α- and γ-BCP cause a decrease in cyclic AMP, although experiments have demonstrated that these peptides also have actions that appear to be

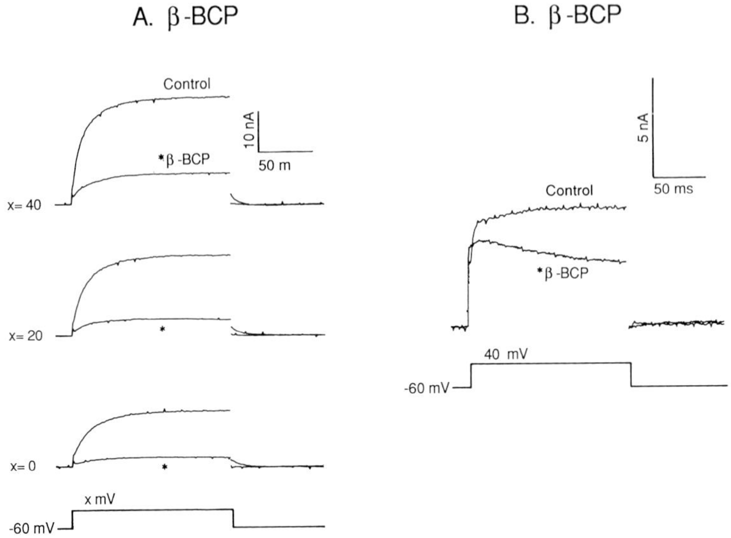

Figure 6. (*A*) Decrease in amplitude of delayed voltage-dependent potassium current after application of β-BCP (10 μM, lower traces) compared to control (upper traces). Currents were measured using whole cell patch clamp of isolated bag cell neurons. EGTA (ethyleneglycol-bis-*N,N,N′,N′*-tetraacetic acid) was included in the patch pipette to eliminate calcium-activated currents. Cells were depolarized from a holding potential of −60 mV. Controls received artificial seawater/protease inhibitors. (*B*) Increase in the rate of inactivation of delayed potassium current after application of β-BCP. Protocol as in *A*.

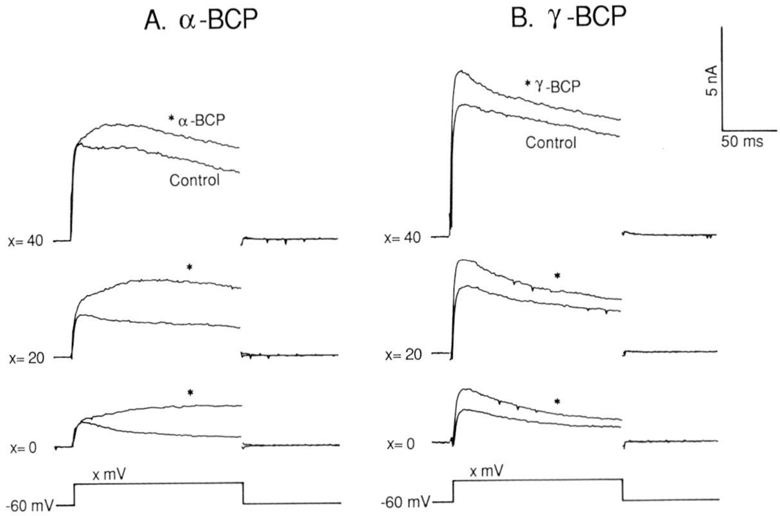

Figure 7. (*A*) Increase in delayed potassium current after application of α-BCP (10 μM, upper traces) compared with control (lower traces). Protocol as in Fig. 6. (*B*) Increase in delayed potassium current after application of γ-BCP (10 μM, upper traces) compared with control (lower traces).

unrelated to their effects on cyclic AMP levels (Brown and Mayeri, 1986; Kauer et al., 1987).

Cyclic AMP has also been shown to modulate characteristics of the bag cell neurons that are not directly related to their excitability. For example, the rate of synthesis of the protein precursor of the bag cell peptides is increased upon elevation of cyclic AMP levels (Bruehl and Berry, 1985). Furthermore, application of cyclic AMP analogues or forskolin/theophylline to isolated bag cell neurons in cell culture causes a prompt acceleration of the movement of secretory granules into the distal tips of growing neurites (Forscher et al., 1987).

It is not yet known whether the three autoactive peptides, α-, β-, and γ-BCP, are coreleased during an afterdischarge or whether differential processing of the peptide precursor can occur to produce a differential release of individual peptides at different sites in the bag cell network, or at different times during an afterdischarge. Experiments using immunocytochemical localization of the bag cell peptides (Kreiner et al., 1986) and studies of vesicle population and precursor processing (Yates and Berry, 1984; Arch et al., 1986; Newcomb and Scheller, 1987; Fisher et al., 1988) suggest that differential processing of the ELH precursor may, indeed, take place, and that α-, β-, and γ-BCP are likely to be found in a different population of granules from those containing ELH. It is not known, however, if the differential processing results in granules containing different amounts of β-BCP compared to α- or γ-BCP.

Finally, it is not yet known if the effects of α-, β-, and γ-BCP on cyclic AMP levels occur through a change in the activity of adenylate cyclase or of a cyclic nucleotide phosphodiesterase. It is possible that the primary function of these peptides is to stimulate the turnover rate of the cyclic AMP system during an afterdischarge and that the apparent opposing effects on net cyclic AMP levels result from differential effects on these two systems.

The Role of Protein Kinase C

The changes in electrical properties of the bag cell neurons after stimulation cannot be attributed to modulation by cyclic AMP levels alone. For example, the major voltage-dependent calcium current is not affected by elevation of cyclic AMP in these neurons (Kaczmarek and Strumwasser, 1984). Calcium currents in bag cell neurons are, however, modified after activation of protein kinase C. Electrical stimulation of the bag cell clusters has been shown to cause an increase in phosphoinositide hydrolysis (Fink et al., 1988). This pathway generates the second messengers inositol 1,4,5-trisphosphate (IP_3) and diacylglycerol (DAG) (Berridge, 1984). IP_3 causes the release of intracellular calcium and the activation of a calcium-dependent potassium current in these neurons (Fink et al., 1988). DAG is a physiological activator of protein kinase C. The application of phorbol esters or synthetic diacylglycerols, which directly activate protein kinase C, leads to the marked enhancement of action potential height (DeReimer et al., 1985). A similar effect is observed on direct microinjection of protein kinase C into isolated bag cell neurons. Fig. 8 *A* illustrates the enhancement of action potentials after application of the phorbol ester TPA (12-O-tetradecanoyl-13-phorbol acetate). This enhancement of action potential height results from a two- to threefold increase in the amplitude of a voltage-dependent calcium current, measured in isolated bag cell neurons using the whole-cell patch-clamp technique (DeRiemer et al., 1985). The mechanism of this increase has been shown to involve the activation of a previously

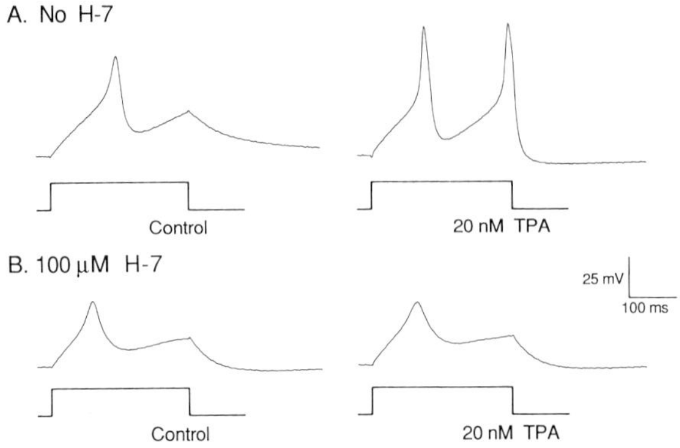

Figure 8. (*A*) Enhancement of action potential height after the addition of TPA (20 nM) to isolated bag cell neurons penetrated with a single microelectrode. Action potentials were elicited with depolarizing current. (*B*) Blockade of the effect of TPA on action potential height by preincubation with the protein kinase C inhibitor, H-7 (100 μM, Conn et al., 1989*a*).

covert calcium channel (Fig. 9) (Strong et al., 1987). Under control conditions, voltage-dependent calcium channels with a mean conductance of 12 pS can be measured using single-channel recording techniques with barium as the charge carrier. This species of calcium channel is found predominantly in clusters on the surface of bag cell neurons. After pretreatment with TPA, a second species of calcium channel can also be recorded. This second channel has a conductance of 24 pS and is distributed relatively uniformly over the surface of the soma. The frequency of occurrence of the 12 pS channels recorded does not change in response to TPA. It is likely, therefore, that the detection of the second channel represents the appearance of a previously covert species of calcium channel, rather than an interconversion of 12 to 24 pS channels.

The development of protein kinase C inhibitors such as sphinganine (Hannun et al., 1986; Merrill et al., 1986) and H-7 (Kawamoto and Hidaka, 1984) have made it possible to provide further tests of whether the effects of phorbol esters and diacylglycerols are, indeed, due to the ability of these drugs to activate protein kinase C. For

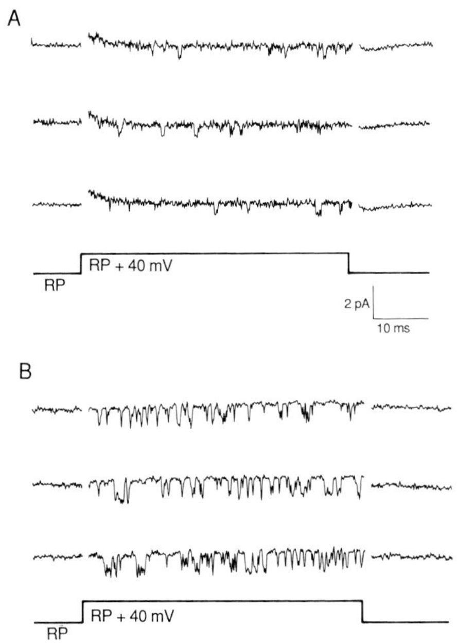

Figure 9. Inward channel activity recorded in cell-attached patches of isolated bag cell neurons with barium-containing pipettes. (*A*) In control cells, one type of inward channel can be seen upon a 40-mV depolarization from the resting potential. This channel has a unitary conductance of ~12 pS. (*B*) In cells treated with 20 nM TPA, a second class of inward channel, shown here, is also detected. This channel has a unitary conductance of ~24 pS (Conn et al., 1989*b*).

example, the enhancement of action potentials by TPA can be blocked in the presence of sphinganine or H-7 (Fig. 8 *B*, Conn et al., 1989*a*, *b*), as can the increase in whole-cell calcium current (Conn et al., 1989*a*, *b*). Moreover, the enhancement of action potential height that occurs during a discharge can be blocked by pretreatment of the clusters within the abdominal ganglion with either H-7 or sphinganine (Conn et al., 1989*b*). The physiological stimulus for the activation of protein kinase C is not known, nor is it known if the receptors for the bag cell peptides are coupled to phosphoinositide hydrolysis. In other systems, such as the muscarinic receptor in hippocampus, agonist-induced decreases in cyclic AMP occur simultaneously with stimulation of phosphoinositide hydrolysis (Ashkenazi et al., 1987; Nicoll, 1988). It will be important, therefore, to determine whether such a mechanism exists for α- and γ-BCP.

Secretion

The enhancement of the action potentials during the afterdischarge provides a possible mechanism for increased peptide secretion. In that the changes in action potential shape occur soon after the onset of the afterdischarge, one who expect that the rate of release of peptides would be maximal relatively early in the afterdischarge. Preliminary data suggests, however, that releases of peptides may be progressively potentiated during an afterdischarge (Loechner et al., 1988). In an intact abdominal ganglion, the bag cell neurons send axonal processes into the vascularized connective tissue sheath covering the abdominal ganglion (Coggeshall, 1967). Peptides released by the bag cell neurons can, therefore, diffuse into the surrounding medium as they would into the body cavity in vivo. In the experiment shown in Fig. 10, the medium surrounding the bag cell clusters within the abdominal ganglion was completely exchanged at 5-min intervals before, during, and after an afterdischarge. Released peptides were separated using reverse-phase high pressure liquid chromatography (HPLC) and post-column derivitization with *o*-phthalic dicarboxaldehyde (Roth, 1971). Each panel represents either 5 min of electrical activity (left) or the HPLC profile of peptides released during that 5-min period (right).

The first trace shows the amount of material released in the 5 min before electrical stimulation of the afterdischarge. The subsequent traces demonstrate that after the onset of an afterdischarge, additional peaks can be detected in the released material. Some of these peaks coelute with synthetic peptide standards, such as the peak seen at 28 min on the HPLC gradient, which coelutes with ELH. ELH has been positively identified by amino acid analysis. Interestingly, several of the peaks increase in height from one chromatogram to the next, suggesting that the rate of release of these peptides increases during the course of the afterdischarge. The progressive increase in release has been seen in afterdischarges of various lengths (10–35 min). Furthermore, after the end of the afterdischarge, the amplitudes of the peaks decline, as is shown in the last set of traces in Fig. 10.

This observed pattern of release is in contrast to that expected from the firing rate of the bag cell neurons, which peaks early in the afterdischarge and subsequently declines. It is possible that the rate of peptide release is, in fact, maximal during the first two or three minutes of the discharge but that their detection in the medium surrounding the ganglion is retarded by binding and diffusion through the vascularized connective sheath. However, a variety of experiments in which the artery to the bag cell clusters was perfused during the afterdischarge, thereby minimizing such diffusion barriers, suggest that this is not a likely interpretation. It appears, therefore, that action potentials occurring later in the afterdischarge may be able to release more peptide than those occurring near its onset. A similar progressive potentiation of release may also occur in other systems of discharging neurons. For example, in the caudodorsal cells of the pond snail *Lymnaea*, a system analogous to that of the bag cell

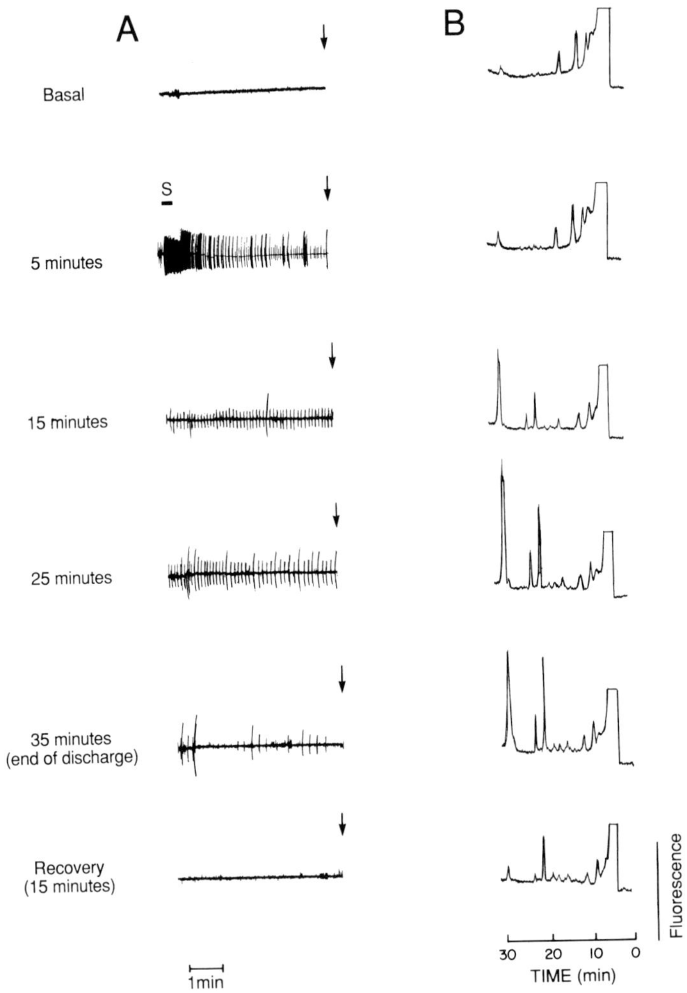

Figure 10. Time course of release of peptides during an afterdischarge. (*Left*) Extracellular recording before, during, and after an afterdischarge. (*Right*) Corresponding HPLC profiles of released peptides for each 5 min of electrical activity. Alternate 5-min samples shown for simplicity. *x*-axis, time on HPLC gradient; *y*-axis, relative fluorescence.

neurons of *Aplysia*, Roubos et al. (1981) have found that the number of exocytotic events measured in the terminals of the caudodorsal cells increases during the course of the caudodorsal cell afterdischarge, and then decreases after its termination. Such ultrastructural analysis of the bag cell neurons during an afterdischarge would provide a further level of analysis of peptide secretion.

Although the mechanism by which release is potentiated is not yet known, it is possible that the initial fast phase of firing at the onset of the afterdischarge serves as a priming stimulus for subsequent changes in the properties of the bag cell neurons. For example, it may allow the initial release of peptides, such as α-, β-, and γ-BCP. These peptides may then stimulate second messenger systems that may remain active throughout the duration of the afterdischarge and play a role in changing the secretory properties of the bag cell neurons. For example, granule mobilization and movement into the terminals may be enhanced, or release sites may be altered, such that subsequent action potentials are capable of triggering greater release.

In conclusion, stimulation of an afterdischarge in bag cell neurons leads to changes in the electrical properties of the neurons and to potentiation of peptide release. The intracellular mechanisms responsible for the modulation of the bag cell neurons include an elevation of cyclic AMP levels and the activation of protein kinase C and may be influenced by autoreceptors for the bag cell peptides that are released during the afterdischarge. These mechanisms allow the bag cell neurons to undergo long-lasting changes in activity that in vivo control the fixed action pattern of egg-laying.

References

Arch, S. 1972. Biosynthesis of the egg-laying hormone (ELH) in the bag cell neurons of *Aplysia californica. Journal of Neurophysiology.* 43:399–519.

Arch, S., A. Linstedt, G. Whitney, P. Teal, and T. Smock. 1986. Neuropeptide routing in the bag cells: kinetic differences in the appearance of newly labeled peptides in transport and secretion. *Journal of Neuroscience.* 6:1545–1552.

Ashkenazi, A., J. W. Winslow, E. G. Peralta, G. L. Peterson, M. I. Schimerlik, D. J. Capon, and J. Ramachandran. 1987. An M2 muscarinic receptor subtype coupled to both adenylyl cyclase and phosphoinositide turnover. *Science.* 238:672–675.

Berridge, M. J. 1984. Inositol triphosphate and diacylglycerol as second messengers. *Biochemical Journal.* 220:345–360.

Berry, R. W. 1981. Proteolytic processing of the neurosecretory egg-laying hormone in *Aplysia.* 1. Precursors, intermediates, and products. *Biochemistry.* 20:6200–6205.

Brown, R. O., and E. Mayeri. 1986. Evidence for excitatory autotransmission in *Aplysia* bag cell neurons mediated by alpha-, beta-, and gamma–bag cell peptides. *Society for Neuroscience Abstracts.* 12:946. (Abstr.)

Bruehl, C. L., and R. W. Berry. 1985. Regulation of synthesis of the neurosecretory egg-laying hormone of *Aplysia:* antagonistic roles of calcium and cyclic adenosine 3′:5′-monophosphate. *Journal of Neuroscience.* 5:1233–1238.

Chiu, A. Y., M. W. Hunkapiller, E. Heller, D. K. Stuart, L. E. Hood, and F. Strumwasser. 1979. Purification and primary structure of neuroactive egg-laying hormone of *Aplysia californica. Proceedings of the National Academy of Sciences.* 76:6656–6660.

Coggeshall, R. E. 1967. A light and electron microscope study of the abdominal ganglion of *Aplysia californica. Journal of Neurophysiology.* 30:1263–1287.

Conn, P. J., J. A. Strong, E. M. Azhderian, A. C. Nairn, P. Greengard, and L. K. Kaczmarek. 1989*a.* Protein kinase inhibitors selectively block phorbol ester- or forskolin-induced changes in excitability of *Aplysia* neurons. *Journal of Neuroscience.* 9:473–479.

Conn, P. J., J. A. Strong, and L. K. Kaczmarek. 1989*b*. Inhibitors of protein kinase C prevent enhancement of calcium current and action potentials in peptidergic neurons of *Aplysia*. *Journal of Neuroscience*. 9:480–487.

DeRiemer, S. A., J. A. Strong, A. Albert, P. Greengard, and L. K. Kaczmarek. 1985. Phorbol ester and protein kinase C enhance calcium current in *Aplysia* neurones. *Nature*. 313:313–316.

Dudek, F. E., and J. E. Blankenship. 1977. Neuroendocrine cells of *Aplysia brasiliana*. I. Bag cell action potentials and afterdischarge. *Journal of Neuroscience*. 40:1301–1311.

Dudek, F. E., J. S. Cobbs, and H. M. Pinsker. 1979. Bag cell electrical activity underlying spontaneous egg laying in freely behaving *Aplysia brasiliana*. *Journal of Neurophysiology*. 42:804–817.

Fink, L. A., J. A. Connor, and L. K. Kaczmarek. 1988. Inositol triphosphate releases intracellularly stored calcium and modulates ion channels in molluscan neurons. *Journal of Neuroscience*. 8:2544–2555.

Fisher, J. M., W. Sossin, R. Newcomb, and R. H. Scheller. 1988. Multiple neuropeptides derived from a common precursor are differentially packaged and transported. *Cell*. 54:813–822.

Forscher, P., L. K. Kaczmarek, J. Buchanan, and S. J. Smith. 1987. Cyclic AMP induces changes in distribution and transport of organelles within growth cones of *Aplysia* bag cell neurons. *Journal of Neuorscience*. 7:3600–3611.

Hannun, Y. A., C. R. Loomis, A. H. Merrill, Jr., and R. M. Bell. 1986. Sphingosine inhibition of protein kinase C activity and of phorbol dibutyrate binding *in vitro* and in human platelets. *Journal of Biological Chemistry*. 261:12604–12609.

Heller, E., L. K. Kaczmarek, M. W. Hunkapiller, L. E. Hood, and F. Strumwasser. 1980. Purification and primary structure of two neuroactive peptides that cause bag cell afterdischarge and egg-laying in *Aplysia*. *Proceeding of the National Academy of Sciences*. 77:2328–2332.

Hokfelt, T. 1987. Neuronal communication through multiple coexisting messengers. *In* Synaptic Function. G. M. Edelman, W. E. Gall, and W. M. Cowan, editors. John Wiley & Sons, New York. 179–211.

Jennings, K. R., L. K. Kaczmarek, R. M. Hewick, W. J. Dreyer, and F. Strumwasser. 1982. Protein phosphorylation during an afterdischarge in peptidergic neurons of *Aplysia*. *Journal of Neuroscience*. 2:158–168.

Kaczmarek, L. K., K. Jennings, and F. Strumwasser. 1978. Neurotransmitter modulation, phosphodiesterase inhibitor effects, and cAMP correlates of afterdischarge in peptidergic neurites. *Proceedings of the National Academy of Sicences*. 75:5200–5204.

Kaczmarek, L. K., K. R. Jennings, and F. Strumwasser. 1982. An early sodium and a late calcium phase in the afterdischarge of peptide secreting neurons of *Aplysia*. *Brain Research*. 238:105–115.

Kaczmarek, L. K., K. R. Jennings, F. Strumwasser, A. C. Nairn, U. Walter, F. D. Wilson, and P. Greengard. 1980. Microinjection of catalytic subunit of cyclic AMP-dependent protein kinase enhances calcium action potentials of bag cell neurons in cell culture. *Proceedings of the National Academy of Sciences*. 77:7427–7491.

Kaczmarek, L. K., and I. B. Levitan. 1987. Neuromodulation: The Biochemical Control of Neuronal Excitability. Oxford University Press, New York.

Kaczmarek, L. K., A. C. Nairn, and P. Greengard. 1984. Microinjection of protein kinase inhibitor prevents enhancement of action potentials in peptidergic neurons of *Aplysia*. *Society for Neuroscience Abstracts*. 10:895. (Abstr.)

Kaczmarek, L. K., and F. Strumwasser. 1981. The expression of long lasting afterdischarge by isolated *Aplysia* bag cell neurons. *Journal of Neuroscience.* 1:626–634.

Kaczmarek, L. K., and F. Strumwasser. 1984. A voltage-clamp analysis of current underlying cyclic AMP-induced membrane modulation peptidergic neurons of *Aplysia. Journal of Neurophysiology.* 52:340–349.

Kauer, J. A., T. E. Fisher, and L. K. Kaczmarek. 1987. Alpha bag cell peptide directly modulates the excitability of the neurons that release it. *Journal of Neuroscience.* 7:3623–3632.

Kauer, J. A., and L. K. Kaczmarek. 1985. Peptidergic neurons of *Aplysia* lose their response to cyclic adenosine 3′:5′-monophosphate during a prolonged refractory period. *Journal of Neuroscience.* 5:1339–1345.

Katz, B., and R. Miledi. 1967. The timing of calcium action during neuromuscular transmission. *Journal of Phsyiology.* 189:535–544.

Kawamoto, S., and H. Hidaka. 1984. 1-(5-isoquinolinesulfonyl)-2-methylpiperazine (H-7) is a selective inhibitor of protein kinase C in rabbit platelets. *Biochemical Biophysics Research Communication.* 125:258–264.

Kreiner, T., W. Sossin, and R. Scheller. 1986. Localization of *Aplysia* neurosecretory peptides to multiple populations of dense core vesicles. *Journal of Cell Biology.* 102:769–782.

Kupferman, I. 1970. Stimulation of egg laying by extracts of neuroendocrine cells (bag cells) of abdominal ganglion of *Aplysia. Journal of Neurophysiology.* 33:877–881.

Kupferman, I., and E. R. Kandel. 1970. Electrophysiological properties and functional interconnections of two symmetrical neurosecretory clusters (bag cells) in abdominal ganglion of *Aplysia. Journal of Neurophysiology.* 33:865–876.

Loechner, K. J., E. Azhderian, R. Dreyer, and L. K. Kaczmarek. 1988. Progressive release of peptides from bag cell neurons of *Aplysia* during a discharge. *Society for Neuroscience Abstracts.* 14:176. (Abstr.)

Loechner, K. J., and L. K. Kaczmarek. 1987. Regulation of isolated bag cell neurons of *Aplysia* by alpha-, beta-, and gamma- bag cell peptides. *Society for Neuroscience Abstracts.* 13:1071. (Abstr.)

Merrill A. H., Jr., A. M. Sereni, V. L. Stevens, Y. A. Hannun, R. M. Bell, and J. M. Kinkade. 1986. Inhibition of phorbol ester-dependent differentiation of human promyelocytic leukemic (HL-60) cells by sphinganine and other long-chain bases. *Journal of Biological Chemistry.* 261:12610–12615.

Newcomb, R., and R. H. Scheller. 1987. Proteolytic processing of *Aplysia* egg-laying hormone and R3-R14 neuropeptide precursors. *Journal of Neuroscience.* 7:854–863.

Nicoll, R. A. 1988. The coupling of neurotransmitter receptors to ion channels in the brain. *Science.* 241:545–551.

Roth, M. 1971. Fluorescence reaction for amino acids. *Analytical Chemistry.* 43:880–882.

Rothman, B. S., E. Mayeri, R. O. Brown, P. Yuan, and J. Shively. 1983. Primary structure and neuronal effects of alpha bag cell peptide, a second candidate neurotransmitter encoded by a single gene in bag cell neurons of *Aplysia. Proceedings of the National Academy of Sciences.* 80:5733–5757.

Roubos, E. W., H. H. Boer, and L. P. C. Schot. 1981. Peptidergic neurons and the control of neuroendocrine activity in the freshwater snail *Lymnaea stagnalis (L.). In Neurosecretion.* D. S. Farner and K. Lederis, editors. Plenum Publishing Corp., New York. 119–127.

Scheller, R. H., J. F. Jackson, L. B. McAllister, B. S. Rothman, E. Mayeri, and R. Axel. 1983. A single gene encodes multiple neuropeptides mediating a stereotyped behavior. *Cell.* 32:7–22.

Strong, J. A., A. P. Fox, R. W. Tsien, and L. K. Kaczmarek. 1987. Stimulation of protein kinase C recruits covert calcium channels in *Aplysia* bag cell neurons. *Nature* 325:714–717.

Stuart, D. K., A. Y. Chiu, and F. Strumwasser. 1980. Neurosecretion of egg-laying hormone and other peptides from electrically active bag cell neurons of *Aplysia. Journal of Neurophysiology.* 43:488–498.

Yates, M. E., and R. W. Berry. 1984. Subcellular sites of processing of precursors to neurosecretory peptides in the bag cells of *Aplysia:* inferences from the effects of monensin, FCCP, and chloroquine. *Journal of Neurobiology.* 15:141–155.

Electrical Phenomena in Secretory Control

Chapter 5

Regulation and Function of ATP-dependent K^+ Channels in Pancreatic B Cells

Gerhard Trube, Jürgen Hescheler, and Klaus Schröter

F. Hoffman-La Roche and Co., ltd., Pharmaceutical Research Department, CH-4002 Basle, Switzerland; II. Physiologisches Institut, Universität des Saarlandes, D-6650 Homburg, Federal Republic of Germany; and the Max-Planck-Institut für Biophysikalische Chemie, Abteilung Zellphysiologie, D-3400 Göttingen, Federal Republic of Germany

Introduction

Pancreatic B cells differ from other secretory cells since their electrical and secretory activity is regulated by small relative changes of the common nutrient glucose, which is normally present at millimolar concentrations in the extra- and intracellular space. B cells maintain a stable resting potential close to the potassium equilibrium potential as long as extracellular glucose stays below a concentration of ~6 mM (Fig. 1 *A*). An increase of glucose to higher concentrations causes a slow depolarization, which initiates action potentials when a threshold of −60 to −50 mV is reached. After an initial period of high action potential frequency, maintenance of a constant level of 10 mM glucose results in a complex burst pattern of action potentials.

Compelling evidence has been accumulated in the past showing that glucose-stimulated insulin release is closely coupled to the initiation and intensity of B cell action potential activity (reviews: Henquin and Meissner, 1984; Petersen and Findlay,

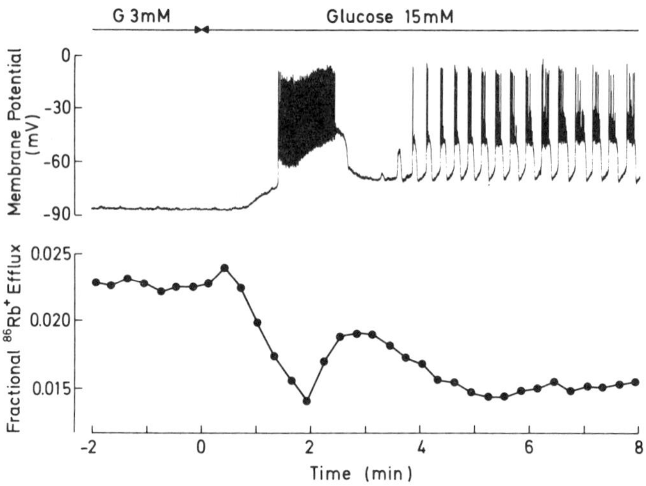

Figure 1. Effects of a rise of glucose concentration from 3 to 15 mM, on the membrane potential of a B cell in a mouse pancreatic islet (*top*) and on $^{86}Rb^{+}$ efflux from a batch of 60 islets (*bottom*). The two representative experiments were performed in different perfusion systems. From Henquin and Meissner, 1984, with permission.

1987). The depolarization initiating the action potentials was found to be caused by a decrease in potassium conductance (monitored as a decrease in $^{86}Rb^{+}$ efflux in Fig. 1 *B*). Studies with glucose substitutes and investigations on the cellular metabolism suggested that the sugar has to be metabolized by the B cell to inhibit potassium conductance and produce the subsequent stimulatory effects (Henquin and Meissner, 1984; Petersen and Findlay, 1987). However, the metabolic product or cofactor inhibiting the potassium permeability of the membrane remained unknown until 1984. Using the patch-clamp method Cook and Hales (1984) then discovered an unexpected type of K^{+} channel in the B cell membrane, which is inhibited by ATP at the intracellular side of the membrane. Similar ATP-sensitive K^{+} channels have also been found in heart cells and skeletal muscles fibers (Kakei et al., 1985; Spruce et al., 1987). The patch-clamp technique revolutionized the electrophysiology of the pancreatic B cell, because its membrane currents could now be identified unequivocally under voltage-clamp conditions. Many experimental results suggest that the ATP-sensitive channel plays a dominant role in mediating the stimulatory influence of glucose on the B cell. This channel is also the target of antidiabetic sulfonylureas, a class of compounds that decrease the K^+ permeability of the B cell membrane and thereby stimulate electrical activity and insulin secretion (Trube et al., 1986; Petersen and Findlay, 1987; Schmid-Antomarchi et al., 1987). In the present communication we will shortly review the effects of ATP and other nucleotides on the ATP-sensitive K^+ channel and we will add a few unpublished results on the action of chloride ions and trypsin. Broader reviews on the properties and function of the ATP-dependent K^+ channel have been published recently (Petersen and Findlay, 1987; Ashcroft, 1988).

Methods and Materials

B cells were isolated from the pancreases of NMRI mice and cultured as previously described (Rorsman and Trube, 1985). RINm5F insulinoma cells, originally obtained from Prof. C. Wollheim (Institut de Biochimie Clinique, University of Geneva, Geneva, Switzerland) were kindly supplied from Prof. E. Neher's tissue culture lab (Max-Planck-Institut für biophysikalische Chemie, Göttingen, FRG). Standard patch-clamp techniques were used for recording currents from inside-out patches of cell membrane (Hamill et al., 1981). Experiments were done at room temperature. The pipette solution contained (concentrations in millimolar): 146 KCl, 1.2 $MgCl_2$, 2.6 $CaCl_2$, and 10 HEPES-KOH (pH 7.4). For the experiments of Fig. 4 the bath solution at the intracellular side of the isolated patches contained 145 glutamic acid, 8 NaCl, 1 $MgCl_2$, 0.2 EGTA, and 10 HEPES. In the bath solutions the pH was always adjusted to 7.15 by adding KOH. Glutamic acid and corresponding amounts of KOH were replaced by KCl (145 mM) as indicated in Fig. 4. ATP (Na_2-salt, Sigma Chemical Co., St. Louis, MO) and equimolar amounts of $MgCl_2$ were also added as specified in the figures. Slightly different bath solutions were used for the experiments of Fig. 5: 130 KCl, 1 $MgCl_2$, 2 $CaCl_2$, 10 EGTA, and 5 HEPES, plus ATP and equimolar $MgCl_2$ as indicated in the figure. Trypsin was purchased from Sigma (type II, no. 8128). Current traces shown in Figs. 4 and 5 were recorded on a chart recorder (40-Hz maximum frequency response). Frequency-vs.-current amplitude histograms (not shown) were obtained by computer analysis to calculate values of the time-averaged current shown in Fig. 4.

Results and Discussion

Identification of a Glucose- and ATP-sensitive K^+ Channel in the B Cell Membrane

Fig. 2 *A* shows single-channel currents recorded from a cell-attached patch on a B cell within a cluster of some hundred islet cells. The conditions of the experiment were designed to allow the detection of K^+ channel openings at the cell's resting potential (~ -70 mV). The pipette solution contained 155 mM K^+ providing ~ -70 mV driving force for inward K^+ currents (from pipette to cytoplasm) at a pipette potential of 0 mV. The downward deflections in the trace of Fig. 2 *A* indicate openings of a channel that, by applying other K^+ concentrations and potentials, was found to be selectively permeable for K^+ (Ashcroft et al., 1984; Findlay et al., 1985*b*; Rorsman and Trube, 1985; Misler et al., 1986; Ashcroft et al., 1988).

Current amplitudes other than those seen in Fig. 2 *A* were only rarely observed at 0-mV pipette potential, suggesting that the channel observed here is the predominant channel active in the resting B cell membrane (Findlay et al., 1985*b*; Ashcroft et al., 1988). The frequency of channel openings was reduced, but not abolished by the addition of glucose in concentrations below the threshold (6 mM) that stimulates action potentials and insulin release in intact islets of Langerhans. Glucose in a concentration of 10–20 mM completely inhibited the channel's activity and finally elicited trains of action potentials (seen as biphasic deflections in Fig. 2 *B*) (Misler et al., 1986; Trube et al., 1986; Ashcroft et al., 1988). Channel activity in the presence of 10–20 mM glucose could not be restored by changing the voltage at the patch

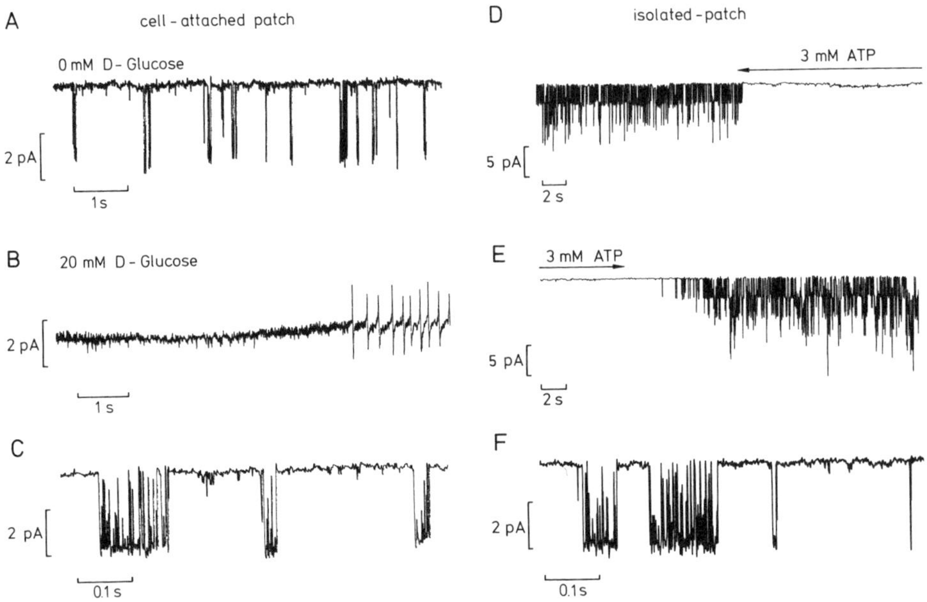

Figure 2. Effects of glucose (*A* and *B*) or ATP (*D* and *E*) on single-channel K^+ currents in a cell-attached patch (*A–C*) and an inside-out patch (*D–E*). (*A–C*) Currents recorded at the cell's resting potential. Trace *B* is the continuation of *A* 6 min after adding 20 mM glucose. Biphasic current deflections due to action potentials are observed at the end of the trace. (*D–F*) Currents from the inside-out patch recorded at −70-mV membrane potential. Traces *D* and *E* show a continuous recording. ATP was applied during the period below the arrows. (*C* and *F*) Examples of the single-channel openings in *A* and *D*, respectively, at an expanded time base. Mouse pancreatic B cell. From Rorsman and Trube, 1985, with permission.

membrane which indicates that the disappearance of the openings was not due to a change of the cell resting potential (Ashcroft et al., 1984, 1988). It should also be noted that glucose was added to the bath solution, but not to the pipette. Therefore, the sugar did not reach the channels observed in Fig. 2 *A* from the extracellular side. It can be concluded that it was taken up into the cell and probably metabolized before inhibiting the K^+ channels. Experiments on excised membrane patches revealed that ATP may be the product of glucose metabolism which regulates the activity of the channel seen in Fig. 2 *A*. Superimposed openings of several identical channels are usually observed when the intracellular face of an inside-out patch is exposed to a bath solution not containing ATP (Fig. 2 *D*, note that time and amplitude scales are different from those in *A* and *B*). These openings are completely blocked during the application of 3 mM ATP (Fig. 2, *D* and *E*, interval below arrows; Cook and Hales, 1984). The conductance (60 pS) and the kinetics of the channels seen in the cell-attached patches and in the inside-out patches were found to be very similar, indicating that the glucose- and ATP-sensitive channels are identical (Rorsman and Trube, 1985). This similarity is also illustrated by the two expanded traces in *C* (cell-attached) and *F* (inside-out) of Fig. 2. Glucose itself has no influence on the channels in inside-out patches, which supports the idea that the sugar has to be metabolized before influencing the electrical activity of the pancreatic B cell (Rorsman and Trube, 1985; Ashcroft et al., 1988).

The ATP-dependent K^+ channel is only weakly influenced by Ca^{2+} at the intracellular side of the membrane (Cook and Hales, 1984; Findlay et al., 1985*b*; Misler et al., 1986). Ca^{2+}-dependent K^+ channels exist in the B cell membrane (Cook et al., 1984; Findlay et al., 1985*a*), but openings of these channels are only rarely observed in cell-attached patches and, if present, are not suppressed by an increase of the extracellular glucose concentration (Findlay et al., 1985*b*; Rorsman and Trube, 1985; Ashcroft et al., 1988). Therefore, the Ca^{2+}-dependent K^+ channel is not any longer believed to be important for the initial glucose-induced depolarization (Petersen and Findlay, 1987). However, some mathematical models of B cell electrical activity suggest that the silent phases between the bursts of action potentials are caused by an activation of the Ca^{2+}-dependent K^+ channel (Chay, 1988).

The ATP-sensitive K^+ channel shows moderate inward rectification at positive potentials and, therefore, has also been called the "inward-rectifier" channel (Findlay et al., 1985*b*). It is, however, not identical with the classical inwardly or anomalously rectifying K^+ channel of skeletal muscle, heart cells, and starfish eggs. ATP-dependent and anomalously rectifying channels coexist in heart cells and skeletal muscle (Kakei et al., 1985; Spruce et al., 1987), but the anomalous rectifier is missing in the B cell.

Openings of the ATP-sensitive K^+ channels are also blocked by the nonhydrolyzable ATP analogues AMP-PCP and AMP-PNP, and by ATP in absence of magnesium. This indicates that the inhibition is a direct effect independent of phosphorylation of the channel protein (Misler et al., 1986; Ohno-Shosaku et al., 1987).

Is ATP the Physiological Regulator of the ATP-dependent K Channel?

The ATP sensitivity of the K^+ channel, which had been found to be indirectly regulated by glucose, suggested that the K^+ permeability of the B cell membrane may be linked to glucose concentration via changes of the intracellular level of ATP. This hypothesis, however, has been questioned since it is difficult to reconcile the relation between ATP concentration and K^+ current in inside-out patches with previously measured intracellular ATP levels. In inside-out patches, ATP in a concentration of 10–20 μM reduces the K^+ channel's open state probability by 50% (filled circles in Figs. 3 and 4 *D*; Cook and Hales, 1984; Ohno-Shosaku et al., 1987). Biochemical assays of the intracellular ATP concentration, on the other hand, always resulted in values above 0.5 mM, and the concentration does not increase more than twofold when glucose is increased from 0 to 20 mM (e.g., Kakei et al., 1986; Malaisse and Sener, 1987; Ribalet and Ciani, 1987). Three types of arguments have been used to explain the discrepancy between the intracellular ATP concentration and the IC_{50} from the experiments on inside-out patches.

(*a*) The density of ATP-dependent K^+ channels in the membrane is quite high and, therefore, it has been concluded that the degree of channel activation must be low (probably $<1\%$) to explain the low K^+ permeability of the intact B cell (Dawson et al., 1983; Rorsman and Trube, 1985; Misler et al., 1985; Ashcroft et al., 1988). Starting from the conclusion that intact B cells work at a very low level of channel activation, Cook et al. (1988) recently estimated the theoretically expected relation between membrane potential and intracellular ATP concentration. Their model predicts an approximately 300-fold shift to higher ATP concentrations of the membrane potential dependency relative to the dose-response curve for channel inhibition.

(*b*) The concentration of ATP in the vicinity of the channels may be lower than the average intracellular concentration of ATP, either because of a high rate of ATP consumption by membrane-bound enzymes or because part of the ATP is confined to intracellular organelles. Evidence for the existence of such ATP gradients has recently been derived from experiments on hepatocytes and pancreatic islets (Jones, 1986; Malaisse and Sener, 1987). Obviously, a method for directly measuring the spatial distribution of intracellular ATP is needed to substantiate the ideal of "functional compartmentalization" of ATP.

(*c*) The presence of unknown cytoplasmic factors, which are lost during the formation of cell-free inside-out patches, may reduce the inhibition of the K^+ channel by ATP in the intact cell. Circumstantial evidence for this assumption arises from the simultaneous measurement of $^{86}Rb^+$ efflux and intracellular ATP levels in the insulinoma cell line, RINm5F, during metabolic inhibition (Schmid-Antomarchi et al., 1987). Glibenclamide-sensitive $^{86}Rb^+$ efflux, which is believed to arise specifically from the ATP-dependent channel, was sigmoidally related to the ATP concentration with half-maximal inhibition at 0.8 mM ATP. This suggests that the K^+ channel in the intact cell may be about 50 times less sensitive to ATP than the channel in the inside-out patch.

Other Modulators of the ATP-dependent Channel

Several recent studies have shown that the activity of the ATP-dependent channel in inside-out patches is also modulated by nucleotides other than ATP. The pyridine nucleotides NAD and NADP, as well as their reduced forms, produce both partially activating and inhibitory effects depending on the concentration and the simultaneous presence of ATP, GTP, and ADP (Dunne et al., 1988). GTP and other guanoside di- and triphosphates induce channel openings on some special conditions (Dunne and Petersen, 1986*a*). The combined effect of the pyridine nucleotides and guanosine phosphates in the intact cell can hardly be predicted, however, since sufficient information about their cytoplasmic concentrations is not available.

ADP in the absence of other nucleotides is a weak channel inhibitor (Ribalet and Ciani, 1987). But when it is present with ATP, probably because of their competition for the same binding site, it is quite potent in reducing the inhibition of the channel (Fig. 3, *A–C*; Dunne and Petersen, 1986*b*; Misler et al., 1986). ADP in a concentration of 2 mM shifts the 50% inhibitory concentration of ATP from 12 to 145 μM (Fig. 3 *D*). It is therefore believed that both an increase of ATP and a decrease of ADP concentration contribute to the inhibition of the "ATP-dependent" K^+ channel in the intact cell after a rise of the glucose concentration (Petersen and Findlay, 1987). In this context, it is important to note that the ATP/ADP ratio varies more strongly than the ATP concentration when the glucose concentration is changed (Kakei et al., 1986; Malaisse and Sener, 1987).

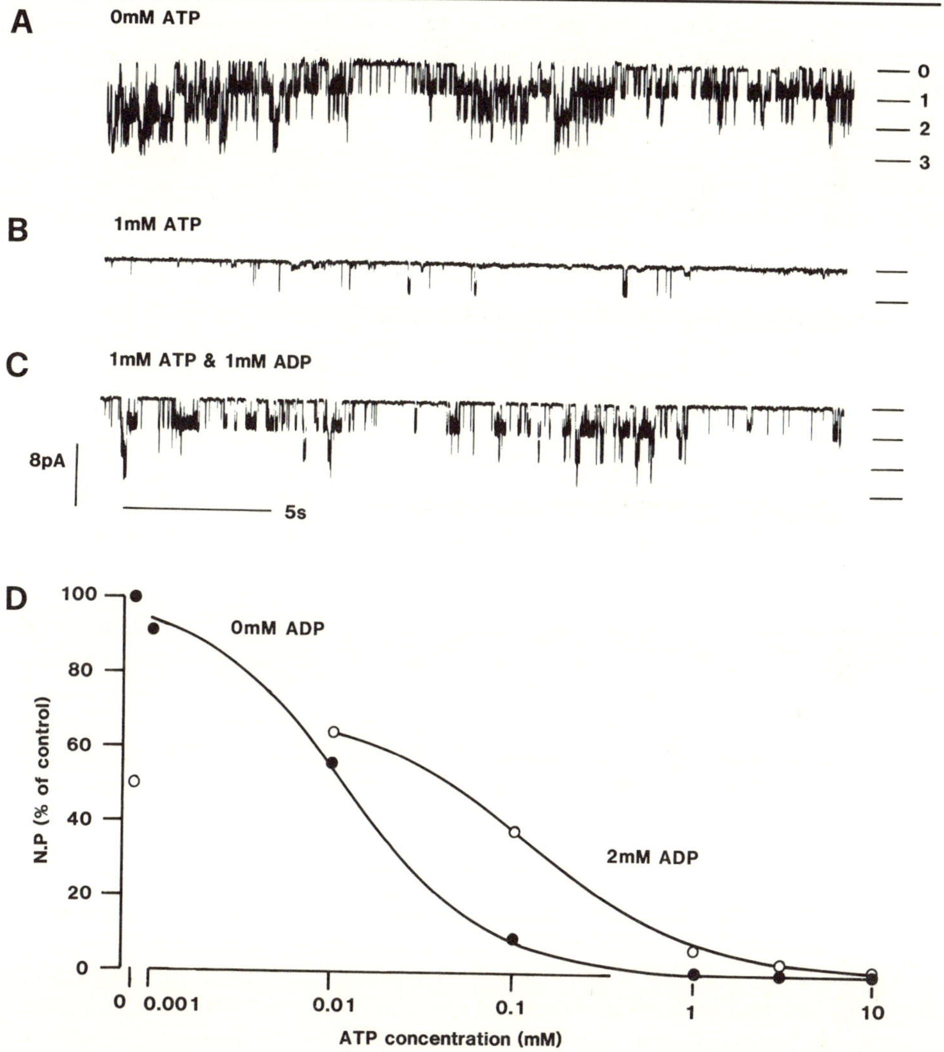

Figure 3. Single-channel K^+ currents recorded at -60 mV from an inside-out patch exposed to control solution (0 mM ATP, *A*) and ~2 s after exposure to an intracellular solution containing 1 mM ATP (*B*) or 1 mM ATP + 1 mM ADP (*C*). The diagram (*D*) shows the effect of 2 mM ADP on the dose-response curve for inhibition of the time-averaged current by ATP. Current in the absence of nucleotides is defined as 100%. (●) 0 mM ADP, (○) 2 mM ADP. The curves are drawn to the equation $y = y_{max}/[1 + (x/IC_{50})^H]$ where x is the ATP concentration, IC_{50} the 50% inhibitory concentration, and H the Hill coefficient. In absence of ADP, $y_{max} = 100$, $IC_{50} = 12\ \mu M$, and $H = 1.1$. In the presence of 2 mM ADP, $y_{max} = 64$, $IC_{50} = 145\ \mu M$ and $H = 1$. Rat pancreatic B cell. From Kakei et al., 1986, with permission.

An influence of intracellular pH on the activity of the ATP-dependent K^+ channel has been overlooked until recently (Gillis and Misler, 1988). The pH-effect seems to be physiologically important, because acidification of the cytoplasm has been shown to facilitate and enhance glucose-induced electrical activity of the B cell, and because glucose is believed to decrease intracellular pH (Pace, 1984). A decrease of intracellu-

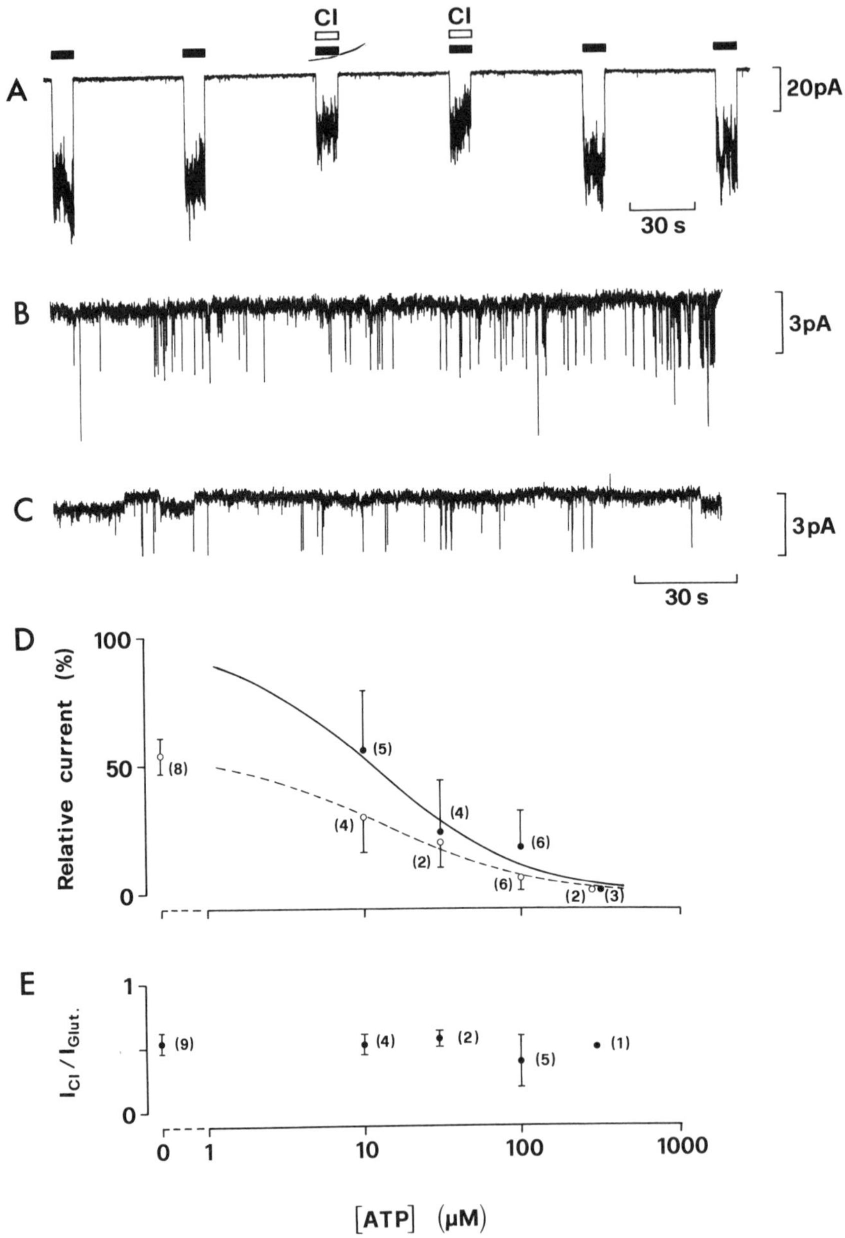

Figure 4. Reduction of activity of ATP-dependent K^+ channels by chloride ions. Inside-out patches taken from RINm5F insulinoma cells. Membrane potential −50 mV. (*A–C*) currents recorded from a patch containing at least 25 channels. (*A*) Solutions without ATP were applied during the intervals labeled by the closed bars (3 mM ATP during the other intervals) and Cl^- was substituted for glutamate as indicated by the open bars. See Methods and Materials for detailed composition of the solutions. (*B*) Current recorded in continuous presence of glutamate (145 mM) and 0.3 mM ATP. (*C*) glutamate replaced by Cl^-, 0.3 mM ATP. (*D*) Relation between time-averaged mean current and concentration of ATP in solutions containing glutamate (*filled circles*) or chloride (*open circles*). Current in presence of glutamate and absence of ATP was defined as 100%. Symbols and bars indicate mean values ± SD from the

lar pH may increase B cell activity by inhibition of both, ATP-dependent and Ca^{2+}-dependent K^+ channels (Cook et al., 1984).

Solutions applied to the intracellular side of the cell membrane in patch-clamp experiments often contain chloride as the main anion. It is usually assumed that this deviation from the composition of the cytoplasm does not influence cation-selective channels. However, in whole-cell recordings from RINm5F insulinoma cells E. Neher (personal communication) recently noticed that ATP (3 mM) was unexpectedly inefficient in blocking the resting K^+ conductance if the intracellular solution contained glutamate instead of chloride. We therefore performed a series of experiments comparing the effects of glutamate and chloride on inside-out patches. Chloride inhibited the ATP-dependent current in the absence and presence of ATP as exemplified by the current traces of Fig. 4, *A–C*, which were taken from a patch containing 20–30 channels. The time-averaged currents in the presence of several concentrations of ATP (10–300 μM) were expressed as a percentage of the maximum ATP-dependent current, i.e., the current in the presence of glutamate and the absence of ATP. The results from 11 patches are plotted vs. the ATP concentration in Fig. 4 *D*. Obviously, the 50% inhibitory concentrations of ATP were similar (12–14 μM) for both conditions, glutamate (closed circles) and chloride (open circles). In any tested concentration of ATP, mean currents in chloride were ~50% smaller than the currents in glutamate. This is better seen in Fig. 4 *E*, which shows the ratios of the current in chloride divided by the current in glutamate. The single-channel conductance was not significantly influenced by the different anions. This is indicated by the similar current amplitudes in Fig. 4, *B* and *C*, and this could also be seen in the absence of ATP when a patch contained only a small number of channels (not shown). Thus, it can be concluded that chloride reduces the open state probability of the ATP-dependent channel. Aspartate was also tested in two patches and resulted in channel activity similar to that in glutamate. The smaller number of superimposing channel openings in presence of chloride (Fig. 4 *A*) suggests that the density of channels may have been underestimated in previous studies using chloride as the main anion of the intracellular solution (Rorsman and Trube, 1985; Misler et al., 1986; Ohno-Shosaku et al., 1987).

In experiments on cardiac cells, intracellular application of trypsin was recently found to increase the amplitude of the calcium current and to reduce the rate of its inactivation (Hescheler and Trautwein, 1988). An additional rise of an outward current was observed in some of these experiments (Hescheler, J., unpublished results). This change might be explained by an activation of ATP-sensitive K^+ channels, which are also present in the heart cell membrane (Kakei et al., 1985). We, therefore, were interested in seeing whether trypsin modifies the activity of the ATP-dependent K^+ channels of pancreatic B cells. Fig. 5, *A* and *B* shows traces recorded from two inside-out patches that were alternatingly exposed to pulses of solutions containing either 0.3 mM ATP (20-s intervals, openings inhibited) or no ATP (10-s intervals,

---

Figure 4 (continued)
number of patches given in parentheses. Curves were fitted to the values by a nonlinear least-squares routine using the equation specified in the legend of Fig. 3. The following values of the fitted parameters were obtained: IC_{50} = 12 μM, H = 1.1 for glutamate and IC_{50} = 14 μM, H = 1.1 for chloride. (*E*) Ratios of the time-averaged K^+ current in chloride divided by the current in glutamate were calculated for each patch and ATP concentration. The panel shows mean values ± SD of the ratios obtained from the number of patches given in parentheses.

channels active). Application of 2 mg/ml trypsin (trace *A*) strongly increased the channel activity in the absence of ATP within 40 s. It turned out that the number of channels in the patch was larger than estimated before the exposure to trypsin (up to four superimposing openings during the first three pulses of zero-ATP solution, but nine superpositions during the fifth pulse). Trypsin also partially relieved channel inhibition during the presence of 0.3 mM ATP. Part of the effect persisted after the removal of trypsin in spite of a slow decline of channel activity, which presumably still represents the run-down seen in the absence of trypsin. At a lower concentration (0.02 mg/ml), trypsin was similarly effective but more time was needed to reach a plateau of

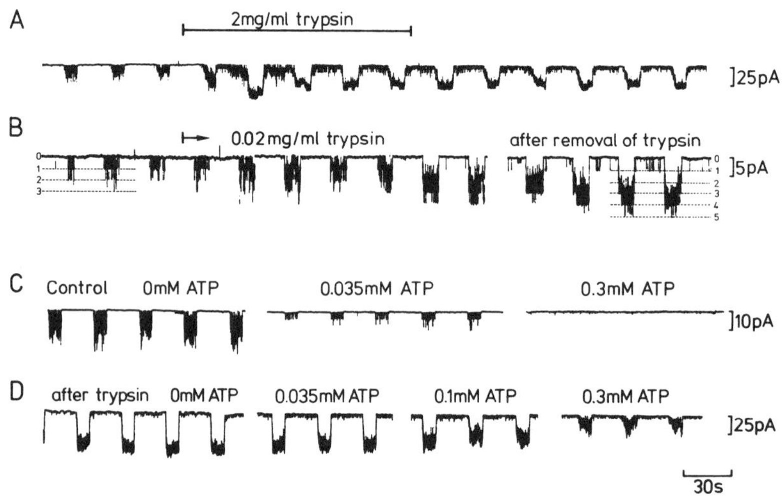

Figure 5. Trypsin increases activity of ATP-dependent K^+ channels. Inside-out patches from mouse pancreatic B cells, −50-mV membrane potential. (*A*) Currents recorded during alternating pulses of solution containing 0.3 mM ATP (channel openings inhibited) or lacking ATP (channels active). Trypsin (2 mg/ml) was added to the solution without ATP during the period labeled by the bar. (*B*) Other patch, same protocol of solution changes. Trypsin (0.02 mg/ml) was added at the time indicated by the arrow and removed 8 min later. High activity of channels persisted after the washout of trypsin as is shown by the second piece of recording. Dashed lines and figures at the traces indicate the number of superimposing channel openings. (*C* and *D*) Other patch. ATP was again applied in a pulse-like fashion, the concentrations (0–0.3 mM) during the invervals of higher channel activity are indicated above the pieces of recording. 1 mM ATP was present during the intermittent periods of (*C*) complete or (*D*) strong channel inhibition. Trypsin (0.05 mg/ml) was applied for 5 min between *C* and *D*.

high channel activity (trace *B*). No additional effect was seen when the concentration was then increased to 2 mg/ml (not shown). The protease (0.05 mg/ml) was without influence when applied together with trypsin-chymotrypsin inhibitor (1 mg/ml; Sigma type I-S; records not shown). Trypsin not only increased the number active channels and/or channel opening probability in the absence of ATP, but also shifted the ATP-sensitivity to higher concentrations. This is seen by comparing the traces in Fig. 5, *C* and *D*, which show responses to solution pulses of different ATP concentrations before (*C*) and 5 min after the application of trypsin (0.05 mg/ml, *D*). Obviously the channels in this patch were >50% inhibited by 35 μM ATP during the control (*C*) which is similar to the previous experiments (Cook and Hales, 1984; Ohno-Shosaku et al., 1984; Figs. 3 and 4). After exposure to trypsin, however, 100 μM ATP had <50% effect (*D*).

Several explanations for the effect of trypsin are possible. It could be concluded that the cleavage of some peptide bonds in the channel and ATP-receptor protein(s) leads to a decrease of intrinsic channel inhibition in the absence of ATP and a decreased affinity for ATP. This might be completely unrelated to the physiological regulation of the channel. Alternatively, one might assume that the channel is normally controlled by another protein which, for example, regulates channel activity and ATP sensitivity by phosphorylation or dephosphorylation. Proteolysis of the regulatory protein might then remove the inhibitory influence of the regulator on the K^+ channel. In experiments on other types of cells, it was recently found that trypsin treatment enlarges the amplitude of Na^+ and Ca^{2+} currents and slows their inactivation (Gonoi and Hille, 1987; Hescheler and Trautwein, 1988). At least the former effect seems to be very similar to the increase in the number of active K^+ channels described above (Fig. 5).

Conclusion

The "ATP-dependent" K^+ channel is modulated by a number of factors including nucleotides, pH, different anion species, and proteolytic processes. Pharmacological interferences by antidiabetic sulfonylureas and related drugs have not been discussed in the present communication although the specificity of their action strongly supports the role of the ATP-dependent K^+ channel in the control of the pancreatic B cell (Trube et al., 1986; Schmid-Antomarchi et al., 1987). The dominant physiological regulator of the channel probably is the ATP/ADP ratio, but the multitude of other modulators indicates that this might not be the only pathway mediating the influence of glucose on the electrical and secretory activity of pancreatic B cells.

Acknowledgments

Previously unpublished experiments described in the present publication were done at the Max-Planck-Institute in Göttingen. The authors thank Profs. B. Sakmann and W. Trautwein for stimulating discussions and Prof. E. Neher for the information about the effect of glutamate.
The work was supported by the Deutsche Forschungsgemeinschaft, SFB 236, project A8.

References

Ashcroft, F. M. 1988. Adenosine 5′-triphosphate-sensitive potassium channels. *Annual Review of Neuroscience.* 11:97–118.

Ashcroft, F. M., S. J. H. Ashcroft, and D. E. Harrison. 1988. Properties of single potassium channels modulated by glucose in rat pancreatic β-cells. *Journal of Physiology.* 400:501–527.

Ashcroft, F. M., D. E. Harrison, and S. J. H. Ashcroft. 1984. Glucose induces closure of single potassium channels in isolated rat pancreatic β-cells. *Nature.* 312:446–448.

Chay, T. R. 1988. Mathematical modeling for the bursting mechanism of insulin secreting β-cell. *Comments on Molecular and Cellular Biophysics.* 4:349–368.

Cook, D. L., and C. N. Hales. 1984. Intracellular ATP directly blocks K^+ channels in pancreatic B-cells. *Nature.* 311:271–273.

Cook, D. L., M. Ikeuchi, and W. Y. Fujimoto. 1984. Lowering of pH_i inhibits Ca^{2+}-activated K^+ channels in pancreatic B-cells. *Nature.* 311:269–271.

Cook, D. L., L. S. Satin, M. L. J. Ashford, and C. N. Hales. 1988. ATP-sensitive K^+ channels in pancreatic β-cells. Spare channel hypothesis. *Diabetes.* 37:495–498.

Dawson, C. M., P. C. Croghan, I. Atwater, and E. Rojas. 1983. Estimation of potassium permeability in mouse islets of Langerhans. *Biomedical Research.* 4:389–392.

Dunne, M. J., I. Findlay, and O. H. Petersen. 1988. Effects of pyridine nucleotides on the gating of ATP-sensitive potassium channels in insulin-secreting cells. *Journal of Membrane Biology.* 102:205–216.

Dunne, M. J., and O. H. Petersen. 1986*a*. GTP and GDP activation of K^+ channels that can be inhibited by ATP. *Pflügers Archiv.* 407:564–565.

Dunne, M. J., and O. H. Petersen. 1986*b*. Intracellular ADP activates K^+ channels that are inhibited by ATP in an insulin-secreting cell line. *FEBS Letters.* 208:59–62.

Findlay, I., M. J. Dunne, and O. H. Petersen. 1985*a*. High-conductance K^+ channel in pancreatic islet cells can be activated and inactivated by internal calcium. *Journal of Membrane Biology.* 83:169–175.

Findlay, I., M. J. Dunne, and O. H. Petersen. 1985*b*. ATP-sensitive inward rectifier and voltage- and calcium-activated K^+ channels in cultured pancreatic islet cells. *Journal of Membrane Biology.* 88:165–172.

Gillis, K., and S. Misler. 1988. Effects of pH_i on the activity of a metabolite-regulated, ATP_i-sensitive K^+ channel [K (ATP)] in rat pancreatic islet cells. *Journal of General Physiology.* 92:6*a*–7*a*. (Abstr.)

Gonoi, T., and B. Hille. 1987. Gating of Na channels. Inactivation modifiers discriminate among models. *Journal of General Physiology.* 89:253–274.

Hamill, O. P., A. Marty, E. Neher, B. Sakmann, and F. J. Sigworth. 1981. Improved patch-clamp techniques for high-resolution current recording from cells and cell-free membrane patches. *Pflügers Archiv.* 391:85–100.

Henquin, J. C., and H. P. Meissner. 1984. Significance of ionic fluxes and changes in membrane potential for stimulus-secretion coupling in pancreatic B-cells. *Experientia.* 40:1043–1052.

Hescheler, J., and W. Trautwein. 1988. Modification of L-type calcium current by intracellularly applied trypsin in guinea-pig ventricular myocytes. *Journal of Physiology.* 404:259–274.

Jones, D. P. 1986. intracellular diffusion gradients of O_2 and ATP. *American Journal of Physiology.* 250:C663–C675.

Kakei, M., R. P. Kelly, S. J. H. Ashcroft, and F. M. Ashcroft. 1986. The ATP-sensitivity of K^+ channels in rat pancreatic B-cells is modulated by ADP. *FEBS Letters.* 208:63–66.

Kakei, M., A. Noma, and T. Shibasaki. 1985. Properties of adenosine-triphosphate–regulated potassium channels in guinea-pig ventricular cells. *Journal of Physiology.* 363:441–462.

Malaisse, W. J., and A. Sener. 1987. Glucose-induced changes in cytosolic ATP content in pancreatic islets. *Biochimica et Biophysica Acta.* 927:190–195.

Misler, S., L. C. Falke, K. Gillis, and M. L. McDaniel. 1986. A metabolite-regulated potassium channel in rat pancreatic B cells. *Proceedings of the National Academy of Sciences.* 83:7119–7123.

Ohno-Shosaku, T., B. J. Zünkler, and G. Trube. 1987. Dual effects of ATP on K^+ currents of mouse pancreatic β-cells. *Pflügers Archiv.* 408:133–138.

Pace, C. S. 1984. Role of pH as a transduction device in triggering electrical and secretory responses in islet B cells. *Federation Proceedings.* 43:2379–2384.

Petersen, O. H., and I. Findlay. 1987. Electrophysiology of the pancreas. *Physiological Reviews.* 67:1054–1107.

Ribalet, B., and S. Ciani. 1987. Regulation by cell metabolism and adenine nucleotides of a K channel in insulin-secreting B cells (RIN m5F). *Proceedings of the National Academy of Sciences.* 84:1721–1725.

Rorsman, P., and G. Trube. 1985. Glucose dependent K^+-channels in pancreatic β-cells are regulated by intracellular ATP. *Pflügers Archiv.* 405:305–309.

Schmid-Antomarchi, H., J. DeWeille, M. Fosset, and M. Lazdunski. 1987. The receptor for antidiabetic sulfonylureas controls the activity of the ATP-modulated K^+ channel in insulin-secreting cells. *Journal of Biological Chemistry.* 262:15840–15844.

Spruce, A. E., N. B. Standen, and P. R. Stanfield. 1987. Studies of the unitary properties of adenosine-5′-triphosphate–regulated potassium channels of frog skeletal muscle. *Journal of Physiology.* 382:213–236.

Trube, G., P. Rorsman, and T. Ohno-Shosaku. 1986. Opposite effects of tolbutamide and diazoxide of the ATP-dependent K^+ channel in mouse pancreatic β-cells. *Pflügers Archiv.* 407:493–499.

Chapter 6

Delay of the Ca Mobilization Response to Muscarinic Stimulation

A. Marty, R. Horn, Y. P. Tan, and J. Zimmerberg

Laboratoire de Neurobiologie, Ecole Normale Supérieure, 75005 Paris, France; Department of Neuroscience, Hoffman La Roche Institute of Molecular Biology, Nutley, New Jersey 07110; Bogaziçi Üniversitesi, BMME PK2, Bebek, Istanbul, Turkey; and the Physical Science Laboratory, Division of Computer Science and Technology, Laboratory of Biochemistry and Metabolism, National Institutes of Health, Bethesda, Maryland 20892

Introduction

Slow kinetics have been recognized for a long time as one of the basic features of muscarinic responses. Thus, Del Castillo and Katz (1955) pointed out that in the frog heart, there is a pronounced lag between the discharge of the vagal nerve and the first sign of hyperpolarization in the muscle fibers. This delay is roughly 1,000 times larger than the corresponding value at the synapse innervating a skeletal muscle fiber. Even though it was felt that the slow rise of muscarinic responses probably reflected some unusual property of the receptors, elucidating the underlying mechanisms has appeared for a long time as a remote, somewhat unrealistic goal. This situation has recently changed with the recognition that several of the many variants of muscarinic responses found in vertebrates use G proteins to link activated receptor and cellular response. This notion rests in part on binding studies showing that GTP modifies agonist-binding curves in a manner that is characteristic of receptors coupled to G proteins (Birdsall et al., 1984). A second, more recent type of evidence supporting a role for G proteins in muscarinic responses comes from patch-clamp experiments showing that GTP and/or GTPγS potentiate muscarinic responses in the heart (Breitweiser and Szabo, 1985; Pfaffinger et al., 1985) and in exocrine glands (Evans and Marty, 1986). These experiments also indicated that the association between muscarinic receptors and G proteins is quite versatile, since the outcome of the system can be an activation of K channels, an inhibition of Ca channels, or a liberation of intracellular Ca. If it turns out that all muscarinic responses involve some kind of G protein, it is evident that this indirect coupling will result in slow kinetics. At the present time, very little is known about the exact nature and the rate constants of the reactions involved. Fortunately, however, G proteins from different systems have been found to follow common rules, so that it is probably legitimate to transpose to the muscarinic receptor some of the conclusions of the extensive kinetic studies that have been carried out on the beta-adrenergic receptor and on the rhodopsin system.

This is the approach followed here. We have focused our attention on the onset of muscarinic responses because, for short times, complicating factors due to the reversibility of slow reactions and feedback controls need not be considered. In the following, a review of available results describing the delay of muscarinic responses in various preparations will be presented. Next, the special case of Ca mobilization will be described. Finally, a kinetic model will be presented to account for the rising phase of this response.

The Delay of Muscarinic Responses in Various Preparations

The response type which has been best studied from a kinetic point of view is the inward rectifying K conductance increase observed in heart muscle cells, and in sympathetic and parasympathetic neurons. Purves (1976) showed that hyperpolarizations recorded in frog interatrial septum follow acetylcholine (ACh) applications with a minimum latency of 0.1 s (at 16°C). It was argued that this delay was not due to diffusion on the basis of geometrical considerations. A minimum latency of 0.1 s (at 23°C) was found in heart muscle fibers and in heart parasympathetic neurons of the mudpuppy (Hartzell et al., 1977) (Fig. 1). Again, it was concluded that the delay was not due to diffusion, mainly because it was far more sensitive to temperature than was expected from any diffusion model. Finally, the K response of bullfrog sympathetic neurons of type C (for which ACh causes an inhibitory postsynaptic potential with an increase in K permeability) was reported to have a latency of 0.16 s at 27°C (Dodd and Horn, 1983).

Thus, a 0.1-s delay at room temperature is a characteristic feature of a first type of muscarinic response exemplified by the K conductance rise of heart muscle. Patch-clamp results suggest that these responses do not involve any soluble intracellular second messenger (Soejima and Noma 1984). Therefore the intermediate G protein is presumably directly linked to the K channel (Brown and Birnbaumer, 1988).

A different situation is found in the case of responses involving Ca mobilization. These have been described mainly in secreting cells of exocrine glands. In the cat submandibulary gland, for instance, there is a delay of 0.7 s at 22°C and 0.3 s at 37°C (Creed and McDonald, 1975). Our results on isolated cells of rat lacrimal glands, to be presented below, give a delay of 1 s at room temperature. A long delay seems therefore to be a feature of Ca-dependent responses. The excitatory muscarinic response of certain smooth muscle cells, which may depend on Ca release, also has a minimum latency of 1 s at room temperature (Bolton, 1976). Other nonmuscarinic responses involving Ca release show a similarly long latency. Thus, the Ca-dependent noradrenergic response of smooth muscles has a minimum latency near 0.5 s at room temperature (Byrne and Large, 1987).

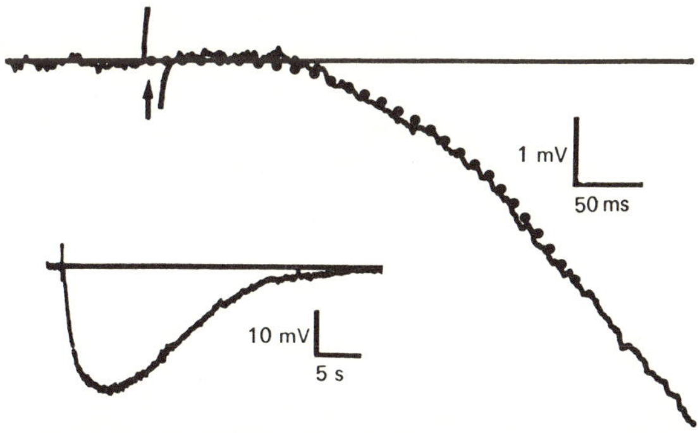

Figure 1. Time course of *Necturus* heart muscle response to iontophoretically applied ACh, recorded with an intracelluar electrode. The arrow shows the time of ACh application (duration of iontophoretic current, 5 ms). (*Dots*) Fit of the rising phase with a t^3 function. Temperature, 23°C. Reproduced with permission from Hartzell et al. (1977).

Results of the latency of other types of muscarinic responses are scarce. The available evidence suggests that M current decrease occurs with a lag of 0.2 s since this is the interval between a presynaptic spike and the ensuing slow excitatory postsynaptic potential that is driven by M current decrease (Adams and Brown, 1982).

To sum up, kinetic studies reveal a distinction between two classes of muscarinic responses: inward rectifying K conductance increases with a relatively short latency, and Ca release with a longer latency. In the following, we will pursue the analysis of kinetic features of these two types of responses.

Interpretation of the Delay of the K Response in the Heart

Let us consider in more detail the rising phase of the K conductance in the heart, which stands as a prime example of a permeability change directly coupled to muscarinic receptors. As already stated, diffusion of the agonist can safely be disregarded as a source of the observed latency. By analogy with permeability systems based on direct receptor-channel coupling, it is most likely that the subsequent agonist binding and conformational change of the muscarinic receptor are also fast on the time scale of the response (see e.g., Hartzell et al., 1977). The events responsible for the response activation may therefore be considered to start with R, the active form of the receptor. Both in heart muscle cells and in parasympathetic neurons, the current rise may be modeled with a t^3 curve, indicating that (at least) three reactions are implicated, none of which are rate limiting (Hartzell et al., 1977). In the light of recent knowledge on the functioning of G proteins, the events linking receptor activation to channel gating may be in fact grouped in three main steps according to the following scheme. Binding of R to the G protein:

$$R + G \xrightarrow{a} RG \tag{1}$$

Dissociation of the RG complex and activation of the G protein:

$$RG \xrightarrow{b} R + G^* \tag{2}$$

Binding to the channel, C, and subsequent opening of the channel:

$$G^* + C \xrightarrow{c} GC^* \tag{3}$$

The observed third-order kinetics would then require that, for short times (e.g., <0.2 s), $a(R) \cdot (G)$, $b(RG)$, and $c(G^*) \cdot (C)$ be of the same order of magnitude. Alternatively, one of the reactions (1–3) could be fast while another one could contain two distinct slow steps. Step 2 is particularly complicated as it involves dissociation of GDP from the RG complex and binding of GTP, plus a conformation change of G which may induce a dissociation of α and $\beta\gamma$ subunits.

The Latency of Ca Rise in Rat Lacrimal Glands

This section deals with recent results obtained in our laboratory on the delay of Ca mobilization in isolated secreting cells from rat lacrimal glands.

Methods

Cells were isolated by enzymatic digestion using sequential treatment with trypsin and collagenase. The preparation was then used within the same day. Experiments were performed at room temperature (20–25°C). ACh was applied using the U-tube method of Krishtal and Pidoplichko (1980), with an electronically controlled valve. This ensured rapid, precisely timed and homogeneous application of the agonist.

Washout of the Response

Upon stimulation with ACh, cells from this preparation respond with a release of Ca from intracellular stores and with a subsequent increase in the opening probability of Ca-dependent K and Cl channels (Marty et al., 1984). In the tight-seal whole-cell recording mode, we usually set the Cl^- equilibrium potential at 0 (with 145 mM Cl^- on both sides of the membrane) and it is then possible to record a pure K response by clamping the cell potential at 0 mV. When this is done, it is found that the K current rises abruptly after a lag of ~1 s (Fig. 2).

A second feature apparent in Fig. 2 is that the delay of the response increases if the cell is exposed several times to ACh. Control experiments showed that this effect is not linked to desensitization. By varying pipette and cell sizes, it was shown that the increase in delay is due to the diffusion of a water-soluble substance of low molecular weight into the recording pipette (Marty and Zimmerberg, 1989). With standard whole-cell recording conditions, this lengthening of the response is observed within a few minutes. In practice, such variations seriously jeopardize kinetic analysis of the response. Subsequently, the amplitude of the response decreases as well, such that after 30 min no response can usually be recorded (Fig. 2).

A quantitative analysis of the time course of the delay increase during whole-cell recording leads to the conclusion that the washout molecule has a diffusion coefficient

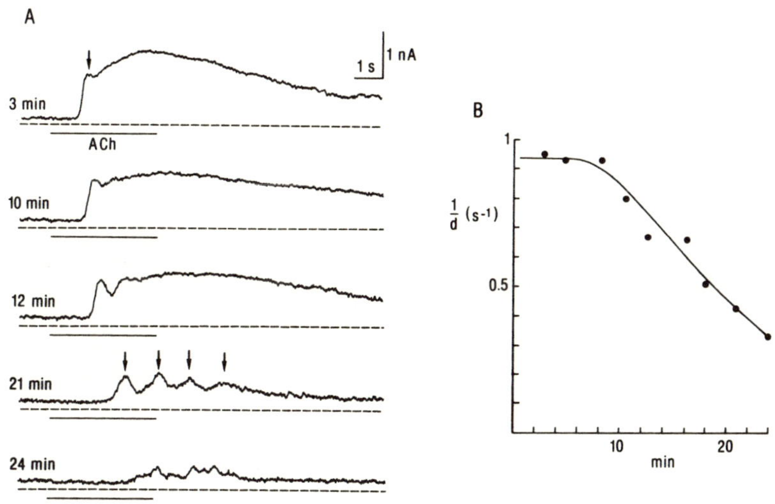

Figure 2. ACh-induced Ca-dependent K current recorded with conventional whole-cell configuration. The cell potential was held at 0 mV, the equilibrium potential for Cl ions, so that the K current response could be obtained in isolation. Upon application of 0.5 μM ACh (bars below records), the K current rose after a delay, ***d***. In *A*, responses are shown at various times after the onset of whole-cell recording, as indicated near each record. At 3 min of whole-cell recording, *d* was close to 1 s. Later on, however, it increased to 2 s and more. A plot of $1/d$ as a function of time displays a short plateau followed by a progressive decline (*B*). The amplitude of the ACh-induced K current simultaneously decreases. Note the initial hump in K current and the multiple transients observed near the end of washout (*arrows*). Reproduced with permission from Marty and Zimmerberg, 1989.

that corresponds to a molecular weight in the range of 200–500. However, we have been unable to stop washout by complementing the pipette solution with various water-soluble compounds falling in this range (such as ATP, cAMP, GTP, or a combination of these three nucleotides). Washed out cells respond normally to the internal application of inositoltrisphosphate ($InsP_3$), but not to GTPγS. This suggests that the washout molecule acts downstream of receptor activation and upstream of Ca liberation from the endoplasmic reticulum. Thus, the washout molecule is likely to act on phospholipase C or on the G protein that controls phospholipase C (Marty and Zimmerberg, 1989).

Perforated Patch Recording

To obviate the limitations set by washout on the quantitative study of the delay, we used a new method similar to that introduced by Lindau and Fernandez (1986) to study degranulation in mast cells. In this method, called perforated patch recording, the polyene antibiotic nystatin (50–100 μg/ml) is introduced into the recording pipette after the original cell-attached configuration is obtained. The cell-pipette connection is then established by insertion of nystatin channels in the membrane patch, rather than by application of suction in the pipette. The development of this conductance is best monitored by following the response to brief voltage pulses applied in the pipette. After the cell-pipette connection is established, such pulses elicit exponentially decaying

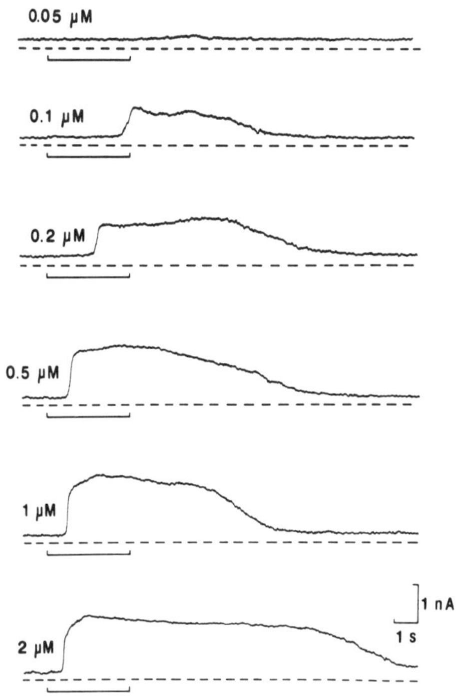

Figure 3. ACh-induced Ca-dependent K current recorded with the perforated patch technique. Cell potential, 0 mV. Responses to increasing ACh concentrations, as indicated next to the records. Reproduced from Horn and Marty, 1988.

current transients reflecting the charging of the cell capacitance. Since the initial current amplitude is inversely related to the access resistance, it is easy to obtain an estimate of this resistance. The values obtained are comparable to those achieved in conventional tight-seal whole-cell recording. However, due to the sieving properties of the nystatin channels, the washout molecules are retained in the cytosol. Most importantly for our present purpose, the delay of the K responses is then stable such that reliable results can be obtained on a given cell (Horn and Marty, 1988).

K Responses in Perforated Patch Recordings

The responses of a cell to ACh concentrations ranging from 0.05 to 2 μM are shown in Fig. 3. The current rises quite suddenly after a comparatively long latency during which absolutely no signal change can be observed. It may also be seen that the delay of the response decreases as the ACh concentration is raised, while the peak amplitude reaches saturation on a narrow concentration range (0.05–0.5 μM).

The results demonstrate a linear relationship between the response delay, *d*, and the inverse of the ACh concentration, $1/A$ (Fig. 4). The intersect of the regression line with the ordinate axis (0.97 s) corresponds to the minimum latency obtained with very high ACh concentrations. It was verified that ACh concentrations as high as 10 μM were unable to shorten this latency further. From the intersect of the regression line with the abscissa axis, a putative value of the dissociation constant of ACh for its receptor is obtained (0.27 μM), as will be discussed further below.

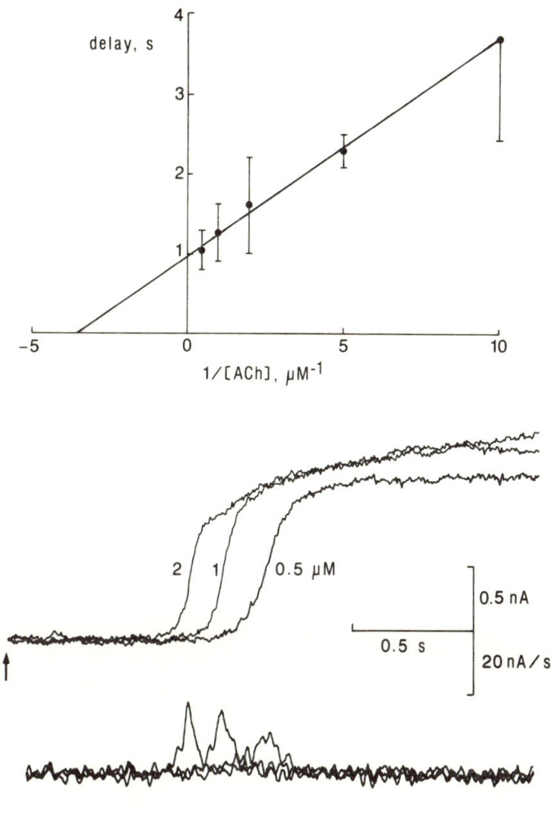

Figure 4. Delay of ACh-induced current as a function of ACh concentration. Individual experiments as in Fig. 3. (*Upper curve*): The delay, *d*, is linearly related to the inverse of the ACh concentration. The intersect with the abscissa axis gives a putative dissociation constant of 0.27 μM for the binding of ACh to its receptor. (*Middle traces*) Superimposed responses to 0.5, 1, and 2 μM (same recordings as in Fig. 3). The arrow shows the onset of the ACh application. (*Lower traces*) Derivatives of the middle traces. Reproduced from Horn and Marty, 1988.

It is useful to compare the kinetics of ACh-induced K responses in the heart (Fig. 1) and in lacrimal glands (Fig. 4). In the first case, the short silent period (0.1 s) merges gradually into a slow rising phase. In the second case, the long silent period (~1 s) is followed by a sharp rising phase, where most of the current excursion occurs in <0.1 s. Our interpretation of the latter behavior, to be discussed later, implies a positive feedback mechanism in the upswing of the Ca signal.

The Problem of Identifying Rate-limiting Steps

The chain of events put into play in the muscarinic response may be described as follows: (0) Binding of ACh to its receptor and subsequent conformation change of the receptor. (1) Binding of the activated receptor to the G protein. (2) Activation of the G protein (either as a whole or of one active subunit). (3) Binding of the activated G protein (or of a subunit) to phospholipase C and subsequent activation of phospholipase C. (4) Accumulation of $InsP_3$. (5) Liberation of Ca from the endoplasmic reticulum. (6) Diffusion of liberated Ca ions to the plasma membrane. (7) Opening of Ca-dependent channels.

Such a list may appear to contain a formidable complexity at first. Fortunately it is possible to restrict the location of rate-limiting steps. As discussed above, step 0 is likely to be fast. Steps 5–7 are not rate limiting since direct application of $InsP_3$ elicits responses with a delay on the order of 0.2 s, five times smaller than that obtained with ACh applications (Fig. 5). We are therefore left with the task of identifying rate-limiting steps in the sequence 1–4. Steps 1–3 are of course essentially the same as those discussed in relation to the heart response.

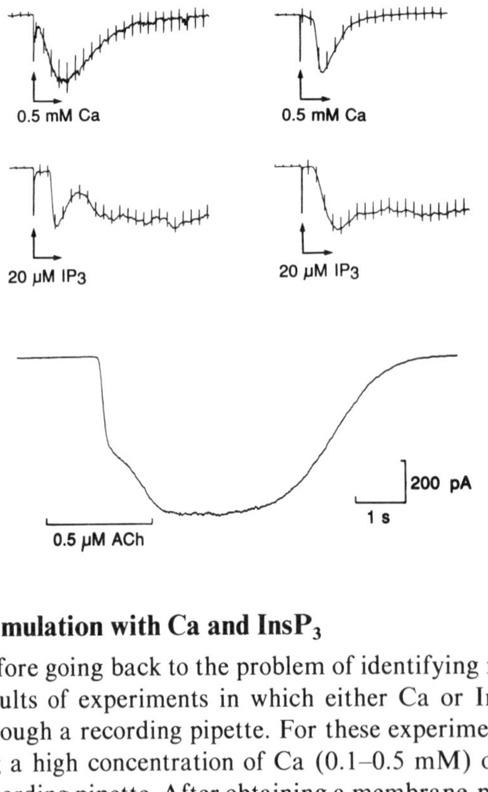

Figure 5. Effects of a local stimulation with $InsP_3$ or with elevated Ca. All recordings were taken at −60 mV (from five different cells). (*Upper traces*) Responses to a sudden application of 0.5 mM Ca through the recording pipette. The pipette-cell connections were established at the vertical arrows. (*Middle traces*) Responses to a sudden application of 20 μM $InsP_3$ through the recording pipette. (*Lower trace*) Example of an ACh-induced Cl current. Note the hump in the current rise. 0.5 μM ACh was applied to the bath during the time indicated below the record.

Stimulation with Ca and $InsP_3$

Before going back to the problem of identifying rate-limiting steps, let us consider the results of experiments in which either Ca or $InsP_3$ are applied abruptly to the cell through a recording pipette. For these experiments, an intracellular solution containing a high concentration of Ca (0.1–0.5 mM) or $InsP_3$ (20 μM) was used to fill the recording pipette. After obtaining a membrane-pipette seal, the pipette-cell connection was established by applying a sharp voltage command to the pipette interior ("zapping"), thus allowing a sudden and local rise of Ca or of $InsP_3$. The responses obtained in either case were remarkably similar (Fig. 5). At −60 mV (a value close to the equilibrium potential for K ions, where mostly Ca-dependent Cl currents are recorded), responses had peak amplitudes of 100–500 pA, and a time to peak close to 0.6 s. Most importantly, there was only a short interval (0.1–0.2 s) between the break-in artifact and the first signs of a response.

The time course of the transients shown in Fig. 5 is puzzling at first. It is much too rapid to reflect the simple diffusion of Ca or of $InsP_3$ into the cell, a process which would take on the order of 1 min as calculated according to simple diffusion models (Pusch and Neher, 1988). Control experiments showed that membrane resealing, which was the rule in the Ca experiments, did not affect the initial transient, and only influenced the later part of the response (as seen in the $InsP_3$ traces for times exceeding 1 s after patch rupture). To explain the transient nature of the responses, one could suppose that Ca and/or $InsP_3$ are rapidly bound or degraded by cellular metabolites. It could thus be imagined that Ca or $InsP_3$ first diffuse into the cell in the favor of a steep concentration gradient giving the peak of the current, and that binding or enzymatic degradation then establish a steady concentration gradient with a low concentration in the cell cytosol, accounting for the return of the Cl current towards baseline. This interpretation is unlikely for two reasons. First, it predicts an asymmetrical response with a sharp rise and a slower decline, whereas the transients illustrated in Fig. 5 are almost symmetrical. Secondly, experiments performed with lower Ca or $InsP_3$ concentrations gave either the same type of response or no response at all. This "all or none" effect, together with the stereotyped kinetics of the response, suggests that intracellular buffering is not the cause of the observed transients, and that a positive feedback mechanism is involved. A plausible molecular mechanism of this kind is the Ca sensitivity of the Ca release channels of the endoplasmic reticulum (Lai et al., 1988), which is presumably the cause of "Ca-induced Ca release." We propose that the responses of Fig. 5 are due to regenerative Ca transients travelling along the endoplasmic reticulum and invading the cell (Marty, A., and Y. P. Tan, manuscript submitted for publication). The finding that Ca and $InsP_3$ elicit the same response is then easily explained by supposing that $InsP_3$ is able to elicit local Ca release and thus initiate the Ca transient.

$InsP_3$ Microdomains and the Origin of the Delay

The conclusions of the Ca/$InsP_3$ stimulation experiments can now be used for interpreting the kinetics of responses to ACh. A key finding is the observation of humps in the rising phase of ACh-induced K or Cl currents (Figs. 2 and 5). These humps are also discernible later in the response under certain circumstances such as partial washout (Fig. 2). At -60 mV, the humps observed in ACh-induced Cl^- currents have an amplitude and a rise time similar to those exhibited by Ca- or $InsP_3$-induced transients (Fig. 5). These results suggest that the initiation of the ACh-induced responses corresponds to the development of a Ca wave across the cell. The delay is then due to the accumulation of $InsP_3$ and Ca up to the point where Ca-induced Ca release is triggered. This raises, however, an immediate difficulty. Why is this buildup not apparent as a small rise of Ca-dependent K current during the delay?

To answer this paradox, we note that $InsP_3$ is rapidly degraded near its production site by an endogeneous phosphatase such that each active phospholipase C molecule is presumably surrounded by a microdomain of elevated $InsP_3$. We suppose that these microdomains are just impinging on the $InsP_3$-sensitive endoplasmic reticulum so that Ca release is kept at a minimum as long as each microdomain is isolated from the others. However, when several of the phospholipase C microdomains overlap, the local Ca concentration reaches a level sufficient to trigger Ca-induced Ca release (Fig. 6).

A quantitative assessment of these hypotheses would require the knowledge of (*a*) the rate constant of $InsP_3$ production by an active phospholipase C molecule, (*b*) the concentration of the $InsP_3$ degrading phosphatase, (*c*) the rate of binding of $InsP_3$ to this enzyme, and (*d*) the diffusion constant of the phosphatase. With these figures, it should be possible to calculate the length constant of the microdomain surrounding each phospholipase C molecule and to compare the result with the distance observed in electron micrographs between the plasma membrane and the neighboring endoplasmic reticulum. Such calculations performed for the case of Ca channels indicate the presence of Ca microdomains in the submicron range around each channel (Chad and Eckert, 1984; Neher, 1986). In the present case, however, none of the above parameters is known and the calculation is impossible. Speculative as it is, the microdomain hypothesis does give an explanation to the paradoxical finding of the long latent period with no observable Ca rise. During the time in which microdomains are isolated, they contribute only very local, undetectable Ca signals.

Delay as a Function of ACh Concentration

According to the model of Fig. 6, Ca_i will rise abruptly as soon as the concentration of active phospholipase C reaches a threshold value, T. In this section, we examine how the time needed to reach T, d, is related to the ACh concentration, A.

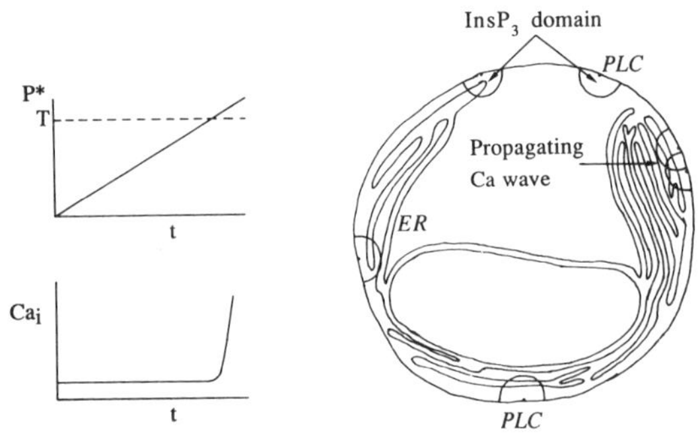

Figure 6. A model for the initiation of Ca release. On the left is depicted the supposed variation of P*, the concentration of active phospholipase C molecules, together with that of Ca_i, starting at the onset of the ACh application. The Ca_i trace takes off when P* reaches a threshold value, T. On the right a secreting cell is drawn. $InsP_3$ microdomains are supposed to build up around each activated phospholipase C molecule. For a given degree of overlap between neighboring domains, a propagating Ca wave is initiated.

It is first assumed that the concentration of active receptor, [R], follows a fast bimolecular reaction such that:

$$[R] = R_T \frac{A}{A + K} \tag{4A}$$

where R_T is the total number of receptor molecules and K the dissociation constant of ACh. Eq. 4 only holds true if RG does not accumulate. In other cases it should be replaced by:

$$[R] = (R_T - [RG]) \frac{A}{A + K} \tag{4B}$$

Next, the activation of the G protein and of phospholipase C are modeled with a sequence of reactions very similar to those used previously in connection with K channels in the heart:

$$R + G \xrightarrow{a} RG \tag{5}$$

$$RG \xrightarrow{b} R + G^* \tag{6}$$

$$G^* + P \xrightarrow{c} GP^* \tag{7}$$

Here P stands for an inactive phospholipase C molecule and GP* for the active complex formed between the G protein and phospholipase C.

Given the above assumptions, various relations between delay (d) and A can be derived. The following discussion is guided by that given in Tolkovsky et al. (1982) to account for the stimulation of adenylate cyclase in turkey erythrocyte membranes.

First reaction rate limiting. Then the rate of formation of GP* is

$$\mathrm{d}\frac{[\mathrm{GP}^*]}{\mathrm{d}t} = a[\mathrm{R}] \cdot [\mathrm{G}] \tag{8A}$$

For short times $[\mathrm{G}] = \mathrm{G_T}$, the total number of G proteins. Because reaction 6 is supposed to be faster than reaction 5, RG does not accumulate and Eq. 4 applies. Integrating Eq. 8A then gives:

$$[\mathrm{GP}^*] = a\mathrm{R_T G_T}\frac{A}{A+K}t. \tag{8B}$$

Since the value of [GP*] obtained for $t = d$ is T,

$$T = a\mathrm{R_T G_T}\frac{A}{A+K}d, \tag{9A}$$

and

$$d = \frac{T}{a\mathrm{R_T G_T}}(1 + K/A). \tag{9B}$$

Second reaction rate limiting. This includes the case where dissociation of GDP from the RG complex is rate limiting. In this case $\mathrm{d[GP]/d}t = b[\mathrm{RG}]$. If $\mathrm{G_T} \gg \mathrm{R_T}$, $[\mathrm{RG}] = \mathrm{R_T}$, (because reaction 5 goes to completion), so that $[\mathrm{GP}^*] = b\mathrm{R_T}t$. Then $d = \mathrm{T}/b\mathrm{R_T}$, independent of A and $\mathrm{R_T}$. If $\mathrm{G_T} \ll \mathrm{R_T}$, then $[\mathrm{RG}] = \mathrm{G_T}$ and $d = \mathrm{T}/b\mathrm{G_T}$, independent of A and $\mathrm{R_T}$.

Third reaction rate limiting. Then $\mathrm{d[GP^*]/d}t = c[\mathrm{G^*}][\mathrm{P}]$. If $\mathrm{G_T} \gg \mathrm{R_T}$ one obtains $d = \mathrm{T}/c\mathrm{R_T P_T}$, where $\mathrm{P_T}$ is the total number of phospholipase C molecules. If $\mathrm{R_T} \gg \mathrm{G_T}$ one obtains $d = \mathrm{T}/c\mathrm{G_T P_T}$, independent of A and $\mathrm{R_T}$.

From the above discussion, it appears that the finding of a linear dependence of d on $1/A$ is readily compatible with the hypothesis that step 5 is rate limiting. Other possibilities cannot be ruled out, however, since the very formulation of phospholipase C activation in steps 5–7 represents at best a gross simplification of the actual kinetic steps at work. We would like to consider two variants of scheme 5–7.

Reaction 5 reversible, reaction 6 rate limiting. Formally, this appears as a way to obtain a slow production of G* with a rate constant tuned by receptor occupancy. Such a model predicts, however, a complicated variation of the system kinetics with G_T which is not born out in the turkey erythrocyte (Tolkovsky et al., 1982). Therefore this model is not considered likely to apply here.

"Collision-coupling" model. Whereas the "shuttle model" described by Eqs. 5–7 describes accurately the coupling of rhodopsin to phosphodiesterase, Levitzki (1988) advocates a "collision-coupling" model for the activation of cAMP-dependent protein kinase in turkey erythrocytes. In this variant, the G protein never dissociates from the target enzyme. Thus our system would read:

$$R + GP \xrightarrow{a} RGP \tag{5'}$$

$$R\,GP \xrightarrow{b} RG^*P \tag{6'}$$

$$RG^*P \xrightarrow{c} GP^* + R \tag{7'}$$

where G* and P* stand for the activated forms of the G protein and of phospholipase C, respectively. For our present purpose, this formalism is equivalent to that of Eqs. 5–7 in the sense that it also requires reaction 5′ to be rate limiting in order to obtain a linear relation of the delay on the inverse ACh concentration.

Use of Delay Vs. ACh Concentration Curves

The main conclusion from the previous section is that kinetic results are consistent with a model in which binding of the activated receptor to the G protein (step 5) is rate limiting. Eq. 9 then describes the variation of the delay as a function of the ACh concentration. In this equation, it appears that the delay should be inversely proportional to R_T, the total number of receptors. Eq. 9 also predicts that the intersect of d vs. $1/A$ curves should yield the value of K, the dissociation constant of ACh for its receptor. Thus, the presence of a threshold and of a regenerative Ca response results in a situation where it is the delay plot, rather than the amplitude plot, that gives useful information on the dissociation constant of the agonist for its receptor. According to Eq. 4, K is the dissociation constant for the free receptor molecule. This is also the low affinity value observed for RG in the presence of guanine nucleotides. In the absence of nucleotides, the RG complex has a higher affinity for the agonist (Gilman, 1987).

Comparison of Muscarinic Current Kinetics in the Heart and in Exocrine Glands

The first steps of the response being formally identical for the heart K conductance and for Ca-dependent responses in exocrine glands it may be assumed that the underlying reactions follow the same type of kinetics. This would imply that the activation of phospholipase C in exocrine glands also obeys t^3 kinetics for times shorter than 0.2 s, but that this is not observed because accumulation of active phospholipase C below the threshold does not show as a signal. At longer times, one of the three reactions would become rate limiting, giving the linear activation of phospholipase C with time, and the hyperbolic relation between delay and ACh concentration. Our hypothesis that reaction 5 is rate limiting during the major part of the delay period is entirely consistent with this proposal. Whereas reaction 5 starts with its maximum speed, reactions 6 and 7 proceed much faster as soon as the concentrations of intermediate species start to rise. Thus, if reactions 5–7 have similar speeds at the very start of the response, reaction 5 is likely to become rate limiting shortly afterwards.

Conclusion

It is our conviction that a careful study of the rising phase of muscarinic responses can give valuable clues on the molecular events after the activation of the receptors. Our efforts to obtain such information in exocrine glands have met with difficulties at first as we found that the kinetics of the ACh-induced currents changed markedly from trial to trial. However, we have shown that this effect is due to the loss of a low molecular weight substance during whole-cell recording (washout). Furthermore, we have shown that washout can be avoided by using a new variant of the tight-seal whole-cell recording method. Using this technique, quantitative results on the kinetics or the Ca rise and on its dependence on ACh concentration have been gathered. The results are unusual in the sense that, contrary to what is found in other types of muscarinic currents, they reveal a long silent period (delay) followed by a rapid upswing. According to our molecular model, the delay corresponds to the accumulation of activated phospholipase C molecules while the ensuing Ca rise corresponds to the initiation of Ca-induced Ca release by the endoplasmic reticulum.

References

Adams, P. R., and D. A. Brown. 1982. Synaptic inhibition of the M-current: slow excitatory postsynaptic potential mechanism in bullfrog sympathetic neurones. *Journal of Physiology*. 332:263–273.

Birdsall, N. J. M., E. C. Hulme, and J. M. Stockton. 1984. Muscarinic receptor heterogeneity. *Trends in Pharmacological Sciences Supplement*. 4–8.

Bolton, T. B. 1976. On the latency and form of the membrane responses of smooth muscle to the iontophoretic application of acetylcholine or carbachol. *Proceedings of the Royal Society of London, B*. 194:99–119.

Breitwieser, G. E., and G. Szabo. 1985. Uncoupling of cardiac muscarinic and β-adrenergic receptors from ion channels by a guanine nucleotide analogue. *Nature*. 317:538–540.

Brown, A. M., and L. Birnbaumer. 1988. Direct G protein gating of ion channels. *American Journal of Physiology*. 254:H401–H410.

Byrne, N. G., and W. A. Large. 1987. Action of noradrenaline on single smooth muscle cells freshly dispersed from the rat anococcygeus muscle. *Journal of Physiology*. 389:513–525.

Chad, J. E., and R. Eckert. 1984. Calcium domains associated with individual channels can account for anomalous voltage relations of Ca-dependent responses. *Biophysical Journal*. 45:993–999.

Creed, K. E., and I. R. McDonald. 1975. The effects of temperature and 2,4-dinitrophenol on transmembrane potentials of submandibular acinar cells. *Journal of Physiology*. 247:521–535.

Del Castillo, J., and B. Katz. 1955. Production of membrane potential changes in the frog's heart by inhibitory nerve impulses. *Nature*. 175:1035.

Dodd, J., and J. P. Horn. 1983. Muscarinic inhibition of sympathetic C neurones in the bullfrog. *Journal of Physiology*. 334:271–291.

Evans, M. G., and A. Marty. 1986. Potentiation of muscarinic and alpha-adrenergic responses by an analogue of guanosine triphosphate. *Proceedings of the National Academy of Sciences*. 83:4099–4103.

Gilman, A. G. 1987. G proteins: tranducers of receptors-generated signals. *Annual Review of Biochemistry*. 56:615–649.

Hartzell, H. C., S. W. Kuffler, R. Stickgold, and D. Yoshikami. 1977. Synaptic excitation and inhibition resulting from direct action of acetylcholine on two types of chemoreceptors on individual amphibian parasympathetic neurones. *Journal of Physiology*. 271:817–846.

Horn, R., and A. Marty. 1988. Use of nystatin to establish whole-cell recordings without wash-out. *Journal of General Physiology*. 92:145–159.

Krishtal, O. A., and V. I. Pidoplichko. 1980. A receptor for protons in the nerve cell membrane. *Neuroscience*. 5:2325–2327.

Lai, F. A., H. P. Erickson, E. Rousseau, Q. Y. Liu, and G. Meissner. 1988. Purification and reconstitution of the calcium release channel from skeletal muscle. *Nature*. 331:315–319.

Levitzki, A. 1988. Transmembrane signalling to adenylate cyclase in mammalian cells and in *Saccharomyces cerevisiae*. *Trends in Biochemical Sciences*. 13:298–301.

Lindau, M., and J. M. Fernandez. 1986. IgE-mediated degranulation of mast cells does not require opening of ion channels. *Nature*. 319:150–153.

Marty, A., Y. P. Tan, and A. Trautmann. 1984. Three types of calcium-dependent channel in rat lacrimal glands. *Journal of Physiology*. 357:293–325.

Marty, A., and J. Zimmerberg. 1989. Diffusion into the patch-clamp recording pipette of a factor necessary for muscarinic current response. *Cellular Signalling*. 1:259–268.

Neher, E. 1986. Concentration profiles of intracellular calcium in the presence of a diffusible chelator. *In* Calcium Electrogenesis and Neuronal Functioning. U. Heinemann, M. Klee, E. Neher, & W. Singer, editors. Springer Verlag; Heidelberg. 80–96.

Pfaffinger, P. J., J. M. Martin, D. D. Hunter, N. M. Nathanson, and B. Hille. 1985. GTP-binding proteins couple cardiac muscarinic receptors to a K channel. *Nature*. 317:536–538.

Purves, R. D. 1976. Function of muscarinic and nicotinic acetylcholine receptors. *Nature*. 261:149–151.

Pusch, M., and E. Neher. 1988. Rates of diffusional exchange between small cells and a measuring patch pipette. *Pflügers Archiv*. 411:204–211.

Soejima, M., and A. Noma. 1984. Mode of regulation of the ACh-sensitive K-channel by the muscarinic receptor in rabbit atrial cells. *Pflügers Archiv*. 400:424–431.

Tolkovsky, A., S. Braun, and A. Levitzki. 1982. Kinetics of interaction between β-receptors, GTP-protein, and the catalytic unit of turkey erythrocyte adenylate cyclase. *Biochemistry*. 79:213–217.

Chapter 7

The Neuronal Growth Cone: Calcium Regulation of a Presecretory Structure

S. B. Kater and P. B. Guthrie

Program in Neuronal Growth and Development, Department of Anatomy and Neurobiology, Colorado State University, Fort Collins, Colorado

Introduction

The neuronal growth cone has been recognized since the time of Ramon y Cajal (1890) as one of the most fascinating structures in the nervous system. It is, without a doubt, the most morphologically dynamic system of interacting components thus far examined. Another prominent structure in the nervous system, the presynaptic terminal, has also been extensively studied in the past, but usually as a neurotransmitter-releasing element. In many ways, the presynaptic terminal and the neuronal growth cone share a variety of characteristic features (Fig. 1). While these structures play very different

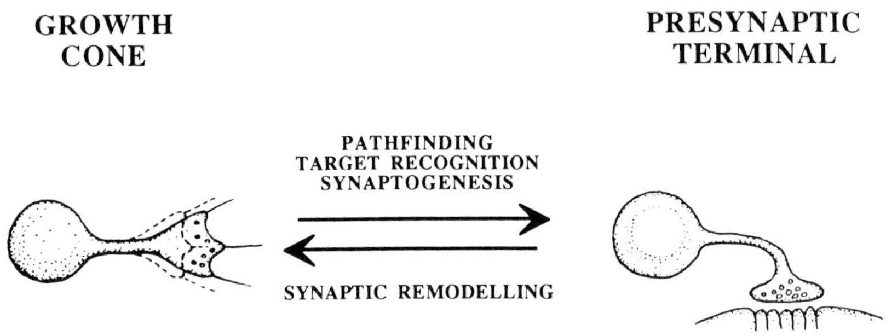

GROWTH CONE	CHARACTERISTIC FEATURE	PRESYNAPTIC TERMINAL
✓	RESOURCES TRANSPORTED FROM CELL BODY	✓
✓	VESICLES INTEGRAL TO FUNCTION	✓
✓	MITOCHONDRIA CONCENTRATED	✓
✓	RELEASE NEUROTRANSMITTERS	✓
✓	FEEDBACK SIGNALS FROM TARGET	✓
✓	NEUROTRANSMITTER RECEPTORS PRESENT	✓
✓	ACTION POTENTIALS TRANSMIT INFORMATION FROM CELL BODY	✓
✓	MEMBRANE IS EXCITABLE	✓
✓	CALCIUM REGULATES PRIME FUNCTION	✓

Figure 1. Growth cones and presynaptic terminals are different morphological states of the same basic structure. The growth cone's primary activities include pathfinding, target recognition, and synaptogenesis. The product of these activities is the mature presynaptic terminal. The mature presynaptic terminal is not a static structure, however. Considerable remodeling is known to occur throughout the lifetime of the neuron. In order for the presynaptic terminal to change its synaptic partner, it must also be capable of pathfinding, target recognition, and synaptogenesis. During remodeling, therefore, the presynaptic terminal may functionally become a growth cone. In line with the preceding arguments, one would expect the growth cone and the presynaptic terminal to share many structural and functional properties. Some of the most obvious of these are listed in the table. Many of these have been amply demonstrated by other workers. In this article, we will provide evidence in support of the last four items.

roles in fulfilling their most obvious functions, it seems likely, from a variety of data, that they may actually represent different morphological states of a continuum of possible nerve terminal forms. To be sure, the presynaptic terminal is specialized for synaptic transmission and the growth cone is specialized for neurite outgrowth. However, as these two structures receive more critical attention, it becomes clear that they may represent dynamically interchangeable morphs because they share a variety of regulatory mechanisms. Our work on the neuronal growth cone has demonstrated the existence of four additional ties between these two entities: (*a*) neurotransmitter receptors are present on both; (*b*) action potentials transmit information from the cell body to both; (*c*) the membranes of both are excitable; and, to a large extent the topic

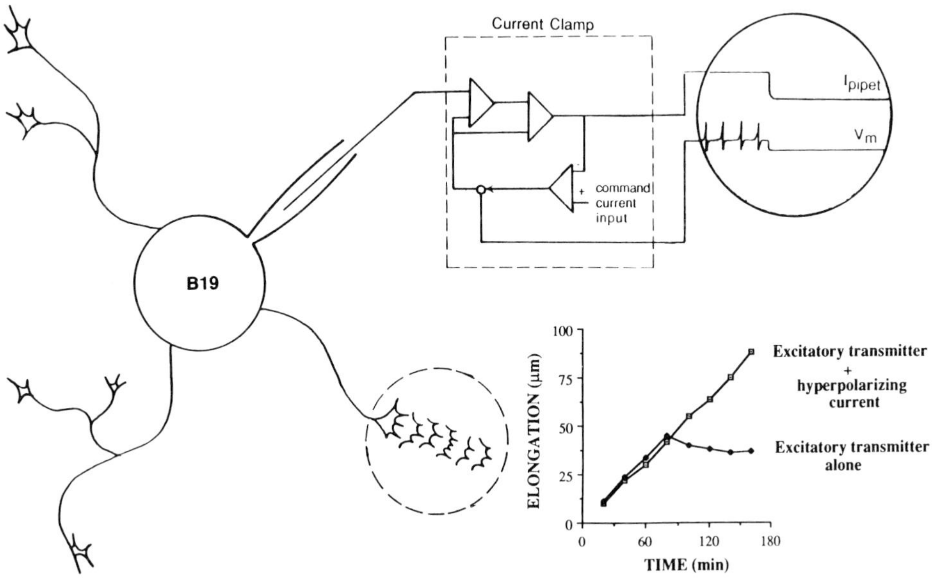

Figure 2. Neurotransmitters regulate growth cone behavior by acting at the level of membrane potential. As shown in schematic form, it is possible to record from or stimulate an identified *Helisoma* neuron, using patch electrodes or microelectrodes, while simultaneously monitoring neurite elongation. Such experiments (McCobb and Kater, 1988) have demonstrated that those neurotransmitters that inhibit neurite elongation also significantly depolarize the neuron. In addition, the outgrowth inhibitory effects of those neurotransmitters can be blocked by an injection of hyperpolarizing current.

of our most recent work, (*d*) calcium regulates the prime function of both the neuronal growth cone and the presynaptic terminal.

The Growth Cone Response to Neurotransmitters

It is well known from studies of presynaptic inhibition that the presynaptic terminal is endowed with specific receptors to neurotransmitters. We have studied the elongation behavior of neuronal growth cones both in molluscan and mammalian preparations and determined that neurotransmitters can significantly affect growth cone behavior. Neurons of the snail *Helisoma* (Haydon et al., 1985), and pyramidal neurons of the

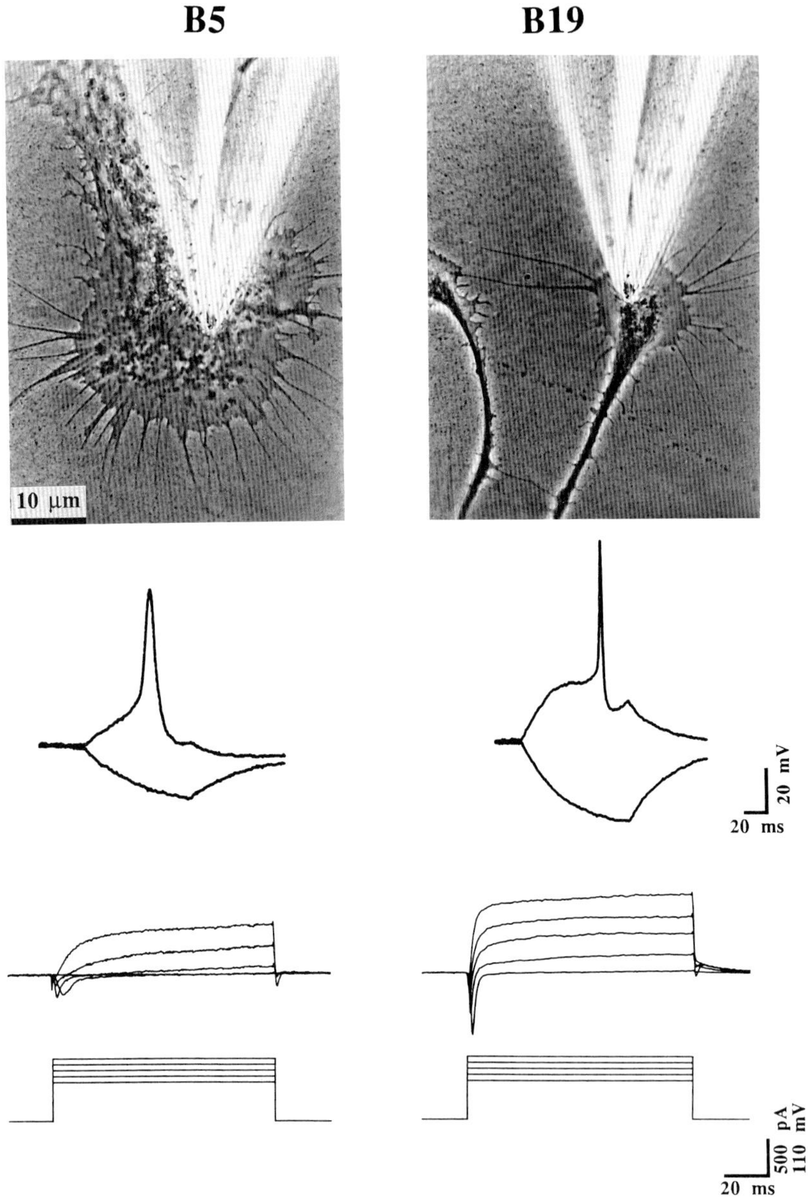

Figure 3. Growth cones from different identified neurons are identifiable, and their membranes possess many of the active properties normally associated with mature neuronal membrane. The phase images on top are typical growth cones from a neuron B5 (*left*) and a neuron B19 (*right*). These growth cones can be identified on purely morphological criteria (Haydon et al. 1985). Such growth cones can be surgically isolated from their associated neurite and patch clamped in either a cell-attached or a whole-cell mode. Under each image are action potentials and whole-growth cone voltage clamp records from isolated growth cones. Isolated growth cones are capable of producing regenerative action potentials. In addition, isolated growth cones from different neurons possess different, but characteristic, membrane properties. In fact, the membrane properties are characteristic for the neuron from which the growth cone was isolated and can serve to identify growth cones.

hippocampal cortex of the rat (Mattson et al., 1988) both show dramatic and consistent responses to neurotransmitters. In *Helisoma,* for example, the application of the neurotransmitter serotonin to any of a group of neurons, of which buccal neuron B19 is exemplary, results in an abrupt termination of elongation, a retraction of filopodia and lamellipodia, and the stabilization of the neuronal growth cone. These effects are seen even when serotonin is applied to a surgically isolated growth cone (Haydon et al., 1984). Similarly, the elongation of hippocampal pyramidal neuron dendrites is inhibited by one of their primary excitatory neurotransmitters, glutamate. In addition, both molluscan and mammalian growth cones respond in a predictable fashion to the simultaneous presentation of pairs of neurotransmitters such that

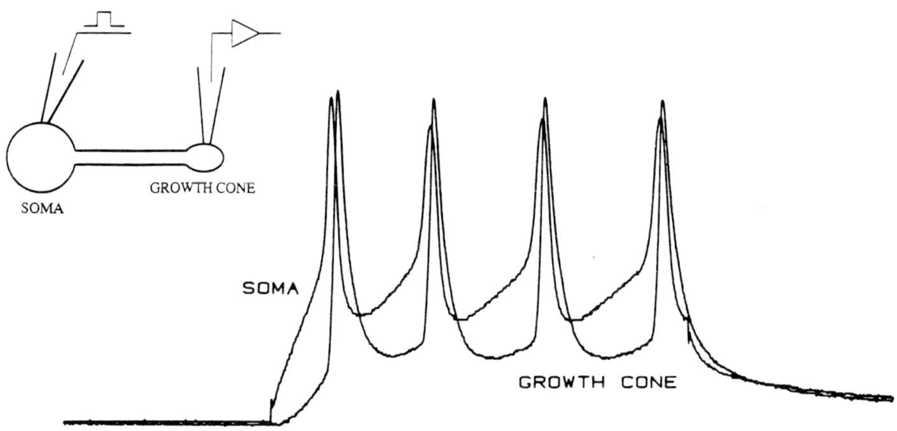

Figure 4. Action potentials generated in the cell body propagate to the growth cones. Patch electrodes were used in the whole-cell patch mode to obtain simultaneous recordings from both the cell body and a distant growth cone. Stimulation of the cell body produced a series of action potentials riding upon a depolarization; overshooting action potentials were also seen in the growth cone recording, following one-for-one with a constant latency. In addition, a depolarization of the growth cone was also seen; the depolarization was smaller and slower than that seen in the cell body, as would be expected by the passive cable properties of the connecting neurite. These data confirm that activity in the cell body will be transmitted to the growth cone at least in the form of action potentials, if not in the form of passive current flow.

acetylcholine can negate the serotonin effects on *Helisoma* growth cones (McCobb et al., 1988) and GABA can negate the effect of glutamate on mammalian growth cones (Mattson and Kater, 1988). Our present point of view is that, irrespective of species, neurotransmitters that produce sustained excitation stabilize growth cones and terminate neurite elongation, whereas neurotransmitters that are able to produce sustained inhibition can negate such effects.

It seems quite reasonable to assume that most of the effects of neurotransmitters are mediated through changes in membrane potential at the growth cone, precisely as neurotransmitters modify membrane potential in classical neuronal integration in adult systems. In a direct test of this idea (McCobb and Kater, 1988), neurotransmitters known to inhibit outgrowth (both serotonin and dopamine) were applied to *Helisoma* growth cones while whole-cell patch recordings were made from the cell

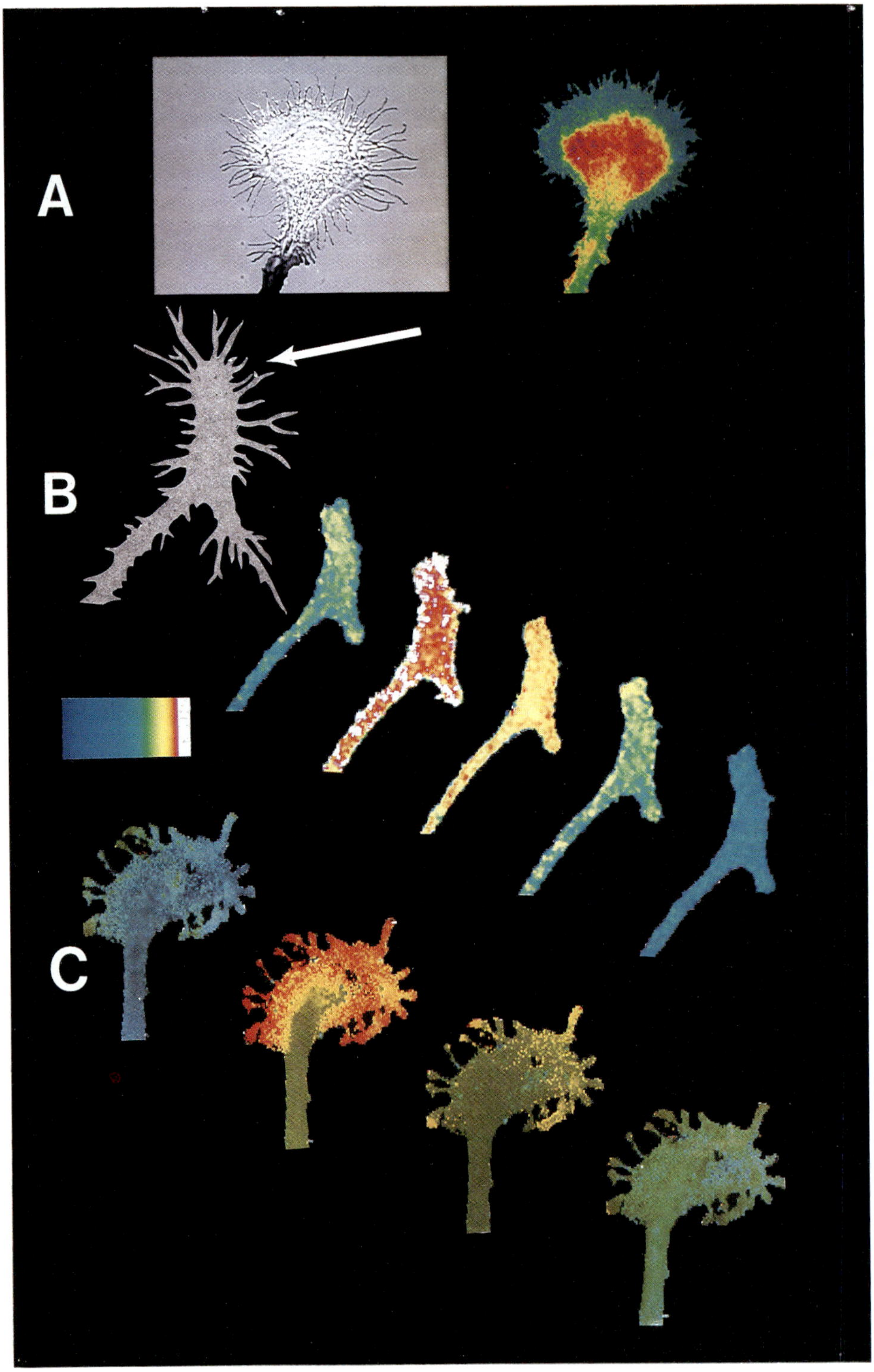
A
B
C

body. The growth cone inhibitory effects of these neurotransmitters could be entirely negated by the application of hyperpolarizing current to buck the excitatory effects of the neurotransmitters (Fig. 2). Thus, it appears that the regulation of the neuronal growth cone resides at the level of voltage-activated channels, as is often the case for the presynaptic inhibition of synaptic transmission.

Growth Cones Are Excitable

It is possible to isolate individual neuronal growth cones by severing them at the growth cone–neurite junction. Whole-cell patch recording techniques can be used to record from and stimulate such isolated growth cones (Guthrie et al., 1988). Growth cones produce regenerative action potentials; under voltage clamp control, distinct inward and outward currents are seen. A most interesting find, in this regard, is that the action potential waveform and the characteristics of the ionic currents are predictable on the basis of the particular identified neuron from which growth cone recordings are made.

Figure 5. Calcium maps of *Helisoma* growth cones. Growth cones were loaded with the fluorescent calcium indicator fura-2 as the /AM (acetoxymethylester) form. Similar results are seen with injection of the salt form of fura-2 (Cohan et al., 1987). Pairs of images of the growth cones were captured (350 ± 10 and 380 ± 10-nm excitation filters with a 505-nm long-pass emission filter) using a Quantex QX7 Image Data Acquisition System. The ratio of the two images is converted to calcium concentration according to the equation $[Ca^{++}] = K_D \cdot [(R - R_{min})/(R_{max} - R)] \cdot F_O/F_S$ (see Grynkiewicz et al., 1985). Calcium concentration is represented as color on the calcium maps with blue being low calcium and red being high calcium. The calibration bar in the middle represents calcium concentration for each panel: (*A*) 0–600 nM; (*B*) 0–300 nM; (*C*) 0–800 nM. (*A*) High resolution image of a neuron B5 growth cone. (*Left*) Phase image of the growth cone showing the filopodia, lamellipodia, main body of the growth cone, and the associated neurite. (*Right*) Calcium map of the growth cone. The important feature of this calcium map is the gradient in calcium concentration with the highest concentration at the leading edge of the main body of the growth cone and a significantly lower concentration at the growth cone-neurite junction. (The blue color of the filopodia and lamelipodia may not represent the true calcium levels; very little fluorescent signal is obtainable from those structures since they are so thin. Further studies are underway to resolve this issue.) (*B*) The concentration of calcium in growth cones is altered by activities that alter growth cone behavior. The left calcium map shows the calcium rest levels in the growth cone ($t = 0$). The color scale has been expanded to permit visualization of major changes in response to external stimuli; therefore, although there is a gradient within the growth cone, it is not as obvious as in section *A*. The growth cone was then gently touched with a glass probe (white arrow in the outline of the growth cone from the phase image). This stimulus is known to cause turning or branching (depending upon the site of contact) in growth cones. Within seconds after contact, the calcium levels in the growth cone have risen significantly (second calcium map from left; $t = 45$ s). The calcium rise is transient, in fact, calcium levels typically fall below the original rest levels after stimulation (subsequent calcium maps; $t = 80$ s; $t = 3$ min; $t = 10$ min). (Courtesy of L. Mills, unpublished and 1988.) (*C*) Stimulation of action potentials in the cell body causes calcium levels to rise in growth cones. (*Left*) Rest calcium levels in the growth cone. The color scale has been expanded to permit visualization of major changes in response to external stimuli; therefore, although there is a gradient within the growth cone, it is not as obvious as in section *A*. The growth cone was then stimulated with a microelectrode to fire eight action potential at 4/s. Calcium concentrations increased throughout the growth cone, but particularly at the leading edge of the main body of the growth cone (second panel; $t = 2$ s). The calcium concentrations decreased after cessation of the stimulation (right panels; $t = 15$ s; $t = 45$ s).

The action potential wave forms of buccal neurons B5 and B19, for instance, are quite different from one another in recordings from the cell body. These same differences are manifest at the level of the growth cone (Fig. 3). Thus, at least at the level of ionic currents, the growth cone is endowed with the identity of its parent cell body, a fact which may prove important to the differential regulation of growth cone behavior.

Action potentials have now been shown to affect growth cone behavior (Cohan and Kater, 1986). During periods of electrical quiescence, normal growth cone behavior can be monitored with whole-cell or cell-attached patch recording electrodes in place. Upon experimentally evoking action potentials, however, growth cone elongation is abruptly terminated. The growth cones remain immobile throughout the period of stimulation and then, upon cessation of stimulation of action potentials, can resume their motile characteristics. The ability of growth cones to generate action potentials is significant, since this permits propagated information from proximal portions of the neuron to reach growth cones at a distance (Fig. 4) and, once there, to depolarize the neuronal growth cone in a fashion that could affect its behavior. Again, as in the case of the presynaptic terminal, we see a striking parallel. Information generated in the form of action potentials can propagate to the growth cone and there regulate its prime function, motility.

Calcium Regulates the Prime Function

Intracellular calcium, the subject of much of this symposium, is a primary regulator of secretion, the central job of the presynaptic terminal. Intracellular calcium is also the prime regulator of growth cone activity for all external stimuli we have thus far studied. It has been possible to examine the distribution of intracellular calcium within growth cones through the use of the fluorescent calcium indicator fura-2 (Grynkiewicz et al., 1985). The digitized images of two excitation wavelengths can be ratioed, and good determinations of intracellular free calcium can be made from structures as small as mammalian growth cones. From our studies of the large molluscan growth cones, we have been able to measure both the steady state distribution of intracellular calcium in the growth cone and the change in this distribution evoked by a variety of stimuli.

The steady state spatial distribution of calcium is strikingly nonuniform in the motile growth cone. Fig. 5 *A* demonstrates a high concentration of intracellular calcium near the leading edge of the growth cone and a significantly lower calcium concentration at the point where the growth cone joins its associated neurite. This pattern of distribution may be crucial to growth cone function. Perhaps different biochemical processes are involved in growth cone motility. For instance, the actin, which is concentrated near the leading edge, may "see" one level of calcium, while the microtubules, which end near the growth cone–neurite junction, may "see" another level of intracellular calcium. Accordingly, with an appropriate distribution of calcium, the associated growth cone behavior propels this organelle in a linear fashion. Alternatively, a change in the distribution pattern of calcium could result in a change in growth cone behavior such as inhibition of motility, branching, or even turning.

A priori, several possible schemes could account for the ability of a growth cone to establish a standing gradient in its motile state. One possibility would involve an influx of calcium near the leading edge of the growth cone and/or an efflux or sequestering of intracellular calcium near the growth cone–neurite junction. At present, our limited data which bear on this possibility come from high voltage electron microscopic images of growth cones, which indicate an abundance of mitochondria, a potential calcium sink, precisely located at a low point in the calcium profile, (Fig. 6). Important future directions for this work, of course, require (*a*) asking if mitochondria are indeed involved in the sequestration of calcium in growth cones, and (*b*) mapping regions of calcium influx (see e.g., Fig. 7).

It is possible to change the observed spatial distribution of calcium with a variety of stimuli. All stimuli that are known to inhibit growth cone motility result in an abrupt

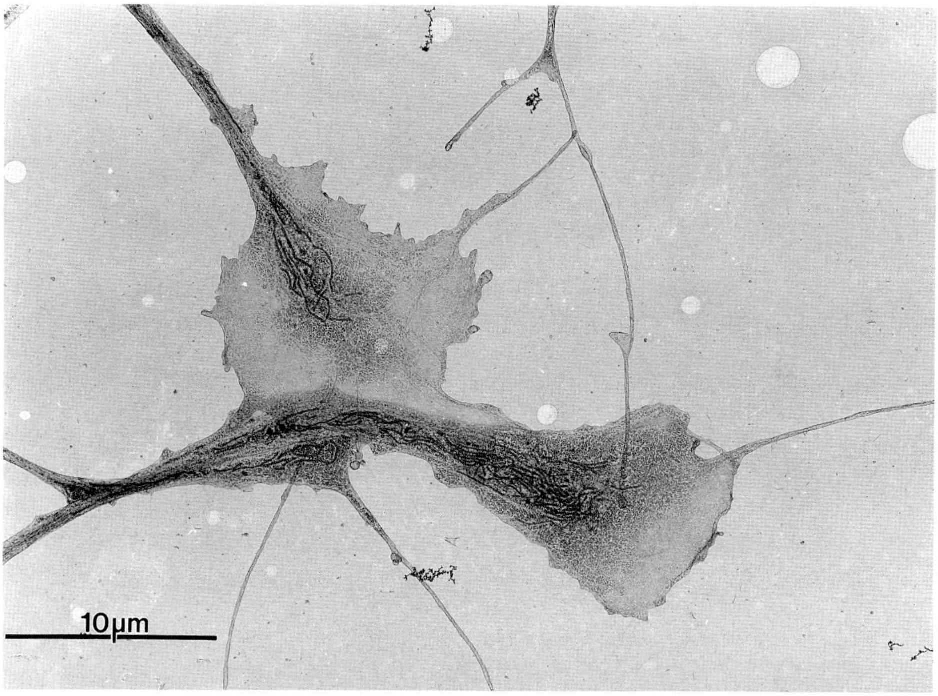

Figure 6. High voltage electron microscopy of *Helisoma* growth cones. Identified neurons were grown on Formvar-coated EM grids, quick frozen, critical-point dried, and viewed at the High Voltage Electron Microscope Facility in Boulder, CO. The distribution of organelles is clearly visible in the growth cone and associated neurite. Most notable is the concentration of mitochondria at the growth cone–neurite junction. This region has a lower calcium concentration than the leading edge of the main body of the growth cone (see Fig. 5 *A*). The lower concentration of calcium in this region may be due to mitochondria acting as calcium sequestration organelles.

and dramatic rise in intracellular calcium. For instance, if the neurotransmitter serotonin is applied to the growth cones of a neuron B19, there is an immediate rise in intracellular calcium and a concomitant inhibition of outgrowth. Similarly, the generation of action potentials (Fig. 5 *C*) which inhibits growth cone motility, also results in a rise in intracellular calcium (Cohan et al., 1987). Even mechanical stimuli (Mills et al., 1988) can alter growth cone behavior; these tactile stimuli also result in an abrupt rise in intracellular calcium (Fig. 5 *B*). Taken together, it would appear that an alteration in the spatial distribution of calcium, in fact, results in an alteration of the motility processes. One test of this has been through the use of the calcium ionophore A23187. Application of A23187 to the growth cones of both *Helisoma* and rat hippocampal pyramidal neurons results in an immediate inhibition of outgrowth and a tremendous rise in intracellular calcium (Kater et al., 1988*b*).

Intracellular calcium changes display important parallels with the behavior of the

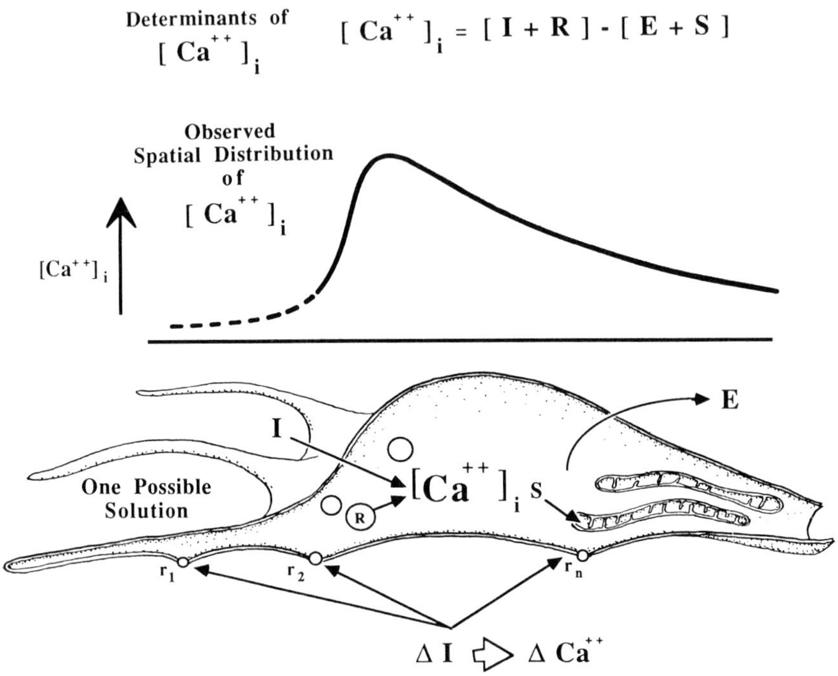

Figure 7. Factors determining calcium concentration within growth cones. The top equation shows the factors determining calcium levels within the growth cone. Calcium levels are increased by influx (I) and release (R) from intracellular stores; levels are decreased by efflux (E)(pumping) and sequestration (S). (*Middle*) The observed calcium rest levels in growth cones (cf. Fig. 5 *A*). (*Bottom*) One possible distribution of mechanisms regulating calcium concentrations which could explain the observed gradient. If influx were localized primarily to the leading edge of the main body of the growth cone, and/or efflux/sequestration were localized primarily to the growth cone–neurite junction, then calcium concentrations would be higher in the leading edge than in the neurite. One important implication of the existence of a gradient is the possibility that anything that would perturb the gradient could have profound effects on the behavior of the growth cone. For example, if neurotransmitter receptors were localized to the leading edge of the growth cone, activation of those receptors could directly increase the influx of calcium at that point, and increase the gradient. Activation of other receptors at the trailing edge of the growth cone (e.g., receptors interacting with extracellular matrix molecules) could reduce the gradient by increasing influx at that point.

neuronal growth cone. While there are, undoubtedly, further complexities to the regulation of the growth cone by intracellular calcium, it is clear that a rise in intracellular calcium can completely change the behavior of the growth cone. We also know from work not discussed here that, if intracellular calcium drops below particular levels, the behavior of the growth cone also changes (Cohan et al., 1987). Thus, our present view of the role of calcium in the growth cone is that there is an optimum range of intracellular calcium concentration correlated with the usual behaviors associated with the neuronal growth cone (Kater et al., 1988*a*).

Conclusions

The growth cone and the presynaptic terminal share a variety of regulatory mechanisms despite an obvious difference in their primary functions. It should be recognized, however, that some key features are shared. The fusion of vesicles is likely to be as important a part of the motility process in the growth cone (Bray, 1973), as the fusion of vesicles is for the secretion of material from the presynaptic ending. Accordingly, as more becomes known about the precise mechanisms of the control of motility of growth cones and the relationship to calcium, it may be seen that these two quite different-appearing organelles are, indeed, representatives of an interchangeable set of morphologies that can exist at the tips of nerve terminals.

References

Bray, D. 1973. Model for membrane movements in the neural growth cone. *Nature*. 244:93–96.

Cohan, C. S., and S. B. Kater. 1986. Suppression of neurite elongation and growth cone motility by electrical activity. *Science*. 232:1638–1640.

Cohan, C. S., J. A. Connor, and S. B. Kater. 1987. Electrically and chemically mediated increases in intracellular calcium in neuronal growth cones. *Journal of Neuroscience*. 7:3588–3599.

Grynkiewicz, G., M. Poenie, and R. Tsien. 1985. A new generation of Ca^{2+} indicators with greatly improved fluorescence properties. *Journal of Biological Chemistry*. 260:3440–3450.

Guthrie, P. B., R. E. Lee, and S. B. Kater. 1989. A comparison of neuronal growth cone and cell body membrane: electrophysiological and ultrastructural properties. *Journal of Neuroscience*. In press.

Haydon, P. G., D. P. McCobb, and S. B. Kater. 1984. Serotonin selectively inhibits growth cone motility and the connectivity of identified neurons. *Science*. 226:561–564.

Haydon, P. G., C. S. Cohan, D. P. McCobb, H. R. Miller, and S. B. Kater. 1985. Neuron-specific growth cone properties as seen in identified neurons of *Helisoma*. *Journal of Neuroscience Research*. 13:135–147.

Kater, S. B., M. Mattson, C. Cohan, and J. Connor. 1988*a*. Calcium regulation of the neuronal growth cone. *Trends in Neuroscience*. 11:315–321.

Kater, S. B., P. B. Guthrie, M. P. Mattson, L. R. Mills, and R. S. Zucker. 1988*b*. Calcium homeostasis in molluscan and mammalian neurons: dynamics of calcium regulation. *Society for Neuroscience Abstracts*. 14:582. (Abstr.)

Mattson, M. P., P. Dou, and S. B. Kater. 1988. Outgrowth-regulating actions of glutamate in isolated hippocampal pyramidal neurons. *Journal of Neuroscience*. 8:2087–2100.

Mattson, M. P., and S. B. Kater. 1988. Excitatory and inhibitory neurotransmitters in the generation and degeneration of hippocampal neuroarchitecture. *Brain Research*. In press.

McCobb, D. P., C. S. Cohan, J. A. Connor, and S. B. Kater. 1988. Interactive effects of serotonin and acetylcholine on neurite elongation. Neuron 1:377–385.

McCobb, D. P., and S. B. Kater. 1988. Membrane voltage and neurotransmitter regulation of neuronal growth cone motility. *Developmental Biology.* In press.

Mills, L. R., M. Murrain, P. B. Guthrie, and S. B. Kater. 1988. Intracellular calcium concentrations in neuronal growth cones change with induced turning and branching behavior. *Society for Neuroscience Abstracts.* 14:583. (Abstr.)

Ramon y Cajal, S. 1890. A quelle epoque apparaissent les expansions des cellules nerveuses de la moelle epiniere du poulet. *Anatomischer Anzeiger.* 5:609–613, 631–639.

Chapter 8

Modulation of Potassium Channels by Dopamine in Rat Pituitary Lactotrophs: A Role in the Regulation of Prolactin Secretion?

Karen A. Gregerson, Leslie Einhorn, Michele M. Smith, and Gerry S. Oxford

Departments of Pediatrics and Physiology, The University of Maryland, Baltimore, Maryland 21201; and the Department of Physiology and The Curriculum in Neurobiology, University of North Carolina, Chapel Hill, North Carolina 27599

Introduction

In recent years, much scientific effort has been spent on investigating the cellular and electrical phenomena associated with secretion in a variety of endocrine cells. From work on neoplastic cell lines, there is evidence that the binding of secretagogues to receptors on the plasma membrane of pituitary cells can lead to a change in the ionic permeability of the cell by opening or closing membrane channels (Taraskevich and Douglas, 1977; Sand et al., 1980; Kaczarowski et al., 1983; Dubinsky and Oxford, 1985). In addition, spontaneous electrical activity, including calcium-dependent action potentials, has been observed in pituitary cells and may be involved in basal secretion (Kidokoro, 1975; Biales et al., 1977; Adler et al., 1983).

Under normal physiological conditions, prolactin (PRL) secretion from the anterior pituitary gland is predominantly under inhibitory control by hypothalamic dopamine (DA) (Gibbs and Neill, 1978; Weiner et al., 1979). DA inhibition of PRL secretion has been demonstrated to be mediated via surface membrane receptors of the D_2 subtype (Caron et al., 1978; Enjalbert and Bockaert, 1983). In addition to cells that release PRL, the mammalian anterior pituitary gland contains several other secretory cell types distributed in a hetereogeneous fashion throughout the gland. Consequently, investigations of the cellular mechanisms of DA inhibition of PRL secretion have until recently been impeded by the lack of a pure preparation of normal PRL-secreting cells (lactotrophs). While the clonal line of GH_3 cells has proven an excellent model for the study of thyrotropin-releasing hormone (TRH)–induced PRL release (cf. Gershengorn, 1986), these cells neither respond to (Faure et al., 1980) nor demonstrate any high affinity binding sites for DA (Cronin et al., 1980).

Recently, the modification and application of an immunological technique, the reverse hemolytic plaque assay (RHPA), has enabled the identification of specific cell types in mixed cell cultures while preserving the viability of the cells (Neill and Frawley, 1983). The procedure uses immunologically triggered complement-induced red blood cell lysis in the vicinity of individual pituitary cells to identify those secreting a particular hormone. Using this technique to unambiguously identify single normal lactotrophs in combination with GΩ seal patch-clamp methods of current and voltage clamp, we have begun to investigate the role of membrane excitability and ion channels in the regulation of PRL secretion by DA at the cellular level.

Methods

Cell Dissociation

Anterior pituitary glands were dissected from adult, female Sprague-Dawley rats on the morning of proestrus (lights on 0600–1800, decapitate between 0830–1000). Cells were dispersed enzymatically using a nontrypsin dissociation protocol modified from that described by Weiner et al. (1983). Briefly, the pituitary tissue was minced in Hank's balanced salt solution, calcium and magnesium free (Hank's CMF). After three washes, the tissue fragments were incubated in Hank's CMF containing 0.3% collagenase (Worthington Biochemicals, Freehold, NJ) 1 mg% DNase I, and 0.1 mg% trypsin inhibitor, for 60 min at 37°C. The fragments were then mechanically dispersed into individual cells by trituration through a siliconized Pasteur pipette and washed twice with Dulbecco's modified Eagle's medium, containing 0.1% bovine serum albumin (DMEM-BSA), to remove the collagenase. The cell suspension was then filtered through nylon mesh (20 μm) and harvested by centrifugation at 150 *g* for 10 min.

Reverse Hemolytic Plaque Assay

The plaque assay was performed essentially as described by Smith et al. (1986) with some modifications to minimize interference with electrophysiological recordings. In brief, pituitary cells suspended in DMEM-BSA (4×10^5/ml) were mixed with an equal volume of an 8% suspension of ovine erythrocytes (oRBC) previously coupled to *Staphylococcus* protein A with chromium chloride hexahydrate. Aliquots of this mixture were infused into modified Cunningham chambers constructed by affixing a polylysine-coated glass coverslip to a microscope slide by means of two parallel pieces of double-stick tape. The chambers were inverted and incubated for 45 min at 37°C in a humid 95% air-5% CO_2 atmosphere to allow cell attachment to the coverslip. Unattached cells were then removed by rinsing the chambers with DMEM-BSA. The assay was initiated by infusion of DMEM-BSA containing antiserum to rat PRL (1:120–1:200 dilution) and plaques were developed 1 h later by infusion of guinea pig complement (1:50–1:120, 30 min).

For identification of lactotrophs to be used in electrophysiological studies, the complement reaction was terminated by repeated rinses with DMEM-BSA. The chambers were then dismantled and the coverslips, with the cells attached, were maintained in culture in DMEM containing 10% defined equine serum (Hyclone Laboratories, Logan, UT) and antibiotics (gentamycin sulfate, 40 μg/ml) until used in electrophysiological studies. In general, these assays were performed on the day of dissociation and the patch-clamp experiments were performed on the subsequent three days.

The PRL antiserum used in the plaque assays was generated against purified rat PRL (PRL I-3, National Institutes of Diabetes and Digestive and Kidney Diseases [NIDDK]) in rabbits. Preimmunized sera from the rabbits elicited no plaque formation. Preabsorption of the antiserum with PRL I-3 (0.5 μg/ml) abolished plaque formation, while preabsorption with purified rat growth hormone (GH I-1, 50 μg/ml) did not diminish plaque formation.

PRL Secretion

For analysis of PRL release from individual cells, plaque assays were performed that included 0.5 mM ascorbic acid with or without DA in the incubation with the antiserum. In these plaque assays, the complement reaction was terminated by infusion of 1% gluteraldehyde and the cells were stained with phloxine (2.5% in distilled water) and 0.05% azure II + 0.02% methylene blue in 0.05% borate buffer.

Radioimmunoassay (RIA) measurements of PRL secretion were performed on cohort pituitary cells plated in sterile, polylysine-treated, multiwell plates (4×10^4 cells/well) and incubated in DMEM containing equine serum and gentamycin. For the release study, the cells were washed with a basic extracellular solution (see below) containing 0.1% BSA to remove medium and serum. After a 30-min preincubation, a 1-h test incubation was performed in 1 ml of solution containing BSA and 0.5 mM ascorbic acid alone (control) or with the various test substances. The solution was then removed, centrifuged at 500 *g* for 10 min, and stored at −20°C until assayed. Secreted PRL was determined by homologous double antibody RIA using reagents provided by

Dr. Albert Parlow through the NIDDK Rat Pituitary Hormone Distribution Program (rPRL RP-2 standard). Samples were assayed in two doses in duplicate. Intra- and interassay variations were <5 and 8%, respectively. Differences between means of groups in the RIA data were analyzed using Student's *t* test.

Electrophysiology

Whole-cell and patch recordings were made according to the GΩ seal patch-clamp technique described by Hamill et al. (1981). Patch electrodes with 2–6 MΩ tip resistances were manufactured from N51A (Drummond Scientific, Broomall, PA) or aluminosilicate (A-M Systems, Everett, WA) capillary glass and coated with Sylgard to reduce stray electrode capacitance. The basic extracellular salt solution contained (in millimolar): 145 NaCl, 5 KCl, 2 $CaCl_2$, 1 $MgCl_2$, 10 glucose, and 10 HEPES. The basic intracellular solution contained in the recording electrode included (in millimolar): 130 K-Asp, 20 KCl, 10 glucose, and 10 HEPES. Intracellular solutions also contained 2 mM MgATP and 2 mM cAMP and a nucleotide regenerating system made up of 20 mM creatine phosphate and 50 U/ml creatine phosphokinase, which minimizes "rundown" of Ca^{2+} currents (Forscher and Oxford, 1985). All solutions were adjusted to pH 7.3–7.4 and 295–305 mosmolal. Variations of these solutions used in the following studies are noted in the corresponding figure legends.

Patch-clamp experiments were performed at room temperature in a Plexiglas recording chamber mounted on the stage of a Nikon Diaphot inverted microscope. External solutions were continuously perfused through the recording chamber, which contained a coverslip with the plaque-identified pituitary cells attached. Application of test drugs (primarily DA, with ascorbic acid to inhibit catecholamine oxidation) and control solutions (external solution with ascorbic acid alone) was accomplished by a "U-tube" device. This device was constructed from a piece of PE10 tubing with a small hole in its wall at the bend of a "U"-shaped loop. Test solution flows through the tubing from a reservoir by hydrostatic pressure. The outlet at the distal end of the tubing is connected to a slight vacuum that maintains the flow through the tubing. When this vacuum is interrupted, test solution flows out the hole in the tubing loop. By positioning this portion of the U-tube next to a cell and manipulating the vacuum at the outlet, test and control solutions can be rapidly applied and withdrawn from a cell during recording with little mechanical disturbance of the cell.

Results

DA Inhibits PRL Secretion and Plaque Formation

Hemolytic plaques, defined as clear zones containing oRBC ghosts completely encircling secretory cells, formed around numerous pituitary cells, and, thus, these plaque-forming cells were considered lactotrophs. Under control conditions (in the absence of secretory modulators), there was a wide range of sizes in the plaques that formed around lactotrophs from the proestrous rats (Fig. 1, left). Since the area of hemolysis is proportional to the amount of PRL release (i.e., cells secreting large amounts of PRL form large plaques while cells secreting much smaller amounts of PRL form small plaques), these control plaques show the wide variation in secretory activity that occurs among individual lactotrophs.

Fig. 1 (right) also illustrates the effect of DA on primary lactotrophs using the plaque assay. DA reduced the amount of PRL released from individual cells, but this inhibition was apparently not uniform. While numerous small plaques still formed in the presence of a maximally effective dose of DA (10^{-6} M) the development of large plaques was completely prevented.

Electrophysiological Response of RHPA-identified Lactotrophs to DA

As illustrated in Fig. 2, patch electrodes could be sealed onto the membrane of plaque-forming cells to record their electrical activity. Seals of 20-GΩ resistance or greater were obtained on these cells and intracellular measurements revealed input

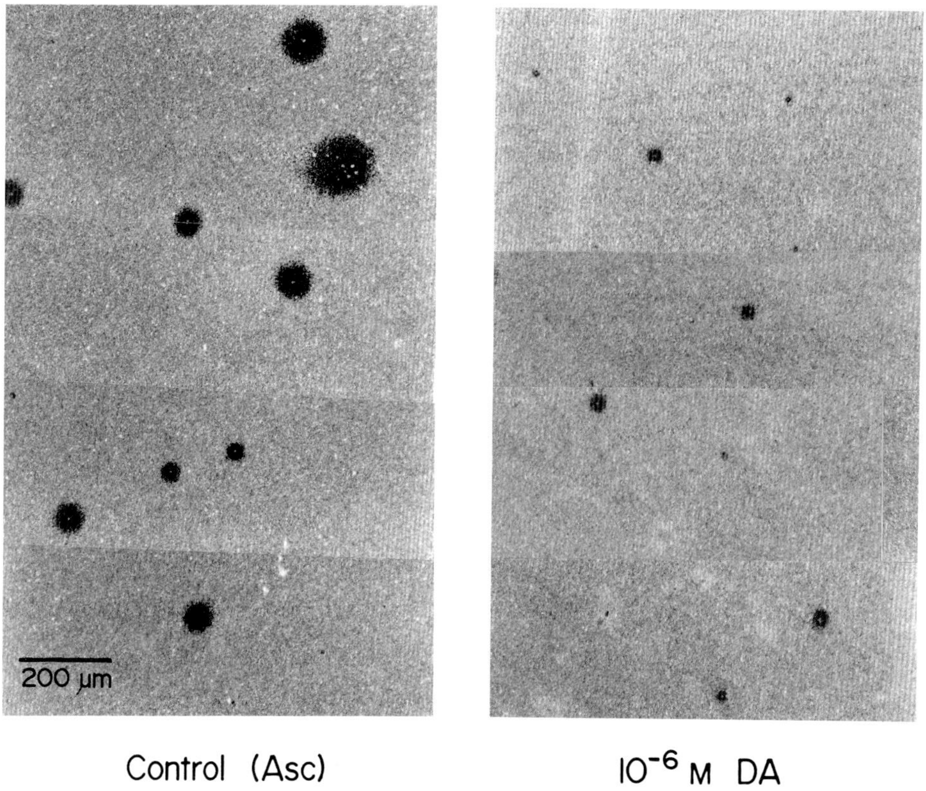

Figure 1. Low magnification photomicrographs of plaques, enhanced by dark-field illumination, formed around PRL-secreting cells. Pituitary cells were derived from proestrous rats. Cells were incubated for 1 h at 37°C with anti-rat PRL in DMEM-BSA with 0.5 mM ascorbic acid alone (control) or in the presence of 10^{-6} M DA. Plaques were developed with complement. Cells were fixed with gluteraldehyde and stained (see Methods).

resistances on the order of 3–7 GΩ. Recordings were made at random on a range of lactotrophs with different cell diameters and forming plaques of different sizes. No differences in electrical properties (e.g., resting membrane potential, spontaneous activity) could be correlated with either cell or plaque size. However, the plaques were developed in assays 1–3 d before electrophysiological experiments and may not reflect the secretory activity of the cells at the time of recording. No overall change in the electrical properties of the lactotrophs was observed in the different times in culture.

The effect of DA on the membrane potential in identified lactotrophs was examined by whole-cell current clamp as illustrated in Fig. 3. The brief application of DA (4 μM) was found to elicit an immediate hyperpolarization of the membrane potential that recovered only slowly. In cells exhibiting spontaneous Ca^{2+} spikes (Fig. 3, upper trace), the hyperpolarization was accompanied by immediate cessation of spiking activity which reinitiated at the control rate with recovery of the resting membrane potential. Application of the vehicle (0.5 mM ascorbic acid in external solution) elicited no response. The rapidly developing, long-lasting hyperpolarization induced by DA was also seen in quiescent cells (Fig. 3, lower trace). The hyperpolarization was coincident with ~50% reduction in the cell input resistance as determined by hyperpolarizing current injections (5 pA, 330-ms duration), indicating an increased membrane conductance in response to DA.

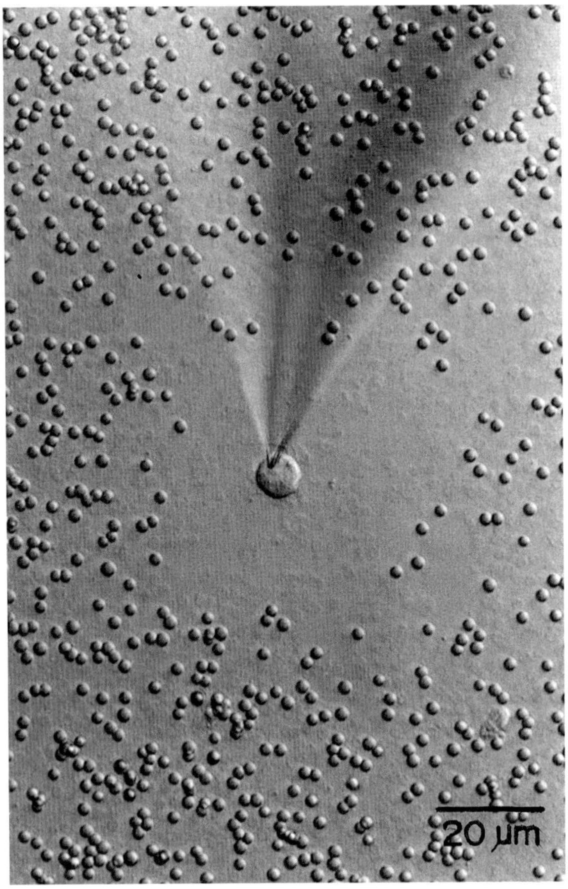

Figure 2. A viable, PRL-secreting cell identified in a plaque assay by the surrounding area of lysed oRBCs. The tip of a patch recording electrode (extending above the focal plane at the top of the picture) is sealed on the membrane surface of the lactotroph. Cells were dissociated from rat anterior pituitary tissue. Photomicrograph of unstained, unfixed cells was taken through Hoffman modulation contrast optics.

Pharmacology of the DA Response

As previously mentioned, DA has been shown to inhibit PRL secretion via the D_2 receptor subtype (Caron et al., 1978; Enjalbert and Bockaert, 1983). To confirm the role of D_2 receptors in the inhibition of PRL release, we investigated the effects of specific D_2 receptor agonist and antagonists. The bar graph in Fig. 4 shows the PRL

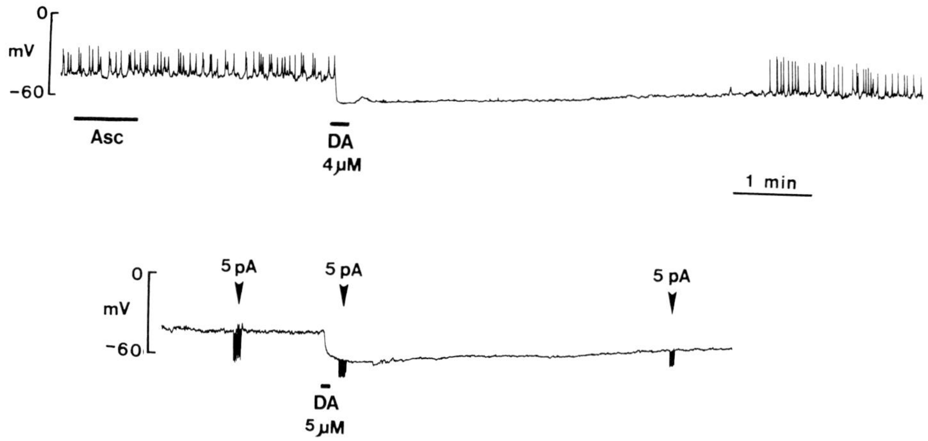

Figure 3. Response of the membrane potential of primary rat lactotrophs in whole-cell current clamp to DA. Brief (duration indicated by bars) application of DA elicited a rapidly developing, long-lasting hyperpolarization in lactotrophs derived from proestrous rats (*Upper trace*) In a spontaneously active cell, the hyperpolarization was accompanied by cessation of spiking activity which resumed after recovery of the resting membrane potential. Application of the vehicle (0.5 mM ascorbate) had no effect. (*Lower trace*) Hyperpolarization elicited by 5 μM DA in a quiescent cell. Attenuated membrane potential responses to the injection of hyperpolarizing current pulses (*arrows;* 5 pA, 330 ms) indicate a decreased membrane resistance during the DA-induced hyperpolarization.

release data for pituitary cells in culture. Incubation with DA decreases the amount of PRL released as compared with vehicle-treated controls. This DA-induced reduction in PRL secretion was blocked by the specific D_2 receptor antagonists, sulpiride (5 μM) and (+)butaclamol (0.1 μM), but was not affected by the inactive isomer (−)butaclamol (0.1 μM). In the absence of DA, these receptor antagonists had no effect on PRL release by themselves. The same results were obtained when the D_2 antagonists were

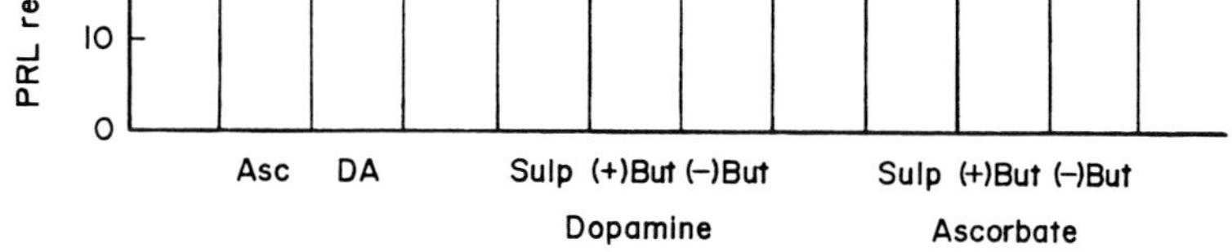

Figure 4. Pharmacology of DA effects on PRL release from primary lactotrophs derived from proestrous rats. Bar graph of PRL release from cultured pituitary cells during 1-h incubation periods with or without DA or D_2 receptor antagonists. Data were derived from four independent experiments, each comprising triplicate wells per treatment group. Data are expressed as the means ± SEM. Values indicated by asterisk (*) differ significantly from ascorbate control group at $P < 0.001$.

TABLE I
Effects of D_2 Receptor Agonists and Antagonists on the Membrane Potential of RHPA-identified Lactotrophs

Compound	Concentration	Hyperpolarizes V_m	Blocks DA effect
	μM		
DA	5.0	Yes	—
	1.0	Yes	—
	0.1	Yes	—
RU 24213	1.0	Yes	—
Sulpiride	5.0	No	Yes
(+)Butaclamol	0.1	No	Yes
(−)Butaclamol	0.1	No	No

tested on PRL release from individual lactotrophs in the reverse hemolytic plaque assay (data not shown).

The pharmacology of the electrophysiological response to DA was found to follow the same pattern as the secretory response and is summarized in Table I. DA induced hyperpolarization of the lactotroph membrane at various concentrations. These have been extended to include DA concentrations as low as 5 and 50 nM (Fig. 5), which are within the range of DA concentrations measured in the hypothalamic-pituitary portal blood (Ben-Jonathan et al., 1977; Plotsky et al., 1978). The specific D_2 receptor agonist, RU24213, also hyperpolarized the lactotroph membrane. As in the secretory response, sulpiride and (+)butaclamol blocked the electrical response to DA while having no inherent effect on the membrane potential themselves. The (−) isomer of butaclamol, again was unable to block the response to DA.

Ionic Basis of the DA-induced Hyperpolarization

The ionic current underlying the hyperpolarizing response most likely involves an increase in either K^+ (outward) or Cl^- (inward) ion movement. We investigated this point using whole-cell current and voltage-clamp measurements during ion substitutions and changes in membrane potential. The peak hyperpolarization response was not altered when the transmembrane gradient for Cl^- ions was reversed by reducing the external Cl^- concentration from 150 to 10 mM (data not shown). Thus, an increase in membrane permeability to K^+, and not Cl^-, likely underlies the DA response of the

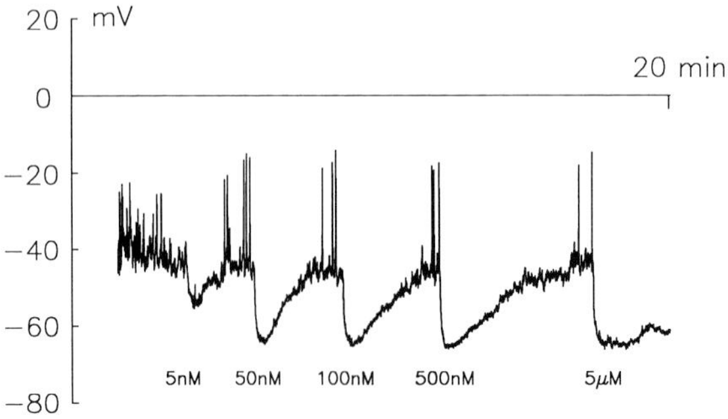

Figure 5. Membrane potential responses to varying concentrations of DA. Whole-cell voltage recordings of a lactotroph showing transient hyperpolarizations of membrane potential in response to consecutive 1-s applications of DA at concentrations from 5 nM to 5 μM as indicated. Maximal voltage change occurred by 50–100 nM DA, yet recovery of the membrane potential became more retarded with increasing doses of DA. Note resumption of spontaneous spiking activity with the recovery of the resting membrane potential after each DA-induced hyperpolarization.

lactotroph membrane potential. To further identify the ion and investigate the ionic currents contributing to the DA-induced change in membrane potential, the reversal potential for this response was determined in both current and voltage clamp (Fig. 6, *A* and *B*). From cells in current clamp, the peak changes in membrane potential elicited by DA (or the DA agonist, RU24213) are plotted against the membrane potential of the cell just prior to drug application (open symbols). Steady hyperpolarizing or depolarizing currents were injected into some cells to poise the membrane potential more negative or positive than the equilibrium potentials of the suspected ions. The reversal potential was also determined from DA-induced whole-cell currents in voltage

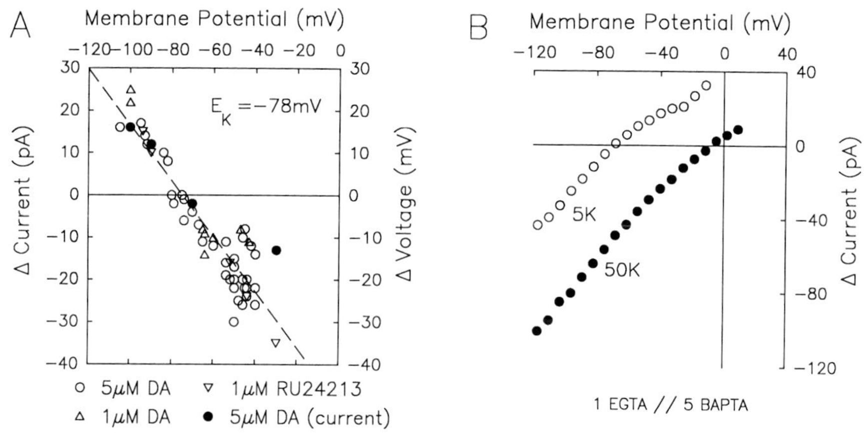

Figure 6. The reversal potential for the effect of DA on primary lactotrophs is consistent with changes in K^+ conductance. (*A*) Changes in membrane potential induced by DA or RU24213 at the indicated concentrations plotted as a function of the "resting" membrane potential in the absence of agonist (*open symbols*). Dashed line is a linear regression fit of the data yielding a voltage intercept of −77.6 mV, which is in agreement with an E_K of −78 calculated from the solution compositions. Also plotted are the current responses of a cell in voltage clamp to 5 μM DA at different holding potentials (*filled circles*). (*B*) Current-voltage relations of the DA-induced current in a lactotroph under voltage clamp. Membrane potential was commanded by voltage ramps from −120 to +10 mV and the current responses to five ramps were averaged. Averaged current responses during a control period were subtracted from those obtained at the peak DA response to yield only the DA-sensitive current. Initial measurements were obtained in a Ca^{2+}-free external solution containing 4 mM $MgCl_2$, 140 mM NaCl, 5 mM KCl, 10 mM glucose, 10 mM HEPES, and 1 mM EGTA (*open circles*). Data were then obtained in a solution containing 50 mM K^+ equimolar substituted for Na^+ (*filled circles*). The internal solution in both cases contained 5 mM K-BAPTA to reduce nominal Ca^{2+} to negligible levels.

clamp (Fig. 6 *A*, closed symbols). The DA response of either membrane potential or current exhibited a reversal potential closely approximating the calculated equilibrium potential of K^+ (E_K).

The reversal potential for the DA-induced ionic current in voltage clamp was sensitive to the external K^+ concentration. As illustrated in Fig. 6 *B*, the current-voltage curve for the DA-induced conductance shifted to more depolarized potentials when the external K^+ concentration was increased from 5 to 50 mM, which is consistent with a primary role for K^+ in the response. Additionally, it is noteworthy that the DA-induced K^+ conductance exhibits minimal rectification about the reversal potential and could be elicited under conditions in which both extracellular and intracellular Ca^{2+} was buffered to negligible levels.

The activation of single K channels associated with DA ligand binding was observed in preliminary experiments in on-cell membrane patches from normal lactotrophs with application of DA to the extrapatch membrane (Fig. 7). 10 μM DA could activate outward directed single-channel currents when both the bath and patch electrode contained standard external solution (5 mM KCl; upper trace of Fig. 7). A transient shift in the baseline of the recording reflects the DA-induced hyperpolarization of the entire cell as it was not under voltage control. The lower trace in Fig. 7 shows inward-directed channel currents in another lactotroph activated by DA when the patch electrode contained 140 mM KCl. The reversed direction of the channel currents when the electrode K^+ concentration was increased is consistent with DA activating a

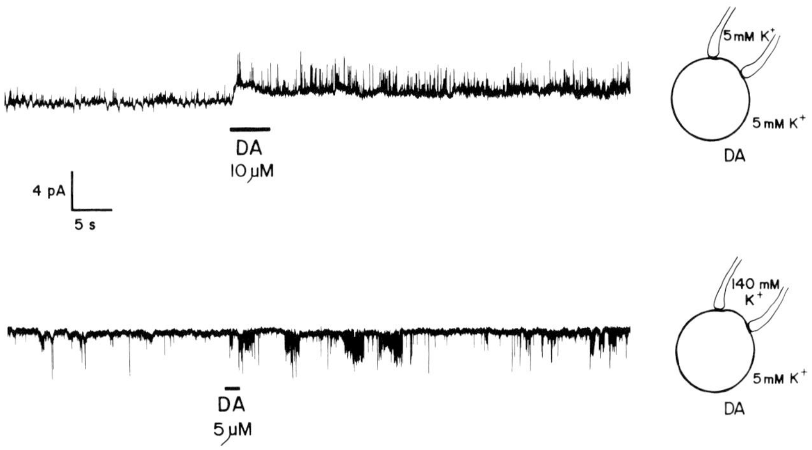

Figure 7. On-cell patch recordings of single-channel activity in lactotrophs before, during, and after application of DA (indicated by bars) to the extrapatch membrane (*Upper trace*) Outward-directed channel currents in cell-attached patch were activated by 10 μM DA. Both the bath and patch electrode contained standard external solution (5 mM K^+) (*Lower trace*) Inward-directed channel currents in cell-attached patch of another lactotroph activated by 5 μM DA. The bathing solution contained 5 mM K^+ while the patch electrode contained 140 mM K^+. In both traces, the patch electrode was held at 0 mV, thus the transpatch potential was the unclamped cell membrane potential.

K conductance. The reversal further indicates the minimal rectification of the DA-induced K conductance.

The on-cell patch recordings from the lactotroph in the upper trace of Fig. 7 are shown with greater time resolution in Fig. 8. Each panel represents randomly chosen segments taken before (control) and after the application of DA. Upward deflections represent channel openings. Before the application of DA only a small conductance outward channel is observed, but after DA exposure a larger conductance channel appears in the records. This difference is not reflective of the change in driving force accompanying the DA-induced hyperpolarization of the cell since one would expect K channel currents to decrease in size as the driving force diminished in this case. Thus, a different channel type gives rise to the larger events following DA application. Since the transmembrane potential of the patch is not known, the single-channel conductances cannot be accurately calculated. Nevertheless, these recordings clearly show an increased opening frequency of channels in response to DA. Moreover, this effect on patch channels by ligand binding to extrapatch receptors suggests the involvement of a second messenger system(s) in the transduction of this response.

Involvement of Guanine Nucleotide Binding Proteins in the DA Response

Previous reports have indicated that DA inhibition of PRL secretion is mediated, at least in part, by a pertussis toxin (PTX)–sensitive guanine nucleotide binding protein (e.g. Cronin et al., 1983). We used radioimmunological and electrophysiological techniques to further investigate the role of guanine nucleotide binding proteins (G proteins) in the signal tranduction pathway for DA inhibition of PRL secretion and activation of a potassium current. Specifically the actions of PTX and nonhydrolyzable analogues of GTP and GDP were examined on lactotroph secretory and electrical responses to DA.

PTX catalyzes the ADP-ribosylation of certain classes of G proteins thereby

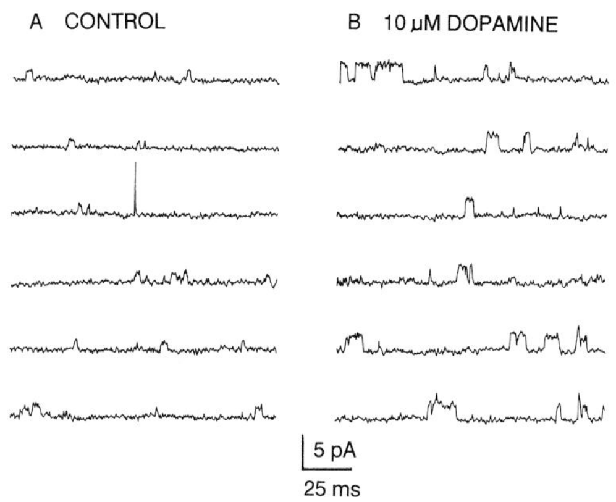

Figure 8. On-cell patch recording from the same lactotroph as shown in the upper trace of Fig. 7, at greater time resolution. Each panel represents randomly chosen segments taken before (*A*) and after (*B*) application of DA (10 μM) to the extrapatch cell membrane. The predominant channel seen under control conditions in this patch exhibited an amplitude of ~2 pA. After DA was applied, a channel of ~4 pA amplitude appears with an increased opening frequency. Data were sampled from analog tape recordings at 14-bit resolution and low-pass filtered at 2.5 kHz.

blocking their activation by agonists (Bokoch et al., 1983; Murayama and Ui, 1983; Katada et al., 1984). As shown in the upper bar graph of Fig. 9, preincubation of primary cultured pituitary cells with PTX (List Biological Laboratories, Campbell, CA; 1 μg/ml, 8 h) blocked the ability of DA to inhibit PRL release. Furthermore, companion whole-cell current clamp studies on RHPA-identified lactotrophs revealed that a similar PTX preincubation (0.1–1 μg/ml, 3–8 h) abolished the ability of DA to hyperpolarize the cells (Fig. 10). Whereas 85% of control cells responded to DA with an average −20 mV hyperpolarization, only 1 out of 18 PTX-treated cells examined responded with a change in membrane potential, and this response to DA was markedly reduced (Fig. 9).

The involvement of G proteins in the DA response was further revealed by experiments in which single cells were internally dialyzed with nonhydrolyzable guanine nucleotide analogues and challenged with agonist. Guanosine 5′-*O*-[2-thiodiphosphate] (GDPβS) is thought to block the actions of G proteins by inhibiting the binding of GTP necessary for G protein activation. Lactotrophs internally dialyzed with GDPβS by including it in the patch electrode during whole-cell recording were substantially less responsive to DA when compared with controls (e.g., Fig. 10). This reduction in responsiveness was dose dependent and reflected both a reduction in the number of cells responding and in the amplitude of hyperpolarization (Fig. 9, middle and bottom).

Guanosine 5′-*O*-[3-thiotriphosphate] (GTPγS) is thought to activate G proteins by displacing GTP from its binding site on the α subunit and thus preventing the

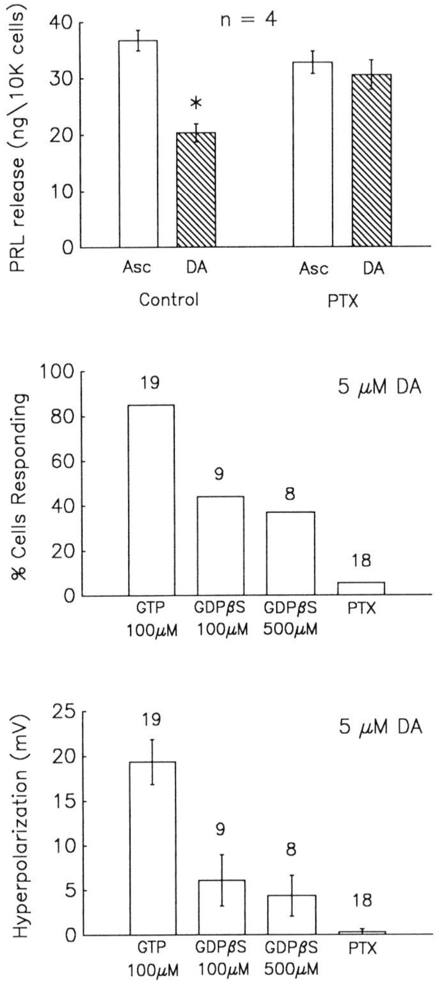

Figure 9. Sensitivity of DA effects to the block of G protein activation. (*Upper panel*) PRL release from cultured pituitary cells during 1 h incubation in basic external solution (+0.1% BSA) with or without DA (1 μM). The two bars on the right are data from cells preincubated for 8 h with 1 μM PTX. Data were derived from four independent experiments, each comprising triplicate wells per treatment group and are expressed as means ± SEM. Values indicated by an asterisk differ significantly from ascorbate control group at $P < 0.001$. (*Middle and lower panels*). Summary of whole-cell current clamp experiments presenting the percentage of $RHPA^+$ proestrous lactotrophs responding to 5 μM DA and the peak hyperpolarization of those responses (means ± SEM). Numbers above bars indicate the number of different cells tested in that group. Control cells had 100 μM GTP in the internal solution. Other cells had 100 or 500 μM GDPβS added to their internal solution. Cells pretreated with PTX (1 μM, 8 h) were examined with 100 μM GTP in the patch electrode solution.

nucleotide hydrolysis that normally terminates G protein activation. In the absence of DA, intracellular dialysis with GTPγS (50 μM) resulted in a slowly developing membrane hyperpolarization that was maximal by 7–10 min after establishing the whole-cell recording configuration (Fig. 10). The GTPγS-induced potential change was comparable to that produced by DA in both amplitude (20 mV) and associated increase in membrane conductance. Furthermore, application of DA to a GTPγS hyperpolarized cell resulted in little to no additional change in membrane potential. In some experiments in GTPγS-dialyzed cells at early times after establishment of the whole-cell recording mode, before a GTPγS effect developed, subsequent DA applications produced rapid and *irreversible* hyperpolarizations (data not shown). While multiple reversible agonist induced hyperpolarizations were observed in control GTP dialyzed cells, a single 1-s application of DA to GTPγS-dialyzed cells was often followed by a hyperpolarization that did not recover during the 20–30 min of the recording session.

In control experiments in which cells were dialyzed with the hydrolyzable GTP, a spontaneous hyperpolarizing change in membrane potential in the absence of DA was never observed, and the potential remained stable for many tens of minutes. In addition, we observed in control cells that the exclusion of GTP from the internal solution did not result in any notable instability of the membrane potential and large DA responses could be routinely elicited. This suggests that despite extensive intracel-

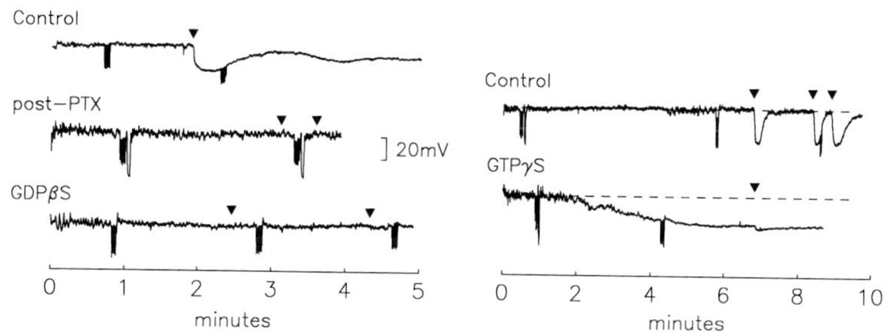

Figure 10. Effects of pertussis toxin and guanine nucleotide analogues on the membrane potential responses of lactotrophs to DA. All voltage recordings are from different single lactotrophs during whole-cell dialysis in the current-clamp mode. Brief DA (100 mM) applications are indicated by the inverted triangles. Occasional brief downward interruptions of the traces represent the responses to a series of −5 pA current injections to assess membrane resistance. (*Left panel*) Recordings from three cells representing a typical hyperpolarizing DA response in a control cell dialyzed with 100 μM GTP alone (upper trace), and the absence of a DA responses in a cell pretreated with PTX and dialyzed with GTP (middle trace), or a cell dialyzed with 500 μM GDPβS. (*Right panel*) Three consecutive DA responses were elicited in a cell dialyzed with GTP alone (upper trace). In another cell from the same coverslip, dialysis with GTPγS (50 μM) resulted in a slow, spontaneous hyperpolarization to a level similar to that seen with DA (lower trace). Subsequent application of DA to this cell resulted in a very small additional hyperpolarization to the identical voltage level seen in the previous cell.

lular dialysis adequate levels of endogenous GTP are maintained during whole-cell recording in pituitary cells as has been reported previously for cardiac muscle (Breitwieser and Szabo, 1988).

Discussion

We have examined the actions of DA in normal rat lactotrophs positively identified using the reverse hemolytic plaque assay. DA, at concentrations that inhibit PRL secretion, alters the electrical activity of normal rat lactotrophs by eliciting a rapid, sustained hyperpolarization of the membrane potential which, in spontaneously active cells, is accompanied by cessation of Ca^{2+}-dependent spiking activity. DA-induced hyperpolarization has previously been reported in human PRL-secreting adenoma cells

(Israel et al., 1985) and in bovine and rat pituitary cell populations enriched for PRL cells (Ingram et al., 1986; Israel et al., 1987).

An increased potassium conductance underlies the hyperpolarization as evidenced by the findings that (*a*) a decrease in membrane resistance accompanies the hyperpolarization, (*b*) the reversal potential for the DA effect on current or voltage was coincident with the calculated E_K, (*c*) the reversal potential was dependent on external $[K^+]$, and (*d*) single channels that conduct K^+ ions were activated by DA agonists binding to D_2 receptors. As regards the latter observation, DA does not appear to directly activate K channels as channels in the patch of membrane isolated by the recording electrode were activated by DA binding to physically remote receptors. This finding implicates a second-messenger signal transduction mechanism interposed between the D_2 receptors and the channels.

The D_2 receptors present on PRL-secreting cells have been shown to mediate the dopaminergic inhibition of PRL secretion (Caron, 1978; Enjalbert and Bockaert, 1983) and are both physically and functionally coupled to a G protein (DeLean, 1982; Kilpatrick, 1983; George, 1985; Senogles et al., 1987). This receptor-G protein coupling appears to play a necessary role in the dopaminergic inhibition of secretion (Fig. 9; Cronin, 1983; Enjalbert et al., 1988; Schofield et al., 1988). Our electrophysiological data provide strong evidence to support the involvement of an analogous stimulus transduction mechanism for a DA-activated and regulated K channel.

PTX treatment leads to the selective ADP-ribosylation of the G protein classes, G_o and G_i, and thus blocks the ability of these proteins to couple to receptors and dissociate into their active subunits (see Birnbaumer et al., 1989). PTX treatment of rat lactotrophs not only blocked DA inhibition of PRL secretion, but also blocked DA-induced electrical responses (Figs. 9 and 10), which is consistent with a role for a common PTX-sensitive G protein in the two phenomena. The inhibition of DA responses in GDPβS-dialyzed cells and the mimicry of the DA response by intracellular GTPγS (Fig. 10) provide additional support for the involvement of a G protein. Furthermore, the observation that DA responses and GTPγS responses were not additive suggests that GTPγS bypasses the D_2 receptor to activate the same G protein(s) in the receptor-channel signal transduction pathway.

Until recently guanine nucleotide binding proteins were thought to be primarily involved in the regulation of traditional second messenger systems such as those involving receptor-activated cyclic nucleotide generation or phospholipid metabolism. Recently it has become evident that G proteins are intimately involved in the regulation of several ion channels (see for example Breitweiser and Szabo, 1985; Pfaffinger et al., 1985; Holz et al., 1986; Lewis et al., 1986; Hescheler et al., 1987; Ewald et al., 1989). In fact, G protein regulation of K channels in clonal pituitary cells has been reported (Yatani et al., 1987; Codina et al., 1987) and DA regulation of K conductance in substantia nigra and *Aplysia* neurons appears to be linked to G proteins (Sasaki and Sato, 1987; Lacey et al., 1988). Although not directly demonstrated in many cases, several investigators have provided evidence that G proteins are "directly" coupled to K channels in heart and endocrine cells (for reviews see Birnbaumer et al, 1989; Clapham et al., 1989).

While we have not yet directly determined the mechanism of coupling between the putative DA-stimulated G protein and the lactotroph K channel, several of our findings impact on this issue. First, we observed that DA, when applied to extrapatch membrane, was able to initiate the opening of K channels in an on-cell membrane patch. Data of this nature are often used to support the involvement of a soluble and diffusible cytosolic messenger in the observed response. We argue that such a finding does not preclude the hypothesis that agonist stimulation of the D_2 receptor leads to activation of a G protein that directly couples to K channels. Because both $G_o\alpha$ and $G_i\alpha$ subunits have been noted to behave as soluble monomers, and because bath application of purified α subunits can potently activate K channels in inside-out membrane patches, some investigators have speculated that the site of action of these subunits may not be solely confined to the plasma membrane (Rodbell, 1985; Sternweis, 1986). It is possible that our observation reflects the soluble and perhaps diffusible nature of the signal transducer subunit.

Secondly, in all of our experiments the level of intracellular cAMP was maintained at 2 mM. The dopaminergic regulation of PRL secretion in rat lactotrophs is known to be coincident with a PTX-sensitive inhibition of adenylate cyclase activity, a characteristic response of D_2 receptor activation involving G_i-type proteins (Cronin et al., 1983; Enjalbert and Bockaert, 1983). Since DA remained capable of eliciting robust hyperpolarizing responses in cells in which the cAMP was effectively "clamped" at 2 mM, one can conclude that the electrical response is independent of the D_2 regulation of adenylate cyclase and corresponding cAMP levels. It is interesting to note that while there is general agreement that D_2 receptor stimulation is linked to inhibition of cAMP, the involvement of this phenomenon in the regulation of PRL secretion is controversial. Correlative studies suggest a relationship between the two events, however, investigators have also demonstrated that inhibition of cAMP is not required for the inhibition of PRL secretion (Swennen and Denef, 1982; Ray and Wallis, 1982; Enjalbert and Bockaert, 1983; McDonald et al., 1984; Delbeke et al., 1986; Lafond et al., 1986; Ray et al., 1986).

Is the DA-induced Activation of K Channels Involved in the Inhibition of PRL Secretion?

While it is not yet possible to provide direct experimental evidence for a causal link between the DA-induced changes in K channel activity and the inhibition of PRL secretion, out observations strongly support a role for K conductance in the process. The correlations between electrical and secretory responses to DA are striking. Both inhibitory responses are (*a*) independent of the intracellular levels of cAMP, (*b*) blocked by pretreatment with PTX, which inhibits certain G proteins, (*c*) pharmacologically associated with D_2 DA receptor action, and (*d*) are associated with decreases in intracellular Ca^{2+} (Margaroli et al., 1987; Mason et al., 1989). A decrease in $[Ca^{2+}]_i$ is likely related to the suppression of spontaneous spiking activity seen in these cells during exposure to DA. The mechanism underlying the suppression of spontaneous activity is not known, however the hyperpolarizations we observe are sufficiently large and occur over a voltage range that would be expected to depress the activity of voltage-dependent Ca^{2+} channels in the lactotrophs (Lingle et al., 1986; DeReimer and Sakmann, 1987). This does not obviate a direct or voltage-independent effect of DA on Ca^{2+} influx. However, Israel et al. (1985) reported that the inhibition of action potential discharge during DA-induced hyperpolarization in prolactinoma cells could be overcome by injection of depolarizing currents.

Thus the most parsimonious mechanism linking D_2 receptor activity to inhibition of PRL secretion would involve G protein activation of a K channel, subsequent membrane hyperpolarization to voltage levels where the probability of Ca^{2+} channel opening would be dramatically reduced, and a consequential decrease in Ca^{2+} influx and intracellular [Ca^{2+}] leading to a decline in PRL secretion. Future experiments will more clearly define the nature of the G proteins, K channels, and their functional coupling to one another.

Acknowledgments

We thank Drs. Marc Caron and Susan Senogles for generously providing dopamine agonists and antagonists for use in these studies and for helpful discussions of the observations.
This work was supported by National Institutes of Health grants NS-18788 (G. S. Oxford) and DK-40336 (K. A. Gregerson).

References

Adler M., B. S. Wong, S. L. Sabol, N. Busis, M. B. Jackson, and F. F. Weight. 1983. Action potentials and membrane ion channels in clonal anterior pituitary cells. *Proceedings of the National Academy of Sciences*. 80:2086–2090.

Ben-Jonathan, N., C. Oliver, H. J. Weiner, R. S. Mical, and J. C. Porter. 1977. Dopamine in hypophyseal portal plasma of the rat during the estrus cycle and throughout pregnancy. *Endocrinology*. 100:452–458.

Biales B., M. A. Dichter, and A. Tischler. 1977. Sodium and calcium action potentials in pituitary cells. *Nature*. 267:172–174.

Birnbaumer, L., A. M. J. Van Dongen, J. Codina, A. Yatani, R. Mattera, R. Graf, and A. M. Brown. 1989. Identification of G protein-gated and G protein-modulated ionic channels. Molecular basis for G protein action. *In* Secretion and Its Control. G. S. Oxford and C. M. Armstrong, editors. Rockefeller University Press, New York.

Bokoch, G. M., T. Katada, J. Northup, E. Hewlett, and A. Gilman. 1983. Identification of the predominant substrate for ADP-ribosylation by islet activating protein. *Journal of Biological Chemistry*. 258:2072–2075.

Breitwieser, G. E., and G. Szabo. 1985. Uncoupling of cardiac muscarinic and adrenergic receptors from ion channels by a guanine nucleotide analogue. *Nature*. 317:538–540.

Breitwieser, G. E., and G. Szabo. 1988. Mechanism of muscarinic receptor-induced K^+ channel activation as revealed by hydrolysis-resistent GTP analogues. *Journal of General Physiology*. 91:469–493.

Caron, M. G., M. Beaulieu, V. Raymond, B. Gagne, J. Drouin, R. Lefkowitz, and F. Labrie. 1978. Dopaminergic receptors in the anterior pituitary gland. *Journal of Biological Chemistry*. 253:2244–2253.

Clapham, D. E., and D. Kim. 1989. G protein activation mechanisms of the cardiac K^+ channel, $I_{K.Ach}$. *In* Secretion and Its Control. G. S. Oxford and C. M. Armstrong, editors. Rockefeller University Press, New York.

Codina, J., D. Grenet, A. Yatani, L. Birnbaumer, and A. Brown. 1987. Hormonal regulation of pituitary GH_3 cell K^+ channels by G_k is mediated by its α-subunit. *FEBS Letters*. 216:104–106.

Cronin, M. J., N. Faure, J. A. Martial, and R. I. Weiner. 1980. Absence of high affinity dopamine receptors in GH_3 cells: a prolactin secreting clone resistant to the inhibitory action of dopamine. *Endocrinology*. 106:718–723.

Cronin, M. J., G. A. Myers, R. M. Macleod, and E. L. Hewlett. 1983. Pertussis toxin uncouples dopamine agonist inhibition of prolactin release. *American Journal of Physiology.* 244:E499–E504.

Delbeke, D., J. G. Scammell, A. Martinez-Campos, and P. S. Dannies. 1986. Dopamine inhibits prolactin release when cyclic adenosine 3′5′-monophosphate levels are elevated. *Endocrinology.* 118:1271–1277.

DeLean, A., B. Kilpatrick, and M. Caron. 1982. Dopamine receptor of the porcine anterior pituitary gland. *Molecular Pharmacology.* 22:290–297.

DeRiemer, S. A., and B. Sakmann. 1987. Two calcium currents in normal rat anterior pituitary cells identified by a plaque assay. *Experimental Brain Research Supplementum.* 14:137–154.

Dubinsky, J. M., and G. S. Oxford. 1985. Dual modulation of K channels by thyrotropin-releasing hormone in clonal pituitary cells. *Proceedings of the National Academy of Sciences.* 82:4282–4286.

Enjalbert, A., and J. Bockaert. 1983. Pharmacological characterization of the D_2 dopamine receptor negatively coupled with adenylate cyclase in rat anterior pituitary. *Molecular Pharmacology.* 23:576–584.

Enjalbert, A., F. Mussett, C. Chenard, M. Priam, C. Kordon, and S. Heisler. 1988. Dopamine inhibits prolactin secretion stimulated by the calcium channel agonist Bay-K-8644 through a pertussis toxin-sensitive G protein in anterior pituitary cells. *Endocrinology.* 123:1:406–412.

Ewald, D., L.-H. Pang, P. Sternweis, and R. Miller. 1989. Differential G protein-mediated coupling of neurotransmitter receptors to Ca^{2+} channels in rat dorsal root ganglion neurons *in vitro. Neuron.* 2:1185–1193.

Faure, N., M. J. Cronin, J. A. Martial, and R. I. Weiner. 1980. Decreased responsiveness of GH_3 cells to the dopaminergic inhibition of prolactin. *Endocrinology.* 107:1022–1026.

Forscher, P., and G. S. Oxford. 1985. Modulation of calcium currents by norepinephrine in internally dialyzed avian sensory neurons. *Journal of General Physiology.* 85:743–763.

George, S. R., M. Watanabe, T. Di Paolo, P. Falardeau, F. Labrie, and P. Seeman. 1985. The functional state of the dopamine receptor in the anterior pituitary is in the high affinity form. *Endocrinology.* 117:690–697.

Gershengorn, M. C. 1986. Mechanism of thyrotropin releasing hormone stimulation of pituitary hormone secretion. *Annual Review of Physiology.* 48:515–526.

Gibbs, D. M., and J. D. Neill. 1978. Dopamine levels in hypophysial stalk blood in the rat are sufficient to inhibit prolactin secretion *in vivo. Endocrinology.* 102:1895–1900.

Hamill, O. P., A. Marty, E. Neher, B. Sakmann, and F. J. Sigworth. 1981. Improved patch clamp techniques for high-resolution current recording from cells and cell-free membrane patches. *Pflügers Archiv.* 391:85–100.

Heschler, J., W. Rosenthal, W. Trautwein, and G. Schultz. 1987. The GTP-binding protein, G_o regulates neuronal calcium channels. *Nature.* 325:445–447.

Holz, G. G., S. G. Rane, and K. Dunlap. 1986. GTP-binding proteins mediate transmitter inhibition of voltage-dependant calcium channels. *Nature.* 319:670–672.

Ingram, C. D., R. J. Bicknell, and W. T. Mason. 1986. Intracellular recordings from bovine anterior pituitary cells: modulation of spontaneous activity by regulators of prolactin secretion. *Endocrinology.* 119:2508–2518.

Israel, J. M., P. Jaquet, and J. D. Vincent, 1985. The electrical properties of isolated human prolactin-secreting adenoma cells and their modification by dopamine. *Endocrinology.* 117:1448–1455.

Israel, J. M., C. Kirk, and J. D. Vincent. 1987. Electrophysiological responses to dopamine of rat hypophysial cells in lactotroph-enriched primary cultures. *Journal of Physiology.* 390:1–22.

Kaczorowski, G. J., R. L. Vandlen, G. M. Katz, and J. P. Reuben. 1983. Regulation of excitation-secretion coupling by thyrotropin-releasing hormone (TRH): evidence for TRH receptor-ion channel coupling in cultured pituitary cells. *Journal of Membrane Biology.* 71:109–118.

Katada, T., J. K. Northup, G. M. Bokoch, M. Ui, and A. G. Gilman. 1984. The inhibitory guanine nucleotide-binding regulatory component of adenylate cyclase. *Journal of Biological Chemistry.* 259:3578–3585.

Kilpatrick, B. F., and M. G. Caron. 1983. Agonist binding promotes a guanine nucleotide reversible increase in the apparent size of the bovine anterior pituitary dopamine receptors. *Journal of Biological Chemistry.* 258:13528–13534.

Kidokoro, Y. 1975. Spontaneous calcium action potentials in a clonal pituitary cell line and their relationship to prolactin secretion. *Nature.* 258:741–742.

Lacey, M. G., N. B. Mercuri, and R. A. North. 1988. On the potassium conductance increase activated by $GABA_B$ and dopamine D_2 receptors in rat substantia nigra neurones. *Journal of Physiology.* 401:437–453.

Lafond, J., J. R. Ducharme, and R. Collu. 1986. Inhibition of prolactin release and blockade of adenophypophyseal cell cyclic AMP accumulation are two dissociable effects of dopaminergic and non-dopaminergic drugs. *Molecular and Cellular Endocrinology.* 44:219–225.

Lewis, D. L., F. F. Weight, and A. Luini. 1986. A guanine nucleotide-binding protein mediates the inhibition of voltage-dependent calcium current by somatostatin in a pituitary cell line. *Proceedings of the National Academy of Sciences.* 83:9035–9039.

Lingle, C. J., S. Sombati, and M. E. Freeman. 1986. Membrane currents in identified lactotrophs of rat anterior pituitary. *Journal of Neuroscience.* 6:2995–3005.

Margaroli, A., L. Vallar, F. R. Elahi, T. Pozzan, A. Spada, and J. Meldolesi. 1987. Dopamine inhibits cytosolic Ca^{2+} increases in rat lactotroph cells. Evidence of a dual mechanism of action. *Journal of Biological Chemistry.* 262:13920–13927.

Mason, W. T., S. K. Sikdar, R. Zorec, S. Akerman, S. Rawlings, T. Cheek, R. Moreton, and M. Berridge. 1989. Ion channels, intracellular calcium, and exocytosis: control of hormone secretion in cultured bovine pituitary lactotrophs. *In* Secretion and Its Control. G. S. Oxford and C. M. Armstrong, editors. Rockefeller University Press, New York.

McDonald, W. M., D. R. Sibley, B. F. Kilpatrick, M. G. Caron. 1984. Dopaminergic inhibition of adenylate cyclase correlates with high affinity agonist binding to anterior pituitary D_2 dopamine receptors. *Molecular and Cellular Endocrinology.* 36:201–209.

Murayama, T., and M. Ui. 1983. Loss of the inhibitory function of the guanine nucleotide regulatory component of adenylate cyclase due to its ADP ribosylation by islet-activating protein, pertussis toxin, in adipocyte membranes. *Journal of Biological Chemistry.* 258:3319–2236.

Neill, J. D., and L. S. Frawley. 1983. Detection of hormone release from individual cells in mixed populations using a reverse hemolytic plaque assay. *Endocrinology.* 113:1135–1137.

Pfaffinger, P. J., J. M. Martin, D. D. Hunter, N. M. Nathanson, and B. Hille. 1985. GTP-binding proteins couple cardiac muscarinic receptors to a K channel. *Nature*. 317:536–538.

Plotsky, P. M., D. M. Gibbs, and J. D. Neill. 1978. Liquid chromatographic-electrochemical measurements of dopamine in hypophysial stalk blood of rats. *Endocrinology*. 102:1887–1894.

Ray, K. P., J. J. Gomm, G. J. Law, C. Sigournay, and M. Wallis. 1986. Dopamine and somatostatin inhibit forskolin-stimulated prolactin and growth hormone secretion but not stimulated cyclic AMP levels in sheep anterior pituitary cell cultures. *Molecular and Cellular Endocrinology*. 45:175–182.

Ray, K. P., and M. Wallis. 1982. Actions of dopamine on prolactin secretion and cyclic AMP metabolism in ovine pituitary cells. *Molecular and Cellular Endocrinology*. 27:139–155.

Rodbell, M. 1985. Programmable messengers: a new theory of hormone action. *Trends in Biochemical Science*. 10:461–464.

Sand, O., E. Haug, and K. M. Gautvik. 1980. Effects of thyroliberin and 4-AP on action potentials and prolactin release and synthesis in rat pituitary cells in culture. *Acta Physiologica Scandinavica*. 108:247–252.

Sasaki, K., and M. Sato. 1987. A single GTP-binding protein regulates K^+-channels coupled with dopamine, histamine and acetylcholine receptors. *Nature*. 325:259–262.

Schofield, J. G., A. I. Khan, and A. Wood. 1988. Modification by pertussis toxin of the responses of bovine anterior pituitary cells to acetylcholine and dopamine: effects on hormone secretion and ^{86}Rb efflux *Journal of Endocrinology*. 116:393–401.

Senogles, S. E., J. L. Benovic, N. Amlaiky, C. Unson, G. Milligan, R. Vinitsky, A. M. Spiegel, and M. G. Caron. 1987. The D_2-dopamine receptor of anterior pituitary is functionally associated with a pertussis toxin-sensitive guanine nucleotide binding protein. *Journal of Biological Chemistry*. 262:4860–4867.

Smith, P. F., E. H. Luque, and J. D. Neill. 1986. Detection and measurement of secretion from individual neuroendocrine cells using a reverse hemolytic plaque assay. *Methods in Enzymology*. 124:443–465.

Sternweis, P. C. 1986. The purified α subunits of G_o and G_i from bovine brain require $\beta\gamma$ for associations with phospholipid vesicles. *Journal of Biological Chemistry*. 261:631–637.

Swennen, L., and C. Denef. 1982. Physiological concentrations of dopamine decrease adenosine 3′5′-monophosphate levels in cultured rat anterior pituitary cells and enriched populations of lactotrophs: evidence for a causal relationship to inhibition of prolactin release. *Endocrinology*. 111:398–405.

Taraskevich, P. S., and W. W. Douglas. 1977. Action potentials occur in cells of the normal anterior pituitary gland and are stimulated by the hypophysiotropic peptide thyrotropin-releasing hormone. *Proceedings of the National Academy of Sciences*. 74:4064–4067.

Weiner, R. I., C. L. Bethea, P. Jaquet, J. S. Ramsdell, and D. J. Gospodarowicz. 1983. Culture of dispersed anterior pituitary cells on extracellular matrix. *Methods in Enzymology*. 103:287–293.

Weiner, R. I., M. J. Cronin, C. Y. Cheung, L. Annunziato, N. Faure, and P. C. Goldsmith. 1979. Dopamine: a prolactin inhibitory hormone. *In* Neuroendocrine Correlates in Neurology and Psychiatry. E. E. Muller and A. Agnoli, editors. Elsevier, Amsterdam. 41–55.

Yatani, A., J. Codina, R. D. Sekura, L. Birnbaumer, and A. M. Brown. 1987. Reconstitution of somatostatin and muscarinic receptor mediated stimulation of K^+ channels by isolated G_K protein in clonal rat anterior pituitary cell membranes. *Molecular Endocrinology*. 1:384–389.

Chapter 9

Hormonal and Neurotransmitter Regulation of Ca Channel Activity in Cultured Adenohypophyseal Cells

Gabriel Cota and Marcia Hiriart

Department of Physiology, Biophysics, and Neurosciences, Centro de Investigacion y de Estudios Avanzados del IPN, AP 14-740, Mexico, DF 07000; and the Department of Neurosciences, Instituto de Fisiologia Celular, UNAM, AP 70-600, Mexico, DF 04510

Introduction

Like many other excitable cells, the endocrine cells of the mammalian adenohypophysis express voltage-dependent Ca channels in the plasma membrane (for reviews, see Armstrong, 1986; Ozawa and Sand, 1986). These channels are effective devices to inject Ca ions from the extracellular space into the cell. In this way the Ca channels contribute not only to the generation of the electrical activity of the adenohypophyseal cells (Kidokoro, 1975; Douglas and Taraskevich, 1980; Israel et al., 1983, 1987; Mason and Ingram, 1986), but also to raising the levels of an intracellular messenger of paramount relevance in the cell economy: Ca^{++} ions (Schlegel et al., 1987; Holl et al., 1988; Meier et al., 1988). Numerous cell functions are regulated by the cytosolic free Ca^{++} concentration, and in adenohypophyseal cells these include the exocytotic release of hormones (Moriarty, 1978; see also Tsuruta et al., 1982; Childs et al., 1987; Dave et al., 1987) and hormone synthesis as well (Gick and Bancroft, 1985; Loeffler et al., 1986; Dave et al., 1987; Enyeart et al., 1987).

Regulation of Ca channel activity by hormones and neurotransmitters is a widespread mechanism that controls cell function (reviewed by Tsien, 1987). It is then conceivable that Ca channels are targets for hormones and neurotransmitters that control hormone synthesis and secretion in adenohypophyseal cells. We have been investigating this possibility using mostly primary cultures of the pars intermedia of the rat hypophysis, an excellent preparation introduced to electrophysiologists by Douglas and Taraskevich (1978).

The endocrine cells of the pars intermedia (PI cells or melanotrophs) secrete several biologically active peptides of the melanotropin and endorphin families (Mains and Eipper, 1979). PI cells are electrically excitable, and display spontaneous action potentials (Davis and Hadley, 1976; Douglas and Taraskevich, 1978; Davis et al., 1985), that are mediated by an increase in the membrane permeability to Na^{+} and Ca^{++} ions (Douglas and Taraskevich, 1978, 1980). Furthermore, there is evidence that Ca channels, in addition to contributing to spikes, also participate in the intense subthreshold electrical activity of the PI cells that makes possible the triggering of spontaneous action potentials (Douglas and Taraskevich, 1980, 1985).

Multiple hormones and neurotransmitters, including catecholamines, control the secretory activity of PI cells in vivo (Meunier and Labrie, 1983; Cote et al., 1984*b*). PI cells are directly innervated by dopaminergic neurons whose cell bodies are located in the hypothalamus (Bjorklund and Lindvall, 1984; Holzbauer and Racke, 1985). Dopamine released from the nerve terminals mediates a powerful tonic inhibition on the secretory activity of the endocrine cells (for a review, see Holzbauer and Racke, 1985; see also Davis, 1986). On the other hand, the PI cell activity can be stimulated by adrenaline and noradrenaline from the systemic circulation (Berkenbosch et al., 1981, 1983).

The effects of catecholamines on PI cells are mediated by specific receptors located in the surface membrane of the endocrine cells. Dopamine binds to D-2 dopaminergic receptors. Acute stimulation of these receptors, during seconds or minutes, inhibits the intracellular levels of cyclic AMP (cAMP) (reviewed by Cote et al., 1984*b*). It also decreases the frequency of spontaneous action potentials (Douglas and Taraskevich, 1978, 1982; Davis et al., 1985), and inhibits hormone release (Munemura et al., 1980; Jackson and Lowry, 1983; Davis, 1986). Prolonging the stimulation of the dopamine receptors to hours or days inhibits both hormone synthesis

(Chen et al., 1983; Beaulieu et al., 1984; 1986; Cote et al., 1986), and the expression of the glucocorticoid receptor (Antakley et al., 1987), while decreasing the mitotic index (Chronwall et al., 1987). On the other hand, the stimulatory effects of adrenaline and noradrenaline are mediated by beta-2 adrenergic receptors (Cote et al., 1984*b*). Acute and chronic stimulation of the adrenergic receptors has opposite effects on adenylate cyclase activity, hormone synthesis, and secretion to those induced by stimulation of the dopaminergic receptors (Cote et al., 1980, 1986; Meunier and Labrie, 1982*b*; Tsuruta et al., 1982).

In this chapter, we present electrophysiological evidence for a dual regulation of Ca channel activity by the catecholaminergic receptors in cultured PI cells. This regulation involves both short- and long-lasting effects.

Methods

PI Cell Culture

PI cells were dissociated from the neurointermediate (NI) lobe of the rat pituitary gland and maintained for 1–20 d in primary culture. Briefly, pituitary NI lobes were removed from male Sprague-Dawley or Wistar rats weighing 250–300 g, and then incubated with constant agitation in Ca^{++}-free, Mg^{++}-containing Spinner salt solution (SSS) (Sigma Chemical Co., St. Louis, MO), supplemented with 10 mM HEPES (pH 7.30), 100 U/ml penicillin, 100 μg/ml streptomycin, 0.1% bovine serum albumin (BSA), and 0.3% collagenase (CLS II; Cooper Biochemical Products, Freehold, NJ; 139 U/mg), for 40 min at 21°C or for 20 min at 37°C. PI cells were then released from the fragments of NI lobes by gentle trituration with siliconized flame-polished Pasteur pipettes, collected by centrifugation, washed once in SSS supplemented with BSA, and resuspended in the medium used for culturing (see below). Finally, the cells were plated on poly-*l*-lysine–coated slivers of coverslips in 35-mm plastic petri dishes (one NI lobe equivalent per culture dish; about 200,000 cells per lobe) and cultured in a humidified atmosphere of 95% air-5% CO_2 at 37°C. The culture medium was either Kennett's HY medium (Cell Center, University of Pennsylvania, Philadelphia, PA) or RPMI 1640 medium (Sigma Chemical Co.) containing 5% fetal bovine serum (Gibco Laboratories, Grand Island, NY), 1% L-glutamine, and antibiotics. The culture medium was replaced after 1 d and thereafter every second day. Neither collagenase nor antibiotics had a measurable effect on the Ca channel activity of PI cells.

Recording Technique

The Ca channel activity of cultured, isolated pituitary cells was recorded at 20–21°C using the whole-cell variation of the patch-clamp technique (Hamill et al., 1981). The compositions of the external//internal recording solutions are given in the figure legends, with concentrations in millimolar units unless otherwise noted. In all cases, the recording solutions also contained 10 mM HEPES and their pH was adjusted to 7.30.

Current responses were sampled at 14- or 20-μs intervals, and the linear components of current were subtracted either with a P/4 procedure or using a scaled current response to 20–50 mV hyperpolarizing pulses. Tail currents associated with Ca channel closing were recorded at -80 mV. Useful current measurements started 50–100 μs after initiation of the voltage step by using low-resistance electrodes (see below) and capacity transient compensation. Tail currents were fit by the sum of two exponentials as described in Matteson and Armstrong (1986) and Cota (1986). The amplitude of the fast and slow components in the tail current was taken as the amplitude of the fitted exponentials measured 100 μs after the onset of repolarization. Patch electrodes were fabricated from borosilicate glass capillaries (KIMAX-51; Kimble Div., Owens-Illinois, Inc., Toledo, OH) and had a resistance close to 1 MΩ after fire-polishing the pipette tip to a bullet shape. With these electrodes, the access resistance was usually 1.5–2.5 MΩ.

Cell capicitance was estimated from the increase in the capacitive current, generated by a 20-mV hyperpolarizing voltage-clamp step, that followed the rupture of the membrane patch spanning the pipette opening. Cells selected for recording had nearly a spherical shape. Cell capacitance ranged from 4 to 13 pF for cells 11–20 μm in diameter.

Drug Application

Most of our analysis of the effects of the dopaminergic or adrenergic stimulation on Ca channel activity involved the pretreatment of the pituitary cells with catecholaminergic receptor agonists. To test the "acute" effect of these agonists, a coverslip with cultured cells were transferred to a petri dish containing 2 ml of culture medium supplemented with dopamine, bromocriptine, or isoproterenol at a concentration of 1 μM, and incubated for 10 min at 37°C. The coverslip was then transferred to a new petri dish containing 1 ml of the external recording solution. Records of Ca channel currents were taken within 3–10 min after immersion of the cells in the external recording solution in the continuous presence of the agonist. The "chronic" effect of bromocriptine or isoproterenol was assayed by pretreating the cells with the agonist for 1 h or longer at 37°C. The agonist was then removed from the culture medium and after a recovery interval of 30–60 min at 37°C the Ca channel activity was recorded in the absence of the agonist. In some experiments we applied dopamine in the vicinity of a cell that had been already internally dialyzed during some minutes with the solution contained in the patch electrode. In those cases dopamine was dissolved in the external recording solution at a high concentration (usually 20 μM), and delivered from a glass pipette using a pressure microinjection system (Picospritzer II; General Valve Corp., Fairfield, NJ).

Results

Cultured PI Cells Express Two Types of Voltage-dependent Ca Channels

SD and FD Ca channels. A variety of voltage-dependent ionic channels are present in the plasma membrane of cultured PI cells, including Na channels, K channels and Ca channels (Cota, 1986; Cota and Armstrong, 1988). Ca currents arise from the activity of two distinct types of Ca channels (Cota, 1986), named SD and FD channels, following the nomenclature proposed by Armstrong and Matteson (1985) in their study of the Ca channels in clonal pituitary GH_3 cells.

The functional activity of Ca channels in a cultured PI cell is shown in Fig. 1 under conditions that eliminate Na and K currents. The traces are a family of currents carried by Ba^{++} ions, obtained by pulsing to various membrane potentials. The SD and FD Ca channels differ between each other in several of their functional properties, and the traces in Fig. 1 illustrate two of those differences: the different apparent activation threshold and the different deactivation (closing) kinetics. At −20 mV or lower depolarizations, the Ba current flows almost exclusively through SD channels. These channels close with a relatively slow time course when the membrane is repolarized to the holding potential (−80 mV). Positive to −20 mV, an additional set of Ca channels activates, the FD channels. FD channels close very fast at −80 mV, and that explains the fast component in the Ba^{++} tail after large depolarizations. Closing of either FD or SD channels develops with a time course that is well fit by a single exponential. With Ba^{++} as the permeant ion the FD channels close 15–20 times faster than the SD channels at −80 mV.

The amplitude of the fast and slow components in the tail current is directly proportional to the number of open FD and SD channels at the end of the activating pulse. Since there is a considerable difference in the closing time course of the two

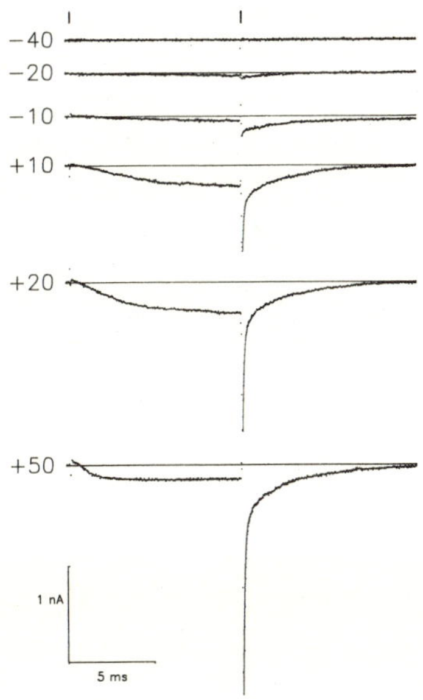

Figure 1. SD and FD Ca channel activity of a cultured PI cell. The traces are Ba currents through voltage-dependent Ca channels. They were induced by 10-ms step depolarizations to various voltage levels from a holding potential of −80 mV, followed by repolarization to −80 mV. Numbers next to each trace correspond to the membrane potential (millivolts) during the steps. Vertical bars on top of the first trace indicate the onset of depolarization and repolarization. The cell was kept in culture for 4 d before this experiment, and the traces were taken within 1.5–2.0 min after break-in with the patch electrode. The composition of the recording solution was (in millimolar) 125 tetraethylammonium chloride (TEA-Cl), 20 $BaCl_2$//30 *N*-methylglucamine (NMG) glutamate, 80 Cs-glutamate, 20 CsCl, 10 EGTA. Experiment Jn1088.

types of Ca channels, analysis of tail currents is a convenient way to separate the contribution of FD and SD channels to the total Ca channel activity. This separation procedure is used in the following sections to determine other functional properties of the SD and FD channels.

Voltage and time dependence of Ca channel conductance. Conductance-voltage curves for the two types of Ca channels, as determined from Ba^{++} tail amplitudes, are given in Fig. 2 *A*. It is clear that FD and SD channels activate over different voltage ranges. Depolarizations to +50 mV, or more positive voltages, activate a maximum number of FD and SD channels in a cell. The half-maximal activation level for SD channels is close to 0 mV, ~20 mV less positive than that for FD channels. Time constants of the fast and slow components in the tail current are not dependent on the extent of Ca channel activation (Fig 2 *B*).

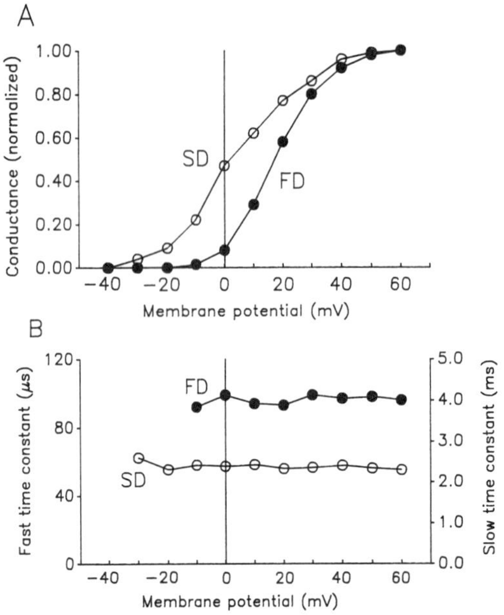

Figure 2. (*A*) Voltage dependence of FD and SD channel activation. Ba tails were recorded at −80 mV after 10-ms activating pulses of various amplitudes. The tail currents were then fit by the sum of two exponentials, and the normalized amplitudes of the fast (FD) and slow (SD) components were plotted as a function of the activating potential. (*B*) Closing kinetics of FD and SD channels do not depend on the amplitude of the activating pulse. Time constants of the fast and slow components in the tail current are plotted as a function of the membrane potential during the activating pulse. Same cell as in Fig. 1.

At −20 mV or smaller depolarizations, current as a function of time mostly reflects the time course of SD channel conductance. The time course of the current during large depolarizations, however, is difficult to interpret because of the significant activation of both channel types, SD and FD. The time course of SD and FD channel conductance at these membrane potentials can be assayed by recording tail currents after activating pulses of various durations. The procedure for activating steps to +20 mV is illustrated in Fig. 3. From visual inspection of the tail currents, it seems that FD channels are almost fully activated after the 1.9-ms pulse and that a substantial fraction of these channels are still open after 100 ms. On the contrary, the SD channels

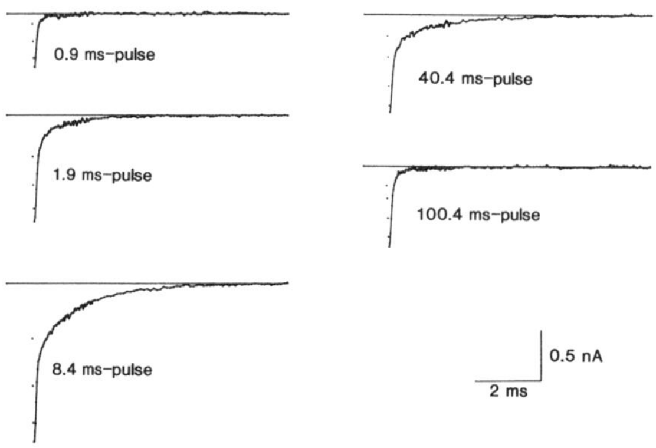

Figure 3. Changes in Ba tail currents as FD and SD channels activate and then inactivate. Tail currents were recorded on returning to −80 mV after activating pulses to +20 mV, for the durations indicated. The time constants for the two tail components, fast and slow, are as follows (pulse durations are given in milliseconds between parentheses): 115 μs, 1.56 ms (0.9); 125 μs, 1.45 ms (1.9); 121 μs, 1.41 ms (8.4); 113 μs, 1.68 ms (40.4); 117 μs, 1.71 ms (100.4). Age of the culture (*T*) = 8 d; time (*t*) after break-in with the patch electrode was 4–5 min. The composition of the recording solution was (in millimolar) 140 NaCl, 10 $BaCl_2$, 1 μM TTX//30 NMG-glutamate, 60 Cs-glutamate, 20 CsF, 20 CsCl, 2 $MgCl_2$, 10 EGTA. Experiment J10385.

activate with a slower time course and then inactivate quickly and completely during the 100-ms pulse.

Changing the holding potential from -80 to -40 mV reduces the number of available SD channels to a larger extent than it does in FD channels, as shown in Fig. 4. The change in holding potential decreases the amplitude of the fast component in the Ba^{++} tail by 35%, and that of the slow component by 80%. The time course and voltage dependence of Ca channel inactivation in PI cells has been explored in detail by Stanley and Russell (1988).

Prolonged intracellular dialysis alters the Ca channel activity. FD and SD channels also differ in their stability under whole-cell mode. Prolonged dialysis induces a loss of functional FD channels and a left voltage shift in the FD conductance-voltage curve. The activity of SD channels is much more resistant to intracellular dialysis (Cota, 1986). Supplementing the pipette solution with 3 mM Mg-ATP helps to

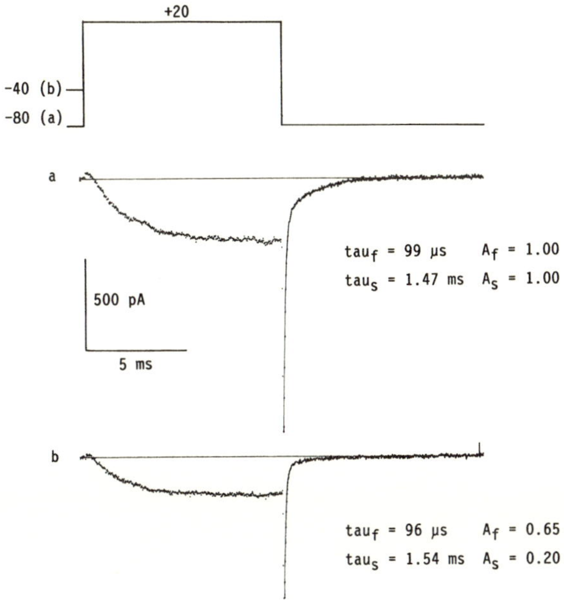

Figure 4. Partial inactivation of FD and SD channels by changing the holding potential from -80 to -40 mV. Trace *a* is the Ba current induced by a step to $+20$ mV from a holding potential of -80 mV, as indicated by the pulse protocol diagrammed on top. The holding potential was then changed to -40 mV, and 15 s later pulsing to $+20$ mV induced the current in *b*. Time constants and relative amplitudes of the fast (f) and slow (s) components in the tail current are given next to each trace. T = 11 d, t = 2.0–2.5 min. Recording solutions as in Fig. 1. Experiment Jn1788.

stabilize the FD channel activity. However, Mg-ATP also reduces the amplitude of the FD current, an effect that is probably explained by a blocking action of internal Mg^{++} ions on the Ca channels (see Byerly and Yazejian, 1986).

Acute Stimulation of the Catecholaminergic Receptors Modulates the Activity of FD Channels

Experiments using a pressure ejection system. To study the effect of acute stimulation of the D-2 dopamine receptors on the Ca channel activity of cultured PI cells, we have used two different approaches. Initially, we applied dopamine from a pipette connected to a pneumatic picopump. The tip of the delivery pipette was located close to the cell under study, usually one cell diameter away from the cell surface. Neither Na currents nor the fast component in Ca tail currents were significantly affected by pulses of dopamine, but we did observe a reduction of the amplitude of the slow component in the Ca tail in most cells investigated. This effect was reversible, and complete recovery took ~1 min. This result would suggest that dopamine inhibits SD channel activity. However, the reduction of the slow Ca tail turned out to be artifactual: it was also observed in control experiments, by pressure-ejecting external solution with no dopamine added.

In most of the cells we did not observe any significant change in the activity of the FD channels in response to dopamine. In some cells (~20% of the total), however, the exposure to dopamine induced a clear, transient reduction of the fast component in the Ca tail. The maximum reduction ranged from 20 to 60% of the control value and was attained 20–30 s after the pulse of dopamine. The FD current usually recovered within 1–2 min. A reversible reduction of the FD current amplitude was never observed in control experiments. This indicates that the acute stimulation of the dopamine

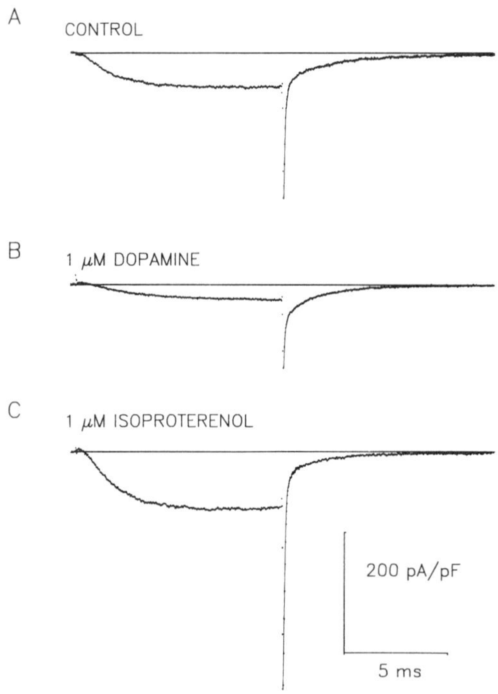

Figure 5. Dual modulation of Ca channel activity in PI cells by acute stimulation of catecholaminergic receptors. The traces are representative Ba currents obtained from (*A*) control cells and (*B*) cells preincubated for ~20 min with 1 μM dopamine or (*C*) 1 μM isoproterenol. Currents were induced by 10-ms test pulses to +20 mV followed by repolarization to the holding potential (−80 mV). The catecholaminergic agonists were continuously present in the bath solution during the electrophysiological recordings. Current amplitude has been normalized by cell capacitance (C_m). C_m was (*A*) 7.0 pF, (*B*) 9.5 pF, and (*C*) 8.5 pF. In every cell, the records were taken within 2 min after break-in with the patch electrode. T = 11–15 d. Recording solutions were as in Fig. 1. Experiments Jn1788 (*A* and *C*) and Jn2088 (*B*).

receptors can inhibit the FD channel activity in some PI cells and that this effect is rapidly reversible.

Dual modulation of FD channels by catecholaminergic receptors. A possible explanation for the variability in the response of FD channels to dopamine in the PI cells is that the intracellular dialysis of these relatively small cells quickly disrupts the coupling of the dopaminergic receptors with the Ca channels. To explore this possibility as well as to avoid the use of pressure ejection we decided to treat the cells with dopamine agonists before the electrophysiological recording. Effects of acute stimulation of beta-2 adrenergic receptors on Ca channel activity were assayed in a similar way.

Fig. 5 presents representative traces of Ba currents obtained at +20 mV from three different sets of cells: (*A*) control cells and (*B*) cells pretreated for 20 min with 1 μM dopamine or (*C*) 1 μM isoproterenol. It seems clear that dopamine reduces and isoproterenol increases the current amplitude through Ca channels, and that these effects are due to an almost selective change in the activity of the FD channels.

Average values for the amplitude of the fast and slow components in the Ba^{++} tail after large activating pulses to +50 or +60 mV are shown in Fig. 6. Dopamine reduces by ~55% the amplitude of the fast component, but does not significantly affect the amplitude of the slow component. Bromocriptine, a specific and potent D-2 dopaminergic agonist, mimics the effect of dopamine. On the other hand, isoproterenol

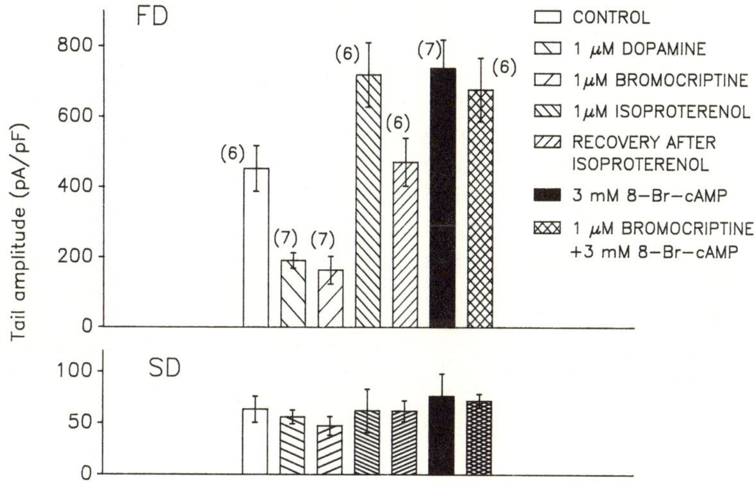

Figure 6. Modulation of Ca channel activity in PI cells by catecholaminergic receptor agonists and 8-Br-cAMP. Ba tails were recorded at −80 mV after 10-ms activating pulses to +50 or +60 mV, from control cells and cells pretreated for ~20 min with dopamine, bromocriptine, isoproterenol, 8-Br-cAMP, or bromocriptine and 8-Br-cAMP, at the concentrations indicated. Ba tails were fit by the sum of two exponentials. The bars indicate mean ± SEM (with the number of cells in parentheses) of the amplitudes of the fast (FD)and slow (SD) components in the tail current, after normalizing the individual traces by cell capacitance. The procedure to obtain the data indicated as "recovery after isoproterenol" was as follows: cells were pretreated for 20 min with 1 μM isoproterenol and the beta-adrenergic agonist was then removed from the bath solution; records were taken within 5–10 min after the end of the beta-adrenergic stimulation. T = 11–15 d. Recording conditions were as in Fig. 5. Experiments Jn1788, Jn2088, and Jn2188.

increases selectively the amplitude of the FD current by ~60%. These changes in FD channel activity are rapidly reversible. For example, Fig. 6 also shows that the channel activity returns to the control level within 5–10 min after the end of the adrenergic stimulation.

Neither dopamine agonists nor isoproterenol induced obvious changes in the closing kinetics of Ca channels. However, the acute stimulation of the catecholaminergic receptors modified the position along the voltage axis of the conductance-voltage curve for FD channels. With isoproterenol the curve was shifted left by ~10 mV, while with dopamine agonists the curve was shifted to the right by ~5 mV.

Effect of external 8-Br-cAMP on Ca channel activity. Stimulating the D-2 dopamine receptors in PI cells inhibits the activity of the adenylate cyclase through the inhibitory GTP-binding protein (G_i) (Cote et al., 1982*b*, 1984*a*; Meunier and Labrie, 1982*a*). Just the opposite effect is obtained by the stimulation of the beta-2 adrenergic receptors, an effect mediated by the stimulatory GTP-binding protein (G_s) (Cote et al., 1982*a*; Meunier and Labrie, 1982*b*). Thus, a relevant question is if the modulation of FD channel activity by catecholaminergic receptors occurs as a consequence of changes in the intracellular levels of cAMP or if it is mediated by a more direct coupling of the receptors with the Ca channels (i.e., via a direct interaction of the G proteins with the channels). As a first attempt to address this question, we have studied the effect of 8-Br-cAMP, a permeable analogue of cAMP, on Ca channel activity when it is added to the external medium.

Pretreatment of PI cells during 20 min with a high 8-Br-cAMP concentration (3 mM) increases by 60%, on average, the FD channel activity, without affecting in a significant way the SD channel activity, as shown in Fig. 6. Furthermore, 8-Br-cAMP shifted the activation curve for FD channels 5–10 mV to the left. Moreover, bromocriptine does not significantly reduce the Ca channel activity stimulated by 8-Br-cAMP (Fig. 6). Thus, 8-Br-cAMP mimics the effect of acute stimulation of the adrenergic receptors and prevents the dopaminergic inhibition of Ca channel activity.

Chronic Stimulation of Catecholaminergic Receptors Induces Long-lasting Changes in Ca Channel Activity

Development of Ca channel activity with time in culture. Our interest in studying the effect of chronic stimulation of the dopamine receptors on Ca channel activity arose with the observation that the amplitude of the current carried by Ca channels in cultured PI cells increases markedly with the age of the culture (Cota, 1986; Cota and Armstrong, 1987). The time course over which the Ca channel activity develops in culture is shown in Fig. 7. The amplitude of the FD current in cells cultured during 9 d or longer is 3–4 times larger than after 1 d in culture. The FD current is at 50% of its maximum by day 3. The SD current also grows with time in culture, but it develops with a slower time course and the maximum increase in current size is less dramatic. The average diameter (14–16 μm) of PI cells does not change with time in culture. Furthermore, there are no significant changes in cell capicitance associated with the increase in Ca channel activity.

The functional properties of Ca channels do not depend on the age of the culture. Closing kinetics of FD and SD channels remained practically unchanged with time in culture. This was also true for other properties of the Ca channels, including the voltage and time dependence of channel conductance, the Ba/Ca conductivity, the acute effects of dopamine agonists, and the sensitivity to dihydropyridines (see below). Thus, the increase of the amplitude of the current carried by Ca channels with time in culture may be due to the increase in the surface density of Ca channels in the plasma membrane. In fact, the development of Ca channel activity seems to require the continual synthesis of proteins. Actinomycin D (40 μM), an inhibitor of mRNA synthesis, added to the culture medium on day 1 prevents the increase of Ca channel activity usually observed on day 3, without modifying the macroscopic kinetic properties of the channels.

Chronic stimulation of catecholaminergic receptors regulates the development of Ca channel activity. The biosynthetic and secretory activity of PI cells in vivo is tonically inhibited by dopaminergic neurons (see Introduction). It is therefore possible that dopamine also mediates a tonic inhibition on Ca channel expression in PI cells. When the endocrine cells are denervated and placed in culture the dopaminergic inhibition is lost and this probably accelerates Ca channel production. To test this possibility we have studied the effect of chronic stimulation of the dopaminergic receptors on the Ca channel activity of cultured PI cells. What we have found is that this treatment prevents and even reverses the increase in Ca channel activity. On the

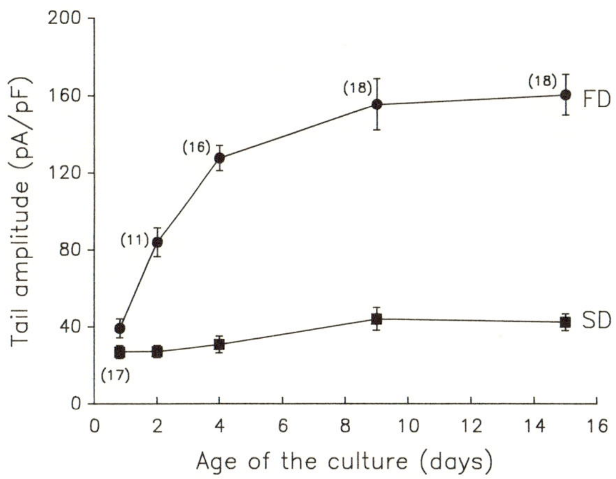

Figure 7. Development of Ca channel activity in cultured PI cells. Ca tails were recorded at −80 mV after 15-ms pulses to +60 mV, from cells kept in culture for the indicated number of days (±8 h). The records were taken within the first minute after establishment of the whole-cell recording. The amplitudes of the fast (FD) and slow (SD) components in the Ca tail were determined by exponential fitting and then normalized by cell capacitance. Tail amplitudes (mean ± SEM, number of cells in parentheses) are plotted as a function of time in culture. The composition of the recording solutions was (in millimolar) 115 NaCl, 10 $CaCl_2$, 1 μM TTX//30 NMG-glutamate, 60 Cs-glutamate, 20 CsF, 20 TEA-Cl, 10 EGTA. PI cell cultures Oc2186 and No1286.

other hand, chronic adrenergic stimulation speeds up the development of Ca channel activity with time in culture.

Fig. 8 illustrates the reversal of Ca channel activity by chronic exposure to bromocriptine. The plots show the average amplitude of the fast and slow components in the Ca^{++} tail in cells kept in culture for 10 d or more, when the Ca channel activity in control experiments had reached a maximum value. The first point in each plot corresponds to the control value. Other points were obtained from cells pretreated with 1 μM bromocriptine. The Ca current records were taken in the absence of bromocriptine in the external recording solution. Actually, the dopamine agonist was removed 30–60 min before the electrophysiological recording to distinguish between short- and long-lasting effects of the dopaminergic stimulation. Under these conditions, pretreatment with bromocriptine for 1 h does not induce any significant change in Ca channel activity. However, prolonging the stimulation of the dopamine receptors to 1 d or longer gradually decreases the amplitude of the Ca current. The activity of both FD and SD channels is reduced with time, but the FD channels are affected earlier; 50% reduction of FD channel activity is observed after 1–2 d of treatment with bromocriptine, while a comparable reduction of the SD current requires 3–4 d.

The reduction of Ca current amplitude in these experiments does not seem to be associated with an obvious change in the kinetic properties of the Ca channels.

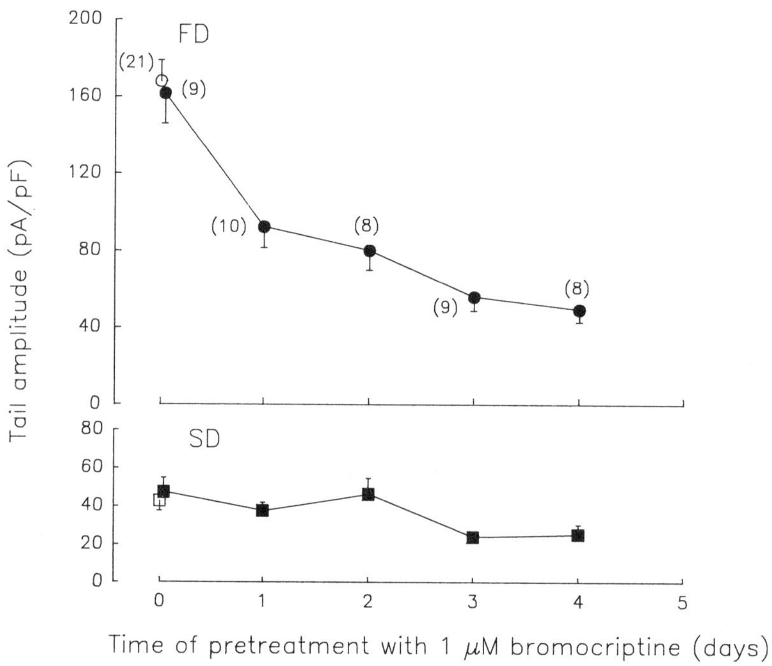

Figure 8. Inhibition of Ca channel activity in PI cells by chronic stimulation of the D-2 dopaminergic receptors. Ca tails were recorded as in Fig. 7 from control cells and from cells cultured in the presence of 1 μM bromocriptine during 1 h, 1 d, or several days. The amplitudes (mean ± SEM) of the fast (FD) and slow (SD) components in the Ca tail are plotted as a function of the time of pretreatment with the dopaminergic agonist. Open symbols correspond to the tail values in control cells. The records were taken in the absence of bromocriptine, 30–60 min after the end of the stimulation of the dopaminergic receptors. T = 10–15 d. PI cell culture Ma2387.

Furthermore, if actinomycin D (40 μM) is added to the culture medium, the FD channel activity decreases with a time course similar to that observed with bromocriptine. This suggests that chronic stimulation of the D-2 dopamine receptors reduces the total number of Ca channels in the plasma membrane.

Adding bromocriptine to the culture medium on day 1 prevents the increase of Ca channel activity, as shown in Fig. 9. The continuous presence of bromocriptine in the culture medium from day 1 to 6 keeps the Ca channel activity at a low level. The channel activity increases with a slow time course after the removal of bromocriptine.

Thus, a delay in the development of Ca channel activity with time in culture can be introduced by chronic stimulation of the dopaminergic receptors.

In contrast to the chronic effects of bromocriptine, if PI cells are cultured in the presence of isoproterenol (1 μM), the Ca channel activity develops with a faster time course (Fig. 10). A similar effect is observed when 8-Br-cAMP (3 mM) is added to the culture medium (Fig. 10). Neither isoproterenol nor 8-Br-cAMP modified the maximum level of Ca channel activity. For example, the addition of 8-Br-cAMP to the culture medium on day 6 did not significantly increase the Ca channel activity, relative to that in control cells, measured 2 d later. However, the treatment with 8-Br-cAMP was able to block the reversal of Ca channel activity induced by chronic stimulation of the dopamine receptors.

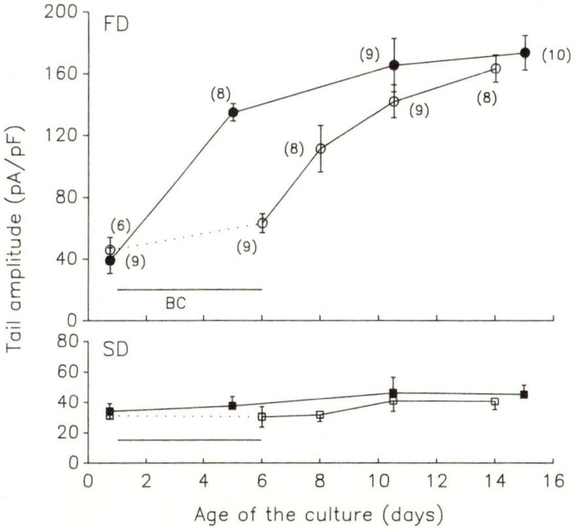

Figure 9. Chronic stimulation of the D-2 dopaminergic receptors prevents the development of Ca channel activity in cultured PI cells. Ca tails were recorded as in Fig. 7 from two different sets of cells kept in culture during the indicated number of days. The plots show the amplitudes (mean ± SEM) of the fast (FD) and slow (SD) components in the Ca tail. Filled symbols correspond to data from control cells, and open symbols correspond to the tail current amplitudes obtained from cells treated with 1 μM bromocriptine (BC) during the interval indicated by the horizontal bar (from day 1 to day 6). PI cell culture as in Fig. 8.

Sensitivity of Pituitary Ca Channels to Dihydropyridines

We have studied the effect of dihydropyridines on the Ca channel activity of PI cells (Torres-Escalante et al., 1989). This information is useful when comparing the properties of the SD and FD channels in the pituitary cells with those present in other excitable cells. It also may help to interpret the effect of dihydropyridines on the synthetic and secretory activity of PI cells (Loeffler et al., 1986; Taraskevich and Douglas, 1986) and other adenohypophyseal cells.

We tested the effect of nifedipine and BAY K 8644 on the Ca channel activity using a holding potential of -80 mV and with 20 mM Ba present in the external solution. Nifedipine (1 μM) added to the bath solution partially blocked the current carried by both FD and SD channels. Nifedipine reduced the amplitude of the fast component in the Ba^{++} tail by >50%, and reduced the amplitude of the slow component by 25%. Reduction of Ca channel activity induced by nifedipine did not depend on the amplitude of the activating pulse. On the contrary, BAY K 8644 (0.5–1 μM) increased the amplitude of the slow component in the Ba tail by ~120%, and that of the fast component by only 10%. Ba tail kinetics was affected by BAY K 8644, but not by nifedipine. The tail time constant measured at −80 mV after 10-ms steps to +60 mV was 94 ± 5 μs (mean ± SEM) for the fast component and 2.1 ± 0.2 ms for the slow component (n = 11) in the control condition. In the presence of BAY K 8644, the time constants were 127 ± 10 μs and 1.5 ± 0.1 ms, respectively.

Conductance-voltage curves for both tail components in the presence of BAY K 8644 had the same position along the voltage axis, and about the same shape, as in control conditions. However, this dihydropyridine drastically modified the time course

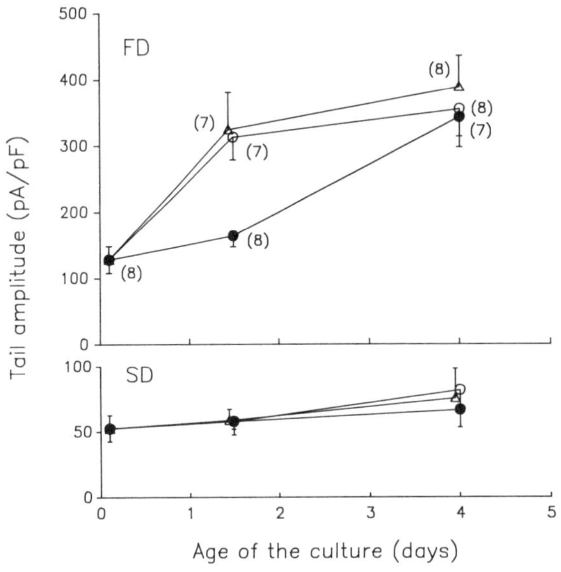

Figure 10. Chronic stimulation of the beta-2 adrenergic receptors or chronic treatment with 8-Br-cAMP speeds up the development of Ca channel activity with time in culture. Ba tails were recorded at −80 mV after 10-ms pulses to +50 or +60 mV from PI cells after various times in culture. The amplitudes of the two tail components are plotted as a function of the age of the culture. Filled circles, controls cells; open circles, cells cultured in the presence of 1 μM isoproterenol; open triangles, cells cultured in the presence of 3 mM 8-Br-cAMP. Isoproterenol and 8-Br-cAMP were removed from the culture medium 30–60 min before the electrophysiological experiments. In all cases, records were taken within the first 2 min after the rupture of the membrane patch. Recording solutions were as in Fig. 1. PI cell culture Jn0688.

of Ca channel conductance. Besides increasing current amplitude, BAY K 8644 prolonged the rising phase of the current, particularly during small depolarizations, and the current at these voltages decayed much less than in the control condition. Analysis of Ba tails after steps to −20 mV of various durations indicated that the slow component inactivates with a slower time course in the presence of BAY K 8644.

BAY K 8644 also affected the voltage dependence of steady-state inactivation for the slow component in the Ba tail. A change in holding potential from −80 to −40 mV reduced by only 20% the amplitude of the slow component. This contrasts sharply with the behavior of normal SD channels (Fig. 4).

It is possible that in the presence of BAY K 8644 a component of the current through FD channels activates at lower depolarizations than usual and deactivates at −80 mV so slowly that its macroscopic activity overlaps almost perfectly with that of normal SD channels, except that it does not inactivate. Alternatively, BAY K 8644 may increase the current carried by SD channels, remove or impede their fast inactivation gating, and speed up in some degree their deactivation kinetics at −80 mV, without altering the voltage dependence of FD and SD channel activation.

Discussion

Two Types of Voltage-dependent Ca Channels in Endocrine Cells

Armstrong and Matteson (1985; Matteson and Armstrong, 1986) demonstrated for the first time, and in an elegant way indeed, the existence of two independent sets of voltage-gated Ca channels in an adenohypophyseal cell type, the GH_3 cell line. The two types of Ca channels, named FD and SD (fast deactivating and slow deactivating), were distinguished on the basis of their quite different rates of deactivation, the voltage range over which they activate, their activation time course, inactivation properties, Ca/Ba conductivity, and stability under whole-cell recording.

Cultured PI cells express two types of voltage-dependent Ca channels that are similar to those present in GH_3 cells (Cota, 1986; Figs. 1–4). The two types of Ca channels, FD and SD, are also expressed by normal prolactin-secreting cells (Hiriart et al., 1989; see also DeRiemer and Sakmann, 1986; Lingle et al., 1986).

FD and SD channels are probably associated with different functions, as discussed by Matteson and Armstrong (1986). The SD channels seem to be well suited to participate in the generation of the subthreshold electrical activity of the pituitary cells, and in this way they probably contribute to the pacemaking mechanism that determines the triggering of spontaneous action potentials. Besides this primarily electrical function, Ca entry through SD channels may be relevant for the secretory activity of adenohypophyseal cells in the absence of action potentials. For example, the "basal" secretory activity of PI cells appears to be independent of action potentials since it is not reduced by tetrodotoxin (TTX), which blocks the spiking activity of the cells (Tomiko et al., 1984). However, TTX does not block the subthreshold fluctuations in membrane potential (Douglas and Taraskevich, 1985), and these fluctuations may depend on SD channel activity. On the other hand, FD channels seem to be well adapted to inject brief, but intense pulses of Ca into the cytoplasm once an action potential has been triggered (Matteson and Armstrong, 1986). Ca entry through FD channels thus may be important to induce the exocytotic release of hormones during conditions of stimulated spiking activity. Long-term activity of the Ca channels may contribute to Ca-dependent hormone synthesis (Enyeart et al., 1987).

The coexistence of two sets of voltage-dependent Ca channels in GH_3 cells and related subclones has been confirmed by other authors on the basis of single-channel characteristics and the sensitivity of the Ca channels to dihydropyridines (Armstrong and Eckert, 1987; Cohen and McCarthy, 1987). Besides GH_3 cells, PI cells and normal prolactin-secreting cells, evidence for two components of Ca current has been obtained in electrophysiological studies of other endocrine cell types of the adenohypophysis (DeRiemer and Sakmann, 1986; Marchetti et al., 1987). The FD channels are also called "high-threshold" or "non-inactivating" channels, and the SD channels are also known as "low-threshold," "inactivating," or "transient" channels. FD and SD Ca channels have also been found in pancreatic beta cells (Hiriart and Matteson, 1988) and adrenocortical endocrine cells (Tabares et al., 1988).

So far, only one study in an endocrine cell, performed on cultured porcine PI cells, argues for three components of voltage-gated Ca current (Taleb et al., 1986). Besides the low-threshold, inactivating component (named I_{CaT}), Taleb et al. presented evidence for two components in the high-threshold Ca current, a sustained component (I_{CaS}) and a transient component (I_{CaN}). Changing the holding potential from −100 to −40 mV eliminates I_{CaT} and I_{CaN}, and raising the internal Ca^{++} concentration from 5 to 50 nM reduces the amplitude of I_{CaS}, but does not significantly affect the amplitude of I_{CaN}. Taleb et al. supposed that I_{CaS} and I_{CaN} are carried by distinct Ca channels. However, interpretation of their results is ambiguous. The partial inactivation of the high-threshold current is equally consistent with two types of channels that differ in their inactivation properties, or with a single type of channel that inactivates only partially (see also Stanley and Russell, 1988). Results obtained by varying the free Ca^{++} concentration in the patch pipette are also difficult to interpret because of the known complex manner in which the cytoplasmic Ca^{++} activity responds to internal perfusion (Byerly and Moody, 1984).

Our study of the Ca channel activity of PI cells indicates that there is only one component in the fast tail current associated with deactivation of high-threshold (FD) Ca channels. The deactivation kinetics of the FD channels do not change as channels activate and then partially inactivate during progressively longer voltage steps (Fig 3). Neither variations in the amplitude of activating pulses nor the change in holding potential from −80 to −40 mV revealed more than one component in the tail current being carried through high-threshold channels (Figs. 2 and 4). Moreover, a similar conclusion, that the high-threshold, FD component of current is carried by a single class of Ca channels, has been reached by Swandulla and Armstrong (1988) in chick sensory neurons, another preparation in which evidence for two types (called "L" and "N") of high-threshold Ca channels has been presented (Nowycky et al., 1985; Fox et al., 1987).

Selective Modulation of FD Channels by Acute Stimulation of Catecholaminergic Receptors

Stimulation of the beta-2 adrenergic receptors of PI cells increases cAMP formation and enhances hormone release (Cote et al., 1984*b*). In contrast, stimulation of the D-2 dopaminergic receptors decreases the capacity of the PI cells to synthesize cAMP and inhibits hormone release (Cote et al., 1984*b*). The present study shows that stimulation of the catecholaminergic receptors is able to modulate the voltage-dependent Ca channel activity of these endocrine cells. The amplitude of the current through Ca channels is decreased by dopamine agonists and increased by isoproterenol, and these changes are caused by a selective modulation of the activity of one type of Ca channel, the FD (Figs. 5 and 6). It is reasonable to suppose that this modulation of Ca channel activity contributes to the catecholaminergic control of hormone release.

"Direct" modulation of FD channels might not be the only way to alter Ca influx by acute stimulation of catecholaminergic receptors in PI cells. As a matter of fact, acute exposure to dopamine also stimulates the K outward current in PI cells (Cota and Armstrong, 1987). Modulation of K channel properties is expected to change the cell electrical activity and thus alter indirectly the voltage-dependent activity of both SD and FD channels. Furthermore, dopamine agonists are known to induce a rapid and long-lasting hyperpolarization of prolactin-secreting cells, an effect that is explained by the increased activity of K channels (see chapter 8 by Gregerson et al. in this book; Israel et al., 1985, 1987; Mason and Ingram, 1986).

Modulation of FD channel activity seems to be in series with changes in the intracellular levels of cAMP induced by the catecholaminergic receptors. The action of the adrenergic stimulation is mimicked by the addition of 3 mM 8-Br-cAMP to the external medium, and dopamine agonists are incapable of reducing the Ca channel activity in the presence of this high concentration of 8-Br-cAMP (Fig. 6). Evidence for modulation of the activity of the FD-type channel by a cAMP-dependent phosphorylation process has been presented in GH_3 cells (Armstrong and Eckert, 1987). It is then tempting to suggest that catecholaminergic receptors in PI cells modulate the activity of FD channels by altering their level of phosphorylation, in a similar way to the proposed mechanism for adrenergic stimulation and muscarinic inhibition of the high-voltage–activated Ca channel activity in cardiac muscle cells (Kameyama et al., 1985, 1986; Fischmeister and Hartzell, 1986; Hescheler et al., 1986). SD-type Ca channels in cardiac muscle cells are insensitive to beta-adrenergic stimulation (Bean, 1985; Hagiwara et al., 1988; Tytgat et al., 1988), as they are in PI cells (Figs. 5 and 6).

An increased Ca channel activity induced by beta-adrenergic stimulation has also been observed in skeletal muscle (Schmid et al., 1985; Garcia et al., 1986; Arreola et al., 1987), and this appears to be mediated in part by a direct interaction of the alpha subunit of the stimulatory GTP-binding protein (G_s) with the Ca channel (Brown and Birnbaumer, 1988). Further investigation is required to examine if this cAMP-independent mechanism contributes to the isoproterenol-induced increase of FD channel activity reported here. In particular, it will be interesting to know if the actions of isoproterenol and 8-Br-cAMP are additive.

Chronic Stimulation of Catecholaminergic Receptors May Regulate Ca Channel Expression

There are several examples of short-term modulation of the functional properties of voltage-gated Ca channels in excitable cells by hormones, neurotransmitters, and drugs (see Tsien, 1987), including the one presented here. In contrast, little is known about the mechanisms that regulate the long-term activity of Ca channels. Our study provides some clues regarding the long-term regulation of Ca channel activity in PI cells.

When PI cells are denervated and placed in culture, there is a marked increase of the current amplitude through Ca channels as a function of time in culture (Fig. 7). The increased Ca channel activity is not associated to a detectable change in the functional properties of the channels. Full development of Ca channel activity takes several days, and seems to depend on an increased mRNA and protein synthesis. The activity of both FD and SD channels increases with time in culture, but the FD channel activity augments more dramatically and develops with a faster time course than the activity of the SD channels. Chronic stimulation of dopamine receptors prevents or even reverts the development of Ca channel activity (Figs. 8 and 9). On the other hand, a faster development of Ca channel activity is obtained by chronic stimulation of the adrenergic receptors (Fig. 10). This does not alter the maximum level of Ca channel activity.

The results suggest that the surface density of Ca channels in the plasma membrane of cultured PI cells can be regulated by catecholaminergic receptors. This regulation seems to be mediated by chronic changes in the intracellular levels of cAMP, since culturing the cells in the presence of 8-Br-cAMP mimics the effect of chronic stimulation of the adrenergic receptors (Fig. 10), and blocks the reversal of Ca channel activity induced by chronic stimulation of the dopamine receptors.

Our results further suggest that the dopaminergic neurons of the hypothalamus, that directly innervate the PI cells in vivo, exert a tonic inhibition on Ca channel expression in the endocrine cells. We propose, in a speculative way, that regulation of Ca channel expression may be a general mechanism for the long-term control of the biosynthetic and secretory activity of the adenohypophyseal cells by the nervous system.

Acknowledgments

This study was started while one of the authors (G. Cota) was a postdoctoral fellow in the laboratory of Dr. Clay M. Armstrong. We would very much like to thank Dr. Armstrong for providing invaluable support during the course of these experiments. Experiments using the picospritzer were done in collaboration with Dr. Robert H. Chow. We thank Mr. Angel Marin for programming the computer, Mrs. Araceli Navarrete for her help with the cell cultures, and Mr. Raul Gonzalez for expert technical assistance.

References

Antakly, T., S. Mercille, and J. P. Cote. 1987. Tissue-specific dopaminergic regulation of the glucocorticoid receptor in the rat pituitary. *Endocrinology.* 120:1558–1562.

Armstrong, C. M. 1986. Electrophysiological studies in endocrine cells. *In* Ionic Channels in Cells and Model Systems. R. Latorre, editor. Plemun Publishing Corp., New York. 243–254.

Armstrong, C. M., and D. R. Matteson. 1985. Two distinct populations of calcium channels in a clonal line of pituitary cells. *Science.* 227:65–67.

Armstrong, D., and R. Eckert. 1987. Voltage-activated calcium channels that must be phosphorylated to respond to membrane depolarization. *Proceedings of the National Academy of Sciences.* 84:2518–2522.

Arreola, J., J. Calvo, M. C. Garcia, and J. A. Sanchez. 1987. Modulation of calcium channels of twitch skeletal muscle fibres of the frog by adrenaline and cyclic adenosine monophosphate. *Journal of Physiology.* 393:307–330.

Bjorklund, A., and O. Lindvall. 1984. Dopamine-containing systems in the CNS. *In* Handbook of Chemical Neuroanatomy. Vol. 2. A. Bjorklund and T. Hokfelt, editors. Elsevier Science Publishers, New York. 55–122.

Bean, B. P. 1985. Two kinds of calcium channels in canine atrial cells. *Journal of General Physiology.* 86:1–30.

Beaulieu, M., R. Felder, and J. W. Kebabian. 1986. D-2 dopaminergic agonists and adenosine 3′,5′-monophosphate directly regulate the synthesis of alpha-melanocyte-stimulating hormone-like peptides by cultured rat melanotrophs. *Endocrinology.* 118:1032–1039.

Beaulieu, M., M. E. Goldman, K. Miyazaki, E. A. Frey, R. L. Eskay, J. W. Kebabian, and T. E.

Cote. 1984. Bromocriptine-induced changes in the biochemistry, physiology, and histology of the intermediate lobe of the rat pituitary gland. *Endocrinology.* 114:1871–1884.

Berkenbosch, F., F. J. H. Tilders, and I. Vermes. 1983. Beta-adrenoceptor activation mediates stess-induced secretion of beta-endorphin-related peptides from intermediate but not anterior pituitary. *Nature.* 305:237–239.

Berkenbosch, F., I. Vermes, R. Binnekade, and F. J. H. Tilders. 1981. Beta-adrenergic stimulation increases the plasma levels of immunoreactive alpha-MSH, beta-endorphin, ACTH and corticosterone. *Life Sciences.* 29:2249–2256.

Brown, A. M., and L. Birnbaumer. 1988. Direct G protein gating of ion channels. *American Journal of Physiology.* 254:H401–H410.

Byerly, L., and W. J. Moody. 1984. Intracellular calcium ions and calcium currents in perfused neurones of the snail, *Lymnaea stagnalis. Journal of Physiology.* 352:637–652.

Byerly, L., and B. Yazejian. 1986. Intracellular factors for the maintenance of calcium currents in perfused neurons from the snail, *Lymnaea stagnalis. Journal of Physiology.* 370:631–650.

Chen, C. L. C., F. T. Dionne, and J. L. Roberts. 1983. Regulation of the proopiomelanocortin mRNA levels in rat pituitary by dopaminergic compounds. *Proceedings of the National Academy of Sciences.* 80:2211–2214.

Childs, V., C. Marchetti, and A. M. Brown. 1987. Involvement of sodium channels and two types of calcium channels in the regulation of adrenocorticotropin release. *Endocrinology.* 12:2059–2069.

Chronwall, B. M., W. R. Millington, W. S. T. Griffin, J. R. Unnerstall, and T. L. O'Donohue. 1987. Histological evaluation of the dopaminergic regulation of proopiomelanocortin gene expression in the intermediate lobe of the rat pituitary, involving in situ hybridization and [^{3}H]thymidine uptake measurement. *Endocrinology.* 120:1201–1211.

Cohen, C. J., and R. T. McCarthy. 1987. Nimodipine block of calcium channels in rat anterior pituitary cells. *Journal of Physiology.* 387:195–225.

Cota, G. 1986. Calcium channel currents in pars intermedia cells of the rat pituitary gland. *Journal of General Physiology.* 88:83–105.

Cota, G., and C. M. Armstrong. 1987. Long term stimulation of dopamine receptors inhibits calcium channel activity in cultured pituitary cells. *Society for Neuroscience Abstracts.* 13:794. (Abstr.)

Cota, G., and C. M. Armstrong. 1988. Potassium channel "inactivation" induced by soft-glass patch pipettes. *Biophysical Journal.* 53:107–109.

Cote, T. E., R. Felder, J. W. Kebabian, R. D. Sekura, T. Reisine, and H.-U. Affolter. 1986. D-2 dopamine receptor-mediated inhibition of pro-opiomelanocortin synthesis in rat intermediate lobe. *Journal of Biological Chemistry.* 261:4555–4561.

Cote, T. E., E. A. Frey, and R. D. Sekura. 1984*a*. Altered activity of the inhibitory guanyl nucleotide-binding component induced by pertussis toxin. *Journal of Biological Chemistry.* 259:8693–8698.

Cote, T. E., E. A. Frey, R. D. Sekura, and J. W. Kebabian. 1984*b*. Dual regulation of adenylate cyclase activity and hormone release in the intermediate lobe of the rat pituitary gland. *Advances in Cyclic Nucleotide and Protein Phosphorylation Research.* 19:151–167.

Cote, T. E., C. W. Grewe, and J. W. Kebabian. 1982*a*. Guanyl nucleotides participate in the beta-adrenergic stimulation of adenylate cyclase activity in the intermediate lobe of the rat pituitary gland. *Endocrinology.* 110:805–811.

Cote, T. E., C. W. Grewe, K. Tsuruta, J. C. Stoof, R. L. Eskay, and J. W. Kebabian. 1982*b*. D-2 dopamine receptor-mediated inhibition of adenylate cyclase activity in the intermediate lobe of the rat pituitary gland requires GTP. *Endocrinology*. 110:812–819.

Cote, T. E., M. Munemura, R. L. Eskay, and J. W. Kebabian. 1980. Biochemical identification of the beta-adrenoceptor and evidence of the involvement of a cAMP system in the beta-adrenergically induced release of alpha-MSH in the intermediate lobe of the rat pituitary gland. *Endocrinology*. 107:108–116.

Dave, J. R., L. E. Eiden, D. Lozovsky, J. A. Washeck, and R. L. Eskay. 1987. Calcium-independent and calcium-dependent mechanisms regulate corticotropin-releasing factor-stimulated proopiomelanocortin peptide secretion and messenger ribonucleic acid production. *Endocrinology*. 120:305–310.

Davis, M. D. 1986. The hypothalamo-hypophyseal rat explant in vitro: endocrinological studies of the pars intermedia dopaminergic neural input. *Journal of Physiology*. 370:381–393.

Davis, M. D., H. L. Haas, and W. Lichtensteiger. 1985. The hypothalamohypophyseal system in vitro: electrophysiology of the pars intermedia and evidence for both excitatory and inhibitory inputs. *Brain Research*. 334:97–104.

Davis, M. D., and M. E. Hadley. 1976. Spontaneous electrical potentials and pituitary hormone (MSH) secretion. *Nature*. 261:422–423.

DeRiemer, S. A., and B. Sakmann. 1986. Two calcium currents in normal rat anterior pituitary cells identified by a plaque assay. *Experimental Brain Research Series*. 14:139–154.

Douglas, W. W., and P. S. Taraskevich. 1978. Action potentials in gland cells of rat pituitary pars intermedia: inhibition by dopamine, an inhibitor of MSH secretion. *Journal of Physiology*. 285:171–184.

Douglas, W. W., and P. S. Taraskevich. 1980. Calcium component to action potentials in rat pars intermedia cells. *Journal of Physiology*. 309:623–630.

Douglas, W. W., and P. S. Taraskevich. 1982. Slowing effects of dopamine and calcium-channel blockers on frequency of sodium spikes in rat pars intermedia cells. *Journal of Physiology*. 326:201–211.

Douglas, W. W., and P. S. Taraskevich. 1985. The electrophysiology of adenohypophyseal cells. *In* The Electrophysiology of the Secretory Cell. A. M. Poisner and J. M. Trifaro, editors. Elsevier/North-Holland, New York. 63–91.

Enyeart, J. J., S.-S. Sheu, and P. M. Hinkle. 1987. Dihydropiridine modulators of voltage-sensitive Ca channels specifically regulate prolactin production by GH_4C_1 pituitary tumor cells. *Journal of Biological Chemistry*. 262:3154–3159.

Fischmeister, R., and H. C. Hartzell. 1986. Mechanism of action of acetylcholine on calcium current in single cells from frog ventricule. *Journal of Physiology*. 376:183–202.

Fox, A. P., M. C. Nowycky, and R. W. Tsien. 1987. Kinetic and pharmacological properties distinguishing three types of calcium current in chick sensory neurones. *Journal of Physiology*. 394:149–172.

Garcia, J., G. Cota, and E. Stefani. 1986. Modulation of fast and slow calcium currents in frog skeletal muscle fibres. *Society for Neuroscience Abstracts*. 12:992. (Abstr.)

Gick, G. G., and F. C. Bancroft. 1985. Regulation by calcium of prolactin and growth hormone mRNA sequences in primary cultures of rat pituitary cells. *Journal of Biological Chemistry*. 260:7614–7619.

Hagiwara, N., H. Irisawa, and M. Kameyama. 1988. Contribution of two types of calcium currents to the pacemaker potentials of rabbit sino-atrial node cells. *Journal of Physiology*. 395:233–253.

Hamill, O. P., A. Marty, E. Neher, B. Sakmann, and F. J. Sigworth. 1981. Improved patch-clamp techniques for high-resolution current recording from cells and cell-free membrane patches. *Pflügers Archiv*. 391:85–100.

Hescheler, J., M. Kameyama, and W. Trautwein. 1986. On the mechanism of muscarinic inhibition of the cardiac Ca current. *Pflügers Archiv*. 407:182–189.

Hiriart, M., and D. R. Matteson. 1988. Na channels and two types of Ca channels in rat pancreatic B cells identified with the reverse hemolytic plaque assay. *Journal of General Physiology*. 91:617–639.

Hiriart, M., A. Navarrete, and G. Cota. 1989. FD and SD Ca channels in normal prolactin-secreting cells. *Biophysical Journal*. 55:298*a*. (Abstr.)

Holl, R. W., M. O. Thorner, and D. A. Leong. 1988. Intracellular calcium concentration and growth hormome secretion in individual somatotropes: effects of growth hormone–releasing factor and somatostatin. *Endocrinology*. 122:2927–2932.

Holzbauer, M., and K. Racke. 1985. The dopaminergic innervation of the intermediate lobe and of the neural lobe of the pituitary gland. *Medical Biology*. 63:97–116.

Israel, J. M., C. Denef, and J. D. Vincent. 1983. Electrophysiological properties of normal somatotrophs in culture. *Neuroendocrinology* 37:193–199.

Israel, J. M., P. Jaquet, and J. D. Vincent. 1985. The electrical properties of isolated human prolactin-secreting adenoma cells and their modification by dopamine. *Endocrinology*. 117:1448–1455.

Israel, J. M., C. Kirk, and J. D. Vincent. 1987. Electrophysiological responses to dopamine of rat hypophysial cells in lactotroph-enriched primary cultures. *Journal of Physiology*. 390:1–22.

Jackson, S., and P. J. Lowry. 1983. Secretion of pro-opiocortin peptides from isolated perfused rat pars intermedia cell. *Neuroendocrinology*. 37:248–257.

Kameyana, M., J. Hescheler, F. Hofmann, and W. Trautwein. 1986. Modulation of Ca current during the phosphorylation cycle in the guinea pig heart. *Pflügers Archiv*. 407:123–128.

Kameyana, M., F. Hofmann, and W. Trautwein. 1985. On the mechanism of beta-adrenergic regulation of the Ca channel in the guinea-pig heart. *Pflügers Archiv*. 405:285–293.

Kidokoro, Y. 1975. Spontaneous Ca action potentials in a clonal pituitary cell line and their relationship to prolactin secretion. *Nature*. 258:741–742.

Lingle, C. J., S. Sombati, and M. E. Freeman. 1986. Membrane currents in identified lactotrophs of rat anterior pituitary. *Journal of Neuroscience*. 6:2995–3005.

Loeffler, J. P., N. Kley, C. W. Pittius, and V. Hollt. 1986. Calcium and cyclic adenosine 3′,5′-monophosphate regulate proopiomelanocortin messenger ribonucleic acid levels in rat intermediate and anterior pituitary lobes. *Endocrinology*. 119:2840–2847.

Mains, R. E., and B. A. Eipper. 1979. Synthesis and secretion of corticotropins, melanotropins, and endorphins by rat intermediate pituitary cells. *Journal of Biological Chemistry*. 254:7885–7894.

Marchetti, C., G. V. Childs, and A. M. Brown. 1987. Membrane currents of identified isolated rat corticotropes and gonadotropes. *American Journal of Physiology*. 252:E340–346.

Mason, W. T., and C. D. Ingram. 1986. Techniques for studying the role of electrical activity in control of hormone secretion by normal anterior pituitary cells. *Methods in Enzymology*. 124:207–242.

Matteson, D. R., and C. M. Armstrong. 1986. Properties of two types of calcium channels in clonal pituitary cells. *Journal of General Physiology*. 87:161–182.

Meier, K., W. Knepel, and C. Schofl. 1988. Potassium depolarization elevates cytosolic free calcium concentration in rat anterior pituitary cells through 1,4-dihydropyridine-sensitive, omega-conotoxin-insensitive calcium channels. *Endocrinology*. 122:2764–2770.

Meunier, H., and F. Labrie. 1982*a*. The dopamine receptor in the intermediate lobe of the rat pituitary gland is negatively coupled to adenylate cyclase. *Life Sciences*. 30:963–968.

Meunier, H., and F. Labrie. 1982*b*. Specificity of the beta-2-adrenergic receptor stimulating cAMP accumulation in the intermediate lobe of the rat pituitary gland. *European Journal of Pharmacology*. 81:411–420.

Meunier, H., and F. Labrie. 1983. Multiple hormonal control of pars intermedia cell activity. *Canadian Journal of Biochemistry and Cell Biology*. 61:516–531.

Moriarty, C. M. 1978. Role of calcium in the regulation of adenohypophyseal hormone release. *Life Sciences*. 23:185–194.

Munemura, M., R. L. Eskay, and J. W. Kebabian. 1980. Release of alpha-melanocyte–stimulating hormone from dispersed cells of the intermediate lobe of the rat pituitary gland: involvement of catecholamines and adenosine 3′,5′-monophosphate. *Endocrinology*. 106:1795–1803.

Nowycky, M. C., A. P. Fox, and R. W. Tsien. 1985. Three types of neuronal calcium channel with different calcium channel agonist sensitivity. *Nature*. 316:440–443.

Ozawa, S., and O. Sand. 1986. Electrophysiology of excitable endocrine cells. *Physiological Reviews*. 66:887–952.

Schlegel, W., B. P. Winiger, P. Mollard, P. Vacher, F. Wuarin, G. R. Zahnd, C. B. Wollheim, and B. Dufy. 1987. Oscillations of cytosolic Ca in pituitary cells due to action potentials. *Nature*. 329:719–721.

Schmid, A., J. F. Renaud, and M. Lazdunski. 1985. Short term and long term effects of beta-adrenergic effectors and cyclic AMP on nitrendipine-sensitive voltage-dependent Ca^{2+} of skeletal muscle. *Journal of Biological Chemistry*. 260:13041–13046.

Stanley, E. F., and J. T. Russell. 1988. Inactivation of calcium channels in rat pituitary lobe cells. *Brain Research*. 476:64–72.

Swandulla, D., and C. M. Armstrong. 1988. Fast-deactivating calcium channels in chick sensory neurons. *Journal of General Physiology*. 92:197–218.

Tabares, L., K. Loesser, C. M. Armstrong, and J. Lopez-Barneo. 1988. The calcium channels in adrenocortical cells. *Journal of General Physiology*. 92:15*a*. (Abstr.)

Taleb, O., J. Trouslard, B. A. Demeneix, and P. Feltz. 1986. Characterization of calcium and sodium currents in porcine pars intermedia cells. *Neuroscience Letters*. 66:55–60.

Taraskevich, P. S., and W. W. Douglas. 1986. Effects of BAY K 8644 and other dihydropyridines on basal and potassium-evoked output of MSH from mouse melanotrophs in vitro. *Neuroendocrinology*. 44:384–389.

Tomiko, S. A., P. S. Taraskevich, and W. W. Douglas. 1984. Effects of veratridine, TTX and other drugs that alter electrical behaviour on secretion of MSH from melanotrophs of the pituitary pars intermedia. *Neuroscience*. 12:1223–1228.

Torres-Escalante, J. L., A. Marin, and G. Cota. 1989. Sensitivity of Ca channels to dihydropyridines in pituitary cells. *Biophysical Journal*. 55:297*a*. (Abstr.)

Tsien, R. W. 1987. Calcium currents in heart cells and neurons. *In* Neuromodulation. L. K. Kaczmarek and I. B. Levitan, editors. Oxford University Press, New York. 206–242.

Tsuruta, K., C. W. Grewe, T. E. Cote, R. L. Eskay, and J. W. Kebabian. 1982. Co-ordinated action of calcium ion and adenosine 3′,5′-monophosphate upon the release of alpha-melanocyte-stimulating hormone. *Endocrinology.* 110:1133–1140.

Tytgat, J., B. Nilius, J. Vereecke, and E. Carmeliet. 1988. The T-type Ca channel in guinea-pig ventricular myocytes is insensitive to isoproterenol. *Pflügers Archiv.* 411:704–706.

Cell Biology of Secretion

Chapter 10

Sorting of Apical Surface Proteins and Lipids by Epithelial Cells: Is There a Common Mechanism?

Michael Lisanti and Enrique Rodriguez-Boulan

Department of Cell Biology and Anatomy, Cornell University Medical College, New York, New York 10021

Introduction

Physiologists have held a long-standing fascination with the radically different functional properties of apical and basolateral membranes of epithelial cells. Different ion channels and pumps are characteristically polarized to one of the two membranes. How does the cell generate this asymmetry?

This question falls within the realm of a very active field of cell biology: protein targeting. Starting with the realization that proteins follow defined biogenetic pathways in eukaryotic cells, after the description of the secretory pathway by Palade and his co-workers (cf. Palade, 1975), this field erupted with the discovery of the signal sequence in secretory proteins (Blobel and Sabatini, 1971; Milstein et al., 1972; Blobel and Dobberstein, 1975). Since then, it has become clear that proteins carry within their structure information that defines their final localization within the cell. Some of these "signals," those that mediate targeting to mitochondria, the nucleus, peroxisomes, and the endoplasmic reticulum, have been elucidated (see Garoff, 1985; Rodriguez-Boulan et al., 1985 for reviews), but much work is still required to understand precisely the targeting mechanisms.

Plasma membrane glycoproteins, including channels, pumps, receptors, and enzymes, follow a biogenetic pathway similar to that employed by secretory proteins (Rodriguez-Boulan et al., 1985). Recent research by several laboratories has considerably increased our knowledge on the mechanisms by which apical and basolateral epithelial molecules are targeted to and retained by their respective surface domain.

Plasma Membrane Domains in Epithelial Cells: Protein and Lipid Composition

Tight junctions (occluding junctions) are a typical epithelial structure in mammals. They are located at the border between apical and lateral membranes, which they help to seperate (Rodriguez-Boulan, 1983; Simons and Fuller, 1985). Recent evidence from several laboratories (reviewed by van Meer and Simons, 1988) indicates that tight junctions effectively block diffusion of proteins and lipids in the expolasmic leaflet of the bilayer, but not in the cytoplasmic leaflet. Thus, lipid compositions are very different on the exoplasmic side, with glycolipids being the main component in the apical surface, and phosphatidylcholine in the basolateral surface.

The protein composition in apical and basolateral membranes has been studied by cell fractionation and surface labeling procedures. The first method was employed for native intestinal cells (Fujita et al., 1973) but the second procedure (labeling) is preferable for epithelial cell lines in culture, since pure fractions are very difficult to obtain. Labeling studies have revealed that apical and basolateral surfaces of epithelial cells in culture share very few of their proteins; practically all of them are polarized (Richardson and Simons, 1979; Brandli et al., 1988; Sargiacomo et al., 1989). Polarity is not, however, absolute. Whereas some proteins are very well polarized (membrane density ratios >100), other proteins have polarity ratios of 10 times or less (Simons and Fuller, 1985). Lipids are usually poorly polarized, less than eightfold (van Meer and Simons, 1988). An important question is what determines the different polarity ratios of proteins and lipids? Are there several mechanisms acting in concert?

Fig. 1 shows the apical and basolateral proteins of the polarized epithelial cell like MDCK (Main-Darby canine kidney), as identified by labeling the surface proteins of confluent monolayers of cells grown on polycarbonate filter chambers with a biotin analogue, followed by sodium dodecylsulfate-polyacrylamide gel electrophoresis (SDS-PAGE) and blotting with ^{125}I-streptavidin which bonds to biotin (Sargiacomo et al., 1989). The composition of integral and peripheral proteins (identified by their ability to partition into the Triton X-114 detergent phase, is very different in each surface. In the case of MDCK cells, the aqueous phase, which comprises peripheral proteins and integral membrane proteins at low hydrophobicity, contains fivefold more proteins than the detergent phase (hydrophobic integral membrane proteins). Similar results have been obtained in other epithelial cell lines (Le Bivic and Rodriguez-Boulan, unpublished results).

Intracellular Pathways of Epithelial Surface Proteins

What are the biogenetic pathways followed by apical and basolateral proteins and lipids? It appears that the answer to this question is: "it depends on the epithelial cell

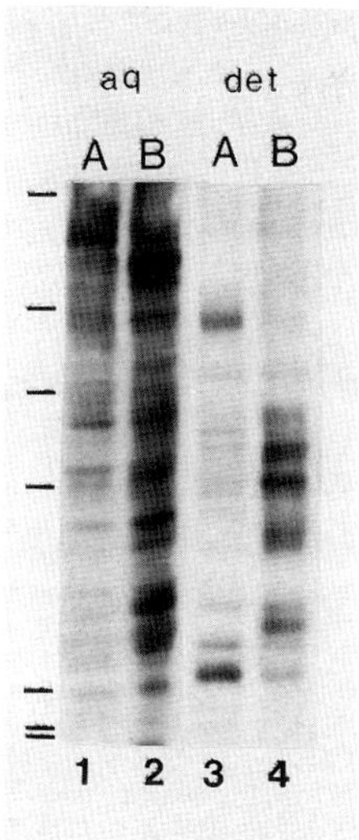

Figure 1. Selective labeling of the apical or basolateral surfaces of MDCK cells. Confluent monolayers of MDCK cells grown on polycarbonate filters were labeled from the apical (*A*) or the basal (*B*) sides with sulfo-*N*-hydroxy-uccinimido-biotin (0.5 mg/ml in phosphate-buffered saline, 4°C), extracted with Triton X-114, and the labeled proteins were fractionated into "peripheral" and "integral" by partition (5 min at 37°C) into aqueous (*aq*) and detergent (*det*) phases. The proteins in each phase were precipitated with cold acetone and analyzed by SDS-PAGE, transfer to nitrocellulose, and ^{125}I-streptavidin blotting. Bars indicate molecular mass standards (starting from the top: 200, 97, 68, 43, 25, 18, and 14). (From Lisanti et al., 1988.)

type" (Hubbard et al., 1989). Studies with enveloped RNA viruses in MDCK cells have shown that some viral glycoproteins are targeted to the apical surface, while others are targeted to the basolateral surface (Rodriguez-Boulan, 1983). Influenza hemagglutinin (HA) is an example of the first case, whereas Vesicular Stomatitis Virus glycoprotein G (VSV G) is an example of the second. In doubly infected cells, HA and G are transported together to the Golgi apparatus and segregated in the *trans*-most cisternae of this organelle into different vesicles, which are then vectorially delivered to the respective surface (Simons and Fuller, 1985; Rodriguez-Boulan and Salas, 1989). In liver cells, however, several apical proteins appear to be delivered first to the basolateral membrane, from where they are removed and translocated to the apical surface (Hubbard et al., 1989). A similar situation has been reported for apical enteropeptidases of native intestinal epithelial cells (reviewed by Hubbard et al., 1989).

Studies in which the fate of newly synthesized glycosphingolipids labeled with a fluorescent ceramide precursor was followed by confocal immunofluorescence microscopy, also showed vectorial delivery of these glycolipids to the apical surface of MDCK cells (van Meer et al., 1987). The question arising is, then, what is the relationship between apical protein and glycolipid sorting in epithelia? Are the two processes related in a causal way (van Meer and Simons, 1988)?

Apical Sorting of Glycosyl-Phosphatidylinositol (GPI)–anchored Proteins

As mentioned above, certain glycolipids are targeted to the apical surface of epithelial cells (van Meer et al., 1987); the hypothesis has been proposed that the apical sorting of glycolipids may play a role in the sorting of apical proteins (van Meer and Simons, 1988). The strongest evidence for the participation of the glycolipid in the sorting of apical proteins is provided by recent results on the novel group of GPI-anchored glycoproteins. These plasma membrane proteins use GPI as a membrane anchor in lieu of the hydrophobic transmembrane sequences used by other conventional integral membrane proteins (Low and Saltiel, 1988). Strikingly, possession of a GPI anchor correlates perfectly with apical localization in epithelial cells (Lisanti et al., 1988). In MDCK cells, a polarity assay that combines the biotin labeling procedure (Fig. 1) with susceptibility to GPI-specific phospholipase C detected six GPI-anchored proteins, all with a preferred apical localization (Fig. 2). The same result was obtained in five other epithelial cell lines of diverse tissue and species origin.[1]

Addition of the GPI anchor occurs immediately after synthesis (1–2 min) and is directed by a COOH-terminal peptide that is cleaved cotranslationally. Transfer of this peptide (by recombinant DNA techniques) to secretory proteins results in the production of chimeric proteins linked to the membrane by GPI. (Caras et al., 1987). When exogenous GPI-anchored proteins are introduced into MDCK cells by transfection of their cDNA genes, they are targeted to the apical surface,[1] which suggests that GPI contains apical targeting information.

A Hypothesis Linking Sorting of Glycolipids and Apical Proteins

Several workers have attempted to localize the sorting information in apically and basolaterally targeted viral glycoproteins by transfecting mutated or chimeric cDNA genes and studying the polarized expression of the protein products. These studies suggest that the apical sorting signals are contained within the ectoplasmic protein domain of influenza HA, an apically targeted molecule. However, there is no information on the exact localization of the sorting information; analysis of the sequences of several apically targeted glycoproteins does not reveal any obvious sequence homology (Stephens and Compans, 1988).

Biophysical evidence suggests that glucosyl-ceramide and other apically targeted glycolipids have a tendency to cluster when they exceed certain concentrations in artificial lipid bilayers (Thompson and Tillack, 1985). Glycosphingolipids are synthesized in the Golgi apparatus and constitute most of the exoplasmic leaflet of the apical membrane (van Meer and Simons, 1988). If clustering occurs in the Golgi, this might

[1]Lisanti, M. P., A. Le Bivic, A. R. Saltiel, and E. Rodriguez-Boulan, manuscript submitted for publication.

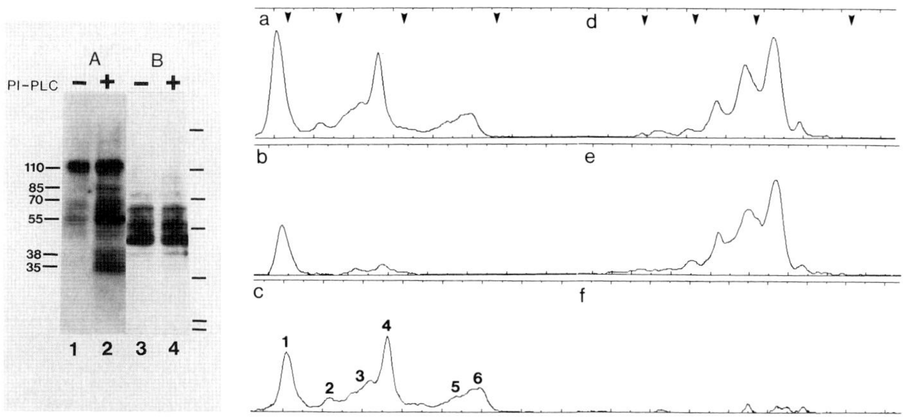

Figure 2. Polarized distribution of GPI-anchored proteins in MDCK cells. MCDK cells grown on polycarbonate filter chambers, labeled from the apical or basal side with sulfo-*N*-hydroxy-uccinimido-biotin, and extracted with Triton X-114. After phase partition at 37°C, the detergent phase containing integral membrane proteins was diluted in buffer, incubated with phosphatidylinositol-specific phospholipase C (*PI-PLC*), and allowed to phase partition for a second time. The aqueous phases, containing the released GPI-anchored proteins, were analyzed by SDS-PAGE and blotting with ^{125}I-streptavidin. (*Left*) autoradiogram of aqueous phases derived from PI-PLC (+) and control samples. (*A*) Monolayers labeled from the apical side; (*B*) monolayers labeled from the basal side. (*Right*) (−) Lanes were subtracted densitometrically from (+) lanes. Scanning densitograms *a*, *b*, *d*, and *e* were obtained by scanning, respectively, lanes *2*, *1*, *4*, and *3* in the left panel. Subtractive densitogram *c* ($a-b$) shows the existence of six apical GPI-anchored proteins. Subtractive densitogram *f* ($d-e$) shows that no GPI-anchored proteins were detected on the basal side. (From Lisanti et al., 1988.)

lead to their segregation and transfer into an apical "carrier vesicle," for transport to the apical surface (Fig. 3). Apical proteins might be sorted into the same vesicle via their (putative) affinity for the glycolipid clusters (van Meer and Simons, 1988). This hypothesis is amenable to experimental testing.

Evidence supporting this model is only circumstantial, presently. The polarity ratios of glycolipids and influenza HA in MDCK cells and comparable (about eightfold more concentrated on the apical surface). Although there is no evidence for

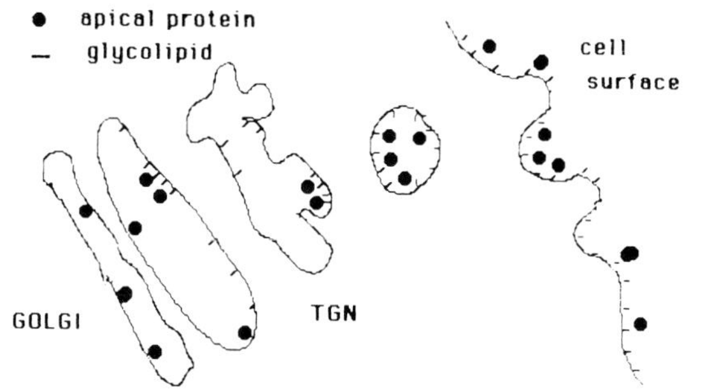

Figure 3. A hypothesis linking sorting of apical proteins and glycolipids in epithelial cells. Some glycolipids are synthesized in the Golgi apparatus and appear to be preferentially incorporated into apically targeted vesicles (van Meer et al., 1987). Apical proteins might have affinity for these glycolipid patches, which would facilitate their incorporation into apically targeted vesicles (van Meer and Simons, 1988). GPI-anchored proteins (not represented) could be equally sorted into the apical "carrier vesicles" through clustering of their GPI anchors. *TGN, trans* elements of the Golgi apparatus.

clustering of GPI, the apical sorting of GPI-anchored proteins provided a tantalizing example of how attachment (in this case covalent) to a glycolipid may result in targeting to the apical surface. Since, in molar terms, glycolipids are much more abundant than apical glycoproteins, "partitioning" into a glycolipid-rich phase may provide a very efficient sorting mechanism. Liposome experiments in which the affinities of apical and basal integral glycoproteins for the glycolipid phase are measured are necessary to investigate this attractive possibility.

Acknowledgments

We thank Ms. Francine Sanchez for typing parts of the manuscript.

This work was supported by grants from National Institutes of Health (GM-34107), New York Heart Association, and an Established Investigator Award from the American Heart Association to E. Rodriguez-Boulan. M. P. Lisanti was supported by the Cornell University Medical College MD/PhD program.

References

Blobel, G., and B. Dobberstein. 1975. Transfer of proteins across membranes. II. Reconstitution of functional rough microsomes from heterologous components. *Journal of Cell Biology*. 67:852–862.

Blobel, G., and D. D. Sabatini. 1971. Ribosome-membrane interaction in eukaryotic cells. *In* Biomembranes. Vol. 2. L. A. Manson, editor. 193–195. Plenum Publishing Co., New York.

Brandli, A. W., G. C. Hansson, E. Rodriguez-Boulan, and K. Simons. 1988. A polarized epithelial cell mutant deficient in translocation of UDP-galactose into the Golgi complex. *Journal of Biological Chemistry*. 263:16283–16290.

Caras, I. W., M. A. Davitz, L. Rhee, G. Weddell, D. W. Martin, and V. Nussenzweig. 1987. Cloning of decay-accelerating to generate two proteins. *Nature*. 325:545–548.

Fujita, M., K. Kawai, S. Asano, and N. Nakao. 1973. Protein components of two different regions of an intestinal plasma epithelial cell membrane. *Biochimica et Biphysica Acta*. 307:141–151.

Garoff, H. 1985. Using recombinant DNA techniques to study protein targeting in the eucaryotic cell. *Annual Review of Cell Biology*. 1:403–445.

Hubbard, A. L., B. Stieger, and J. R. Bartles. 1989. Biogenesis of endogenous plasma membrane proteins in epithelial cells. *Annual Review of Physiology*. 51:755–770.

Lisanti, M. P., M. Sargiacomo, L. Graeve, A. R. Saltiel, and E. Rodriguez-Boulan. 1988. Polarized apical distribution of glycosyl-phosphatidylinositol anchored proteins in a renal epithelial cell line. *Proceedings of the National Academy of Science*. 85:9557–9561.

Low, M., and A. R. Saltiel. 1988. Structural and functional roles of glycosyl-phosphatidylinositol in membranes. *Science*. 239:268–274.

Milstein, C., G. G. Brownlee, T. M. Harrison, and M. B. Matthews. 1972. A possible precursor of immunoglobulin light chains. *Nature New Biology*. 239:117–120.

Palade, G. 1975. Intracellular aspects of the process of protein secretion. *Science*. 189:347–358.

Richardson, J. C. W., and N. L. Simmons. 1979. Demonstration of protein asymmetries in the plasma membrane of cultured renal (MDCK) epithelial cells by lactoperoxidase-mediated iodination. *FEBS Letters.* 105:201–204.

Rodriguez-Boulan, E. 1983. Membrane biogenesis, enveloped RNA viruses and epithelial polarity. *In* Modern Cell Biology. Vol. 1. 119–170. Alan R. Liss, Inc., New York.

Rodriguez-Boulan, E., D. Misek, D. Vega-Salas, P. J. I. Salas, and E. Bard. 1985. Protein sorting in the secretory pathway. *Current Topics in Membranes and Transport.* 24:251–294.

Rodriguez-Boulan, E., and P. J. I. Salas. 1989. External and internal signals for epithelial cell surface polarization. *Annual Review of Physiology.* 51:741–754.

Sargiacomo, M., M. P. Lisanti, L. Graeve, A. LeBivic, and E. Rodriguez-Boulan. 1989. Integral and peripheral protein composition of the apical and basolateral membrane domains in MDCK cells. *Journal of Membrane Biology.* 107:277–286.

Simons, K., and S. D. Fuller. 1985. Cell surface polarity in epithelia. *Annual Review of Cell Biology.* 1:243–288.

Stephens, E. B., and R. W. Compans. 1988. Assembly of animal viruses at cellular membranes. *Annual Review of Microbiology.* 42:489–516.

Thompson, T. E., and T. W. Tillack. 1985. Organization of glycosphingolipids in bilayers and plasma membranes of mammalian cells. *Annual Review of Biophysics and Biophysical Chemistry.* 14:361–386.

van Meer, G., and K. Simons. 1988. Lipid polarity and sorting in epithelial cells. *Journal of Cellular Biochemistry.* 36:51–58.

van Meer, G., E. H. K. Stelzer, R. W. Wijnaendts-van-Resandt, and K. Simons. 1987. Sorting of sphingolipids in epithelial (Madin-Darby canine kidney) cells. *Journal of Cell Biology.* 105:1623–1635.

Chapter 11

A Role for GTP-binding Proteins in Vesicular Transport through the Golgi Complex

Paul Melançon, Benjamin S. Glick, Vivek Malhotra, Peggy J. Weidman, Tito Serafini, Lelio Orci, and James E. Rothman

Department of Molecular Biology, Princeton University, Princeton, New Jersey, 08544; and the Institute of Histology and Embryology, University of Geneva Medical School, 1211 Geneva 4, Switzerland

Introduction

Eukaryotic cells are highly compartmentalized. The distinct lipid and protein composition of the various subcellular compartments is maintained despite extensive vesicular traffic between them. The successful cell-free reconstitution of transport between separate organelles has been an essential step to initiate a biochemical dissection of the complex machinery that must underlie the specificity of this traffic.

Cell-free reconstitution of vesicular transport was first achieved in a system that measures the transfer of proteins between the *cis* and medial compartments of the Golgi apparatus (Balch et al., 1984*a*; Balch and Rothman, 1985). Briefly, the Golgi transport assay measures the transfer of a specific glycoprotein encoded by the Vesicular Stomatitis Virus (VSV-G) between two Golgi-enriched membrane fractions termed donor and acceptor. The donor fraction is prepared from homogenates of a VSV-infected mutant CHO cell line (clone 15B), which lacks the enzyme *N*-acetyl-glucaminosyl transferase 1 (GlcNAc transferase 1). The acceptor fraction, on the other hand, is prepared from homogenates of uninfected wild type CHO cells lacking VSV-G protein but containing GlcNAc transferase I. The transport reaction is carried out in the presence of a radiolabeled substrate, UDP-[^{3}H]GlcNAc; transfer of VSV-G from the donor compartment to the acceptor compartment can then be easily quantitated by measuring how much [^{3}H]GlcNAc has been attached to VSV-G during incubation. In addition to the donor and acceptor membrane fractions, the transport reaction requires the presence of an ATP-regenerating system, and of soluble factors provided as the high speed supernatant of cell extracts. Morphological analysis has established that transport is mediated by a novel type of vesicles coated by material that is morphologically and immunologically different from clathrin (Orci et al., 1986). The vesicular reaction has been further dissected both biochemically and morphologically into a series of intermediates (Balch et al., 1984*b*; Wattenberg et al., 1986; Orci et al., 1989). These include the budding of coated of vesicles from donor membranes, their attachment to the acceptor compartment, and subsequent uncoating and fusion. The Golgi transport assay has already allowed both purification of a first transport component (Block et al., 1988) and characterization of some of its properties (Weidman et al., 1989).

A first link between GTP-binding proteins and vesicular transport was established upon sequencing the SEC4 gene (Salminen and Novick, 1987). Yeast strains harboring a sec4 temperature-sensitive allele accumulate transport vesicles at a nonpermissive temperature, suggesting that the SEC4 protein is involved in the fusion of secretory vesicles with the plasma membrane (Novick et al., 1981). The sequence of SEC4 shows a strong similarity to that of elongation factor EF-Tu and of the oncogene *ras,* particularly within the consensus domains involved in GTP-binding and hydrolysis. Another well-characterized GTP-binding protein, Ypt1p, has been implicated in intracellular transport. Studies with a cold-sensitive allele suggest that Ypt1p is likely to be involved at a stage earlier than Sec4P, possibly within the Golgi (Segev et al., 1988). (Note that mutations in the YPTI gene cause pleiotropic effects that cannot be unambiguously interpreted; Segev and Botstein, 1987, Schmitt et al., 1988.) These results suggest that different families of GTP-binding proteins could be involved at different stages of intracellular transport.

To examine the possibility that GTP-binding proteins play a role in constitutive transport more directly, we have added GTPγS and AlF_4^- (activators of mammalian "G" proteins; Gilman, 1987) to the Golgi transport assay. We find that both compounds inhibit transport. GTPγS blocks a step that corresponds to the processing of vesicles for fusion but not the fusion process itself. Electron microscopy reveals the extensive accumulation of Golgi-associated–coated vesicles under these conditions. Inhibition by GTPγS requires a novel component of cytosol as a cofactor.

Materials and Methods

Materials

Nucleotide analogues were purchased as lithium salts from Boehringer Mannheim Inc., Indianapolis, IN. They were dissolved as stock solutions in 10 mM HEPES-KOH (pH 7.2) and 1 mM dithiothreitol (DTT), frozen in aliquots in liquid nitrogen and stored at −80°C. Each aliquot was used only once.

Donor and acceptor membrane fractions were prepared from VSV-infected 15B CHO cells and uninfected wild type CHO cells, respectively, as before (Balch et al., 1984*a*). They were stored in 1 M sucrose at 0.5–0.8 mg/ml protein. CHO cytosol was prepared from homogenized 15B CHO cells, and gel-filtered as before (Block et al., 1988). The protein concentration was 7.5 mg/ml. Bovine brain cytosol was prepared as described by Wattenberg and Rothman (1986), except that the extract was precipitated with 60% ammonium sulfate, resuspended in one-fifth volume and gel-filtered before use. This modification removed nonspecific inhibitors of transport and increased protein concentration to 37 mg/ml. All other assay materials were prepared as described (Balch et al., 1984*a*).

Transport Assay Conditions Optimal for Inhibition by GTPγS

Unless otherwise indicated in this paper, each 50-μl assay contained donor and acceptor membranes (1.25 μl of each), cytosol (10 μl), 7.5 μl of 1 M sucrose (to achieve a final concentration of sucrose of 0.2 M), 7.3 IU/ml creatine kinase, 2 mM creatine phosphate, 50 μM ATP, 250 μM UTP, and 0.4 μM (0.5 μCi) UDP-[^{3}H]GlcNAc in 25 mM HEPES-KOH (pH 7.0), 15 mM KCl, 2.5 mM $MgCl_2$, and typically 0.2 mM DTT. After 60 min at 37°C, the VSV-G protein was immunoprecipitated, the precipitate was collected and washed on filters, and the filters were counted as described (Balch et al., 1984*a*).

Results

Inhibition of Cell-free Transport by GTPγS and AlF_4^-

The data presented in the first figure demonstrate that nonhydrolyzable analogues of GTP inhibit transport. Half-maximal inhibition ($K_{1/2}$) by GTPγS occurs at ~0.5 μM. GMP-PNP and GMP-PCP also affect transport but are much poorer inhibitors with $K_{1/2}$'s of ~40 and 160 μM. The data presented in Fig. 2 show that the effect is guanine specific and that GTPγS acts by strongly reducing the rate of the transport reaction. Transport assays were carried out in the presence of 10 μM ATPγS, 10 μM GTPγS, or in the absence of nucleotide analogues. The extent of transport was determined at different times after the start of the reaction. Under conditions where ATPγS has virtually no effect on transport, GTPγS reduces the rate of the reaction considerably. ATPγS only inhibits at much higher concentrations, with a $K_{1/2}$ of roughly 0.5 mM. (data not shown).

As shown in Fig. 3, addition of a high concentration of GTP at the start of a

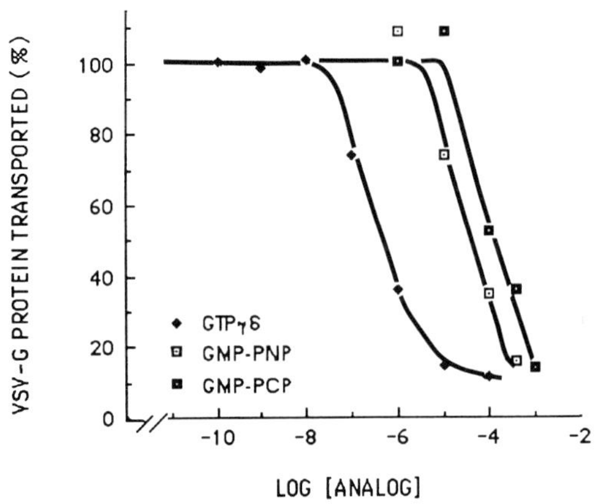

Figure 1. Transport of VSV-G protein is inhibited by GTPγS. A series of transport assays were carried out in the presence of increasing amounts of GTPγS (♦), GMP-PNP (□), or GMP-PCP (■) as described in Materials and Methods. The data are expressed as the fraction (percent) of transport observed in the presence of analogue relative to reactions not containing guanine nucleotide analogues, and are plotted as a function of the analogue concentration.

reaction largely prevents inhibition by GTPγS, most likely because GTP can compete with GTPγS for binding. This experiment further demonstrates that GTPγS binds its target irreversibly. GTP, when added to assays incubated with GTPγS at 37°C for various lengths of time, loses the ability to rescue transport with a $t_{1/2}$ of 2–3 min (Fig. 3). The irreversible nature of the inhibition by GTPγS was also demonstrated by measuring the activity of membranes that had been first incubated either with or without GTPγS, and then reisolated and assayed with fresh cytosol. Such experiments showed that inhibition of transport remained long after GTPγS had been effectively removed (see Melançon et al., 1987).

These results suggest that a GTP-binding protein is involved in the transport reaction. A hallmark of the signal transducing GTP-binding proteins (G proteins) is activation by either GTPγS or fluoride ions. It has been shown that activation by fluoride absolutely requires the presence of aluminum (Sternweis and Gilman, 1982). It appears that the complex anion AlF_4^- binds to the GDP form of G proteins at the position of the γ-phosphate, thereby freezing the protein in the GTP conformation (Bigay et al., 1987). The data in Fig. 4 show the effect of adding increasing amounts of

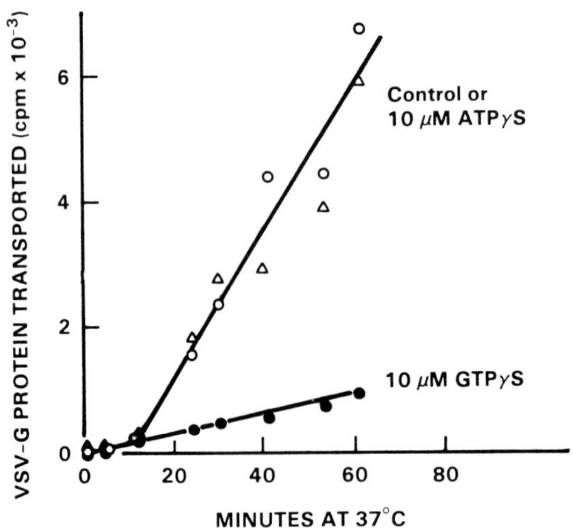

Figure 2. The rate of transport of VSV-G protein is strongly reduced by GTPγS but not by ATPγS. Transport assays were performed as described in Materials and Methods in either the absence of a nucleotide analogue (○), in the presence of 10 μM ATPγS (△), or with 10 μM GTPγS (●). Aliquots from each reaction were removed for immunoprecipitation after the indicated times at 37°C.

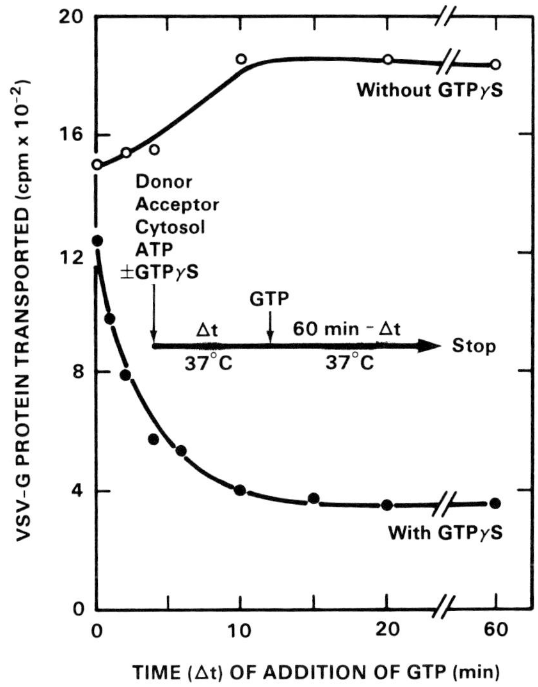

Figure 3. Time course of irreversible inhibition by GTPγS. Transport reactions were carried out either in the absence (○) or the presence of 10 μM GTPγS (●). At various times at 37°C, 2 mM Mg-GTP was added to aliquots, and the incubation was continued for a total of 60 min. The amount of VSV-G protein transported is plotted as a function of the time (Δt) of incubation before the addition of GTP.

Al ions to the transport assay. Titrations were carried out either in the presence of 5 mM KCl or KF. Whereas the addition of fluoride alone has little effect on transport, the presence of both aluminum and fluoride causes strong inhibition. This clear synergy between Al^{3+} and F^- provides further evidence for the involvement of a G protein in transport.

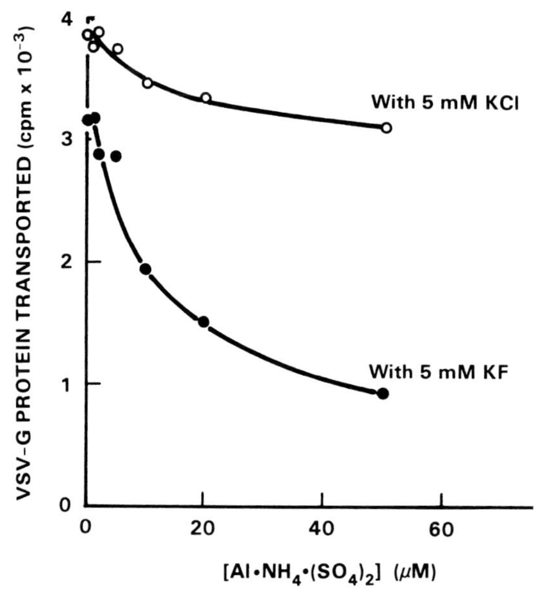

Figure 4. Synergistic inhibition of transport by fluoride and aluminum ions. Transport assays were performed under standard conditions (Balch et al., 1984*a*); 2.5 μl each of donor and acceptor membrane fractions were in a final volume of 25 μl and 0.75 mg/ml cytosol with 2 mM DTT and an increasing concentration of aluminum ions (added as $AlNH_4(SO_4)_2$. The amount of VSV-G protein transported is plotted as a function of the Al^{3+} concentration.

The transport assay is carried out with membrane fractions that are enriched in Golgi membranes but remain contaminated with plasma membrane fragments that will contain classical G proteins (G_i, G_o, G_s, G_n). Controls were therefore carried out to test the possibility that these G proteins or their secondary messengers could be responsible for our observation. Treatments with pertussis and cholera toxins were without effect on transport. The secondary messengers cAMP (1 mM) and cGMP (1 mM), activators of specific kinases were also without effect. Addition of IP_3 (10 μM) and phorbol-myristic acid (1 μM) had no detectable effect on transport. Contrary to what is the case in exocytosis, Ca^{2+} does not seem to play a role in transport through

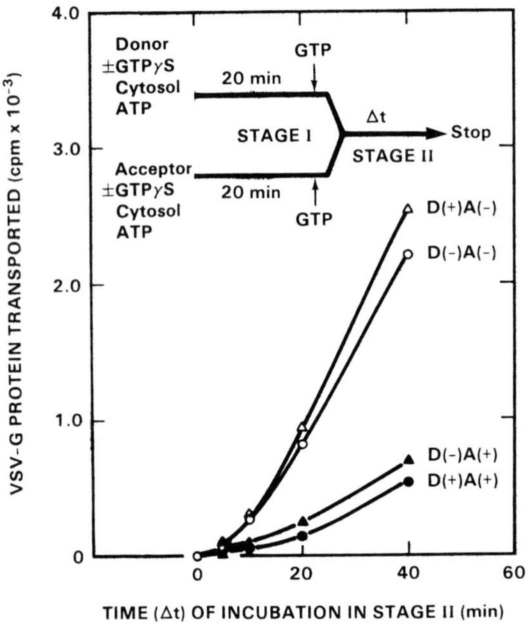

Figure 5. GTPγS inhibits the activity of the acceptor but not that of the donor compartment. Two stage assay. In stage 1, donor (*D*) and acceptor (*A*) membrane fractions were incubated separately under assay conditions described in Materials and Methods. These stage 1 incubations were carried out either in the presence (+) or in the absence (−) of 10 μM GTPγS. After 20 min at 37°C, the stage 1 incubations were chilled on ice and 2 mM Mg-GTP was added to each. Equal volumes of donor and acceptor incubations were mixed pairwise as indicated on the graph. The reactions were returned to 37°C, and aliquots were removed for immunoprecipitation at the indicated times (Δt) after initiation of the stage II incubation at 37°C. The amount of VSV-G protein transported is plotted as a function of the length of incubation in stage II.

the Golgi: the addition of 1 mM EGTA does not affect transport, nor does it prevent inhibition by GTPγS. Lastly, the addition of up to 100 μM of taxol and colchicine neither inhibited transport nor interfered with the effect of GTPγS, therefore demonstrating that stabilization of microtubules is not responsible for our observation (data not shown).

GTPγS Inhibits the Function of the Acceptor but not the Donor Compartment

The donor and acceptor Golgi fractions are chemically very similar, differing principally in the presence of VSV-G protein in the donor and of GlcNAc transferase 1 in the acceptor. However, the two fractions perform very different functions in the transport assay: the donor stack is responsible for budding vesicles, whereas the function of the acceptor is to bind, process, and fuse these vesicles.

To test whether donor or acceptor functions are inhibited by GTPγS, we preincubated the fractions separately with GTPγS and then measured transport after adding the untreated membrane partner, as well as GTP to prevent subsequent

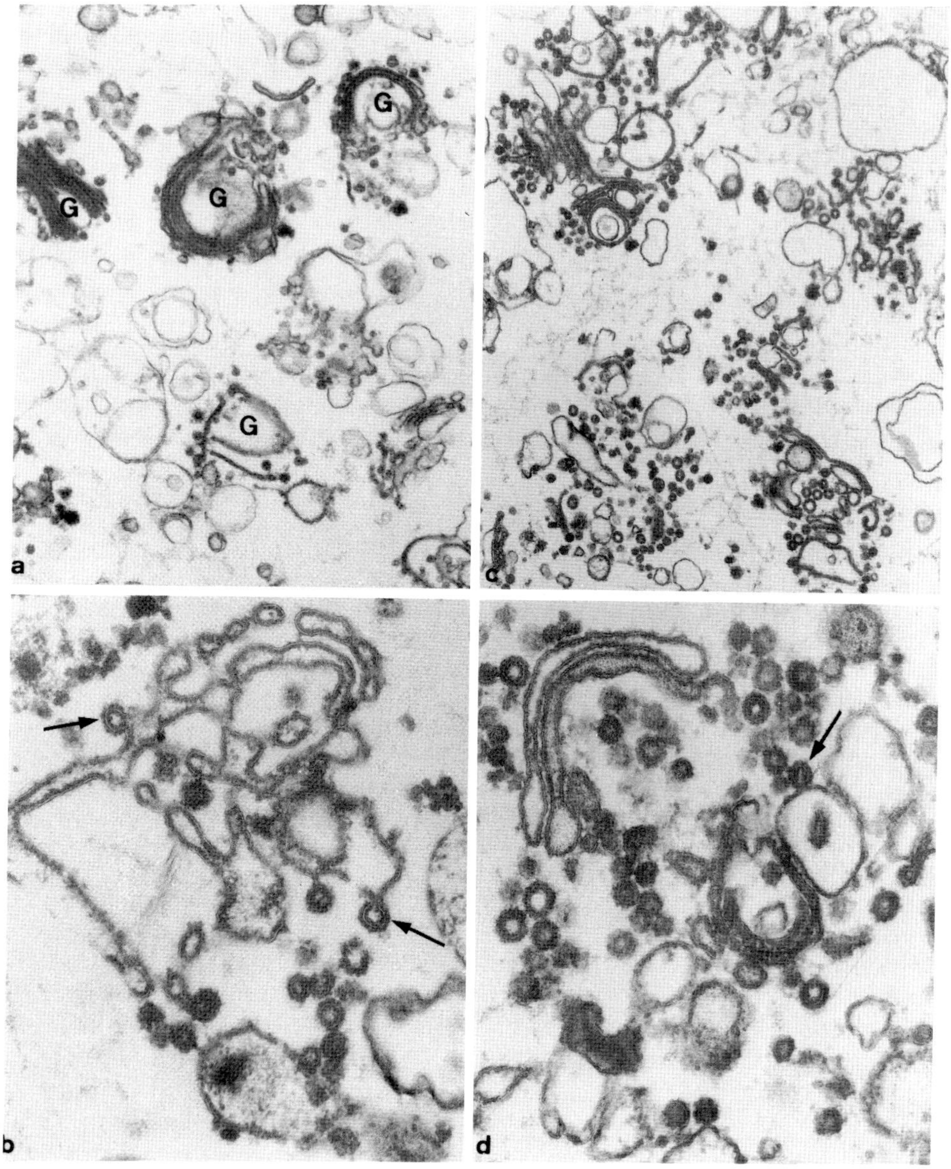

Figure 6. Coated vesicles accumulate in the presence of GTPγS. Golgi fractions were incubated as described in Materials and Methods in the absence (*a* and *b*) or in the presence (*c* and *d*) of 10 μM GTPγS. The pellets of the Golgi fractions were fixed and processed as described (Orci et al., 1986). Magnification: 21,000× (*a* and *b*), 61,000× (*c* and *d*). *G* indicates Golgi area. Arrows indicate examples of budding coated vesicles.

inhibition by GTPγS (Fig. 5). Preincubation with GTPγS clearly inhibited the activity of the acceptor, but had no effect upon the activity of the donor.

Golgi-associated–coated Vesicles Accumulate in the Presence of GTPγS

The preceding experiment suggests that GTPγS does not affect budding of transport vesicles. Given that budding is not affected, but that GTPγS clearly blocks some function of the acceptor, we would predict the accumulation of transport intermediates such as vesicles attached to Golgi cisternae or budding vesicles that pile up behind the block.

Fig. 6 reveals that there is indeed an accumulation of transport vesicles in the presence of GTPγS. Fig. 6 *a* shows the appearance of Golgi stacks incubated in the absence of GTPγS. Numerous vesicles and buds are associated with the stacks. The appearance of such vesicles under transport conditions has been previously demonstrated to occur only under conditions that favor transport; vesicles appear only upon incubation at 37°C, and the presence of both cytosol and ATP is required (Orci et al., 1986). Since VSV-G protein was detected in these vesicles, they were postulated to be transport intermediates responsible for the transfer of VSV-G between the donor and acceptor compartments. Fig 6 *c* shows a much greater number of these vesicular elements upon incubation with GTPγS. While some of these vesicles are clearly budding from cisternae (Fig. 6, *b* and *d*), the majority are fully formed and separate from Golgi cisternae and yet clearly associated in some specific fashion to the Golgi stacks. Using scrape-loading to introduce the analogue into living cells, it has been shown that GTPγS causes accumulation of coated vesicles in vivo as well (Orci et al., 1989).

TABLE I
Quantitation of Golgi Vesicles and Buds in the Presence and Absence of GTPγS

	Density of vesicles plus buds*	Mean circumferential length of vesicles and buds per Golgi‡	Percentage of buds or vesicles with a complete cytoplasmic coat§
	No/μm² Golgi	*μm*	
+GTPγS	54.5 + 1.5	12.9 + 0.8	85.2 + 2
−GTPγS	10.7 + 0.6	2.6 + 0.2	79.4 + 4

Pellets of the Golgi fractions were fixed and processed as described (Orci et al., 1986). Golgi areas were randomly photographed at a calibrated magnification of 38,100× or 82,000×. The following measurements were carried out on photographic prints with the aid of an electronic pen connected to a microprocessor as described by Orci et al. (1986).

*30 Golgi areas were analyzed.

‡The sum of circumferences or arc length of vesicles plus buds divided by the number of distinct Golgi areas counted. 30 Golgi areas were analyzed.

§Either 564 (+GTPγS) or 186 (−GTPγS) vesicles or buds were counted. Shown are the mean values followed by the standard error of the mean.

A quantitative analysis (Table I) of micrographs such as that shown in Fig. 6 reveals that the number of vesicles plus buds per μm^2 Golgi areas increased from 10.7 + 0.6 to 54.4 (i.e., fivefold) in the presence of GTPγS. Similarly, the length of membrane in cross section in vesicles plus buds increases fivefold. Note that 85% of the accumulated vesicles bear a cytoplasmic coat (Fig. 6 and Table I) of the sort previously shown to lack clathrin (Orci et al., 1986). This suggests that GTPγS blocks a step in vesicle processing that occurs before or during the uncoating of the vesicles in preparation for their fusion.

There is a striking correspondence between the GTPγS-dependent accumulation of coated vesicles seen by electron microscopy and the effects of GTPγS on transport measured biochemically. This provides very strong and independent evidence that the coated vesicles do indeed act as carriers for the VSV-G protein between Golgi cisternae, as has been proposed (Orci et al., 1986).

Inhibition by GTPγS Requires a Cytosolic Component

The extent of inhibition caused by the addition of GTPγS varies with the amount of cytosolic proteins present in the transport reaction. Even when the requirement for cytosol in the assay is nearly satisfied (150 μg cytosol), further addition of cytosol greatly increases inhibition by GTPγS (Fig. 7). Similar cytosol titration experiments

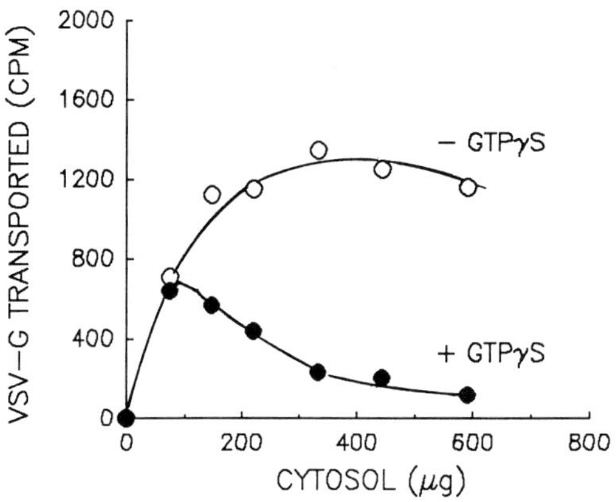

Figure 7. Inhibition by GTPγS as a function of cytosol concentration. Assays were carried out either in the absence (○) or presence (●) of 10 μM GTPγS. Each 50-μl assay contained 2.25 μg membrane protein and included 5 μM palmitoyl coenzyme A. The amount of VSV-G protein transported in a 60-min incubation at 37°C is plotted as a function of the final concentration of bovine brain cytosol.

were carried out at various membrane concentrations. The results, presented in Fig. 8, clearly establish that it is the ratio of cytosolic to membrane proteins and not the cytosolic protein concentration that determines the extent of inhibition. (Note that, unless otherwise indicated, all experiments described in this report were performed under low membrane, high cytosol conditions.) These observations suggest that a component(s) present in limiting amounts in the cytosol is needed to enable inhibition by GTPγS. The importance of the cytosol to membrane ratio most likely indicates that this cytosolic component interacts with a complementary membrane-associated component.

This putative interaction with membranes was tested directly in the experiment shown in Fig. 9. Supernatants were prepared from incubations of cytosol and membranes in the presence or the absence of GTPγS. Transport reactions using as "cytosol" the supernatant from the preincubation with GTPγS were no longer inhibited by GTPγS. In contrast, supernatants from preincubations without GTPγS still permitted inhibition, i.e., these supernatants still contained the inhibitory factor. Removal of the cytosolic factor requires the presence of membranes, and the amount of the factor removed from cytosol increases in direct proportion to the amount of

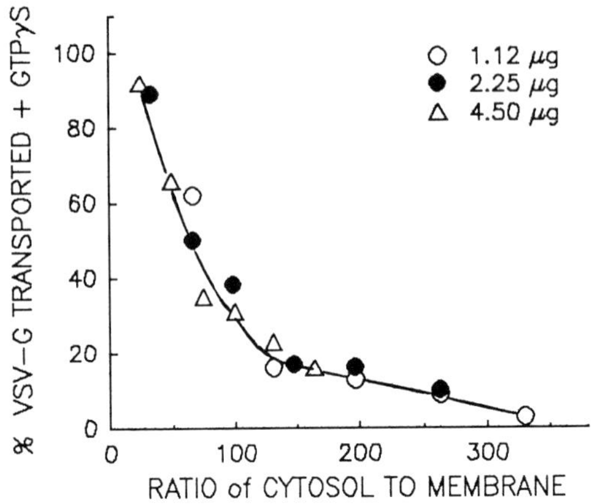

Figure 8. Inhibition by GTPγS as a function of the ratio of cytosol to membrane. Assays, carried out as in Fig. 7, contained either 1.12 (○), 2.25 (●), or 4.5 (△) μg of membrane proteins. The amount of VSV-G protein transported in a 60-min incubation at 37°C was determined both in the absence and presence of 10 μM GTPγS. The fraction of transport (percent) observed in the presence of GTPγS relative to incubations free of analogue is plotted as a function of the ratio of cytosol to membrane proteins in the assay.

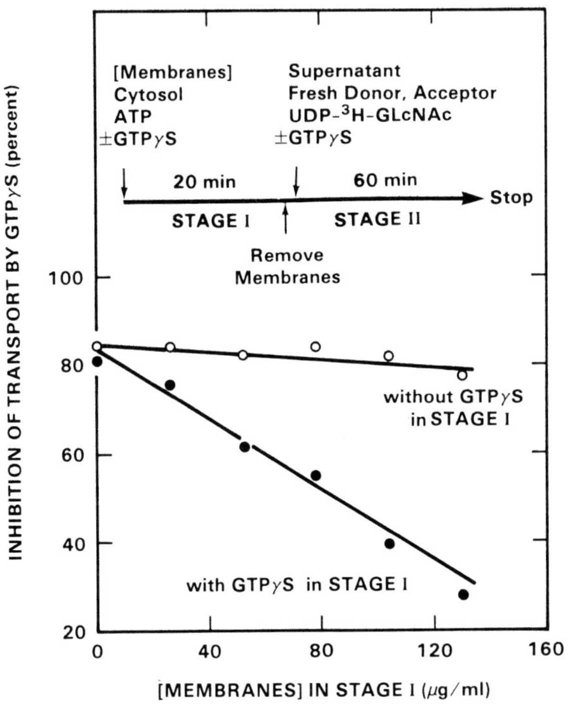

Figure 9. An inhibitory factor is removed from membranes upon incubation with Golgi membranes and GTPγS. Two stage assay. Cytosol was treated to remove inhibitory factor in stage I. The supernatants of stage I were assayed with fresh donor and acceptor membranes in stage II. In stage I, cytosol (2.1 mg/ml) was incubated for 20 min at 37°C in assay buffer with increasing amounts of acceptor membranes either in the absence or presence of 10 μM GTPγS. The final sucrose concentration was maintained at 0.2 M. Membranes were then removed by centrifugation at 400,000 *g* for 10 min. In stage II, the supernatants were assayed by adding fresh membranes and UDP-[^{3}H]GlcNAc to give the final assay conditions described in Materials and Methods. The supernatants from stage I incubations carried out without GTPγS were assayed in Stage II in the absence and presence of 20 μM GTPγS. Supernatants from stage I incubations with GTPγS were assayed in stage II with fresh GTPγS (final concentration, 20 μM). The percentage inhibition by GTPγS was calculated using as the 100% value the amount of transport observed when GTPγS was absent from both stages I and II. The percentage of inhibition measured in stage II is plotted as a function of the membrane concentration in stage I. GTPγS absent in stage I (○); GTPγS present in stage I (●).

membranes added in the preincubation. Additional experiments demonstrated that the GTPγS-dependent depletion occurs with a half-time of ~7 min and is complete within 20 min at 37°C (data not shown). These kinetics are independent of membrane concentration. Depletion does not occur at 0°C. The concentration of GTPγS required to obtain half-maximal depletion ($K_{1/2} < 1$ μM) is similar to that required for inhibition. These observations suggest the formation of a 1:1 complex between the factor and a membrane protein, and they explain why it is the ratio of cytosol to membranes that is critical in determining the extent of inhibition of GTPγS.

The fact that the inhibitory component can be removed without diminishing the transport activity of cytosol suggests that this component may not play an essential role in the transport reaction but rather could be part of a regulatory mechanism that can switch constitutive transport on and off. The possibility that this component is involved in the shutdown of vesicular transport that occurs at mitosis (Warren, 1985) was therefore tested. Cytosol extracts from interphase and mitotic cells were compared in their ability to support transport or its GTPγS-dependent inhibition. All experiments up to this point have failed to show a difference between the two extracts (data not shown).

Discussion

Intracisternal protein transport as assayed in our cell-free system is inhibited by nonhydrolyzable analogues of GTP and by AlF_4^-. Inhibition of transport of GTPγS is irreversible and requires the presence of a factor(s) from cytosol. The cytosolic factor appears to act stoichiometrically rather than enzymatically since (*a*) it is the ratio of cytosolic to membrane proteins rather than cytosol concentration that determines the extent of inhibition and (*b*) the factor is removed from cytosol as it inactivates acceptor membranes in a GTPγS-dependent fashion. The transport pathway seems to be inhibited after the budding and attachment of transport vesicles, but before their fusion with the acceptor cisternae, at or before the step at which the cytoplasmic coat is removed.

Our observation made with the Golgi transport assay has now been extended: specific inhibition by low concentration of guanine analogue and by AlF_4^- has been reported in three different cell-free assays that reconstitute transport between the endoplasmic reticulum and the Golgi apparatus in a yeast (Baker et al., 1988; Ruohola et al., 1988) and a mammalian (Beckers and Balch, 1988) system. This additional evidence clearly supports the notion that GTP-binding proteins are involved in intracellular transport.

Is the putative GTP-binding protein an integral piece of the constitutive transport machinery, or is it a part of a regulatory mechanism that can switch transport on and off? In the "regulatory" model, the addition of GTPγS leads to the activation of a control switch that negatively regulates transport. Such a switch could be used at mitosis where intracellular transport is arrested and where the ER and the Golgi apparatus extensively vesiculate (Warren, 1985; Lucocq et al., 1987). In the "constitutive" model, a cycle of GTP hydrolysis plays an essential role in the transport reaction. When this cycle is blocked with GTPγS or AlF_4^-, transport is also blocked and vesicles accumulate. A good analogy for the constitutive model is the cycling of binding and hydrolysis of GTP by initiation and elongation factors that accompany protein translation (Miller and Weissbach, 1977; Bourne, 1988).

The data at hand do not permit a clear distinction between these two alternatives. As noted in the Results section, our efforts to establish a connection between the effect of GTPγS on transport and mitosis have failed so far. On the other hand, genetic and immunological studies of the GTP-binding protein Sec4p strongly support a role for this protein in the constitutive transport reaction. In particular, Sec4p is clearly associated with secretory vesicles involved in transport to the plasma membrane (Goud et al., 1988). Further support for the constitutive model comes from studies demonstrating that, in addition to Ca^{2+}, guanine nucleotides are required to support secretion from permeabilized cells (Howell et al., 1987).

One can only speculate at this point on the possible function that GTP-binding proteins have in the transport reaction. An intriguing possibility, also raised by several other authors, is that a family of GTP-binding protein may in some way regulate the specificity that underlies vesicle targeting. One such model has been recently discussed in detail in Bourne (1988). The issue will be resolved by the isolation of several of the putative GTP-binding proteins and by the characterization of the proteins and/or factors with which they interact.

Acknowledgments

We wish to thank Professors Alfred Gilman and Felix Wieland for their advice throughout the course of this work. We also thank M. Amherdt and A. Perrelet for their contributions, and M. L. Gleason, F. Freymark, G. Perrelet, G. Negro, and P.-A. Ruttimann for technical assistance.

This research was supported by grants from the National Institutes of Health (to J. E. Rothman), the Swiss National Science Foundation (3.404.86 to L. Orci), by a Medical Research Council (Canada) Postdoctoral Fellowship (P. Melançon), and by a Damon Runyon-Walter Winchell Postdoctoral Fellowship (V. Malhotra).

References

Baker, D., L. Hicke, M. Rexach, M. Schleyer, and R. Scheckman. 1988. Reconstitution of SEC gene product-dependent intercompartmental protein transport. *Cell.* 54:335–344.

Balch, W. E., W. G. Dunphy, W. A. Braell, and J. E. Rothman. Reconstitution of the transport of protein between successive compartments of the Golgi measured by the coupled incorporation of *N*-acetylglucosamine. 1984*a*. *Cell.* 39:405–416.

Balch, W. E., B. S. Glick, and J. E. Rothman. 1984*b*. Sequential intermediates in the pathway of intercompartmental transport in a cell-free system. *Cell.* 39:525–536.

Balch, W. E., and J. E. Rothman. 1985. Characterization of protein transport between successive compartments of the Golgi apparatus: asymmetric properties of donor and acceptor activities in a cell-free system. *Archives of Biochemistry and Biophysics.* 240:413–425.

Beckers, C. J. M., and W. B. Balch. 1989. Calcium and GTP: essential components in vesicular trafficking between the endoplasmic reticulum and Golgi apparatus. *Journal of Cell Biology.* 108:1245–1256.

Bigay, J., P. Deterre, C. Pfister, and M. Chabre. 1987. Fluoride complexes of aluminum or beryllium act on G-proteins as reversibly bound analogues of the γ phosphate of GTP. *EMBO Journal.* 6:2907–2913.

Block, M. R., B. S. Glick, C. A. Wilcox, F. T. Wieland, and J. E. Rothman. 1988. Purification of an *N*-ethylmaleimide-sensitive protein catalyzing vesicular transport. *Proceedings of the National Academy of Sciences*. 85:7852–7856.

Bourne, H. R. 1988. Do GTPases direct membrane traffic in secretion? *Cell*. 53:669–671.

Gilman, A. 1987. G proteins: transducers of receptor-generated signals. *Annual Review of Biochemistry*. 56:615–649.

Goud, B., A. Salminen, N. C. Walworth, and P. J. Novick. A GTP-binding protein required for secretion rapidly associates with secretory vesicles and the plasma membrane in yeast. 1988. *Cell*. 53:754–768.

Howell, T. W., S. Cockcroft, and B. D. Gomperts. 1987. Essential synergy between Ca^{2+} and guanine nucleotides in exocytic secretion from permeabilized rat mast cells. *Journal of Cell Biology*. 105:191–197.

Lucocq, J. M., J. G. Pryde, E. G. Berger, and G. Warren. 1987. A mitotic form of the Golgi apparatus in HeLa cells. *Journal of Cell Biology*. 104:865–874.

Melançon, P., B. S. Glick, V. Malhotra, P. J. Weidman, T. Serafini, M. J. Gleason, L. Orci, and J. E. Rothman. 1987. Involvement of GTP-binding "G" proteins in transport through the Golgi stack. *Cell*. 51:1053–1062.

Miller, D. L., and H. Weissbach. 1977. Molecular Mechanism of Protein Biosynthesis. H. Weissbach and S. Petska, editors. Academic Press, New York. 323–373.

Novick, P. J., S. Ferro, and R. Scheckman. 1981. Order of events in the yeast secretory pathway. *Cell*. 25:171–184.

Orci, L., B. S. Glick, and J. E. Rothman. 1986. A new type of vesicular carrier that appears not to contain clathrin: its possible role in protein transport within the Golgi stack. *Cell*. 46:171–184.

Orci, L., V. Malhotra, M. Amherdt, T. Serafini, and J. E. Rothman. 1989. Dissection of a single round of vesicular transport: sequential intermediates for intercisternal movement in the Golgi stack. *Cell*. 56:357–368.

Ruohola, H., A. Kastan Kabcenell, and S. Ferro-Novick. 1988. Reconstitution of protein transport from the endoplasmic reticulum to the Golgi complex in yeast: the acceptor Golgi compartment is defective in the sec23 mutant. *Journal of Cell Biology*. 107:1465–1476.

Salminen, A., and P. J. Novick. 1987. A ras-like protein is required for a post-Golgi event in yeast secretion. *Cell*. 49:527–538.

Schmitt, H. D., M. Puzicha, and D. Gallwitz. 1988. Study of a temperature-sensitive mutant of the ras-related YPT1 gene product in yeast suggests a role in the regulation of intracellular calcium. *Cell*. 53:635–647.

Segev, N., and D. Botstein. 1987. The ras-like yeast YPT1 gene is itself essential for growth, sporulation, and starvation response. *Molecular and Cellular Biology*. 7:2367–2377.

Segev, N., J. Mulholland, and D. Botstein. 1988. The yeast GTP-binding YPT1 protein and a mammalian counterpart are associated with the secretion machinery. *Cell*. 52:915–924.

Sternweis, P. C., and A. G. Gilman. 1982. Aluminum: a requirement for activation of the regulatory components of adenylate cyclase by fluoride. *Proceedings of the National Academy of Sciences*. 79:4888–4891.

Warren, G. 1985. Membrane traffic and organelle division. *Trends in Biochemical Sciences*. 10:439–443.

Wattenberg, B. W., W. E. Balch, and J. E. Rothman. 1986. A novel prefusion complex formed during protein transport between Golgi cisternae in a cell-free system. *Journal of Biological Chemistry*. 261:2202–2207.

Wattenberg, B. W., and J. E. Rothman. 1986. Multiple cytosolic components promote intra-Golgi protein transport. *Journal of Biological Chemistry*. 261:2208–2213.

Weidman, P. J., P. Melançon, M. R. Block, and J. E. Rothman. 1989. Binding of an *N*-ethylmaleimide-sensitive fusion protein to Golgi membranes requires both a soluble protein and an integral membrane receptor. *Journal of Cell Biology*. 108:1589–1596.

Chapter 12

Protein Secretion by Constitutive and Regulated Pathways

Hsiao-Ping H. Moore, Catherine Brion, Koong-Nah Chung, Leo Lehmicke, Rodolfo Rivas, and David Quinn

Department of Physiology-Anatomy, University of California, Berkeley, California 94720

The Secretory Pathways of Eucaryotic Cells

Secretion is one of the fundamental processes of all cells. Therefore, it is not too surprising that the basic elements involved in secretion are quite conserved throughout evolution. From yeast to mammalian cells, proteins destined for the cell surface follow a similar intracellular pathway that includes sequential transfer of proteins from the rough endoplasmic reticulum to the Golgi to the secretory vesicles and finally to the plasma membrane (Palade, 1975; Farquhar, 1985; Schekman, 1985). This common route suggests that there may be a universal pathway for intracellular transport of all exported proteins. Indeed, studies with yeast cells have shown that the same set of secretory vesicles can mediate transport of a number of secretory and plasma

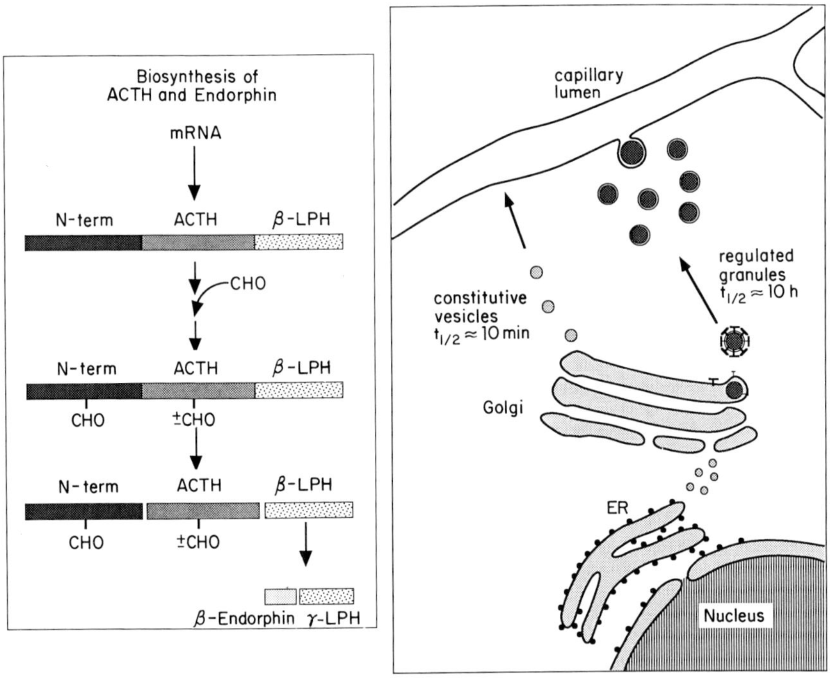

Figure 1. Biosynthesis and intracellular transport of ACTH and β-endorphin in AtT-20 cells. Secretory proteins are segregated into two types of post-Golgi vesicles. The hormones preferentially enter the regulated secretory granules.

membrane proteins (Holcomb et al., 1988). However, this apparently simple scheme is not sufficient to explain recent findings in higher eucaryotic cells. It is now clear that cells with more differentiated secretory functions segregate their secretory products into distinct secretory pathways (Tartakoff et al., 1978; Kelly, 1985; Moore, 1987; Moore et al., 1988). Each pathway allows precise temporal and spatial control of secretion. For example, epithelial cells contain two types of secretory vesicles that shuttle proteins to the apical or the basolateral surfaces (Simons and Fuller, 1985). In this article we will discuss sorting of secretory proteins into the constitutive and the regulated secretory pathways in endocrine cells.

The two secretory pathways have been studied most extensively in the mouse pituitary AtT-20 (Fig. 1). These cells are derived from corticotrophs, and secrete two peptide hormones: adrenocorticotropic hormone (ACTH) and β-endorphin. Both hormones are derived from a single precursor, proopiomelanocortin (POMC; Mains et al., 1977; Roberts and Herbert, 1977). During transport through the Golgi and packaging into secretory granules, the mature hormones are exicsed from their precursors at pairs of basic amino acid residues (Loh et al., 1984; Tooze et al., 1987*b*). Both ACTH and β-endorphin are packaged at high concentrations within dense-core secretory granules, which accumulate in the cytoplasm until secretion is triggered by external stimuli. Secretion by this pathway is controlled by second messenger systems and is called "regulated secretion" (Mains and Eipper, 1981; Gumbiner and Kelly, 1982). In contrast, export of many plasma membrane proteins and extracellular matrix materials from these cells involves a constitutive secretory pathway wherein transport vesicles budding from the Golgi fuse directly with the plasma membrane without prior storage (Gumbiner and Kelly, 1982; Moore et al., 1983*b*). Their half life in the cytosol is on the order of 10 min, compared with 7–10 h for hormone-containing granules. Exocytosis of these vesicles is not influenced by external stimuli. Experimentally, the two pathways can be readily distinguished by their distinct kinetics of secretion, their differential sensitivities to secretagogues, and their unique intracellular localization, which can be analyzed by subcellular fractionation or immunoelectron microscopy.

Intracellular Sorting of Secretory Proteins

Our understanding of molecular sorting is greatly facilitated by studies using DNA transfection and in vitro mutagenesis techniques. A variety of foreign hormones, including human proinsulin (Moore et al., 1983*c*), human growth hormone (Moore and Kelly, 1985), and mouse prolactin (Powell, S., D. Linzer, and H. P. Moore, unpublished data), can be transfected into AtT-20 cells where they are sorted correctly into dense-core granules. Even proteins of exocrine origin, such as rat trypsinogen, are also sorted into the regulated pathway of the pituitary AtT-20 cells (Burgess et al., 1985). However, the sorting machinery does discriminate against other secretory proteins that normally do not enter the regulated secretory pathway. Immunoglobulin light chains (Matsuuchi et al., 1988) and a viral membrane glycoprotein devoid of its membrane anchor (Moore and Kelly, 1985) are both secreted constitutively when transfected into AtT-20 cells. Thus, the AtT-20 cell provides a reliable surrogate system for studying sorting between the constitutive and the regulated secretory pathways.

Where do the two secretory pathways diverge? In principle, proteins could be sorted at early stages of transport; one possibility is that different classes of secretory proteins could contain separate signal peptides that are recognized by receptors residing within district domains of the endoplasmic reticulum membrane. Alternatively, they could be sorted during later stages of transport. Construction of chimeric proteins with swapped presequences demonstrates that the signal peptides of constitutive and regulated proteins are functionally equivalent. Fig. 2 shows the secretory pathway taken by a hybrid protein in which the NH_2-terminal signal sequence of trypsinogen (TG), a constitutively secreted viral glycoprotein, was replaced with that of a regulated protein, human growth hormone. This hybrid polypeptide (hGHsp-TG) is targeted efficiently to the secretory apparatus of transfected cells, but is not sorted into storage granules. Instead, it follows the pathway of the viral protein and is secreted constitutively. Thus, the presequence of human growth hormone can substitute for the signal peptide of TG but it contains no additional information to target it to storage granules. The converse experiment was carried out by Burgess et al. (1987) who demonstrated that replacing the signal peptide of a regulated protein trypsinogen with that of a constitutively secreted protein immunoglobulin light chain has no effect on its intracellular localization (Fig. 3). Taken together, these experiments show that targeting of secretory proteins must take place in at least two steps. Both classes of secretory proteins are separated from cytosolic, nuclear, and mitochondrial proteins via functionally equivalent signal peptides. At a later stage during transport, constitutive and regulated proteins are further sorted into distinct secretory pathways via sorting signals internal to the mature coding sequences.

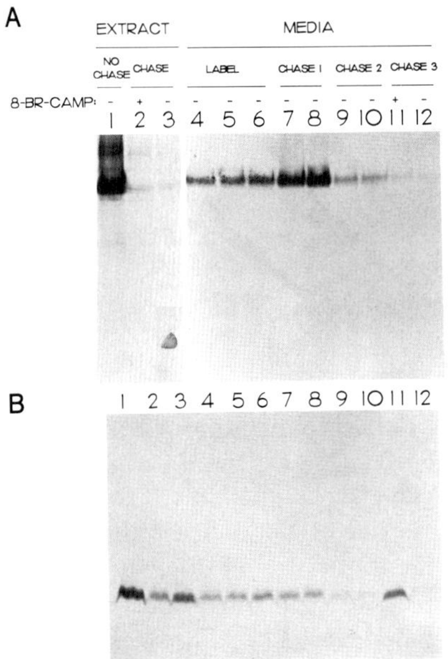

Figure 2. Transport and secretion of a chimeric protein consisting of the signal peptide of a regulated secretory protein (human growth hormone) fused to the mature coding sequence of a constitutively secreted protein TG (Moore and Kelly, 1985). (*A*) Identical dishes of hGHsp-TG12, and AtT-20 cell line transfected with the chimeric gene, hGHsp-TG, were labeled with [^{35}S]methionine for 16 h followed by three 3-h chase periods. During the last chase period one dish of cells was stimulated with 8-Br-cAMP. Media and cell extracts were immunoprecipitated with anti-VSV G antisera. A duplicate control dish was immunoprecipitated before the chase. Lane *1*, cell extract before chase; lanes *2* and *3*, cell extracts after 9 h chase from stimulated and unstimulated dishes, respectively; lanes *4*, *5*, and *6*, media samples collected from cells during the last hour of labeling period; lanes *7* and *8*, media samples during 0–3 h chase; lanes *9* and *10*, media samples during 3–6 h chase; lanes *11* and *12*, media samples during 6–9 h chase from stimulated and unstimulated cells, respectively. (*B*) The identical experiment was carried out with hGH313, an AtT-20 cell line transfected with the wild type hGH DNA. Media and extracts were immunoprecipitated with anti-hGH antisera.

Double immunoelectron microscopy was used to directly identify the sorting compartment (Orci et al., 1987). Stably transfected AtT-20 cells expressing human insulin were infected with influenza virus. The viral glycoprotein hemagglutinin (HA) was used as a marker for the constitutive pathway, and insulin was used as a marker for the regulated pathway. The two proteins were found to be intermixed within all cisternae of the Golgi stacks, as well as in the *trans*-most cisternae where peptide hormones begin to condense. HA and insulin are clearly segregated in post-Golgi vesicles: HA in 100–300 nm clear vesicles and insulin in dense-core granules. Thus, sorting of a constitutively externalized membrane protein and a regulated secretory protein occurs in the *trans*-Golgi compartment. Sorting of proteins into apically or basolaterally directed vesicles in epithelial cells also involves the same compartment

(Griffiths and Simons, 1986). Perhaps the *trans*-Golgi cisternae can be considered as the grand sorting station for all surface-bound proteins.

Physiology of Constitutive and Regulated Secretion

As discussed earlier, segregation of proteins into constitutive and regulated secretory pathways enables the cell to differentially regulate the discharge of different proteins. In addition to such temporal regulation, the two pathways also offer the opportunity to segregate proteins spatially (Rivas and Moore, 1989). AtT-20 cells grown in laminin form long processes, and regulated granules, traveling on microtubules, accumulate at the tips of these processes (Tooze and Burke, 1987*a*; Matsuuchi et al., 1988). Are constitutively secreted membrane proteins also inserted at the process tips or somewhere else? To answer this, AtT-20 cells were infected with ts-045, a temperature-sensitive mutant of vesicular stomatitis virus in which transport of the surface glycoprotein G is conditionally blocked in the endoplasmic reticulum. Upon switching

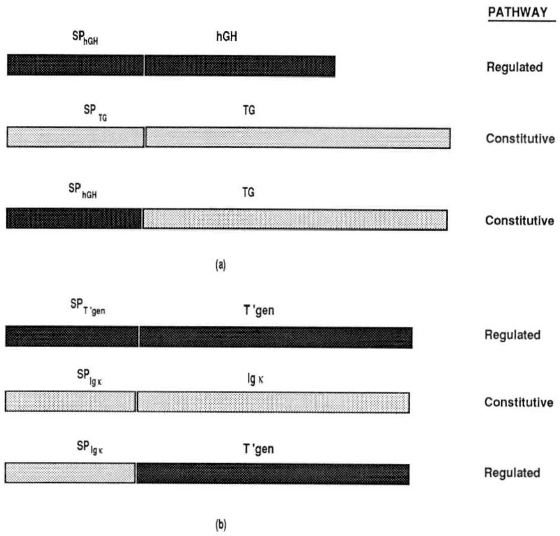

Figure 3. Effects of signal peptide substitutions on the pathways of secretion. In (*a*), the signal peptide of a constitutively secreted protein is replaced with that of a regulated secretory protein (see Fig. 2). In *B*, the signal peptide of a regulated secretory protein is replaced with that of a constitutive one. In both cases, the secretory pathway of the chimeric protein is dictated by its mature coding sequences, not by the signal peptide. Data in *b* are taken from Burgess et al., 1987.

to the permissive temperature, insertion of G protein was detected at the cell body, not at the process tips. Disruption of microtubules by colchicine had no effect on constitutive secretion, but blocked the accumulation of regulated granules at special release sites. The differential effect of microtubule disruption suggests that regulated and constitutive vesicles differ in their ability to associate with microtubules. One intriguing possibility is that regulated granules possess unique membrane receptors for a microtubule-based motor such as kinesin (Vale et al., 1985), but constitutive vesicles lack such receptors and thus do not associate with the motor (Fig. 4). Differential insertion of proteins into distinct membrane domains may be particularly important in neuronal cells, where it can give rise to the regional differences in structure and function between the cell body and axons.

Differential targeting of proteins into distinct vesicles also helps to regulate the activities of the final secreted products. Proteolytic processing and activation of prohormones often occurs in maturing granules after they are targeted to the correct secretory pathway (Orci et al., 1985; Tooze et al., 1987*b*).

Assembly and Biogenesis of Secretory Vesicles

How do the different types of vesicles form in the *trans*-Golgi region, and how are proteins sorted into different kinds of vesicles? One possibility is that the ionic environment in an area of the *trans*-Golgi region is conducive for formation of aggregates of those proteins destined for the same vesicle type. For example, the slightly acidic milieu of the *trans*-Golgi region may cause those proteins with isoelectric points (pI) similar to the pH of Golgi lumen to precipitate, and thereby become segregated away from other proteins within the same cisternae. However, comparison of pIs of secretory proteins to their intracellular locations shows little

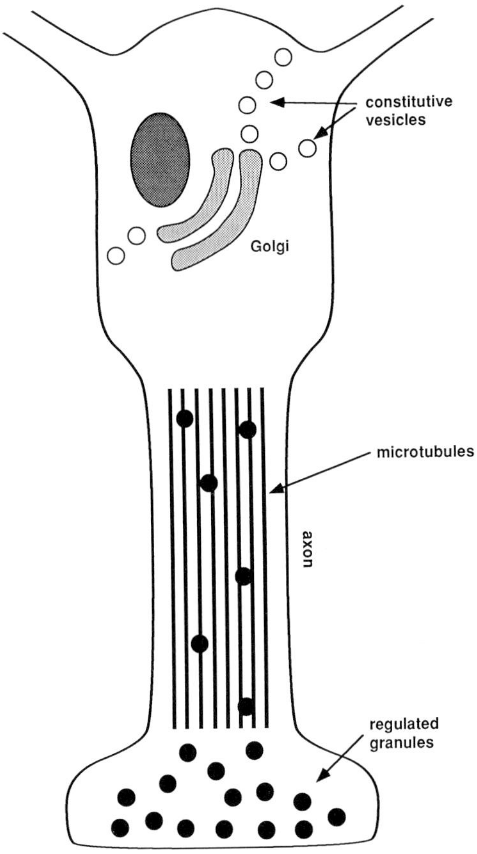

Figure 4. Spatial segregation of constitutive and regulated secretion in AtT-20 cells. Regulated granules selectively attach to microtubules and are transported down to the process tips. Constitutive vesicles, on the other hand, insert into plasma membrane near the cell body.

correlation between sorting efficiencies and their pI values: proteins with pIs within the range of the pH of Golgi lumen can be sorted either extremely efficiently into regulated granules, or not at all (Fig. 5). Thus, segregation of different classes of proteins most likely involves more selective sorting processes.

Although the exact mechanisms for sorting and vesicle formation are not well understood, the molecular processes governing the assembly of constitutive vesicles appear to be fundamentally different from those involved in forming regulated secretory granules. Proteins destined for regulated granules are actively sorted, whereas transport to the constitutive vesicles appears to be bulk-flow. Two lines of evidence support this conclusion. The first of these comes from analysis of AtT-20 cells

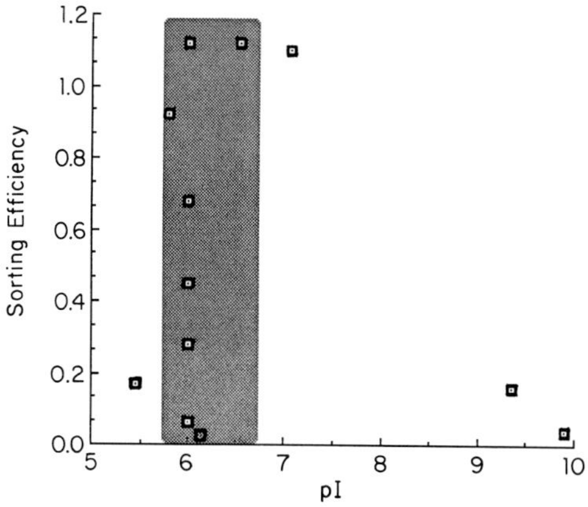

Figure 5. Plot of sorting efficiency as a function of the isoelectric points of transported proteins. The sorting efficiency (Moore and Kelly, 1985), which measures the efficiency with which a protein is targeted into regulated secretory granules, was analyzed for 12 foreign proteins transfected into AtT-20 cells. Shaded areas indicate the estimated lumenal pH of the *trans*-Golgi cisternae. Proteins whose pIs are in this range can be sorted either very efficiently or very poorly.

transfected with a hybrid protein consisting of the coding sequences of the cytoplasmic protein globin fused to the signal peptide of β-lactamase (Fig. 6). As shown by Lingappa and his colleagues (1984), the hybrid protein is inserted into the endoplasmic reticulum and its signal sequence is subsequently cleaved. The cleaved product, globin, contains no sorting signals, but is secreted preferentially by the constitutive pathway from AtT-20 cells. These results demonstrate that once the protein is targeted to the endoplasmic reticulum, secretion by the constitutive pathway does not further require

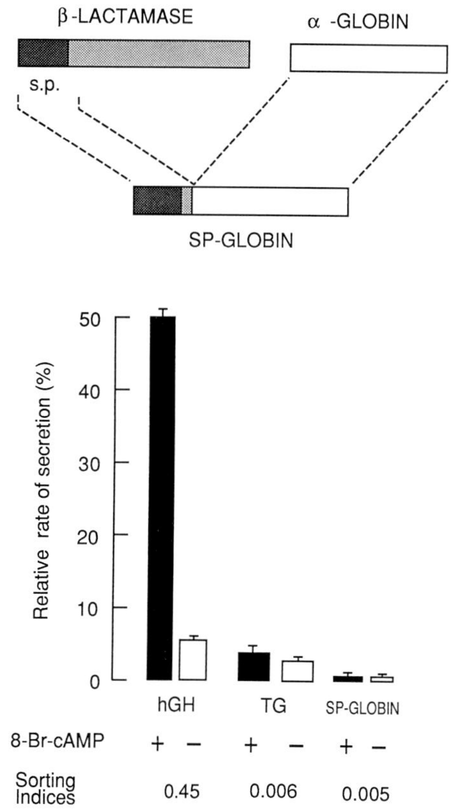

Figure 6. Secretion of a signal peptide-globin fusion protein from AtT-20 cells. The fusion protein consists of the signal peptide (SP) of β-lactamase and the coding sequences of alpha-globin. The construct was kindly provided to us by Dr. V. Lingappa. Sorting indices were measured according to published procedures (Moore and Kelly, 1985). Cells were labeled with [^{35}S]methionine until steady state. The rate of secretion was then measured (this rate = 100%). The labeled cells were then chased for 6 h. During 6–9 h of chase, 8-Br-cAMP was added to the medium to stimulate release from the regulated secretory pathway. Plot shown is the relative rate of secretion from stimulated cells compared with that from unstimulated cells. The fusion protein behaves like the constitutively secreted TG, rather than the regulated secretory protein hGH.

an obligatory sorting signal. The second line of evidence is that fusion between a constitutively secreted protein and a regulated secretory protein results in a protein that is targeted to the regulated pathway (Moore and Kelly, 1986). Thus, the signal for the regulated pathway appears to be dominant over that for the constitutive pathway.

The assembly of regulated and constitutive vesicles also differs in their requirements for new protein synthesis. While constitutive vesicles continue to bud from the Golgi in the absence of protein synthesis, formation and/or maturation of regulated granules requires a continuous supply of newly synthesized proteins (Brion, C., and H.-P. H. Moore, manuscript submitted for publication). To determine whether

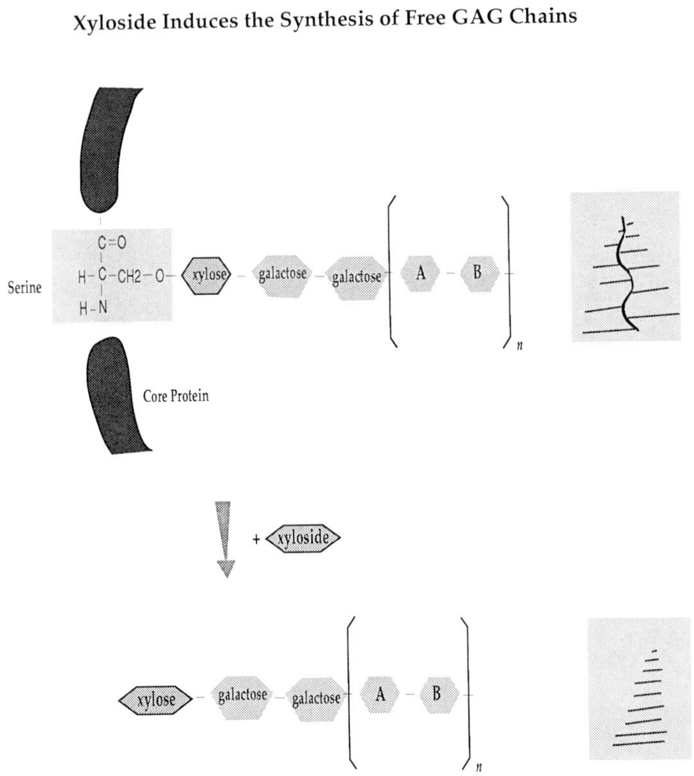

Figure 7. Induction of GAG chain synthesis by xyloside treatment GAG chains in proteoglycans are linked to serine via xylose residues. β-D-xyloside can substitute for xylose, thus promoting synthesis of free GAG chains.

secretory vesicles can form in the absence of protein synthesis, we needed a vesicle content marker that was not a protein. Burgess and Kelly (1984) showed that treating AtT-20 cells with the drug 4-methyl umbelliferyl-β-D-xyloside induces the synthesis of free glycosaminoglycan (GAG) chains. The xyloside acts as an acceptor for GAG chain synthesis (Fig. 7). Unlike a 25-kD proteoglycan, which is efficiently packaged into regulated secretory granules, the free GAG chains are secreted by both the constitutive and the regulated pathways. These findings suggest that the protein core of the proteoglycan contains a granule-localization signal; when not attached to such a signal, the GAG chains synthesized in the Golgi become passively incorporated into both constitutive and regulated vesicles (Fig. 8). Thus, they serve as convenient fluid phase tracers for both pathways. GAG chains are sulfated and can easily be detected when cells are labeled with [^{35}S]sulfate. The basic experimental design thus involves labeling xyloside-treated cells with [^{35}S]sulfate, and following the secretion of labeled GAG chains in the presence of cycloheximide. We found that cessation of protein synthesis arrests transport of GAG chains to the regulated secretory pathway, but has no effect on their trafficking through the constitutive secretory pathway (Fig. 9). Thus, either budding of regulated secretory granules is directly triggered by their content proteins, or proper maturation of granules requires a short-lived factor that becomes rapidly depleted upon inhibition of protein synthesis. In contrast, constitutive vesicles bud from the Golgi at a fixed rate, and this is not dependent on the presence of any cargo proteins.

Carrier-mediated Targeting of Peptide Hormones

DNA transfection studies suggest that sorting of proteins between constitutive and regulated secretory pathways may be mediated by cellular carriers that recognize

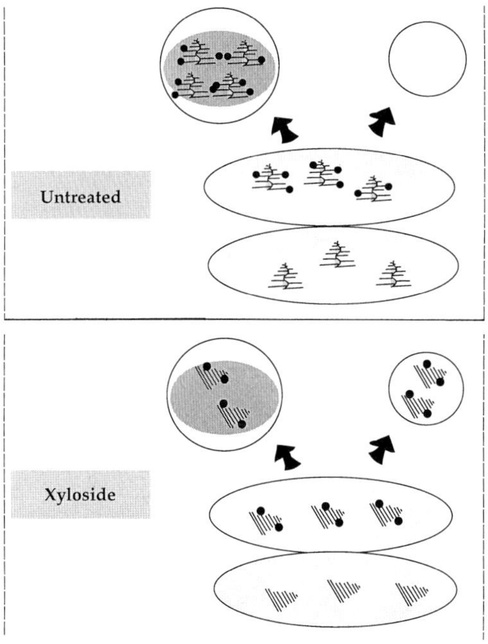

Figure 8. Use of GAG chains as fluid trace markers to measure secretion from the constitutive and regulated pathways. Untreated AtT-20 cells synthesize proteoglycans which are sorted into regulated, dense-core secretory granules. Treatment with xyloside induces synthesis of free GAG chains. These chains act as fluid tracer and are incorporated into both constitutive and regulated secretory pathways. Both proteoglycans and GAG chains are sulfated (indicated as filled circles) in the Golgi, and can be easily labeled with [^{35}S]sulfate.

regulated but not constitutive proteins. What is the molecular nature of these carriers? To look for transport carriers, we used affinity chromatography to isolate proteins that bind regulated secretory proteins (Chung et al., 1989). Canine Golgi membranes were solubilized with a nonionic detergent Nikkol, and the extracts were passed over an affinity column containing regulated secretory proteins. A set of proteins with molecular weights in the range of 25 kD specifically bind to the column and can be eluted at low pH. These proteins bind to the regulated hormones, prolactin and insulin, but do not show detectable affinities for immunoglobulin, hemoglobin, myoglobin, and bovine serum albumin. Both immunoglobulin and albumin are natural constitutive secretory proteins whereas globin, though normally a cytoplasmic protein, can be secreted constitutively if fused to a signal sequence (see above). Thus, the proteins exhibit the expected ligand specificity. Furthermore, the levels of expression of the 25-kD proteins correlate with the amount of storage granules manufactured. In cells induced to store regulated secretory granules, the cellular level of HBP25 is greatly elevated. The 25-kD proteins are immunolocalized to the perinuclear Golgi region of expressing cells. Fig. 10 shows our current working model for sorting of peptide hormones. Sorting signals on peptide hormones are recognized by the 25-kD carriers, which segregate them into immature secretory granules. The low pH lumen of the compartment causes dissociation of the complexes, and the carriers are recycled back

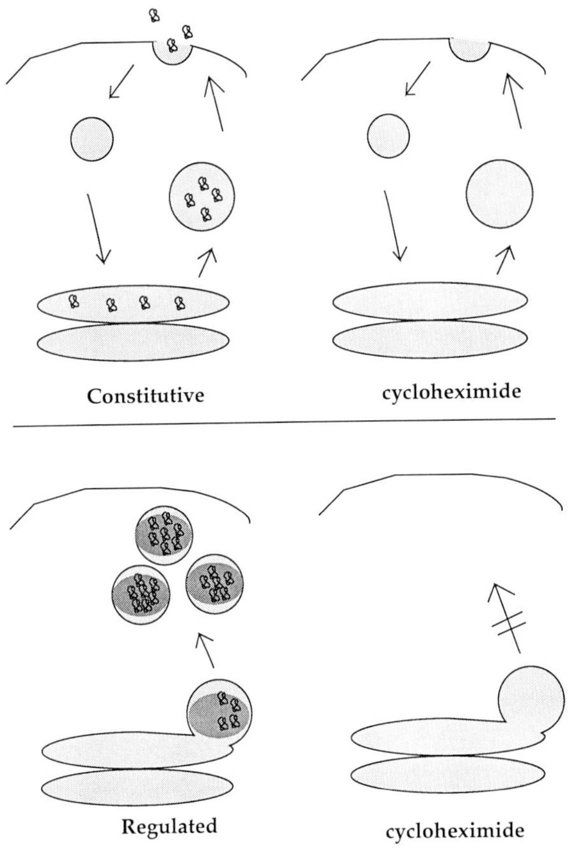

Figure 9. Dependence of granule assembly on protein synthesis. Inhibition of protein synthesis by cycloheximide blocks transport to the regulated pathway, but has no effect on constitutive secretion. A possible explanation is that budding of regulated granules from the Golgi is triggered by cargo proteins, whereas budding of constitutive vesicles is not triggered by their cargo.

to *trans*-Golgi for another round of transport. This model predicts that raising intracompartmental pH experimentally should prevent targeting of hormones to the granules, as it is for lysosomal targeting (Sly and Fischer, 1982). This is indeed the case, as shown for targeting of peptide hormones (Moore et al., 1983*a*), and for granular localization of von Willebrand factor (Wagner et al., 1986).

Conclusion

The protein sorting events described here for AtT-20 cells are probably representative of many other endocrine, exocrine, and neural cells. Recent experiments show that

PC-12 cells (Schweitzer and Kelly, 1985), GH_3 (Green and Shields, 1984), and endothelial cells (Sporn et al., 1986) also sort their secretory proteins into a constitutive and a regulated secretory pathway. It should be noted that the two pathways described so far are probably not the only pathways present in these cells. Many neural cells contain different classes of dense-core secretory granules (Fisher et al., 1988). It is possible that they also contain more than one type of constitutive vesicles in order to segregate channels and receptors to distinct dendritic or axonal regions of the plasma membrane. Much remains to be learned about the types of secretory pathways in mammalian cells, and their sorting mechanisms.

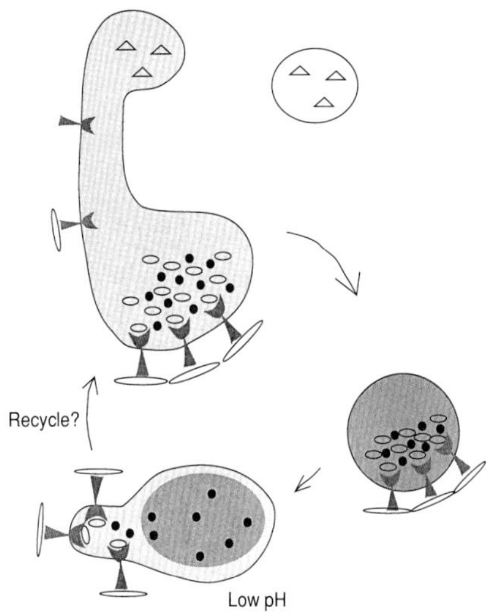

Figure 10. Model for targeting of peptide hormones in AtT-20 cells. Regulated secretory proteins (●) are sorted away from constitutive proteins at the *trans*-Golgi cisternae by binding to the 25-kD carriers (o). The carriers are depicted as partially soluble and partially membrane-associated by binding to a transmembrane protein (⧖). The soluble carriers can cross-link hormones in the lumen. The resulting aggregates then associate with the membranes of incipient granules. As granules pinch off from the Golgi, the lumen becomes acidified. Since purified carriers bind hormones only at neutral pH, the acidity of the granule lumen could cause dissociation of hormones from the carriers. The empty carriers then recycle back to the Golgi by association with the transmembrane protein. △ = constitutively secreted protein; ⬭ = cytoskeletal element or Clathrin.

Acknowledgments

We thank Dr. G. F. Oster for critical reading of the manuscript.

This work was supported by National Institutes of Health grant GM-35239, National Science Foundation Presidential Young Investigator Award DCB-8451636, a Juvenile Diabetes Foundation Research Grant 187381 (to H. P. Moore), a NIH postdoctoral fellowship 1F32GM1178 (to K.-N. Chung), a 1967 Science and Engineering Scholarship from the Natural Sciences and Engineering Research Council of Canada (to C. Brion), and an Alfred P. Sloan Research Fellowship (BR2497; to H.P. Moore).

References

Burgess, T., C. Craik, and R. B. Kelly. 1985. The exocrine protein trypsinogen is targeted into the secretory granules of an endocrine cell line: studies by gene transfer. *Journal of Cell Biology*. 101:639–645.

Burgess, T., C. Craik, L. Matsuuchi, and R. B. Kelly. 1987. In vitro mutagenesis of trypsinogen: role of the amino terminus in intracellular targeting to secretory granules. *Journal of Cell Biology*. 105:659–668.

Burgess, T. L., and R. B. Kelly. 1984. Sorting and secretion of adrenocorticotropin in a pituitary tumor cell line after perturbation of the level of a secretory granule-specific proteoglycan. *Journal of Cell Biology*. 99:2223–2230.

Chung, K. N., P. Walter, G. W. Aponte, and H.-P. H. Moore. 1989. Molecular sorting in the secretory pathway. *Science*. 243:192–197.

Farquhar, M. G. 1985. Progress in unraveling pathways of Golgi traffic. *Annual Review of Cell Biology*. 1:447–488.

Fisher, J. M., W. Sossin, R. Newcomb, and R. H. Scheller. 1988. Multiple neuropeptides derived from a common precursor are differentially packaged and transported. *Cell*. 54:813–822.

Green, R., and D. Shields. 1984. Somatostatin discriminates between the intracellular pathways of secretory and membrane proteins. *Journal of Cell Biology*. 99:97–104.

Griffiths, G., and K. SImons. 1986. The trans Golgi network: sorting at the exit site of the Golgi complex. *Science*. 234:438–443.

Gumbiner, B., and R. B. Kelly. 1982. Two distinct intracellular pathways transport secretory and membrane glycoproteins to the surface of pituitary tumor cells. *Cell*. 28:51–59.

Holcomb, C., W. Hansen, T. Etcheverry, and R. Schekman. 1988. Secretory vesicles externalize the major plasma membrane ATPase in yeast. *Journal of Cell Biology*. 106:641–648.

Kelly, R. B. 1985. Pathways of protein secretion in eukaryotes. *Science*. 230:25–32.

Lingappa, V., J. Chaidez, C. Yost, and J. Hedgpeth. 1984. Determinants for protein localization: beta-lactamase signal sequence directs globin across microsomal membranes. *Proceedings of the National Academy of Sciences*. 81:456–460.

Loh, Y. P., M. J. Brownstein, and H. Gainer. 1984. Proteolysis in neuropeptide processing and other neural functions. *Annual Review of Neuroscience*. 7:189–222.

Mains, R. E., and B. A. Eipper. 1981. Coordinate, equimolar secretion of smaller peptide products derived from pro-ACTH/endorphin by mouse pituitary tumor cells. *Journal of Cell Biology*. 89:21–28.

Mains, R. E., B. A. Eipper, and N. Ling. 1977. Common precursor to corticotropins and endorphins. *Proceedings of the National Academy of Sciences*. 74:3014–3018.

Matsuuchi, L., K. M. Buckley, A. W. Lowe, and R. B. Kelly. 1988. Targeting of secretory vesicles to cytoplasmic domains in AtT-20 and PC-12 cells. *Journal of Cell Biology*. 106:239–251.

Moore, H.-P. H. 1987. Factors controlling packaging of peptide hormones into secretory granules. Annals of the New York Academy of Science. 493:50–61.

Moore, H.-P. H., B. Gumbiner, and R. B. Kelly. 1983*a*. Chloroquine diverts ACTH from the regulated secretory pathway to a constitutive one in AtT-20 cells. *Nature*. 302:434–436.

Moore, H.-P. H., B. Gumbiner, and R. B. Kelly. 1983*b*. A subclass of proteins and sulfated macromolecules secreted by AtT-20 (mouse pituitary tumor) cells is sorted with adrenocorticotropin into dense secretory granules. *Journal of Cell Biology*. 97:810–817.

Moore, H.-P. H., M. Walker, F. Lee, and R. B. Kelly. 1983*c*. Expressing a human proinsulin cDNA in a mouse ACTH secretory cell. Intracellular storage, proteolytic processing and secretion on stimulation. *Cell*. 35:531–538.

Moore, H.-P. H., and R. B. Kelly. 1985. Secretory protein targeting in a pituitary cell line: differential transport of foreign secretory proteins to distinct secretory pathways. *Journal of Cell Biology*. 101:1773–1781.

Moore, H.-P. H., and R. B. Kelly. 1986. Rerouting of a secretory protein by fusion with human growth hormone sequences. *Nature*. 321:443–446.

Moore, H.-P. H., L. Orci, and G. F. Oster. 1988. The biogenesis of secretory granules. *In* Protein Transfer and Organelle Biogenesis. Academic Press, Inc., New York. 521–561.

Orci, L., M. Ravazzola, A. Perrelet, S. K. Powell, D. L. Quinn, and H.-P. H. Moore. 1987. The trans-most cisternae of the Golgi complex: a compartment for sorting of secretory and plasma membrane proteins. *Cell*. 51:1039–1051.

Orci, L., M. Ravazzola, M. Amherdt, O. Madsen, J.-D. Vassalli, and A. Perrelet. 1985. Direct identification of prohormone conversion site in insulin-secreting cells. *Cell*. 42:671–681.

Palade, G. 1975. Intracellular aspects of the process of protein secretion. *Science*. 189:347–58.

Rivas, R. R., and H.-P. Moore. 1989. Spatial segregation of the regulated and constitutive secretory pathways. *Journal of Cell Biology*. In press.

Roberts, J. L., and E. Herbert. 1977. Characterization of a common precursor to corticotropin and β-lipotropin: identification of β-lipotropin peptides and their arrangement relative to corticotropin in the precursor synthesized in a cell-free system. *Proceedings of the National Academy of Sciences*. 74:5300-5304.

Schekman, R. 1985. Protein localization and membrane traffic in yeast. Annual Review of Cell Biology. 1:115–143.

Schweitzer, E. S., and R. B. Kelly. 1985. Selective packaging of human growth hormone into synaptic vesicles in a rat neuronal (PC12) line. *Journal of Cell Biology*. 101:667–675.

Simons, K. and S. D. Fuller. 1985. Cell surface polarity in epithelia. *Annual Review of Cell Biology*. 1:243–288.

Sly, W. S., & H. D. Fischer. 1982. The phosphomannosyl recognition system for intracellular and intercellular transport of lysosomal enzymes. *Journal of Cellular Biochemistry*. 18:67–85.

Sporn, L. A., V. J. Marder, and D. D. Wagner. 1986. Inducible secretion of large biologically potent von Willebrand factor multimers. *Cell* 46:185–190.

Tartakoff, A., P. Vassalli, and M. Detraz. 1978. Comparative studies of intracellular transport of secretory proteins. *Journal of Cell Biology*. 79:694–707.

Tooze, J., and B. Burke. 1987*a*. Accumulation of adrenocorticotropin secretory granules in the midbody of telophase AtT-20 cells: evidence that secretory granules move anterogradely along microtubules. *Journal of Cell Biology*. 104:1047–1057.

Tooze, J., M. Hollinshead, R. Frank, and B. Burke. 1987*b*. An antibody specific for an endoproteolytic cleavage site provides evidence that proopiomelanocortin is packaged into secretory granules in AtT20 cells before its cleavage. *Journal of Cell Biology*. 105:155–162.

Vale, R., T. Reese, and M. Sheetz. 1985. Identification of a novel force-generating protein, kinesin, involved in microtubule-based motility. *Cell*. 42:39-50.

Wagner, D. D., T. Mayadas, and V. J. Marder. 1986. Initial glycosylation and acidic pH in the Golgi apparatus are required for multimerization of von Willebrand factor. *Journal of Cell Biology*. 102:1320–1324.

Calcium and Calcium Channels in Secretion

Chapter 13

Fingering the Trigger for Neurotransmitter Secretion: Studies on the Calcium Channels of Squid Giant Presynaptic Terminals

George J. Augustine, JoAnn Buchanan, Milton P. Charlton, Luis R. Osses, and Stephen J. Smith

Section of Neurobiology, Department of Biological Sciences, University of Southern California, Los Angeles, California 90089-0371; Section of Molecular Neurobiology, Howard Hughes Medical Institute, Yale University Medical School, New Haven, Connecticut 06510; Department of Physiology, University of Toronto, Toronto, Ontario M5S 1A8, Canada; and the Marine Biological Laboratory, Woods Hole, Massachusetts 02543

Introduction

The role of Ca ions in triggering secretion of neurotransmitters at chemical synapses is well established (Katz, 1969; Augustine et al., 1987). In fact, unlike secretion from certain nonexcitable cells (Penner and Neher, 1988), there is no doubt that Ca is an essential second messenger in initiating the process of synaptic transmitter secretion.

Secretion of neurotransmitters is widely believed to be a consequence of the exocytotic discharge of the contents of transmitter-filled synaptic vesicles from presynaptic terminals (Heuser et al., 1979; Torri-Tarelli et al., 1985). The triggering of exocytosis by Ca ions can be divided into three steps: (*a*) Depolarization-induced opening of voltage-gated Ca channels present in the presynaptic membrane, resulting in Ca influx into the presynaptic terminal. (*b*) Diffusion of the entering Ca ions through presynaptic cytoplasm, producing a transient elevation of Ca concentration at

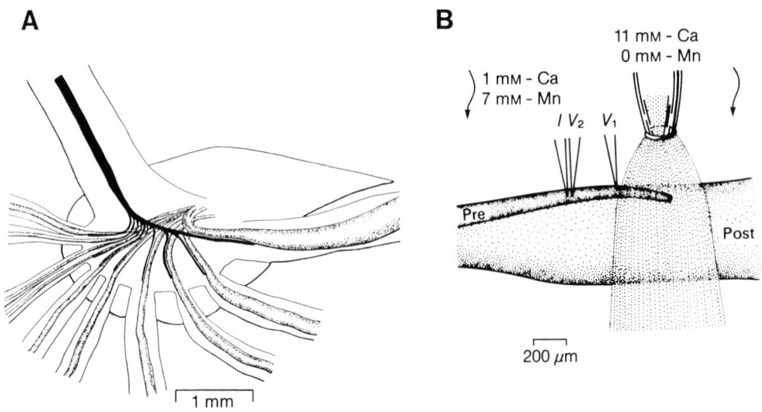

Figure 1. (*A*) Anatomy of the squid stellate ganglion. A large neuronal axon (*black*) enters the stellate ganglion and forms a series of finger-like branches. Each of these digits is a giant presynaptic terminal that innervates an even larger postsynaptic axon (*stippled*). After Young (1939). (*B*) A method for local application of Ca to voltage-clamped presynaptic terminals (*Pre*) of squid. A stellate ganglion is bathed in a low Ca saline containing Mn to block presynaptic Ca influx. Ca influx was then restored to the well-clamped region of the most distal giant synapse by application of Ca-containing saline from a blunt pipette. V_1, V_2, and I indicate the electrodes used to measure presynaptic current with the three-microelectrode voltage-clamp method. The postsynaptic cell is indicated by *Post*. From Augustine et al. (1985*a*).

sites of transmitter secretion. We term this signal the presynaptic Ca transient. (*c*) Initiation of exocytosis by the presynaptic Ca transient is due to increased Ca binding to a receptor molecule. The identity of this Ca receptor and the means by which it triggers exocytosis are unknown (Augustine et al., 1987).

This paper will focus on the first step in this scheme by summarizing what is known about the Ca channels responsible for triggering release at the "giant" synapse of squid. The reason we concentrate on this particular synapse is that it has the largest presynaptic terminal in the entire animal kingdom (Fig. 1 *A*). The large size of this terminal ($\sim 10^6$ times larger in volume than a typical synaptic terminal in a vertebrate brain) has made it possible to characterize in considerable detail the functional properties of its Ca channels and to relate influx of Ca through these channels to the secretion of synaptic transmitter (Llinas, 1982). In this paper we would like to summarize the insights that have emerged from studies of the presynaptic Ca channels of the squid giant nerve terminal. Our goal is to identify some general principles useful in thinking about how Ca channels trigger exocytosis from secretory cells.

Our description of the functional attributes of squid presynaptic Ca channels is divided into several sections. First we will discuss the activation of these channels, that is, how they are gated by the membrane potential of the presynaptic terminal. This will be followed by a brief consideration of the permeation of Ca and other ions through the channels. Following the discussion of these biophysical attributes of the presynaptic Ca channels we will examine their sensitivity to various pharmacological agents and thereby attempt a classification of these channels based on schemes devised to categorize the diversity of Ca channels in different cells. We will then consider the localization of Ca channels in the presynaptic membrane, and conclude with a brief consideration of the role of these channels in the initiation of secretion by examining the quantitative relationship between Ca entry and neurotransmitter release.

Activation of Presynaptic Ca Channels

There are several experimental approaches that have been used to determine the functional properties of presynaptic Ca channels. To study the gating of squid presynaptic Ca channels we have used the three-microelectrode voltage-clamp method (Adrian et al., 1970). When combined with the use of pharmacological agents to block other presynaptic channels, this method permits measurement of currents flowing through presynaptic Ca channels at high temporal resolution (Llinas et al., 1981*a*; Charlton et al., 1982; Augustine and Eckert, 1984*a*; Augustine et al., 1985*a*). While this method reduces problems associated with potential decrement along the lengthy presynaptic terminal, space-clamp problems are still evident when attempting to measure transmitter release from these terminals (Augustine et al., 1985*b*). We therefore developed the method of local Ca application shown in Fig. 1 *B* to allow spatially-controlled measurement of both presynaptic Ca current and postsynaptic responses resulting from transmitter secretion (Augustine et al., 1985*a*, *b*). Unless otherwise indicated, all voltage-clamp results illustrated here were obtained with the combination of this method and the three-microelectrode voltage clamp. This combination yields high-fidelity measurements of presynaptic Ca influx, as revealed by a close correspondence between Ca current measurements and independent optical measurements of Ca accumulation (Augustine et al., 1985*a*).

Examples of Ca currents recorded from squid presynaptic terminals are shown in Fig. 2. These currents were elicited by brief depolarizations to a wide range of membrane potentials (indicated in the upper traces), and have several interesting features. One key feature is the time course of the currents. Upon depolarization Ca currents activate after a delay and reach a maximum within a few milliseconds. The time course of the activation of the currents is voltage dependent (Fig. 3 *A*) and presumably due to a voltage-sensitive rate of transition through multiple closed states to an open, conducting conformation (Llinas et al., 1981*a*; Tsien, 1983). At the end of a depolarization the Ca current rapidly deactivates to produce a large, transient "tail" current. The rapid, more or less exponential, time course of the tail current reflects the transition of the Ca channel from its open configuration through one or more closed, nonconducting states.

These kinetic features of Ca channel gating are important for triggering neurotransmitter secretion. The slow activation of presynaptic Ca channels introduces a delay between production of the presynaptic action potential and the postsynaptic response (Llinas et al., 1981*b*; Augustine et al., 1985*b*, see also Fig. 9 below). This delay probably is one of the main determinants of the "synaptic delay" characteristic of chemical synapses (Katz, 1969). In addition, the rapid deactivation of the Ca channels allows Ca influx to terminate rapidly after the presynaptic action potential and thereby helps to limit the duration of transmitter release (Llinas et al., 1982). Thus the kinetics of Ca channel gating helps determine the kinetics of transmitter secretion.

The magnitude of the Ca currents elicited during and after depolarization are determined by the presynaptic membrane potential. The voltage dependence of the

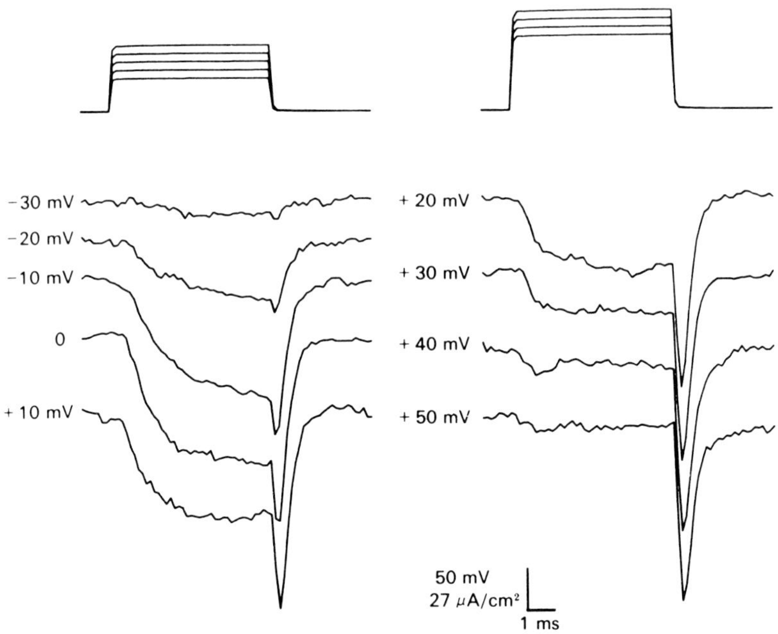

Figure 2. Ca currents recorded from the squid giant presynaptic terminal. Ca currents (*lower traces*) were elicited by depolarizations (*upper traces*) to the potentials indicated to the left of the current records. Ca currents were corrected for leakage, capacitive, and Ca-independent currents, and were recorded in the presence of tetrodotoxin, tetraethylammonium, and 3-, 4-diaminopyridine to eliminate currents flowing through Na and K channels. From Augustine et al. (1985*a*).

current generated during depolarizing pulses is shown graphically in Fig. 3 *B*. In the presence of physiological Ca concentrations this current reaches a maximum density of $\sim 100\ \mu A/cm^2$ at -10 to 0 mV and is minimal at potentials negative to -60 mV or positive to $+60$ mV. This bell-shaped voltage dependence is a consequence of two opposing factors: as the membrane potential becomes more positive the number of open Ca channels increases, due to voltage-dependent activation, but the number of Ca ions flowing through each open Ca channel is reduced by a decreasing electrochemical driving force. The former action, namely the recruitment of Ca channels by depolarization, can be seen in the sigmoidal voltage dependence of Ca tail current amplitude

(Fig. 3 *C*). This curve reveals that the presynaptic Ca channels are all opened at potentials positive to +20 mV and have a 50% probability of opening at ~−15 mV. The reduction in electrochemical driving force with depolarization is evident as a reduction in tail current amplitude when these currents are measured at more positive potentials (Llinas et al., 1981*a*). Unfortunately, it has not yet been possible to record

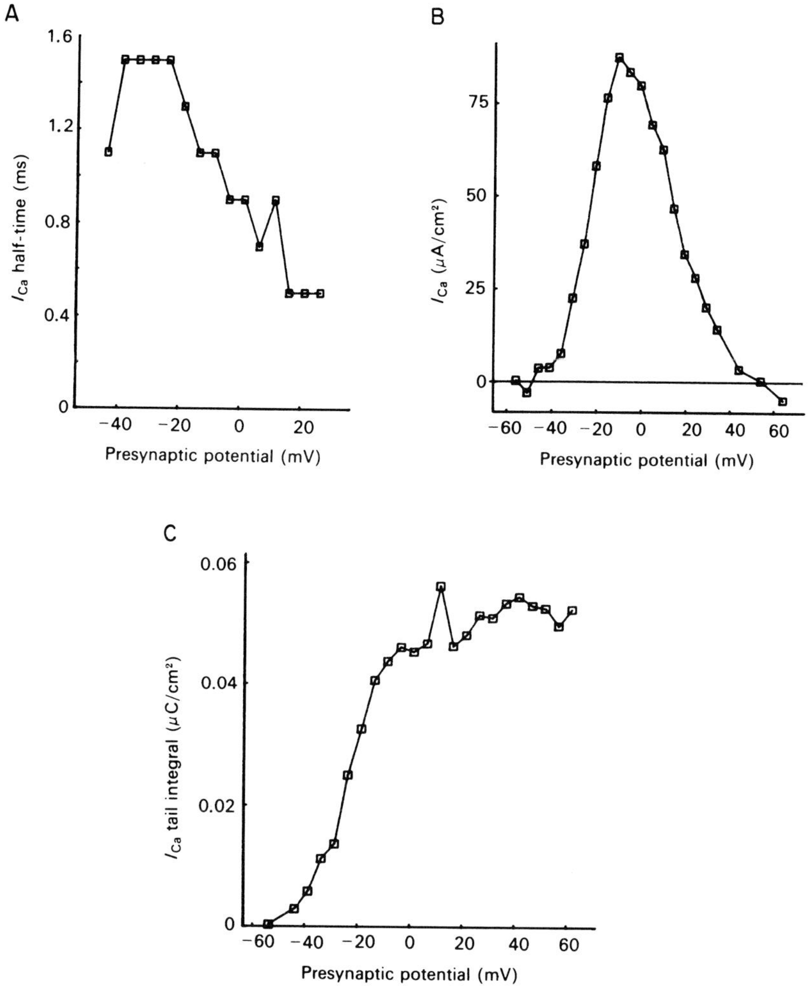

Figure 3. Gating parameters of presynaptic Ca channels. (*A*) Time course of Ca current activation, measured as the time required for the current to reach half-maximum amplitude at the indicated potentials. (*B*) Voltage dependence of Ca current magnitude, measured at the end of 6-ms long depolarizations to the indicated potentials. (*C*) Voltage dependence of Ca channel activation, assessed by measuring the integral of tail currents produced following 6-ms long depolarizations to the indicated potentials. From Augustine et al. (1985*a*).

currents flowing through single presynaptic Ca channels (e.g., Fenwick et al., 1982) to directly determine their unitary conductance and gating properties.

These features of the voltage dependence of Ca current are very important determinants of transmitter release from the presynaptic terminal. One obvious consequence is that the probability of Ca channel opening is greatly enhanced by a presynaptic action potential, thus permitting the coupling of action potential invasion to evoked release of transmitter. Further, the decreased electrochemical driving force at positive potentials, when combined with the delayed activation of Ca channels in response to a depolarization, means that Ca influx primarily occurs during the repolarizing phase of a presynaptic action potential (Llinas et al., 1981*b*, 1982). This makes transmitter release very sensitive to the time course of action potential repolarization (e.g., Benoit and Mambrini, 1970; Jan et al., 1977; Kandel and Schwartz, 1982; Augustine, G.J., manuscript submitted for publication).

During brief depolarizations, such as those shown in Fig. 2, the current flowing through presynaptic Ca channels is sustained. However, during prolonged depolarizations, lasting 1 s or longer, the current relaxes due to inactivation of the Ca channel (Augustine and Eckert, 1982, 1984*b*). This inactivation depends upon the magnitude of Ca entry, because it increases with larger quantities of Ca influx and is reduced when Ba ions are substituted for Ca (Augustine and Eckert, 1984*b*, Augustine, G. J., and R. Eckert, manuscript in preparation). These features are diagnostic of current-dependent inactivation (Eckert and Chad, 1984) and suggest that an elevation of the Ca concentration within the presynaptic terminal is responsible for Ca channel inactivation. Experiments on extrasynaptic Ca channels have led to the proposal that such Ca-dependent inactivation is due to a Ca-regulated dephosphorylation of Ca channels (Chad and Eckert, 1986; Armstrong and Eckert, 1987). It remains to be determined whether this is true for squid presynaptic Ca channels.

Regardless of its molecular mechanism, it is not yet clear whether the Ca-dependent inactivation of the presynaptic Ca channel exerts much influence on the transmitter-releasing capabilities of the presynaptic terminal. In principle, such an arrangement allows intracellular Ca ions to have opposing effects on release: a rapid stimulation of exocytosis and a slower inhibition of Ca influx, which would attenuate exocytosis. This could provide a partial explanation for the observation that elevated presynaptic Ca concentrations inhibit evoked release (Kusano, 1970; Charlton et al., 1982; Adams et al., 1985). In practice, the amount of the Ca influx required for inactivating Ca channels is so large, and the apparent time course of inactivation so slow, that it seems unlikely that this mechanism plays much of a role in regulating transmitter release evoked by presynaptic action potentials, at lease in the case of the squid synapse.

Ion Permeation through Presynaptic Ca Channels

Relatively little work has been done on the permeation properties of presynaptic Ca channels, in squid giant terminals or elsewhere. The magnitude of the presynaptic Ca current depends upon the external concentration of Ca ions (Llinas et al., 1981*a*; Augustine and Charlton, 1986) and appears to saturate at very high external Ca levels (Fig. 4). At the squid synapse the apparent affinity constant for external Ca, in the presence of external Mg ions, is ~60 mM (Augustine and Charlton, 1986). A similar kind of saturation has been reported for Ca fluxes into rat brain synaptosomes (Nachshen and Blaustein, 1982; Suszkiw et al., 1986). By analogy with studies on extrasynaptic Ca channels, the saturation of Ca influx at high Ca concentrations indicates that the presynaptic Ca channel has one (or more) sites to which Ca must bind in order to flow through the channel (Hagiwara and Takahashi, 1967; Almers and McClesky, 1984; Byerly et al., 1985; Hess et al., 1986).

It is possible to view other studies on the permeation properties of squid presynaptic Ca channels in the context of such a binding-site model. For example, it appears that certain other divalent ions are able to permeate the channel and therefore presumably bind to this site. Such ions include Ba, which carries ~20% more current than a comparable concentration of Ca, and Sr, which is roughly equivalent to Ca in current-carrying ability (Augustine and Eckert, 1984*a*). Still other divalent ions block the presynaptic Ca channel and carry no measurable current. The best-studied example is Cd (Llinas et al., 1981*a*; Charlton et al., 1982; Augustine and Eckert, 1984*a*; and see footnote 1), which blocks presynaptic Ca current with an affinity constant of ~100 μM at 0 mV (Fig. 5). Co is another example of a potent and impermeant blocker (Augustine and Charlton, 1986). Both Cd and Co presumably bind to the ion-binding site of the Ca channel in order to block, but may be unable to

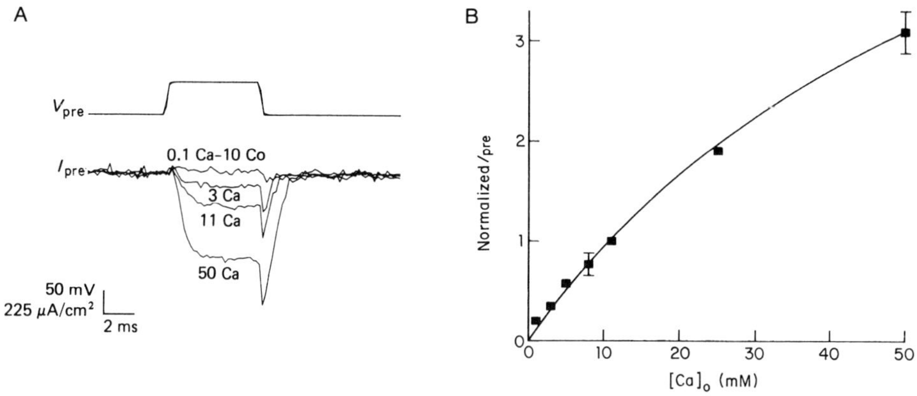

Figure 4. Dependence of presynaptic Ca current on extracellular Ca concentration. (*A*) Depolarization of the presynaptic membrane potential (V_{pre}) to 0 mV elicited inward Ca currents (I_{pre}) whose magnitude depended upon the extracellular Ca concentration. Inward currents were largest in saline containing 50 mM Ca and were eliminated by saline containing 0.1 mM Ca and 10 mM Co. Ca currents were measured as described for Fig. 2 except that the local Ca application method was not used. (*B*) Dependence of presynaptic Ca current (I_{pre}) upon extracellular Ca concentration ($[Ca]_o$). Currents were elicited by depolarizations to −10 mV and were normalized relative to the current recorded in saline containing 11 mM Ca. Points represent the mean current recorded from 1 to 12 different preparations, and bars indicate ± 1 SEM. From Augustine and Charlton (1986).

permeate because their affinity for the binding site is too high (Lansman et al., 1986). The trivalent ion La is a potent blocker of evoked transmitter release (Miledi, 1971; Stanley and Adelman, 1984) and presynaptic Ca spikes (Miledi, 1971) at the squid synapse, and thus are presumably in this category as well. Mn ions block current carried by Ca ions (Llinas et al., 1982) but also appear to carry small presynaptic currents (Augustine et al., 1985*a*). Thus Mn appears to be intermediate between the two extremes of ions that bind to the channel with a very high affinity and block it, and ions that bind to the channel with a lower affinity and are conducted through it.

Mg ions are a classical competitive inhibitor of evoked neurotransmitter release (reviewed in Silinsky, 1985), but appear to be in a different category than the other divalent cations discussed above. Mg carries no current through presynaptic Ca channels, even when present (65 mM) as the only extracellular divalent cation (Augustine and Eckert, 1984*a*). These results suggest that Mg is a blocker of the Ca channel, yet has a much lower binding affinity than Cd, Mn, or other blocking ions. This affinity apparently is too low to even allow Mg to carry current through the Ca channel. Such behavior has been documented in cardiac muscle cells and explained in terms of a relatively slow binding of Mg to Ca channels (Lansman et al., 1986).

In summary, less is known about the permeation properties of presynaptic Ca channels than their extrasynaptic counterparts. What little is known seems to conform to expectations based on studies of extrasynaptic Ca channels. From the standpoint of triggering transmitter secretion the most critical permeation feature of presynaptic Ca channels is that these channels are able to selectively import Ca ions despite the fact that these ions are a relatively minor constituent of the extracellular medium. This presumably is a consequence of the association of these cations with specific binding site(s) within the channel.

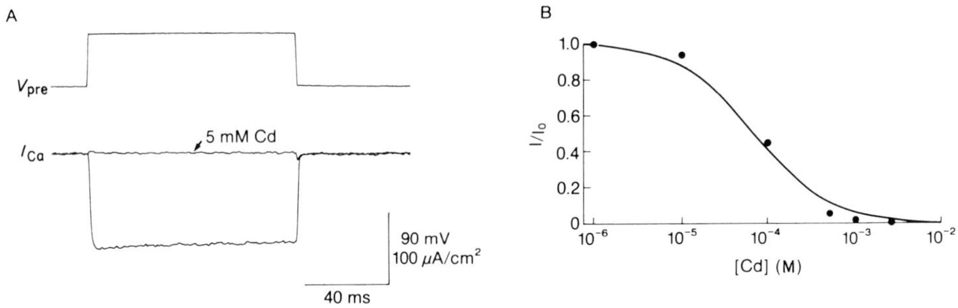

Figure 5. Blockade of presynaptic Ca current by Cd. (*A*) Cd (5 mM) produced a complete blockade of Ca current elicited by a 100-ms long depolarization to 0 mV. (*B*) Concentration dependence of Cd block. Points represent relative reduction (I/I_o) in Ca currents recorded from a single presynaptic terminal treated with the indicated concentrations of Cd ([Cd]). Line indicates the relationship expected if Cd blocked the channel with a 1:1 stoichiometry and had an affinity constant of 120 μM.[1]

Presynaptic Ca Channel Classification

Comparison of the functional and pharmacological properties of Ca channels in a number of different cell types suggests that there is more than one type of Ca channel (Hagiwara and Byerly, 1981). This diversity of Ca channel types was put into a coherent classification scheme by the work of Nowycky et al. (1985), which showed that single neurons from chick dorsal root ganglia (DRG) contain three distinct types of Ca channels. These three channel types, designated T, N, and L, were distinguished on the basis of the criteria shown in Table I. Numerous subsequent studies have found that Ca channels corresponding to these three general classes are found in a wide range of cells (Tsien et al., 1988).

We have asked whether the Ca channels in the squid presynaptic terminal fit into this classification scheme and have found that they do not.[1] The presynaptic Ca channels activate at potentials too negative to be N or L type, they inactivate too slowly to be T type, and their sensitivity to pharmacological agents also does not correspond to

[1] Charlton, M. P., and G. J. Augustine, manuscript submitted for publication.

that of T, N, or L-type Ca channels (summarized in Table I). Thus they seem to represent a different type of Ca channel than those found in chick DRG neurons.

Studies on the sensitivity of Ca channels in other presynaptic terminals to the diagnostic Ca channel blockers listed in Table I have suggested that N, L, or both types of Ca channels mediate transmitter secretion in these various systems (Miller, 1987; Smith and Augustine, 1988). These results, when combined with the results for presynaptic Ca channels of squid, indicate that a variety of types of Ca channels are capable of mediating neurotransmitter secretion. Thus far, the only common feature of all the presynaptic Ca channel types identified is that they inactivate slowly, perhaps as a means of permitting sustained secretion during prolonged depolarization (e.g., Charlton and Atwood, 1977; Blight and Llinas, 1980; Augustine and Eckert, 1984*a*). Other common (and perhaps unique) features of presynaptic Ca channels may emerge as studies of their properties progress beyond their present, rather primitive, state.

TABLE I
Properties of Four Types of Ca Channels

	T	N	L	Squid
Activation range	>-70 mV	>-10	>-10	>-60
Inactivation rate	Fast	Variable	Slow	Slow
Sensitivity to Cd	Low	High	High	Low
Sensitivity to dihydropyridines	Low	Low	High	Low
Sensitivity to ωCgTX	Low	High	High	Low

Localization of Presynaptic Ca Channels

Another important consideration regarding presynaptic Ca channels is their distribution in the membrane of the presynaptic neuron. Perhaps the best way to address this question would be to use molecular probes, such as toxins that bind specifically to presynaptic Ca channels, as histochemical labels to identify the location of these channels. Since development of such labels has progressed rather slowly we instead have used the fluorescent Ca indicator, fura-2 (Grynkiewicz et al., 1985), and digital imaging technology (Tsien, 1988) to identify sites of Ca influx in squid presynaptic neurons. The approach was to open voltage-gated Ca channels and use the resultant increases in intracellular Ca concentration to pinpoint the location of the channels.[2]

In these experiments presynaptic Ca channels were opened by electrically stimulating squid presynaptic neurons with brief trains of action potentials. The resulting changes in dye fluorescence were digitally processed to determine the relative change in fura-2 fluorescence compared with resting fluorescence. In Fig. 6 this ratio has been encoded into pseudocolors, with hotter colors indicating larger changes in fluorescence and, presumably, larger changes in presynaptic Ca concentration. We have resisted calibrating these signals in terms of absolute Ca concentration because the equilibrium calibrations normally used for this purpose (Grynkiewicz et al., 1985) are not readily applied to the rapid and localized Ca concentration changes that we are considering here (Neher, 1986).

[2] Smith, S. J., J. Buchanan, L. Osses, M. P. Charlton, and G. J. Augustine, manuscript submitted for publication.

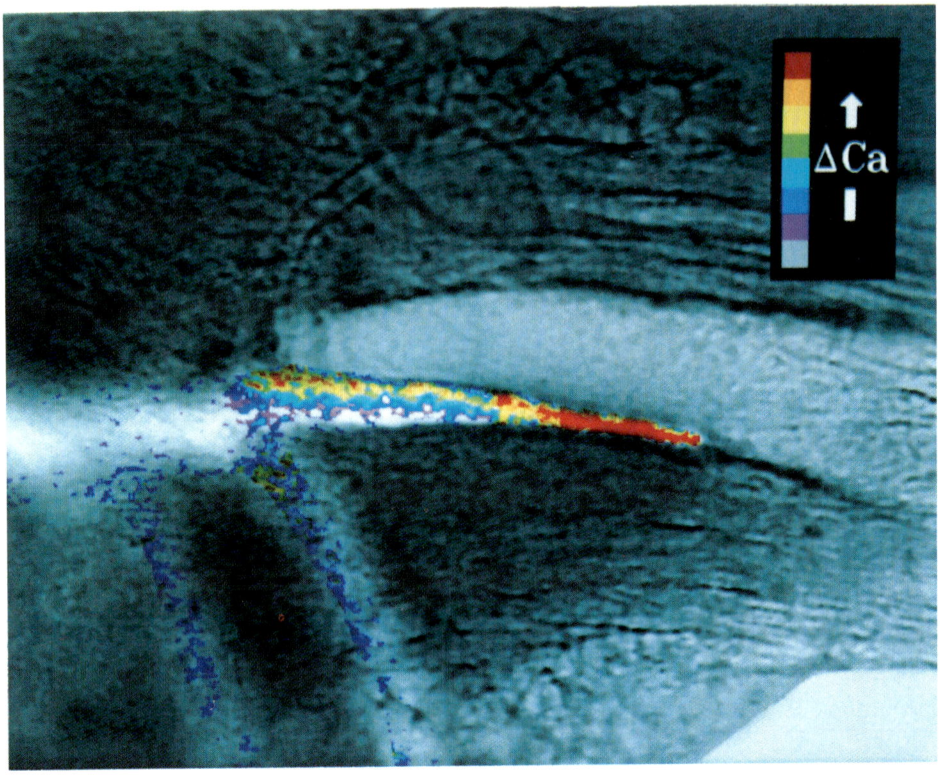

Figure 6. Digital imaging of changes in fura-2 fluorescence induced by trains of action potentials in the presynaptic terminal. (*Left*) Low-magnification image of the fluorescence change induced by a brief train (80 Hz, 0.5 s) of presynaptic action potentials. The magnitude of the regional fluorescence changes have been encoded with the pseudocolor scale shown at right. These fluorescence changes have been superimposed on a brightfield image of the stellate ganglion to illustrate that the changes are most prominent in the terminal digits of the presynaptic cell. (*Right*) Higher-magnification image of the fluorescence changes shown in the top panel, taken 0.5 s after terminating the stimulus. At this time and magnification it can be seen that the fluorescence changes initially are restricted to the portion of the presynaptic terminal closest to the postsynaptic cell.[2]

The change in fura-2 fluorescence produced by a brief train of action potentials is shown at a low magnification in Fig. 6 *left*. One striking feature of these fluorescence changes is that their distribution is longitudinally nonuniform, being abruptly restricted to terminal regions of the presynaptic cell. Fluorescence changes in the axonal region of the presynaptic neuron are at least 10-fold smaller than in the most distal terminal digit. Such results indicate that the intracellular Ca concentration changes and, by inference, the presynaptic Ca channels, are concentrated in the terminals rather than in the axon. This is consistent with the early work of Katz and Miledi (1969), which showed that Ca spikes in this presynaptic neuron are larger in the terminal than in the axon. It also is consistent with previous, lower-resolution measurements of the spatial distribution of presynaptic Ca transients in this neuron (Miledi and Parker, 1981; Llinas, 1984) and in the barnacle photoreceptor neuron (Stockbridge and Ross, 1984).

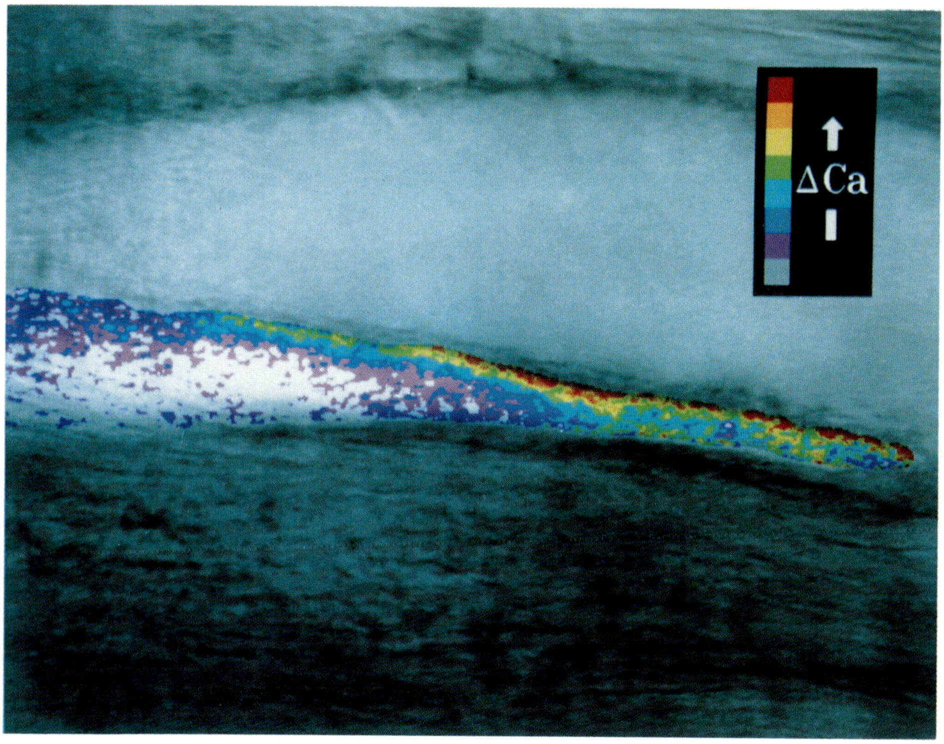

Fig. 6 cont'd.

We have attempted to localize the presynaptic Ca channels at an even higher level of resolution to test the prediction that Ca channels are clustered at specializations of the presynaptic cell known as active zones. Freeze-fracture studies have revealed that exocytosis occurs at active zones and that these active zones also include distinctive arrays of large intramembranous particles that are immediately adjacent to sites of vesicle fusion (Heuser et al., 1978; Torri-Tarelli et al., 1985). These large intramembranous particles, which are found only at active zones, have been proposed to be presynaptic Ca channels (Heuser et al., 1974; Pumplin et al., 1981; Walrond and Reese, 1985; Probst and Ko, 1987). If this is true then Ca influx should be restricted to presynaptic regions that contain active zones.

The organization of the giant presynaptic terminal of squid permits a direct test of this prediction. In this terminal, as in most presynaptic terminals (Heuser and Reese, 1977), active zones are concentrated in the region of the presynaptic membrane most closely apposed to the postsynaptic cell (Fig. 7). Squid synapses are organized such that the giant presynaptic terminal sometimes lies along the side of the postsynaptic cell (Martin and Miledi, 1986). In such synapses the large size of the giant terminal makes it possible to use the fluorescence microscope to distinguish between the region of the terminal nearest the postsynaptic cell and that away from the postsynaptic cell. Since active zones are much more concentrated on the side of the terminal nearest the postsynaptic cell, Ca influx sites also should be concentrated in this region if Ca channels are restricted to active zones.

An example of a high-magnification image of stimulus-induced changes in fura-2 fluorescence from such a specimen is shown in Fig. 6 *right*. At early times after initiating the presynaptic stimulus the changes in fura-2 fluorescence are indeed much larger on the side of the presynaptic terminal that is closest to the postsynaptic cell. At later times the Ca signal is more evenly distributed, presumably due to diffusion of Ca away from sites of entry.[2] From such images of highly lateralized Ca accumulation at early times after the opening of Ca channels, we conclude that the sites of Ca influx,

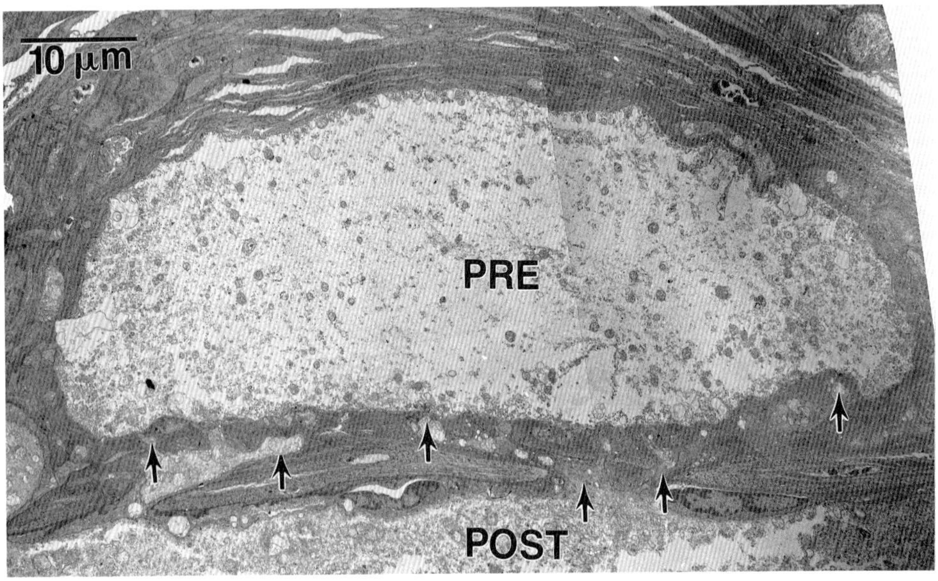

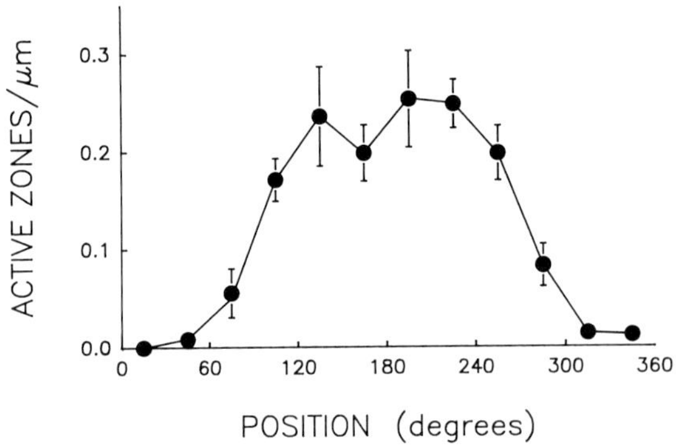

Figure 7. Location of active zones within the squid giant presynaptic terminal. (*Top*) Electron micrograph illustrating a cross section through the presynaptic terminal (*pre*). Active zones are indicated by the arrows and are most abundant in the portion of the presynaptic terminal closest to the postsynaptic cell (*post*). (*Bottom*) Distribution of active zones around the perimeter of the giant presynaptic terminal. The density of active zones was determined by dividing the terminal perimeter into segments based on polar coordinates, with 0 degrees being the point of the terminal that was farthest from the postsynaptic cell. Points represent mean active zone density for 12 cross sections cut through a single presynaptic terminal at 50 μm intervals. Bars indicate ± 1 SEM.[2]

like the active zones, are located in the portion of the presynaptic membrane closest to the postsynaptic cell. This colocalization of Ca influx and active zones provides strong evidence supporting the hypothesis that Ca channels are located preferentially at active zones.

To take this test of the hypothesis one step further, it would be ideal to make these measurements of Ca influx at single active zones. Unfortunately, this has not yet been possible because of the limited spatial and temporal resolution of conventional fluorescence microscopy methods. In an attempt to circumvent these limitations, we have recently begun using the newly developed confocal laser scanning microscope (White et al., 1987) to image Ca transients in squid presynaptic terminals. Such a microscope permits optical sectioning of the presynaptic terminal because illumination only reaches a thin (few micrometers thick) slice of the terminal that is in focus. This results in better spatial resolution, as shown in Fig. 8. Preliminary images obtained with this method (Smith, S. J., M. P. Charlton, and G. J. Augustine, unpublished observations) reveal that sites of Ca influx appear to be located directly opposite postsynaptic dendritic processes (Young, 1973), which is where active zones are also found (Pumplin and Reese, 1978). Thus, these preliminary measurements apparently extend the correlation between sites of Ca influx and active zones to the level of single active zones.

In summary, sites of Ca influx are tightly correlated with the location of active zones, which probably means that Ca channels are clustered at active zones. This lends

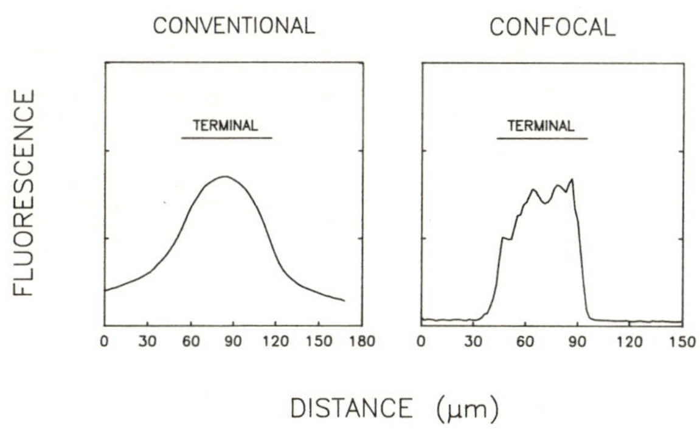

Figure 8. Comparison of the spatial resolution of conventional (*left*) and confocal (*right*) fluorescence microscopy for imaging applications at the squid giant presynaptic terminal. Graphs illustrate the profile of fluorescence scanned across the width of two different presynaptic terminals injected with either fura-2 (left) or fluo-3 (right). The actual width of the terminals, as assessed by bright-field microscopy, is indicated by the lines above the graphs.

further support to the proposal that the large intramembranous particles found only at active zones are, in fact, the Ca channels responsible for triggering transmitter secretion.

The clustering of presynaptic Ca channels at active zones has at least two significant implications for transmitter release. First, locating the Ca channels directly at release sites permits rapid triggering of exocytosis. The synaptic delay, which is the interval between a presynaptic action potential and the postsynaptic response that it produces, is only ~1 ms at this and many other chemical synapses (see below). Because of the relatively slow diffusion of Ca within cytoplasm (Hodgkin and Keynes, 1957; Rose and Loewenstein, 1975), it is necessary to have the channels close to release sites to minimize diffusion distances to produce such a brief delay.

Second, clustering yields a high local density of Ca channels at active zones. This means that there is more potential for Ca entering from single channels (Chad and Eckert, 1984; Simon and Llinas, 1985; Zucker and Fogelson, 1986) to sum together, potentially producing very high Ca concentrations at release sites within the active zones (Smith and Augustine, 1988). Such a large and rapid Ca transient probably is needed to produce the high rates of transmitter release produced by a presynaptic action potential. This also is interesting from the standpoint of understanding the molecular mechanism responsible for triggering exocytosis, because it suggests that the Ca receptor that initiates exocytosis could have a low affinity for Ca (Smith and Augustine, 1988).

Initiation of Transmitter Secretion by Ca Influx

The ability to measure the properties of squid presynaptic Ca channels presents an opportunity to directly evaluate the influence of Ca channel properties on transmitter release. We and others have performed studies to correlate Ca influx through these channels with the resultant secretion of neurotransmitters (Llinas et al., 1981*b*, 1982; Charlton et al., 1982; Augustine et al., 1985*b*; Smith et al., 1985; Augustine and Charlton, 1986). As an assay for neurotransmitter secretion we have measured the electrical responses resulting from interaction of released neurotransmitter with its receptors on the membrane of the postsynaptic cell. While this assay is not as direct as chemical measurements of secreted transmitters or electrical measurements of synap-

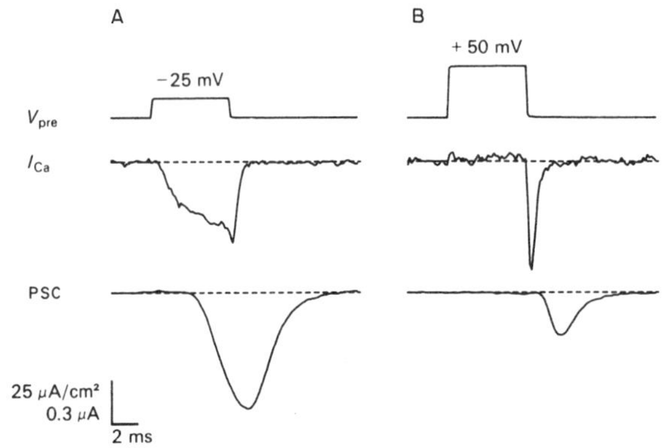

Figure 9. Presynaptic Ca currents (I_{Ca}) and postsynaptic currents (*PSC*) elicited by depolarizations of the presynaptic membrane potential (V_{pre}) to −25 mV (*left*) and +50 mV (*right*). Dotted lines indicate baselines measured before the depolarizations. From Augustine et al. (1985*b*).

tic vesicle fusion (Jaffe et al., 1978; Gillespie, 1979; Neher and Marty, 1982), it does offer rapid temporal resolution and high sensitivity. Our correlation between presynaptic Ca currents and transmitter-induced postsynaptic currents was performed by combining the use of the local Ca application method (Fig. 1 *B*), a three-microelectrode voltage clamp to measure presynaptic Ca current and another, two-microelectrode voltage clamp to measure transmitter-induced currents in the postsynaptic cell. This combination of methods ensures optimal measurement of both Ca current and transmitter release from a nearly isopotential region of the presynaptic terminal (Augustine et al., 1985*a*, *b*).

Representative examples of records obtained under these conditions are shown in Fig. 9. This figure illustrates responses elicited by depolarizations to two different presynaptic membrane potentials, −25 and +50 mV. The depolarization to −25 mV elicited a slowly activating Ca current that was followed, after a delay of roughly 2 ms, by a smoothly rising postsynaptic current response. The depolarization to +50 mV was sufficiently large to suppress the influx of Ca, so that there was neither a presynaptic Ca current nor a postsynaptic response during the depolarization. However, upon cessation of the depolarization there was an immediate and large Ca tail current, which produced a postsynaptic response after a delay of only ~0.5 ms.

Experiments of the sort illustrated in Fig. 9 permit a number of inferences about the molecular basis of exocytosis (e.g., Augustine et al., 1988), but a discussion of this subject is beyond the scope of the present discourse. However, these observations also provide two informative hints about the role of presynaptic Ca channels in triggering neurotransmitter secretion.

One hint can be derived from an examination of the temporal relationship between presynaptic Ca current and postsynaptic response. Because of the finite rate of Ca diffusion in cytoplasm, the brief (0.5–2 ms) delay between Ca entry into the terminal and resultant transmitter secretion indicates that the Ca channels must be located very near the sites of exocytosis (Parsegian, 1977). This reinforces the notion that Ca channels are clustered at active zones, as discussed above.

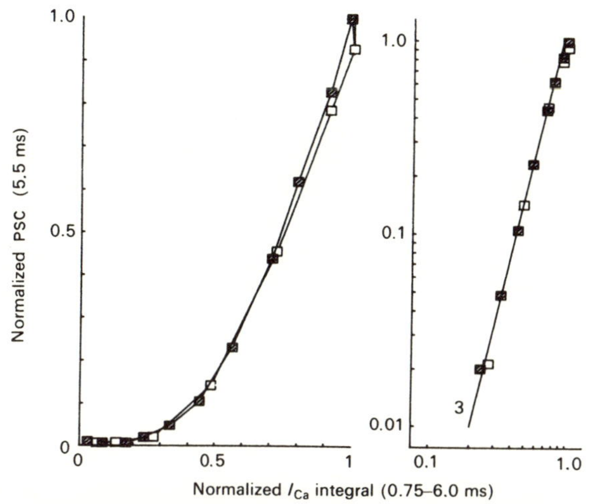

Figure 10. Synaptic transfer curves relating magnitudes of Ca currents (I_{Ca}) and postsynaptic responses (*PSC*) elicited by variable-amplitude presynaptic depolarizations. Ca currents were integrated from 0.75 to 6 ms after the onset of the 6-ms long depolarizations and postsynaptic responses were measured from 5 to 6 ms after the onset of the depolarizations. The same data are plotted on linear coordinates on the left and on logarithmic coordinates on the right. Open symbols represent responses elicited by depolarizations to potentials to −8 mV or more negative, while filled symbols represent responses elicited by depolarizations to potentials more positive than −8 mV. The straight line on the right graph is a power function with an exponent of 3. From Augustine et al. (1985*b*).

A second hint results from consideration of the quantitative relationship between the magnitude of presynaptic Ca currents and resultant postsynaptic responses. Such measurements can be made by varying the magnitude of the presynaptic depolarization over a wide range of values and obtaining a series of Ca currents and postsynaptic responses that are graded in magnitude. These responses can then be used to define a "synaptic transfer curve" relating the magnitude of presynaptic Ca influx to the magnitude of the resultant postsynaptic response. Such curves have been found to be power functions (Fig. 10) with exponents as great as 4 (Augustine et al., 1985*b*; Augustine and Charlton, 1986). Further, very similar transfer curves are obtained regardless of whether responses are elicited by small depolarizations (open symbols in Fig. 10) or by larger depolarizations (filled symbols in Fig. 10). This indicates that transmission has very little voltage dependence beyond that imparted by the gating of Ca channels (Smith et al., 1986).

These two observations on synaptic transfer curves provide information about the range over which Ca acts after entering the presynaptic terminal. Previous theoretical studies have predicted that Ca performs its intracellular second messenger functions by acting over highly localized "domains" centered over individual Ca channels (Chad and Eckert, 1984; Simon and Llinas, 1985; Zucker and Fogelson, 1986). Mathematical models of transmitter release at the squid synapse predict transfer curves with a high exponent power function form and voltage-independent properties only if the domains generated by individual presynaptic Ca channels overlap (Simon and Llinas, 1985; Zucker and Fogelson, 1986). Thus, transfer curves provide another indication that Ca channels are clustered closely together in the squid giant presynaptic terminal and that the resultant high density of Ca channels causes exocytosis at a given release site to be triggered by Ca ions contributed by multiple, adjacent channels.

Conclusions

The Ca channels of the giant presynaptic terminal of squid are tailored for the task of triggering the exocytotic secretion of neurotransmitter. Binding sites within these channels appear to have a high selectivity for Ca over other cations present in the extracellular environment, allowing these channels to deliver only this messenger cation to the presynaptic cytoplasm. The gating kinetics of these channels cause the channels to open slowly and close quickly in response to a presynaptic action potential, properties that shape the temporal characteristics of action-potential–evoked neurotransmitter release. These channels also inactivate only over seconds to permit sustained release in response to prolonged presynaptic depolarization. This combination of kinetic properties, when coupled with the highly nonlinear sensitivity of the exocytotic process to incoming Ca ions (Fig. 10), means that evoked transmitter release changes substantially in response to small changes in action potential waveform, presynaptic resting potential, or to more direct changes in Ca channel current. Thus the terminal is poised for regulation by the sorts of modulatory influences that are likely to be important in plastic changes in synaptic function (Kandel and Schwartz, 1982).

Three independent lines of evidence point toward the clustering of presynaptic Ca channels near sites of transmitter secretion within active zones. The brief delay between Ca influx and transmitter secretion indicates that Ca channels must be very close to release sites. The high-order, relatively voltage-insensitive synaptic transfer function suggests that Ca channels are present at such a high density that Ca from adjacent channels sums to trigger secretion from a given release site. Finally, combined fluorescence and electron microscopy reveals that sites of Ca influx colocalize with active zones. Clustering of Ca channels at active zones is likely to result in a very large and rapid presynaptic Ca transient when an action potential invades the presynaptic terminal.

To conclude we will return to the three-step model of Ca-triggered exocytosis that we invoked at the beginning of this paper. While our understanding of the first step—Ca influx into the presynaptic terminal—is incomplete, it is clear that we know far more about this step than the two that follow. It now appears that our appreciation of the properties of presynaptic Ca channels is sufficiently advanced to permit future studies to concentrate on the last two steps, thereby defining the specific spatiotemporal properties of the presynaptic Ca transient and elucidating the molecular basis for the initiation of exocytosis by this Ca transient.

Acknowledgments

We thank G. M. Armstrong and G. S. Oxford for inviting us to participate in the 1988 Symposium of the Society of General Physiologists and C. K. Augustine, L. Byerly, and S. D. Hess for commenting on this manuscript. This paper was written at the Max-Planck-Institut für biophysikalische Chemie, Göttingen, FRG, and we thank E. Neher for his hospitality and support of this effort.

The research described here was supported by National Institutes of Health grant NS-21624 to G. J. Augustine, Howard Hughes Medical Institute funds to J. Buchanan and S. J. Smith, a Medical Research Council (Canada) grant to M. P. Charlton, and a Muscular Dystrophy Association fellowship to L. R. Osses.

References

Adams, D. J., K. Takeda, and J. A. Umbach. 1985. Inhibitors of calcium buffering depress evoked transmitter release at the squid giant synapse. *Journal of Physiology*. 369:145–159.

Adrian, R. H., W. K. Chandler, and A. L. Hodgkin. 1970. Voltage clamp experiments in striated muscle fibers. *Journal of Physiology*. 208:607–644.

Almers, W., and E. W. McClesky. 1984. Non-selective conductance in calcium channels of frog muscle: calcium selectivity in a single-file pore. *Journal of Physiology*. 353:585–608.

Armstrong, D., and R. Eckert. 1987. Voltage-activated calcium channels that must be phosphorylated to respond to membrane depolarization. *Proceedings of the National Academy of Sciences*. 84:2518–2522.

Augustine, G. J., and M. P. Charlton. 1986. Calcium dependence of presynaptic calcium current and post-synaptic response at the squid giant synapse. *Journal of Physiology*. 381:619–640.

Augustine, G. J., M. P. Charlton, and S. J. Smith. 1985a. Calcium entry into voltage-clamped presynaptic terminals of squid. *Journal of Physiology*. 367:143–162.

Augustine, G. J., M. P. Charlton, and S. J. Smith. 1985*b*. Calcium entry and transmitter release at voltage-clamped nerve terminals of squid. *Journal of Physiology*. 367:163–181.

Augustine, G. J., M. P. Charlton, and S. J. Smith. 1987. Calcium action in synaptic transmitter release. *Annual Review of Neuroscience*. 10:633–693.

Augustine, G. J., M. P. Charlton, and S. J. Smith. 1988. Toward a molecular understanding of synaptic transmitter release: physiological clues from the squid giant synapse. *In* Calcium and Ion Channel Modulation. A. D. Grinnell, D. Armstrong, and M. B. Jackson, editors. Plenum Publishing Corp., New York. 157–168.

Augustine, G. J., and R. Eckert. 1982. Calcium-dependent potassium current in squid presynaptic nerve terminals. *Biological Bulletin*. 163:397. (Abstr.)

Augustine, G. J., and R. Eckert. 1984*a*. Divalent cations differentially support transmitter release at the squid synapse. *Journal of Physiology*. 346:257–271.

Augustine, G. J., and R. Eckert. 1984*b*. Calcium-dependent inactivation of presynaptic calcium channels. *Society for Neuroscience Abstracts*. 10:194. (Abstr.)

Benoit, P. R., and J. Mambrini. 1970. Modification of transmitter release by ions which prolong the presynaptic action potential. *Journal of Physiology*. 210:681–695.

Blight, A. R., and R. Llinas. 1980. The non-impulsive stretch-receptor complex of the crab: A study of depolarization-release coupling at a tonic sensorimotor synapse. *Philosophical Transactions of the Royal Society of London, B*. 290:219–276.

Byerly, L., P. B. Chase, and J. R. Stimers. 1985. Permeation and interaction of divalent cations in calcium channels of snail neurons. *Journal of General Physiology*. 85:491–518.

Chad, J. E., and R. Eckert. 1984. Calcium domains associated with individual channels can account for anomalous voltage relations of Ca-dependent responses. *Biophysical Journal*. 45:993-999.

Chad, J. E., and R. Eckert. 1986. An enzymatic mechanism for calcium current inactivation in dialysed *Helix* neurones. *Journal of Physiology*. 378:31–51.

Charlton, M. P., and H. L. Atwood. 1977. Slow release of transmitter at the squid giant synapse. *Neuroscience Letters*. 5:165–169.

Charlton, M. P., S. J. Smith, and R. S. Zucker. 1982. Role of presynaptic calcium ions and channels in synaptic facilitation and depression at the squid giant synapse. *Journal of Physiology*. 323:173–193.

Eckert, R., and J. E. Chad. 1984. Inactivation of Ca channels. *Progress in Biophysics and Molecular Biology*. 44:215–267.

Fenwick, E. M., A. Marty, and E. Neher. 1982. Sodium and calcium channels in bovine chromaffin cells. *Journal of Physiology*. 331:599–635.

Gillespie, J. T. 1979. The effect of repetitive stimulation on the passive electrical properties of the presynaptic terminal of the squid giant synapse. *Proceedings of the Royal Society of London, B*. 206:293-306.

Grynkiewicz, G., M. Poenie, and R. Y. Tsien. 1985. A new generation of Ca^{2+} indicators with greatly improved fluorescence properties. *Journal of Biological Chemistry*. 260:3440-3450.

Hagiwara, S., and L. Byerly. 1981. Calcium channel. *Annual Review of Neuroscience*. 4:69–125.

Hagiwara, S., and K. Takahashi. 1967. Surface density of calcium ions and calcium spikes in the barnacle muscle fiber membrane. *Journal of General Physiology*. 50:583–601.

Hess, P., J. B. Lansman, and R. W. Tsien. 1986. Calcium channel selectivity for divalent and monovalent cations. Voltage and concentration dependence of single channel current in ventricular heart cells. *Journal of General Physiology*. 88:293–319.

Heuser, J. E., and T. S. Reese, 1977. Structure of the synapse. *In* Handbook of Physiology, Section 1: The Nervous System. E. R. Kandel, editor. *American Physiological Society*, Bethesda, MD. 161–194.

Heuser, J. E., T. S. Reese, M. J. Dennis, Y. Jan, L. Jan, and L. Evans. 1979. Synaptic vesicle exocytosis captured by quick freezing and correlated with quantal transmitter release. *Journal of Cell Biology*. 81:275–300.

Heuser, J. E., T. S. Reese, and D. M. D. Landis. 1974. Functional changes in frog neuromuscular junctions studied with freeze-fracture. *Journal of Neurocytology*. 3:109–131.

Hodgkin, A. L., and R. S. Keynes. 1957. Movements of labelled calcium in squid giant axons. *Journal of Physiology*. 138:253–281.

Jaffe, L. A., S. Hagiwara, and R. T. Kado. 1978. The time course of cortical vesicle fusion in sea urchin eggs observed as membrane capacitance changes. *Developmental Biology*. 67:243–248.

Jan, Y. N., L. Y. Jan, and M. J. Dennis. 1977. Two mutations of synaptic transmission in *Drosophila. Proceedings of the Royal Society of London, B*. 198:87–108.

Kandel, E. R., and J. H. Schwartz. 1982. Molecular biology of learning: modulation of transmitter release. *Science*. 218:433–443.

Katz, B. 1969. The Release of Neural Transmitter Substances. Liverpool University Press, Liverpool, UK. 60 pp.

Katz, B., and R. Miledi. 1969. Tetrodotoxin-resistant electrical activity in presynaptic terminals. *Journal of Physiology*. 203:459–487.

Kusano, K. 1970. Influence of ionic environment on the relationship between pre- and postsynaptic potentials. *Journal of Neurobiology*. 1:435–457.

Lansman, J. B., P. Hess, and R. W. Tsien. 1986. Blockade of current through single calcium channels by Cd^{2+}, Mg^{2+}, and Ca^{2+}. Voltage and concentration dependence of calcium entry into the pore. *Journal of General Physiology*. 88:321–347.

Llinas, R. R. 1982. Calcium in synaptic transmission. *Scientific American*. 247:56–65.

Llinas, R. R. 1984. The squid giant synapse. *Current Topics in Membrane Transport*. 22:519–546.

Llinas, R., I. Z. Steinberg, K. Walton. 1981*a*. Presynaptic calcium currents in squid giant synapse. *Biophysical Journal*. 33:289–322.

Llinas, R., I. Z. Steinberg, and K. Walton. 1981*b*. Relationship between presynaptic calcium current and postsynaptic potential in squid giant synapse. *Biophysical Journal*. 33:323–352.

Llinas, R., M. Sugimori, and S. M. Simon. 1982. Transmission by presynaptic spike-like depolarization in the squid synapse. *Proceedings National Academy of Sciences*. 79:2415–2419.

Martin, R., and R. Miledi. 1986. The form and dimensions of the giant synapse of squids. *Philosophical Transactions of the Royal Society of London, B*. 312:355–377.

Miledi, R. 1971. Lanthanum ions abolish the "calcium response" of nerve terminals. *Nature*. 229:410–411.

Miledi, R., and I. Parker. 1981. Calcium transients recorded with arsenazo III in the presynaptic terminal of the squid giant synapse. *Proceedings of the Royal Society of London, B*. 212:197–211.

Miller, R. J. 1987. Multiple types of calcium channels and neuronal function. *Science*. 235:46–52.

Nachshen, D. A., and M. P. Blaustein. 1982. Influx of calcium strontium and barium in presynaptic nerve endings. *Journal of General Physiology*. 79:1065–1087.

Neher, E. 1986. Concentration profiles of intracellular Ca^{++} in the presence of a diffusible chelator. *Experimental Brain Research*. 14:80–96.

Neher, E., and A. Marty. 1982. Discrete changes of cell membrane capacitance observed under conditions of enhanced secretion in bovine adrenal charomaffin cells. *Proceedings of the National Academy of Sciences*. 79:6712–6716.

Nowycky, M. C., A. P. Fox, and R. W. Tsien, 1985. Three types of neuronal calcium channel with different calcium agonist sensitivity. *Nature*. 316:440–443.

Parsegian, V. A. 1977. Considerations in determining the mode of influence of calcium on vesicle-membrane interaction. *In* Society for Neuroscience Symposia. Vol. 2. W. M. Cowan and J. A. Ferrendelli, editors. Society for Neuroscience, Bethesda, MD. 161–171.

Penner, R., and E. Neher. 1988. The role of calcium in stimulus-secretion coupling in excitable and non-excitable cells. *Journal of Experimental Biology*. 139:329–345.

Probst, J. W., and C.-P. Ko. 1987. Correlations between active zone ultrastructure and synaptic function studied with freeze-fracture of physiologically identified neuromuscular junctions. *Journal of Neuroscience*. 7:3654–3664.

Pumplin, D. W., and T. S. Reese. 1978. Membrane ultrastructure of the giant synapse of the squid *Loligo pealei*. *Neuroscience*. 3:685–696.

Pumplin, D. W., T. S. Reese, and R. Llinas. 1981. Are the presynaptic membrane particles the calcium channels? *Proceedings of the National Academy of Sciences*. 78:7210–7213.

Rose, B., and W. R. Loewenstein. 1975. Calcium ion distribution visualized by aequorin: diffusion in cytosol restricted by energized sequestering. *Science*. 190:1204–1206.

Silinsky, E. M. 1985. The biophysical pharmacology of calcium-dependent acetylcholine secretion. *Pharmacology Reviews*. 37:81–132.

Simon, S. M., and R. R. Llinas. 1985. Compartmentalization of the submembrane calcium activity during calcium influx and its significance in transmitter release. *Biophysical Journal*. 48:485–498.

Smith, S. J., and G. J. Augustine. 1988. Calcium ions, active zones and synaptic transmitter release. *Trends in Neuroscience*. 11:458–464.

Smith, S. J., G. J. Augustine, and M. Charlton. 1985. Transmission at voltage-clamped giant synapse of squid: evidence for cooperativity of presynaptic calcium action. *Proceedings of the National Academy of Sciences*. 82:622–625.

Smith, S. J., M. P. Charlton, and G. J. Augustine. 1986. The voltage dependence of transmitter release. *Journal de Physiologie*. 81:332–339.

Stanley, E. F., and W. J. Adelman. 1984. Direct access of ions to the squid stellate ganglion giant synapse by aortic perfussion: effect of calcium-free medium, lanthanum, and cadmium. *Biological Bulletin*. 167:467–476.

Stockbridge, N., and W. N. Ross. 1984. Localized Ca^{2+} and calcium-activated potassium conductances in terminals of a barnacle photoreceptor. *Nature*. 309:266–268.

Suszkiw, J. B., M. E. O'Leary, M. M. Murawsky, and T. Wang. 1986. Presynaptic calcium channels in rat cortical synaptosomes: fast-kinetics of phasic calcium influx, channel inactivation, and relationship to nitrendipine receptors. *Journal of Neuroscience*. 6:1349–1357.

Torri-Tarelli, F., F. Grohovaz, R. Fesce, and B. Ceccarelli. 1985. Temporal coincidence between synaptic vesicle fusion and quantal secretion of acetylcholine. *Journal of Cell Biology*. 101:1386–1399.

Tsien, R. W. 1983. Calcium channels in excitable cell membranes. *Annual Review of Physiology*. 45:341–358.

Tsien, R. W., D. Lipscombe, D. V. Madison, K. R. Bley, and A. P. Fox. 1988. Multiple types of neuronal calcium channels and their selective modulation. *Trends in Neuroscience*. 11:431–437.

Tsien, R. W. 1988. Fluorescence measurement and photochemical manipulation of cytosolic free calcium. *Trends in Neuroscience*. 11:419–424.

Walrond, J. P., and T. S. Reese. 1985. Structure of axon terminals and active zones at synapses on lizard twitch and tonic muscle fibers. *Journal of Neuroscience*. 5:1118–1131.

White, J. G., W. B. Amos, and M. Fordham. 1987. An evaluation of confocal versus conventional imaging of biological structures by fluorescence light microscopy. *Journal of Cell Biology*. 105:41–48.

Young, J. Z. 1939. Fused neurons and synapstic contacts in the giant nerve fibres of cephalopods. *Philosophical Transactions of the Royal Society of London, B*. 229:465–503.

Young, J. Z. 1973. The giant fibre synapse of *Loligo*. *Brain Research*. 57:457–460.

Zucker, R. S., and A. L. Fogelson. 1986. Relationship between transmitter release and presynaptic calcium influx when calcium enters through discrete channels. *Proceedings of the National Academy of Sciences*. 83:3032–3036.

Chapter 14

Ion Channels, Intracellular Calcium, and Exocytosis: Control of Hormone Secretion in Cultured Bovine Pituitary Lactotrophs

W. T. Mason, S. K. Sikdar, R. Zorec, S. Akerman, S. R. Rawlings, T. Cheek, R. Moreton, and M. Berridge

Department of Neuroendocrinology, Agricultural and Food Research Council Institute of Animal Physiology and Genetics Research, Babraham, Cambridge CB2 4AT; the Agricultural and Food Research Council Invertebrate Chemistry and Zoology Unit, Department of Zoology, University of Cambridge, Cambridge, CB2 3EJ United Kingdom; and the Institute of Pathophysiology, 61105 Ljubljana, Yugoslavia

Introduction

The anterior pituitary gland contains a number of cell types secreting a variety of physiologically important hormones. Central to the control of circulating hormone levels is the mechanism by which the action of blood-borne factors is transduced at the single-cell level into the secretion of a hormone. In many cases the binding of peptides to specific cell surface receptors results in changes in the cell membrane ionic permeability presumably via the opening and/or closing of specific ion channels (Sand et al., 1980; Thorner et al., 1980; Ingram et al., 1986). The resulting changes in ionic flux have been proposed to lead to changes in hormone secretion (see Mason et al., 1988*a*, for review). The presence of these channel types gives the potential for complex interactions between membrane voltage, intracellular messengers, and ionic permeability. Although the intracellular pathways linking receptor activation to the modulation of exocytosis is still poorly understood, the requirement for a rise in intracellular free calcium concentration ($[Ca^{2+}]_i$) for the secretion of a hormone appears to be the case in many endocrine cells (Douglas, 1968).

Our work has centered on normal, as opposed to neoplastic, anterior pituitary cells which can be isolated into cell culture while retaining many properties believed to characterize normal cells. We have concentrated on three anterior pituitary cell types: the bovine lactotroph, the bovine somatotroph, and the ovine gonadotroph. In the case of the bovine cell preparation, the cells are enzymatically dissociated and enriched for a particular cell type (somatotrophs and lactotrophs) by separation on the basis of their relevant densities using Percoll density gradient centrifugation (Ingram et al., 1988). Gonadotrophs however, are cultured from the ovine pars tuberalis, which contain gonadotrophs as the only secretory cell type (Gross et al., 1984).

This paper will discuss our present knowledge of the modulation of intracellular calcium and hormone secretion in anterior pituitary cells with particular emphasis on bovine lactotrophs as a model secretory system.

The Lactotroph Cell and Prolactin Secretion: A General Overview

The hormone prolactin (PRL) is primarily contained in a cell type termed a lactotroph, although there appears to be a certain heterogeneity in this cell population (Winiger et al., 1987; Lewis et al., 1988). PRL is stored within the cell in discrete secretory granules and it is believed that hormone can be released from the cell by fusion of the granule with the cell membrane in response to an appropriate stimulus. PRL is physiologically important in milk production in mammals.

Under physiological conditions release of PRL is controlled by a variety of factors. In particular, neurohormones released from the median eminence of the hypothalamus and transported by the hypophysial portal system to the anterior pituitary gland play a major role in controlling release. These hypothalamic factors include thyrotropin-releasing hormone (TRH) and vasoactive intestinal polypeptide (VIP) (Abe et al., 1985), which stimulate release, and dopamine (DA), which is the major inhibiting factor for PRL secretion (Ben-Jonathan, 1985).

These factors bind to specific cell surface receptors, and in a number of cases activate membrane-bound enzymes which catalyze the production of various intracellular messengers that modulate the response of the cell. For example, VIP binding to its cell surface receptor activates a guanine nucleotide-binding protein (G-protein; G_s), which stimulates the enzyme adenylate cyclase (AC) to produce the intracellular messenger cyclic adenosine-3′,5′-monophosphate (cAMP) (Bjoro et al., 1987), and leads ultimately to the secretion of PRL. DA exerts its action through the activation of an inhibitory G-protein (G_i), which inhibits the production of cAMP via an action on the AC enzyme. In contrast, binding of the TRH receptor stimulates the phospholipase C enzyme, which catalyzes the hydrolysis of phosphatidylinositol-4,5-bisphosphate (PIP_2) to form the intracellular messengers D-myo-inositol-1,4,5-trisphosphate (IP_3) and diacylglycerol (DAG) (Gershengorn, 1986). It has been proposed that there is a complex interplay between these different intracellular factors to produce the final cellular response (e.g., Gourdji et al., 1979).

In addition, the lactotroph cell possesses a variety of ion channels in its cell membrane, whose opening is modulated not only by chemical factors, but also by the voltage across the membrane (DeRiemer and Sakmann, 1986; Lingle et al., 1986; Cobbett et al., 1987*a, b*). These include: (*a*) a voltage-activated, TTX-sensitive Na^+ channel, (*b*) two types of voltage-activated Ca^{2+} channels (T and L type channels), and (*c*) at least two types of voltage-activated K^+ channels with different sensitivities to calcium.

In this review we shall discuss results using calcium imaging techniques to record dynamic changes in $[Ca^{2+}]_i$, and cell membrane capacitance measurements (a sensitive method for assaying exocytosis), and we shall attempt to relate this to data obtained from measurements of electrical properties and hormone secretion from these cells. In the following sections we will deal with three aspects of PRL secretion from bovine lactotrophs, and the involvement of Ca^{2+} ions. These will be the basal release of PRL, the stimulation of PRL release both by the hypophysiotropic factor TRH and depolarization, and the inhibition of PRL release by the neurotransmitter DA. We will discuss the results that we have obtained using a number of different techniques.

Basal Secretion of PRL and Calcium Flux

Much of our present knowledge on hormonal release mechanisms has come from measurements by radioimmunoassay of hormone content in the bathing or superfusing medium. PRL is released from cultured bovine lactotrophs under nonstimulated (basal) conditions (Fig. 1). However, if extracellular Ca^{2+} is removed by use of EGTA buffers, or if 4 mM Co^{2+} ions are added to the extracellular medium, basal secretion of PRL is inhibited by 30–70% (Ingram et al., 1986). However, basal PRL secretion is unaffected by extracellular Na^+ replacement with isotonic choline, or if high concentrations of TTX (sufficient to block Na^+ action potential firing in these cells) are added to the medium (Cobbett et al., 1987*a*). Hence neither Na^+/Ca^{2+} exchange nor Na^+ entry through voltage-gated Na^+ channels (such as might occur during an action potential) appear to play a prominent role in basal secretion of hormone.

Monitoring the Granular Fusion Events Related to Basal Release

Although the fact that basal secretion indeed occurs may not be surprising, it is important to be able to identify and quantitate the exocytotic process itself. The final intervening step before the hormone within the secretory granule is liberated out of the cell is the fusion of the secretory granule with the plasma membrane. The granular fusion and retrieval would lead to changes in the cell surface area due to incorporation of the vesicle membrane into the plasma membrane, and this can be monitored as a change in cell capacitance. The cell capacitance itself was monitored under sinusoidal excitation with a two phase lock-in amplifier (Neher and Marty, 1982) and with the aid of a computer (Lindau and Neher, 1988). Under voltage clamp conditions, at a holding potential of ~−70 mV, we could observe step jumps in cell capacitance between 2 and 20 fF (Mason et al., 1988*b*), which correlates well with the measured secretory granule diameter in lactotrophs (Ingram et al., 1988). This suggests that spontaneous fusion of secretory granules with the plasma membrane does occur at the resting membrane potential of the cell, and probably contributes to basal release. Since the fusion process itself is affected by intracellular Ca^{2+} concentration (as discussed in a later section) it is also important to know the intracellular Ca^{2+} concentration profile of the cells at rest.

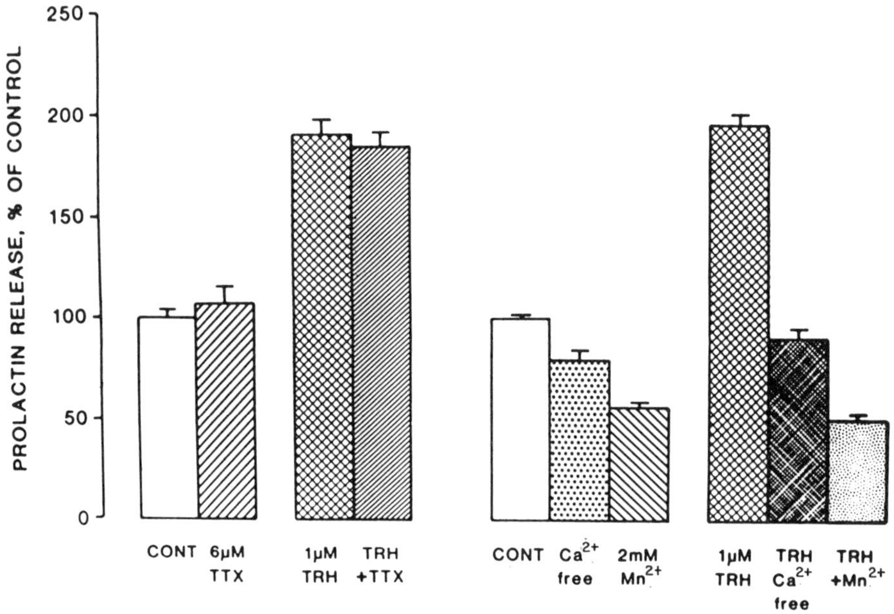

Figure 1. Effect of Na^+ channel and Ca^{2+} channel blocking agents on release of PRL. PRL release was measured by standard radioimmunoassay techniques. Both basal and 1 μM TRH-stimulated PRL release were unaffected by application of TTX at 10 μM. However, these data clearly show that both basal and TRH-stimulated PRL release are potently suppressed by removal of Ca^{2+} from the extracellular medium, or by extracellular application of 2 mM Mn^{2+}.

Intracellular Ca^{2+} Levels at Rest

To measure intracellular Ca^{2+} concentration ($[Ca^{2+}]_i$) lactotrophs were loaded with the Ca^{2+}-sensitive dye (fura-2 A/M) at room temperature (see legend to Fig. 5 for details). Basal $[Ca^{2+}]_i$ appeared to be dependent to some extent on the flow of Ca^{2+} from the extracellular compartment, since the mean $[Ca^{2+}]_i$ in cells bathed in 1.8 mM extracellular calcium (47 nM) was about half that measured in 5 mM extracellular calcium (68 nM). This evidence is consistent with the idea that calcium entry pathways exist that are activated even when the cell is resting under basal conditions. Below, we shall consider the possible ionic mechanisms controlling Ca^{2+} entry at rest.

Ionic Mechanisms Controlling Ca^{2+} Entry and Basal Secretion at Rest

Normal lactotroph cells as well as the clonal pituitary tumor lines GH_3 and GH_4 are known to spontaneously fire action potentials in 15–30% of cells in the population (Fig.

2, *A* and *B*). The depolarization caused by an action potential probably opens voltage-activated Ca^{2+} channels, and firing of a single action potential in the GH_3 cell has been reported to cause an observable spike of $[Ca^{2+}]_i$ as measured by fura-2 technology (Schlegel et al., 1987). In the bovine lactotrophs, basal secretion is unaffected by application of TTX under conditions that block voltage-activated Na^+

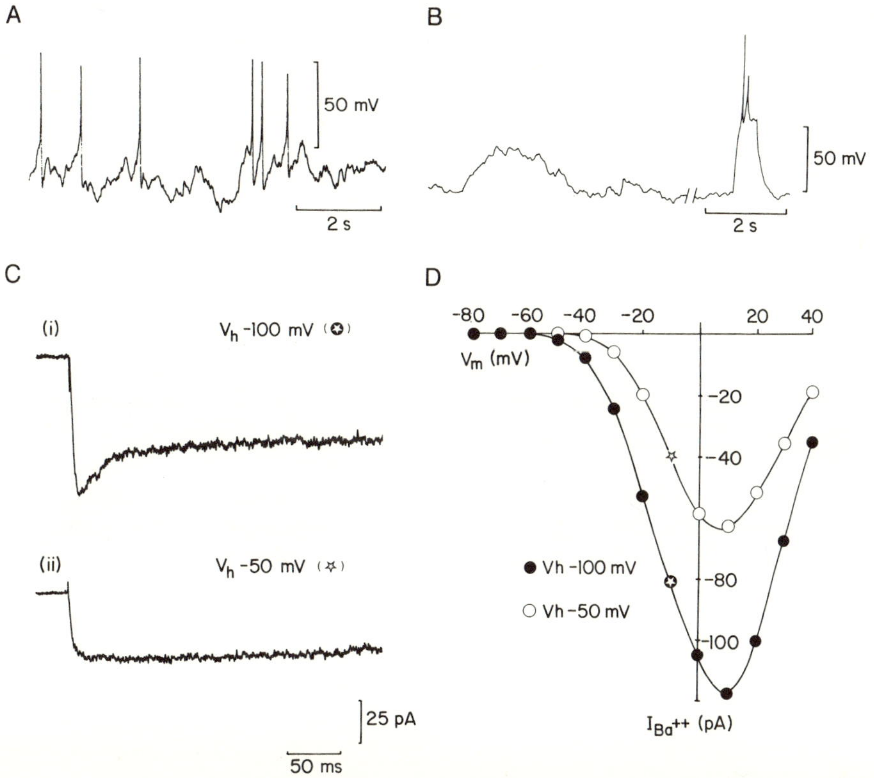

Figure 2. Calcium currents and general electrical properties of cultured bovine lactotrophs. (*A*) Using the current-clamp mode of the whole-cell technique (see Hamill et al., 1981), spontaneous fluctuations in membrane potential were seen, which rarely give rise to spontaneous action potentials. However, positive current pulses (5 pA, 600 ms) elicited a marked depolarization and firing of between one and three action potentials. (*B*) These action potentials were blocked by addition of 10 μM TTX (not shown). Internal solution (in millimolar): 150 KCl, 10 EGTA, 10 HEPES, pH 7.3. External solution: 135 NaCl, 2 KCl, 5.6 glucose, 1 $MgCl_2$, 10 HEPES, pH 7.3. (*C*) Whole-cell voltage-clamp recordings of isolated calcium currents, using 30 mM Ba^{2+} as the charge carrier. (*i*) From a holding potential (V_h) of -100 mV, a pulse to membrane potential (V_m) of -10 mV gave rise to two types of current; a rapidly activating and inactivating current ("*T*" current) and a noninactivating current ("*L*" current). (*ii*) However, from a Vh of -50 mV, a pulse to a V_m of -10 mV shows a predominance of the noninactivating current. The currents are averages of five sweeps, cancelled for linear leak with suitably scaled hyperpolarizing pulses. Internal solution (in millimolar): 140 CsCl, 5 EGTA, 10 HEPES, 1 $MgCl_2$, 10 glucose, pH 7.3. External solution: 120 choline-Cl, 30 TEA, 10 HEPES, 10 glucose, 30 $BaCl_2$, pH 7.3. *D* shows the current-voltage curve for the peak current obtained for pulses from the two holding potentials shown in *C*. The currents seen in *C* are marked with stars as they appear on the I-V curve. From a V_h of -100 mV, the peak current activated between -60 and -50 mV, a potential that is frequently attained by the potential fluctuations demonstrated in *A*.

currents and firing of action potentials (Fig. 1). However, recordings under current clamp show significant fluctuations in the cell membrane potential around the resting potential (see Fig. 2, *A* and *B*). These fluctuations can be up to 30 mV in amplitude (Cobbett et al., 1987*a*) and thus may account for the opening of voltage-activated Ca^{2+} channels and Ca^{2+} influx in the absence of an action potential spike.

Using the whole-cell recording configuration and manipulation of the intracellular and extracellular media, we recorded isolated currents flowing through Ca^{2+} channels in the plasma membrane typically using either Ca^{2+} or Ba^{2+} as a charge carrier (see the legend of Fig. 2 for details). Under these conditions, positive voltage pulses from negative holding potentials cause current flow through voltage-activated Ca^{2+} channels in the lactotroph membrane (Fig. 2 *C*). If we use a very negative holding potential (such as -100 mV) and apply positive voltage pulses, it can be clearly seen that voltage-activated Ca^{2+} channels can be activated around the resting membrane potential of these cells (Fig. 2 *D*). These experiments clearly show that in lactotrophs as well as other pituitary cells as much as 2–10% of the total current evoked at strong depolarized potentials (-20 to $+20$ mV) occurs in the range of -70 to -40 mV. Typically this current may be on the order of 1–5 pA. Such current flow close to the resting potential may be quite significant, particularly in the case of a small cell. For instance, 1 pA of current represents the entry of about five million divalent cations per second.

The current flow occurs almost certainly through at least two types of Ca^{2+} channel. One is a rapidly activating and inactivating channel and the second is a Ca^{2+} channel that does not significantly inactivate even during test pulses of many seconds (Cobbett et al., 1987*a*). This noninactivating current is metabolically sensitive in that it runs down during dialysis of the cell interior with ATP-free media, whereas the transient inactivating current is less sensitive to the absence of ATP. Also, the noninactivating current can be partially blocked by dihyropyridine Ca^{2+} antagonists such as nifedipine. Both of these currents can be blocked by addition of Co^{2+} ions to the extracellular medium.

Further electrophysiological evidence for a tonic flow of Ca^{2+} ions at rest comes from the fact that with 1.8–5 mM Ca^{2+} in the external medium, application of 4 mM Co^{2+} elicits a small decrease in the inward holding current of the cell, as well as a decrease in membrane conductance when held at ~ -70 mV.

Are Ionic Currents Activated by Spontaneous Action Potentials Important in Promoting Exocytosis?

Whole-cell recordings were made from lactotrophs under conditions in which the cell interior was dialyzed with a physiological-like intracellular medium, containing K-gluconate (140 mM) and low buffering capacity EGTA (100 μM) with 1.8 or 5 mM extracellular Ca^{2+}. The results showed that significant increases in cell capacitance occur during long periods of time after dialysis of the cell interior (Fig. 3). When intracellular Ca^{2+} was buffered to <10 nM with EGTA solutions, the capacitance increases were not observed. The capacitance change is clearly seen in Fig. 4, which also shows that depolarizing pulses that are used to activate voltage-dependent ion channels (from a holding potential around rest) significantly increase the rate of this capacitance change. Addition of TTX and TEA to the external medium, to block voltage-activated Na^{+} and K^{+} channels, respectively, did not prevent the increased

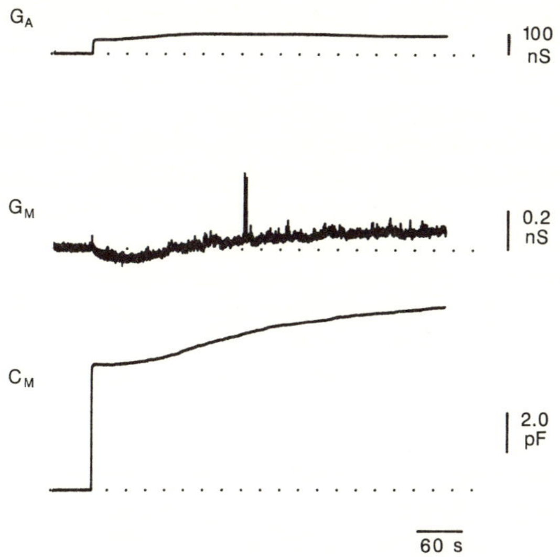

Figure 3. Increase in membrane capacitance after dialysis of the cell with internal solution containing 1 $\mu M\ Ca^{2+}$. The pipette contained 0.5 mM EGTA + 3.5 Ca-EGTA in K-gluconate, which diffused into the cell when the whole-cell configuration was achieved (indicated by the capacitance jump). The capacitance plot was derived from a computer-aided reconstruction of the cell parameters: cell capacitance (C_m), membrane conductance (G_m), and the access conductance (G_a) in series with the parallel resistance-capacitance circuit of the cell, from the two-phase lock-in measurements (see Lindau and Neher, 1988; Pusch and Neher, 1988). The large initial increase in C_m and G_a indicates the point where the whole-cell configuration was achieved. After this, the cell capacitance increased sigmoidally, and the maximal percentage change in C_m from rest was 48%; the half maximal capacitance was reached in 208 s.

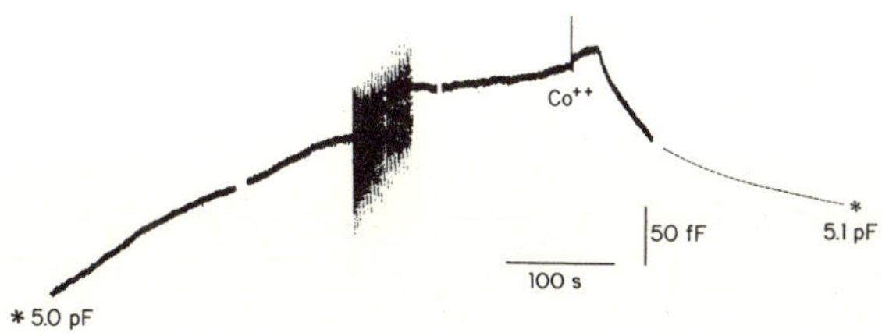

Figure 4. Spontaneous increase in membrane capacitance after dialysis of the cell with 0.1 μM EGTA-containing internal solution (in millimolar): 10 NaCl, 140 K-gluconate, 2 $MgCl_2$, 10 HEPES, 0.1 μM EGTA, 3 ATP, pH 7.2 with Trizma. External bathing solution contained: 131.8 NaCl, 5 KCl, 2 $MgCl_2$, 0.5 NaH_2PO_4, 5 $NaHCO_3$, 1.8 $CaCl_2$, 10 HEPES, 10 glucose. During the application of a train of pulses (+80 mV, 410 ms every 2 s, 18 pulses) indicated by the artifacts on the record, there was a capacitance increase of 45 fF, and the rate of C_m change increased from 25.7 fF/50 s to 41 fF/50 s. The membrane capacitance increased from 5.0 to 5.22 pF after breaking the patch for whole-cell recording. Application of Co^{2+} into the bath (final concentration 2 mM) caused a decrease in capacitance 24 s after application, and the capacitance decreased to 5.1 from 5.22 pF. Holding potential, −72 mV. Membrane capacitance was recorded using a two-phase lock-in amplifier incorporated into a patch-clamp amplifier (Henigman et al., 1987) by the compensated method (for details see Lindau and Neher, 1988). The calibrated out-of-phase output was 1 pF/V (sinusoidal voltage amplitude, peak to peak 1 mV, frequency 1,600 Hz). The capacitance record was appropriately scaled using a correction factor, and the cross-talk was checked for interactions between the resistive and capacitative components due to inappropriate phase setting performed by turning the C-slow compensation knob to mimic cell capacitance change, and the phase-shift circuit of the lock-in amplifier was adjusted so that 200–300 fF displacements in the capacitance signal caused no deflection in the conductance trace.

rate of change of membrane capacitance (Zorec, R., S. K. Sikdar, and W. T. Mason, unpublished data).

This increase in capacitance is most likely associated with the entry of Ca^{2+} ions through voltage-activated Ca^{2+} channels in the cell membrane. In addition, the spontaneous increases in capacitance before and after the depolarizing voltage pulses are also probably due to the influx of Ca^{2+} ions through the membrane. This is supported by the fact that spontaneous increases in membrane capacitance can be reversed by the addition of Co^{2+} ions (2–8 mM), which are known to block voltage-activated Ca^{2+} channels (Fig. 4). One other explanation for this effect is that Co^{2+} directly inhibits the process of exocytosis by blocking interactions at the cell membrane–granule fusion site. However, this possibility is unlikely since Co^{2+} ions applied to cells in which exocytosis was forced by dialysis of the cytosol with high Ca^{2+} -containing solutions, did not block the capacitance increase (data not shown).

Increased rate of exocytosis after electrical stimulation of the cell depends upon the accumulation of Ca^{2+} within the cell from the influx of Ca^{2+} from the extracellular space. The efficiency of depolarizing voltage pulses in facilitating capacitance changes (Fig. 4) in the bovine lactotroph thus differs from that reported for bovine adrenal chromaffin cells (Clapham and Neher, 1984), although both cells are excitable.

Thus, three lines of evidence using very different technologies suggest that pathways exist for calcium entry into lactotrophs under basal conditions, and that these pathways may modulate intracellular calcium, which directly interacts with the exocytotic event. In addition, currents activated during spontaneous action potentials (primarily Ca^{2+}, but not Na^+) may also play a small role in basal PRL release.

Stimulation of PRL Release by TRH and Depolarization

The hypophysiotrophic peptide TRH stimulates PRL release from bovine lactotrophs. This release is insensitive to the voltage-activated Na^+ channel blocker TTX, but it is significantly blocked by 4 mM Co^{2+} (Fig. 1). In addition, elevated extracellular K^+ (56 mM) will also promote PRL release, and this in turn is reduced by the Ca^{2+} channel blocker Co^{2+}. Hence both these methods of stimulating PRL release appear to be dependent on the influx of extracellular Ca^{2+}. Much of our work has sought to investigate the hypothesis that stimulated secretion of PRL is dependent on action potentials, a concept originally developed to explain the link between stimulus and secretion.

Using intracellular recordings with high-resistance micropipettes, it was found that only 15% of normal lactotrophs (Ingram et al., 1986) fire action potentials spontaneously (as in Fig. 5). TRH applied at 100 nM to 1 μM consistently causes a hyperpolarization lasting typically 5–10 s (Ozawa, 1985), which is followed by a small depolarization to a value of 2–5 mV more positive than that recorded before TRH application (Fig. 5 *A*). In many of the spiking cell types, this elevation is sufficient to cause a small increase in the frequency of action potential firing. In some "quiet" cells firing is also observed after TRH application, such as that shown in Fig. 5. In both cell types application of TTX abolishes spiking induced by TRH as well as abolishing action potential firing in the absence of TRH. Application of high extracellular K^+, usually at 56 mM, depolarizes the cell, but it also abolishes firing of action potentials as they are strongly inactivated during depolarization of the cells.

One of the interesting aspects that we wished to examine was whether the biphasic electrophysiological response to TRH via hyperpolarization followed by depolarization is reflected in the intracellular Ca^{2+} change.

Intracellular Ca^{2+} Response to TRH

Application of 1 μM TRH to cells in the presence of 5 mM extracellular Ca^{2+} causes a rapid, but transient, increase in the level of intracellular ionized Ca^{2+} after a delay of 2–5 s (Fig. 5 *B*). This decays back to near-basal levels within 2–5 min. This transient rise in intracellular Ca^{2+} is not blocked by Co^{2+} in the external medium or by the

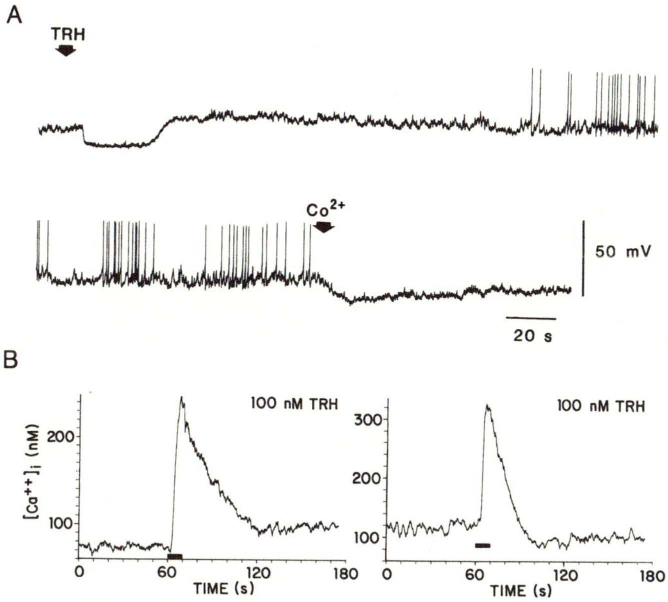

Figure 5. (*A*) TRH electrophysiology. Continuous intracellular recording of bovine lactotroph when TRH was applied at a concentration of 1 μM. TRH caused an initial hyperpolarization followed by a small depolarization and action potentials in ~20% of the recorded cells. Co^{2+} caused a significant hyperpolarization of the cell, and spiking ceased. (*B*) The effect of TRH on intracellular free calcium in two lactotroph cells. Cells were loaded with fura-2 A/M and alternatively excited with 340 and 380 nM light using the method of Grynkiewicz et al., 1985. The resultant emission at 510 nM was collected by an intensified CCD camera, ratioed, and stored on video tape. The traces were produced by calculating the mean ratio value for a cell and converting to calcium values. In the first trace a 10-s application of 100 nM TRH caused a transient elevation of intracellular free calcium, followed by a decay to a level higher than that at rest. In the second trace a similar transient increase is observed but the post-response level is below that found prior to TRH application.

removal of extracellular Ca^{2+} with EGTA. Both treatments would affect flow through voltage-activated Ca^{2+} channels in the plasma membrane.

The effects of TRH strongly suggest rises in intracellular Ca^{2+} occurring through its release from intracellular stores that are likely due to the action of compounds in the phosphoinositide pathway (see Gershengorn, 1989, this volume). There is little evidence to suggest that these intracellular Ca^{2+} transients are in any way voltage dependent. Of more interest is that Ca^{2+} imaging experiments have shown us that recharging of intracellular Ca^{2+} stores after discharge by TRH is dependent on extracellular Ca^{2+} entry. Thus, repetitive responses of a single lactrotroph to TRH can be generated if extracellular Ca^{2+} is present in the range of 1.8–5 mM. Removal of extracellular Ca^{2+} prevents multiple responses to TRH pulses. In view of the fact that no lowering of $[Ca^{2+}]_i$ was observed before the transient rise and decay and that the response was monophasic in contrast with the biphasic electrophysiological response indicates that the changes in the electrophysiological properties are independent of the $[Ca^{2+}]_i$ change.

Intracellular Ca^{2+} and Exocytosis

We were unable to see changes in cell capacitance after extracellular application of TRH in the whole-cell configuration presumably because the cells were being dialyzed

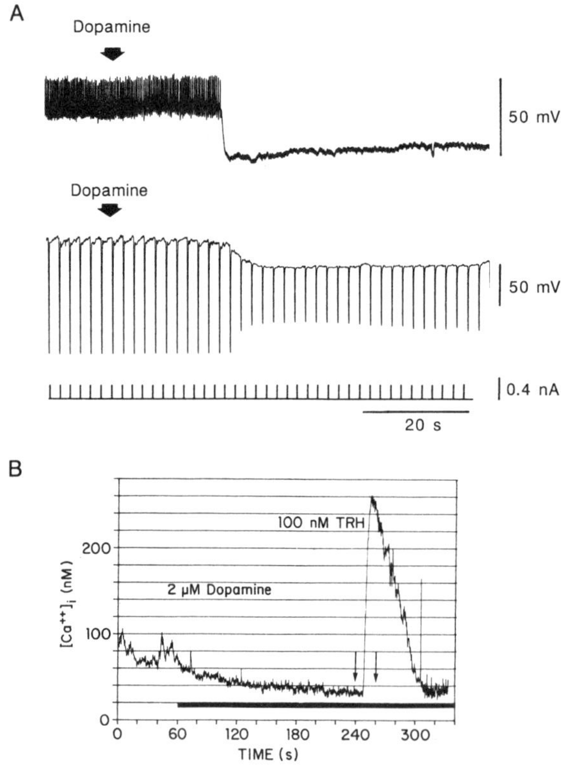

Figure 6. (*A*) DA electrophysiology. Top trace shows an intracellular voltage recording from a lactotroph firing action potentials. Application of 1 μM DA in the bath caused a strong, sustained hyperpolarization of ~25 mV in this cell. The bottom trace also shows that such hyperpolarization is accompanied by a large increase in membrane conductance, measured with hyperpolarizing pulses, the current for which is shown below the record. (*B*) The effects of DA and TRH on intracellular free calcium in a single lactotroph cell. 2 μM DA causes intracellular free calcium to decrease dramatically to below 50 nM. A 10-s application of TRH, however, still induced a typical transient response even in the continued presence of DA.

and the biochemical factors transducing the TRH effect may have washed out. We are presently attempting to use nystatin-permeabilized patches (Horn and Marty, 1988) for monitoring cell capacitance, wherein the integrity of the intracellular biochemical milieu is kept intact and the ionic conductivity is maintained.

The whole-cell configuration, however, affords the possibility of dialyzing voltage-clamped cells with putative intracellular messengers and examining their role in the exocytotic process independent of voltage. Our preliminary studies on the involvement of G-proteins in exocytosis have indicated that inclusion of nonhydrolyzable GTP analogues such as GTP-γ-S does not significantly promote exocytosis on its own. This contrasts with the GTP-γ-S–induced degranulation reported in dialyzed rat peritoneal mast cells (Neher, 1988; Penner and Neher, 1989, this volume).

Compelling evidence against a requirement for Na^+-dependent action potentials in the mediation of secretion comes from the PRL release data. Release of PRL stimulated by either TRH or high K^+-induced depolarization is completely insensitive to TTX and to the complete replacement of Na^+ with choline in the extracellular medium (Fig. 1). Similar findings have been made in other pituitary cells such as somatotrophs (Mason and Rawlings, 1988) and gonadotrophs (Mason and Sikdar, 1988). In all these cells, which possess voltage-activated Na^+ channels, hormone secretion is insensitive to TTX.

DA: An Inhibitor of PRL Secretion

Release experiments performed in a number of different laboratories have shown that basal PRL secretion can be inhibited by the neurotransmitter DA (see Ben-Jonathan, 1985; Gregerson et al., 1989, this volume). Intracellular recording (Fig. 6 *A*) from bovine lactotrophs shows that DA strongly hyperpolarizes the lactotroph cell membrane (Ingram et al., 1986), and produces a concomitant increase in cell conductance. This will have the effect of closing voltage-activated Ca^{2+} channels, and in addition it will reduce the excitability of the cell to further stimuli. Application of 1 μM DA to lactrotrophs under basal conditions induces a pronounced lowering of intracellular calcium (from 67 to 40 nM measured at 5 mM extracellular Ca^{2+}). Similar results were observed when Co^{2+} (2–5 mM) was applied to the cells. However, the presence of DA does not inhibit the TRH-induced transient rise in $[Ca^{2+}]_i$, although it will lower $[Ca^{2+}]_i$ in the continued presence of TRH (Fig. 6 *B*). Its action appears to be similar to that of Co^{2+} application under basal conditions or removal of extracellular Ca^{2+}.

Conclusions

From the results obtained using these diverse techniques, we can draw some conclusions that relate to the control of intracellular Ca^{2+}, and PRL release from these cells.

PRL Secretion Is a Ca^{2+}-dependent Process

The cell capacitance work clearly demonstrates that high intracellular Ca^{2+} levels promote exocytosis in lactotrophs, and this process can be significantly slowed by application of buffers containing low ionized $[Ca^{2+}]_i$ ($[Ca^{2+}]_i < 10$ nM) inside the cell. This important result then raises the question as to the source of the Ca^{2+} (intracellular vs. extracellular store).

Basal and TRH-induced PRL Secretion Is Not Dependent on Na^+ Action Potentials

One of the most interesting pieces of data that has come from the release experiments is that both basal and stimulated PRL release can occur in the presence of the voltage-activated Na^+ channel blocker TTX, or in the complete absence of any extracellular Na^+. However, if we consider the large fluctuations in membrane potential in silent cells that constituted a majority in the lactotroph population and the intactness of the voltage-activated Ca^{2+} channels in these cells, this is not surprising. The only reasonable way of interpreting the TTX insensitivity of immunoassayed basal PRL release and TRH-stimulated hormone release is that the voltage-activated Na^+

current is not required for secretion. In fact, the small capacitance change to a train of depolarizing pulses can be observed even when the Na^+ channels are blocked by TTX, which suggests that exocytosis does not require Na^+.

There Is a Tonic Influx of Ca^{2+} Under Resting Conditions

The most direct evidence of this comes from the Ca^{2+}-imaging work which showed that cells bathed in 5 mM extracellular Ca^{2+} had significantly higher resting $[Ca^{2+}]_i$ levels than cells bathed in 1.8 mM Ca^{2+}. Electrophysiological recording has demonstrated that Ca^{2+} channels may be open around the resting membrane potential of these cells. Furthermore, the bovine lactotroph shows significant fluctuations in membrane potential (up to 30 mV) around the mean resting potential, and these are probably sufficient to maintain the opening of Ca^{2+} channels at rest. In addition, application of the Ca^{2+} channel blocker Co^{2+} under resting conditions reduces the conductance of the cell, implicating a block of a standing current at rest.

The Ca^{2+}-imaging work has also shown that application of the PRL release-inhibitor DA to the cell at rest results in a decrease in $[Ca^{2+}]_i$. Since other work has implicated a mechanism of action of DA that hyperpolarizes the cell and thus reduces the probability of opening of voltage-activated Ca^{2+} and reducing influx of Ca^{2+}; this implies that there is a tonic influx of Ca^{2+} at rest. This is supported by the fact that Co^{2+} will block the basal release of PRL (Fig. 1) and the spontaneous increases in membrane capacitance (Fig. 4).

TRH Invokes a Transient Rise in $[Ca^{2+}]_i$, Probably Via Ca^{2+} Release from Intracellular Stores

Ca^{2+}-imaging data (Fig. 5) clearly show that a single application of TRH (shown in release studies to stimulate PRL release) causes a transient rise in $[Ca^{2+}]_i$, a response that is unaffected by blockade of cell membrane Ca^{2+} channels with Co^{2+}. This supports work done in other laboratories that suggests that TRH probably stimulates the production of the intracellular messenger IP3, which can stimulate Ca^{2+} release from the intracellular pool.

Acknowledgments

The bovine lactotroph cell culture was originally established in our department by Dr. John Bicknell and Dr. Colin Ingram. Dr. Peter Cobbett was involved in some aspects of the work. We are indebted to Miss Anna Tibbs for providing the cell cultures, and Richard Bunting and John Hoyland for setting up the Ca^{2+}-imaging analysis in our department.

Various aspects of this work have been generously supported by the Meat and Livestock Commission, Medical Research Council, Lalor Foundation, and the Research Council of Slovenia.

References

Abe, H., D. Engler, M. E. Molitch, J. Bollinger-Gruber, and S. Reichlin. 1985. Vasoactive intestinal peptide is a physiological mediator of prolactin release in the rat. *Endocrinology*. 116:1383–1390.

Ben-Jonathan, N. 1985. Dopamine: a prolactin inhibiting hormone. *Endocrine Review*. 6:564–589.

Bjoro, T., B. C. Ostberg, O. Sand, J. Gordeladze, J-G. Iverson, P. A. Torjesen, K. M. Gautvik, and E. Haug. 1987. Vasoactive intestinal peptide and peptide with N-terminal histidine and C-terminal isoleucine increase prolactin secretion in cultured rat pituitary cells (GH_4C_1) via a cAMP-dependent mechanism which involves transient elevation of intracellular Ca^{2+}. *Molecular and Cellular Endocrinology*. 49:119–128.

Clapham, D. E., and E. Neher. 1984. Trifluoperazine reduces inward ionic currents and secretion by separate mechanisms in bovine chromaffin cells. *Journal of Physiology*. 353:541–564.

Cobbett, P. C., C. D. Ingram, and W. T. Mason. 1987*a*. Sodium and potassium currents involved in action potential propagation in normal bovine lactotrophs. *Journal of Physiology*. 392:273–299.

Cobbett, P. C., C. D. Ingram, and W. T. Mason. 1987*b*. Voltage-activated currents through Ca^{2+} channels in normal bovine lactotrophs. *Neuroscience*. 23:661–667.

DeRiemer, S. A., and B. Sakmann. 1986. Two Ca^{2+} currents in normal rat anterior pituitary cells identified by a plaque assay. *Experimental Brain Research*. 14:139–154.

Douglas, W. W. 1968. Stimulus-secretion coupling: the concept and clues from chromaffin and other cells. *British Journal of Pharmacology*. 34:451–474.

Gershengorn, M. C. 1986. Mechanism of thyrotropin-releasing hormone (TRH) stimulation of pituitary hormone secretion. *Annual Review of Physiology*. 48:515–526.

Gershengorn, M. C. 1989. Role of inositol lipid second messengers in regulation of secretion: studies of thyrotropin-releasing hormone action in pituitary cells. *In* Secretion and Its Control. G. S. Oxford and C. M. Armstrong, editors. Rockefeller University Press, New York.

Gourdji, D., D. Bataille, N. Vauclin, D. Grouselle, G. Rosselin, and A. Tixier-Vidal. 1979. Vasoactive intestinal peptide (VIP) stimulates prolactin (PRL) release and cyclic AMP production in a rat pituitary cell line (GH_3B_6). Additive effects of VIP and TRH on PRL release. *FEBS Letters*. 104:165–168.

Gregerson, K., L. Einhorn, M. Smith, and G. S. Oxford. 1989. Modulation of potassium channels by dopamine in rat pituitary lactotrophs: a role in the regulation of prolactin secretion? *In* Secretion and Its Control. G. S. Oxford and C. M. Armstrong, editors. Rockefeller University Press, New York.

Gross, D. S., J. L. Turgeon, and D. W. Waring. 1984. The ovine pars tuberalis: a naturally occurring source of pure gonadotropes. *Endocrinology*. 114:585–588.

Grynkiewicz, G., M. Poenie, and R. Y. Tsien. 1985. A new generation of calcium indicators with greatly improved fluorescence properties. *Journal of Biological Chemistry*. 260:3440–3450.

Hamill, O. P., A. Marty, E. Neher, B. Sakmann, and F. J. Sigworth. 1981. Improved patch-clamp techniques for high-resolution current recording from cells and cell-free membrane patches. *Pflügers Archiv*. 391:85–100.

Henigman, F., M. Kordas, and R. Zorec. 1987. An inexpensive head stage for the patch clamp apparatus. *Journal of Physiology*. 391:11P. (Abstr.)

Horn, R., and A. Marty. 1988. Muscarinic activation of ionic currents measured by a new whole-cell recording method. *Journal of General Physiology*. 92:145–159.

Ingram, C. D., R. J. Bicknell, and W. T. Mason. 1986. Intracellular recordings from bovine anterior pituitary cells: modulation of spontaneous activity by regulators of prolactin secretion. *Endocrinology*. 119:2508–2518.

Ingram, C. D., P. D. Keefe, F. B. P. Wooding, and R. J. Bicknell. 1988. Morphological characterization of lactotrophs separated from the bovine pituitary by a rapid enrichment technique. *Cell and Tissue Research*. 252:655–659.

Lewis, D. L., M. B. Goodman, P. A. St. John, and J. L. Barker. 1988. Calcium currents and fura-2 signals in fluorescence-activated cell sorted lactotrophs and somatotrophs of rat anterior pituitary. *Endocrinology*. 123:611–621.

Lindau, M., and E. Neher. 1988. Patch-clamp techniques for time-resolved capacitance measurements in single cells. *Pflügers Archiv*. 411:137–146.

Lingle, C. J., S. Sombati, and M. E. Freeman. 1986. Membrane currents in identified lactotrophs of rat anterior pituitary. *Journal of Neuroscience*. 6:2995–3005.

Mason, W. T., and S. R. Rawlings. 1988. Whole-cell recordings of ionic currents in bovine somatotrophs and their involvement in growth hormone secretion. *Journal of Physiology*. 405:577–593.

Mason, W. T., S. R. Rawlings, P. Cobbett, S. K. Sikdar, R. Zorec, S. N. Akerman, C. D. Benham, M. J. Berridge, T. Cheek, and R. B. Moreton. 1988*a*. Control of secretion in anterior pituitary cells: linking ion channels, messengers and exocytosis. *Journal of Experimental Biology*. 139:287–316.

Mason, W. T., S. K. Sikdar, and R. Zorec. 1988*b*. Ca^{2+}-induced cell membrane capacitance increase in bovine lactotrophs in vitro. *Journal of Physiology*. 407:88P.

Mason, W. T., and S. K. Sikdar. 1988. Characterisation of voltage-gated sodium channels in ovine gonadotrophs: relationship to hormone secretion. *Journal of Physiology*. 399:493–517.

Neher, E. 1988. The influence of intracellular calcium concentration on degranulation of dialysed mast cells from rat peritoneum. *Journal of Physiology*. 395:193–214.

Neher, E., and A. Marty. 1982. Discrete changes of cell membrane capacitance observed under conditions of enhanced secretion in bovine adrenal chromaffin cells. *Proceedings of the National Academy of Sciences*. 79:6712–6716.

Ozawa, S. 1985. TRH-induced membrane hyperpolarization in rat anterior pituitary cells. *American Journal of Physiology*. 248:E64–E69.

Penner, R., and E. Neher. 1989. Stimulus-secretion coupling in mast cells. *In* Secretion and Its Control. G. S. Oxford and C. M. Armstrong, editors. Rockefeller University Press, New York.

Pusch, M., and E. Neher. 1988. Rates of diffusional exchange between small cells and a measuring patch pipette. *Pflügers Archiv*. 411:204–211.

Sand, O., E. Haug, and K. M. Gautvik. 1980. Effects of thyroliberin and 4-aminopyridine on action potentials and prolactin release and synthesis in rat pituitary cells in culture. *Acta Physiologica Scandinavica*. 108:247–252.

Schlegel, W., P. B. Winiger, P. Mollard, P. Vacher, F. Wuarin, G. R. Zahnd, C. B. Wollheim, and B. Dufy. 1987. Oscillations of cytosolic Ca^{2+} in pituitary cells due to action potentials. *Nature*. 329:719–721.

Thorner, M. O., J. T. Hackett, F. Murad, and R. M. MacLeod. 1980. Calcium rather than cyclic AMP as the physiological intracellular regulator of prolactin release. *Neuroendocrinology*. 31:390–402.

Winiger, B. P., F. Wuarin, G. R. Zahnd, C. B. Wollheim, and W. Schlegel. 1987. Single cell monitoring of cytosolic calcium reveals subtypes of rat lactotrophs with distinct response to dopamine and thyrotropin-releasing hormone. *Endocrinology*. 121:2222–2228.

Chapter 15

Calcium Channels That Regulate Neurosecretion

Kathleen Dunlap, George G. Holz, Clark A. Lindgren, and John W. Moore

Department of Physiology, Tufts Medical School, Boston, Massachusetts 02111; and Department of Neurobiology, Duke University Medical Center, Durham, North Carolina 27710

Introduction

It has been acknowledged for many years that calcium ions play a central role in the control of neurotransmitter release and synaptic transmission. The pioneering work of Katz and Miledi demonstrated that calcium ions must be present at the time of presynaptic depolarization if transmitter release is to occur (Katz and Miledi, 1967) and that the size of the postsynaptic potential (a measure of the quantity of neurotransmitter released) is a function of the amount of calcium entering presynaptically through voltage-dependent channels (Katz and Miledi, 1973). Recent developments suggest that there are several different types of calcium channel and raise questions as to their functional significance; in other words, do different types of calcium channel subserve different cellular functions? It is of particular interest in neurons to determine which type of channel (or channels) controls exocytosis and whether different channel types regulate the release of separate transmitter pools. In this paper we will focus on these two issues through a discussion of our work on the electrically-stimulated release of (*a*) substance P (SP) from dorsal root ganglion neurons and (*b*) acetylcholine from vertebrate motoneurons. Results from these two studies will be further discussed in the larger framework of the published literature.

Types of Voltage-dependent Calcium Channel in Vertebrate Neurons

Calcium current recorded from nerve cell somata is kinetically and pharmacologically complex, which is consistent with the presence of more than one type of voltage-dependent calcium channel (for review, see Fox et al., 1988). In some nerve cells, for example dorsal root ganglion neurons in vitro, at least three biophysically distinct calcium channel types are thought to coexist (Carbone and Lux, 1984*a*, *b*; Nowycky et al., 1985*a*; Fox et al., 1987*a*; *b*; Kostyuk et al., 1988). Single-channel studies of these neurons demonstrate that step depolarizations of the membrane from negative holding potentials to near 0 mV evoke inward currents that fall into three distinct amplitude (or conductance) classes. These current types can be further characterized by their inactivation kinetics and pharmacological profiles.

The largest of these is a 25-pS channel, termed the L-type (Nowycky et al., 1985*a*) or HVA (Carbone and Lux, 1984*a*, *b*) channel, which exhibits a high threshold for activation (near-20 mV), inactivates slowly under conditions of low intracellular calcium concentration, and demonstrates virtually complete removal of inactivation at holding potentials near rest (Fox et al., 1987*a*, *b*). The two other calcium channel types, termed T (or LVA, Carbone and Lux, 1984*a*) and N (Nowycky et al., 1985*a*, Fox et al., 1987*a*, *b*), have smaller single-channel conductances (8 and 13 pS, respectively) and are largely inactivated at the resting potential. The T channel can be further distinguished in that it exhibits a threshold some 40 mV lower (-60 mV) than that of the other calcium channel types.

Pharmacological differences between the channel types also exist. The L-type channel is a target for dihydropyridines (DHPs). The probability for channel opening decreases in the presence of antagonists such as nifedipine while DHP agonists such as Bay K 8644 promote L-type channel opening (Nowycky et al., 1985*a*, *b*; Fox et al., 1987*a*, *b*). In contrast, the function of T and N channels in dorsal root ganglion neurons is reportedly unaffected by the DHPs. Although other pharmacological agents are known to inhibit some or all of these channels (e.g., cadmium or omega-conotoxin),

DHPs are the only agents presently thought to be specific for a single-channel type. They have, therefore, been useful tools for evaluating the physiological roles of L-type calcium channels. In this paper we describe results of experiments in which DHPs have been used to determine if L-type calcium channels are involved in the control of neurosecretion. Our results suggest that these channels play a dominant role in the release of both classical neurotransmitters (acetylcholine from motoneurons at the vertebrate nerve muscle junction) and peptides (SP from embryonic dorsal root ganglion neurons in vitro).

The Good and the Bad of DHPs

Although DHPs can be useful tools in such studies, the results from our laboratories, as well as from those of others, underscore the need for cautious interpretation of data, particularly when the experimental conditions do not allow direct measurement of DHP action on calcium current. The voltage- and time-dependent characteristics of the DHP-induced inhibition of calcium channels allow the block to be manifest only under particular experimental conditions. These properties of the DHPs were first observed in the experiments of Kass, Bean, and colleagues, who demonstrated that the

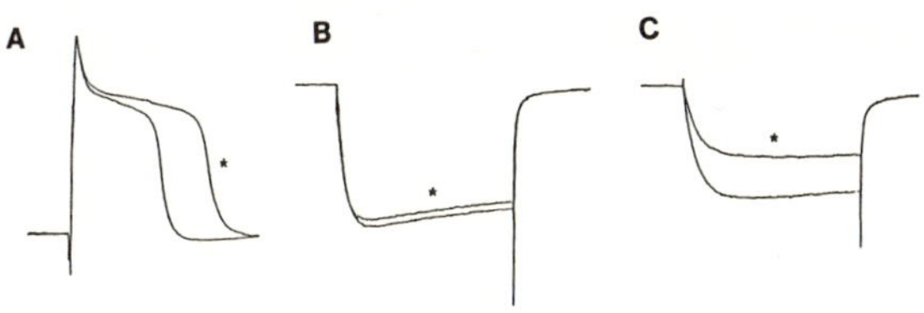

Figure 1. Effect of DHPs on dorsal root ganglion neuron calcium channels. (*A*) Action potential recorded before and during (*) application of 5 μM Bay K 8644. (The duration of the action potential is dependent on calcium current.) (*B*) Calcium current recorded after 60-ms step depolarizations to 0 mV from a holding potential of −70 mV before and during (*) application of 5 μM nifedipine. (*C*) Nifedipine effect (*) on calcium current recorded after 60-ms test depolarizations to 0 mV which were preceded by 5-s conditioning pulses to −30 mV. Peak control currents were 3.5 (*B*) and 1.8 (*C*) nA. Currents were recorded in an external solution containing 1 mM Ba and 2 mM Ca, the same solutions as those used in the release experiments described below.

high threshold calcium current in cardiac muscle cells was insensitive to DHP antagonists when the current was evoked by step depolarizations from holding potentials near the resting potential (Sanguinetti and Kass, 1984; Bean, 1984). Depolarizing holding potentials, on the other hand, promoted the effectiveness of the DHPs on this current. In contrast to these antagonist properties, the calcium channel agonist Bay K 8644 was found to be effective in potentiating calcium currents evoked by step depolarizations from potentials near rest (Sanguinetti et al., 1986).

Similar actions of the DHPs have been observed in dorsal root ganglion neurons as well (Rane et al, 1987; Holz et al., 1988). Bay K 8644 increases high threshold, L-type calcium currents and calcium-dependent action potentials recorded from these neurons (Fig. 1*A*). Likewise, nifedipine, a DHP antagonist, inhibits L-type calcium current but only under conditions in which the cells are chronically depolarized as shown in Fig. 1, *B* and *C*. Calcium currents (or calcium-dependent action potentials) evoked by depolarization from a normal resting potential are insensitive to DHP antagonists. DHP effectiveness is further complicated in these cells by the time dependence of its block. After depolarization of the neurons to a potential permissive for antagonist DHP action, no significant inhibition of current is observed for at least 1 s in dorsal root ganglion neurons. Thus, the conditional nature of DHP-induced inhibition of high threshold calcium currents mitigates, to a certain extent, the usefulness of these drugs as pharmacological probes for calcium channel function. However, at present they are the only known agents specific for a single class of calcium channel and are, thus, our most effective tool for studying the involvement of different channel types in exocytosis.

Peptide Release from Embryonic Dorsal Root Ganglion Neurons In Vitro

The peptide SP is synthesized, packaged, and released by dorsal root ganglion neurons in vitro as it is in vivo (Mudge et al., 1979; Holz et al., 1988). To determine whether L-type channels play a role in regulating its release, we have developed a method to stimulate secretion electrically. Current is passed between two extracellular electrodes placed in contact with the culture solution and the cells in the culture are thereby stimulated to produce propagating action potentials (Fig. 2 *A*). These action potentials stimulate the release of SP into the bathing medium where it is collected and assayed with standard radioimmunoassay techniques using an antibody specific for SP (Kream et al., 1985).

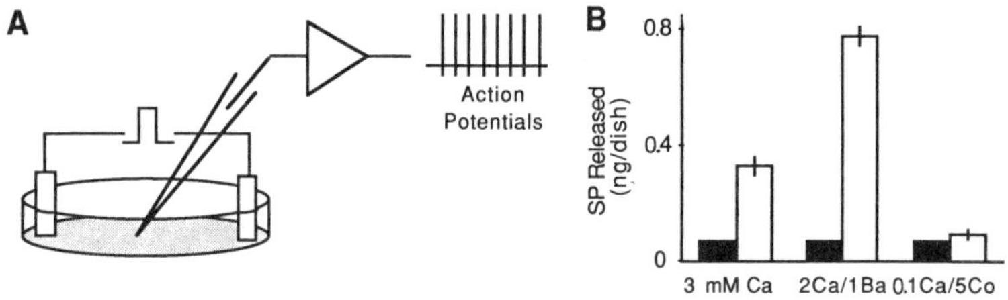

Figure 2. (*A*) Method for stimulating cultured neurons through a change in the electrical field. (*B*) Histogram of SP release in solutions containing different divalent cation concentrations. Stimulated release (*open bars*).

As shown in Fig. 2 *B*, the release is blocked by cobalt, and potentiated by the addition of a small amount of barium to the external solution (which increases calcium influx by prolonging the action potential duration), indicating the involvement of calcium channels.

To test for a role of L-type Ca channels in the release of SP, we have studied the action of the DHP agonist Bay K 8644 and antagonist nifedipine. Neurons were stimulated under three conditions: in control solutions, or in solutions containing either nifedipine or Bay K 8644. As can be seen in Fig. 3, Bay K enhanced the release of SP but nifedipine was ineffective. Taken in isolation, this result suggests that the L-type Ca channel may not play a very important role in SP release. However, considering the conditional effectiveness of the DHP antagonists produced by their time and voltage dependence, L-type calcium channels would not be expected to be sensitive to the DHPs under the conditions of the experiment. That is, the action potentials responsible for the evoked release of SP are generated from a normal resting potential and last only 5–10 ms, conditions which would prevent the DHP-induced inhibition of the current.

To investigate the properties of SP release under conditions conducive to antagonist DHP action, the release was evoked by chronic depolarization. External solutions containing 60 mM KCl (in place of NaCl) were applied to the neurons for 5 min and the release of SP was measured. Exocytosis under these conditions has beer shown previously to be calcium dependent (Mudge et al., 1979). In contrast to electrically stimulated release, K-stimulated release was sensitive to nifedipine (Fig. 3). The DHP antagonist at 5 μM blocked the release by 60%, suggesting that L-type calcium channels plays a significant role in triggering depolarization-induced exocytosis of the peptide from dorsal root ganglion neurons.

Do L-type channels play an equally important role in the release of classical neurotransmitters? The work of Miller, Tsien, and colleagues would suggest perhaps not. The K-stimulated release of norepinephrine from rat sympathetic neurons is not sensitive to DHP antagonists although Bay K potentiated the release from these cells (Hirning et al., 1987). In addition, these investigators do find a DHP-sensitive release process of SP in rat dorsal root ganglion neurons using the same experimental protocol (Perney et al., 1986). Based on the differences observed for these two peripheral neurons, they and others have suggested that peptides might be released by a mechanism involving L-type channels while more classical neurotransmitters are released by DHP-insensitive channels, perhaps N-type (Perney et al., 1986; Hirning et al., 1987; Miller, 1987). The results reviewed below suggest that, if true, this dichotomy between peptide and classical neurotransmitters is not universally the case.

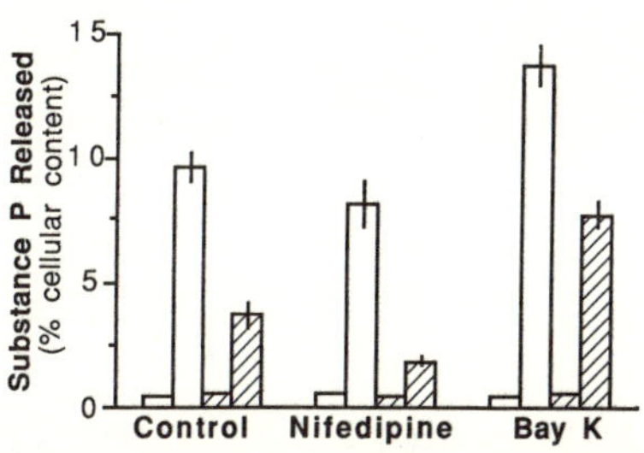

Figure 3. SP release was evoked either by electrical field stimulation (*open bars*) or by 60 mM potassium stimulation (*cross-hatched bars*) for 5 min in the presence or absence of DHP antagonist nifedipine (5 μM) or agonist Bay K 8644 (5 μM). Values are means ± SEM (n = 5). Mean cellular content of SP was 14,200 ± 250 pg/culture (n = 30). Baseline values for release precede evoked values.

Classical Neurotransmitter Release at the Vertebrate Nerve Muscle Junction

Extracellular recording techniques have been used to measure the currents underlying the presynaptic action potential and the postsynaptic end-plate potential at the neuromuscular junction of the lizard. The nerve terminals on the ceratomandibularis muscle are characterized by single *en plaque* endings that extend only 30–60 μm along the muscle surface. This compact structure is ideal for extracellular recording of presynaptic and postsynaptic currents after the invasion of the action potential into the nerve terminal. The ceratomandibularis muscle is so thin that it is possible to visualize the nerve endings using high power (640×) Nomarski optics and accurately position an extracellular electrode (inside diameter, ~5 μm) over specific regions of the nerve ending. Presynaptic currents may then be recorded under conditions in which the postsynaptic current (due to the opening of acetylcholine-sensitive channels) is blocked by the reversible acetylcholine antagonist pancuronium. Removal of the antagonist allows postsynaptic currents to be recorded as well. To improve the signal-to-noise ratio, we normally averaged between 25 and 100 current traces per measurement.

Calcium Current

Presynaptic currents recorded extracellularly from the nerve terminal are characterized by two peaks of outward current (Fig. 4 *A*). The first represents the discharging of the membrane capacitance by the passive spread of current into the terminal; the second (I_K) corresponds to the voltage-dependent potassium current that is responsible for action potential repolarization and is blocked by tetraethylammonium (TEA). In the presence of TEA, a small inward current (I_{Ca}) can be recorded whose amplitude is proportional to the calcium concentration in the extracellular solution. This voltage-dependent calcium current provides the nerve terminal with the calcium entry required for neurotransmitter release. It is blocked by cadmium (50 μM) and omega-conotoxin GVIA (5 μM), but not by nickel (100 μM). This pattern of sensitivities is similar to N- and L-type calcium channels described in embryonic chick dorsal root ganglion neurons (Nowycky et al., 1985*a*; Fox et al., 1987*a*, *b*). In an attempt to determine which of these two channel types underlies this presynaptic nerve terminal calcium current, we investigated its DHP sensitivity.

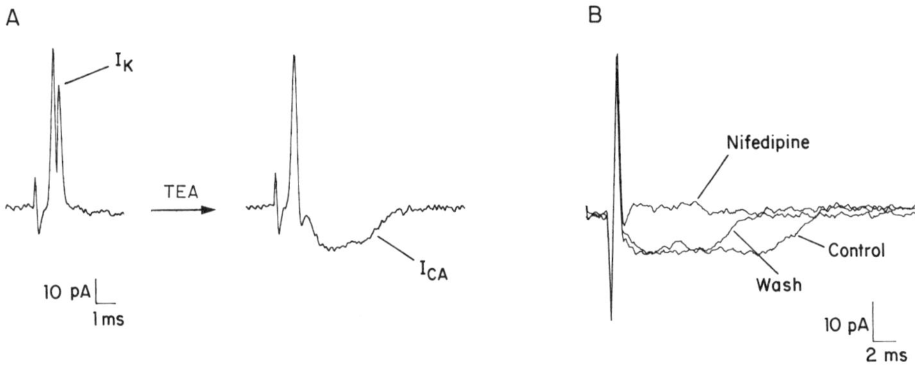

Figure 4. (*A*) Presynaptic calcium current revealed by application of TEA (10 mM) for 7 min. Calcium, 5 mM; pancuronium, 10 μM. Temperature, 18°C. (*B*) DHP sensitivity of calcium current. Superimposed averages of 100 traces obtained before (*Control*) and after the application of nifedipine (10 μM), and after washing the preparation in control saline for 40 min. (*Wash*). Calcium, 3 mM; TEA, 10 mM; pancuronium, 10 μM. Temperature, 23°C.

As mentioned above, the time- and voltage-dependent characteristics of the DHP-induced inhibition of calcium channels require that the cell membrane be depolarized (for at least 1 s in dorsal root ganglion neurons) during application of the DHP antagonist. Unfortunately, this presents a problem at the neuromuscular junction because the measurement of calcium current requires that the nerve impulse propagate into the presynaptic terminal. It is not feasible to maintain the preparation in a chronically depolarized state, as was possible for dorsal root ganglion neurons, because this would inactivate sodium channels and render the motor nerve inexcitable. Instead, we increased the duration of the action potential several fold and stimulated the nerve repetitively during the application of nifedipine to achieve an effect equivalent to chronic depolarization. We prolonged the action potential by increasing the amount of time the preparation was exposed to TEA from the normal 5–10 min to at least 1 h. The extent of action potential broadening could be monitored by measuring the duration of the presynaptic currents, which provides a lower estimate of the length of the action potential.

After expanding the action potential to at least 20 ms (see Fig. 4 *B*, *Control*), we applied nifedipine. During the time the drug was administered, 100 stimuli at 0.5 Hz were delivered to the nerve every 5 min. Using this procedure, the calcium current was progressively reduced until, by 25 min, the current was blocked completely (Fig. 4 *B*, *Nifedipine*). The antagonist was then removed but the stimulation protocol continued. During nifedipine washout, the current gradually returned to its control amplitude although it never reached its initial duration. After 40 min of wash the current pattern had stabilized *(Wash)* and additional washing had no further effect. Thus, as with the DHP sensitivity of high threshold calcium currents in dorsal root ganglion neurons and cardiac muscle cells, the nifedipine-induced inhibition of the calcium current in motor nerve terminals also appears to be voltage and time dependent. Because of the indirect means used to depolarize the motor endings, it is difficult to quantify the time constant of the DHP-induced block of calcium channels. However, we estimate that the nerve terminals must be depolarized for a cumulative total of 10 s (i.e., five groups of 100 action potentials, each 20 ms long) to allow a complete block of the calcium current by nifedipine.

Transmitter Release

The DHP sensitivity of transmitter release was examined under normal conditions and under conditions similar to those in which the presynaptic calcium current was found to be most responsive to nifedipine. To discriminate between changes in transmitter release and postsynaptic sensitivity to transmitter, we monitored the amplitude of both spontaneous miniature (MEPCs) and evoked end-plate currents (EPCs) by reducing the concentration of calcium and elevating that of magnesium in the bath. Since none of the agents tested altered MEPC amplitude, we were able to rule out the possibility of postsynaptic effects and could use the EPC amplitude directly to monitor changes in transmitter release.

Nickel, cadmium, and omega-conotoxin GVIA had the same relative effects on the EPC as they had on presynaptic calcium current (Lindgren et al., 1988). Nifedipine also decreased the evoked release of transmitter; however, as with the calcium current, the drug-induced inhibition was greater with a prolongation of the nerve terminal action potential. In the presence of 1 mM TEA and 0.1 mM 3,4-diaminopyridine (DAP), a potent blocker of the delayed-rectifier potassium current in this preparation, nifedipine completely abolished the EPC (Fig. 5 *A*). In contrast, the calcium channel agonist Bay K 8644 reversibly potentiated evoked transmitter release during a presynaptic action potential of normal duration (Fig. 5 *B*). Thus, unlike the antagonist action of nifedipine, the agonist properties of Bay K 8644 do not require chronic depolarization (Sanguinetti et al., 1986).

Our results are consistent with the hypothesis that the calcium current in motor nerve terminals of the lizard flows through high threshold (L- or HVA-type) calcium channels and that this current is directly associated with the release of the classical neurotransmitter, acetylcholine. We believe that the enhancement of nifedipine's effectiveness that is observed after the prolongation of the action potential duration reflects the voltage and time dependence of nifedipine's binding to calcium channels in the nerve terminal membrane.

Other investigators have failed to observe effects of DHPs at the vertebrate neuromuscular junction. Atchison and O'Leary (1987) found that, although Bay K 8644 increased both evoked and spontaneous release of acetylcholine at the rat phrenic nerve-diaphragm preparation, an action which could be blocked by the DHP antagonist nimodipine, there was no effect of nimodipine itself on transmitter release. Since no attempt was made in these studies to optimize the binding of nimodipine to the presynaptic calcium channels, the difference between their result and ours could be explained by the voltage dependency of the DHP antagonists. However, this cannot account for the results of Penner and Dreyer (1986) who measured calcium current in motor nerve endings on the triangularis sterni muscle of the mouse. They found that the DHP antagonists nitrendipine and nisoldipine had no effect on calcium currents recorded inside the perineurium of motor nerves treated with potassium channel blockers that produced action potentials whose durations equaled or exceeded those in our experiments on the lizard. This result suggests that calcium channels in mamma-

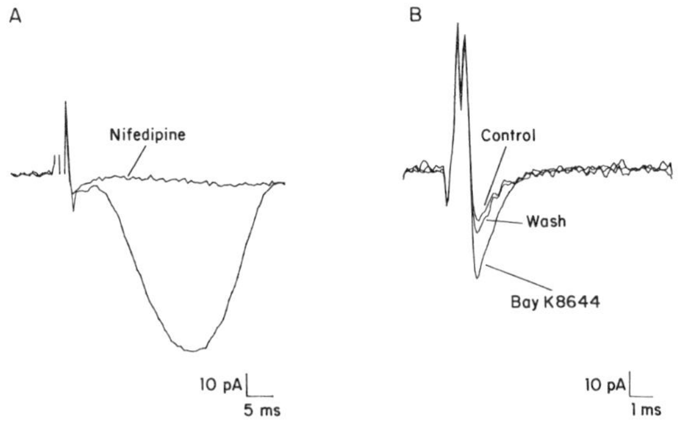

Figure 5. Sensitivity of transmitter release to DHPs. (*A*) Superimposed records before and after the application of 10 μM nifedipine. A preparation was first exposed to 1 mM TEA and 0.1 mM DAP for 2 h to increase the duration of the presynaptic action potential. Because of the large and prolonged EPC, it was necessary to add 5 μM curare to prevent a slight twitching of the muscle in spite of the fact that the bathing solution contained 0.5 mM calcium and 8 mM magnesium. Nifedipine was added after the EPC had been stable for 20 min, and it blocked the EPC completely after ~25 min. This effect was not reversed with washing. (*B*) Superimposed averages of 100 traces recorded before (*Control*) and after the application of 5 μM Bay K 8644 and then after washing the preparation in control saline (*Wash*). All of the solutions contained 0.25 mM calcium and 2 mM magnesium. Temperature, 23°C.

lian motor nerve endings may be quite different from those in reptiles. Another striking difference between lizard and mouse is that calcium channels in mouse motor nerve endings are insensitive to omega-conotoxin (Anderson and Harvey, 1987; Sano et al., 1987). This clearly distinguishes these calcium channels from those at the neuromuscular junctions of the lizard (Lindgren et al., 1988) and the frog (Kerr and Yoshikami, 1984) as well as the L- and N-type channels in chick dorsal root ganglion neurons (McCleskey et al., 1987). These differences suggest that throughout the animal kingdom a heterogeneity of calcium channel types underlie synaptic transmission at the neuromuscular junction.

DHPs and Transmitter Release in Other Systems

Agonist DHP sensitivity of calcium influx and/or transmitter release has been reported in a number of other systems, including secretory cells (Garcia et al., 1984; Chang et al., 1986; Hoshi and Smith, 1986; Kunze et al., 1987), cerebellar (Carbone and Wojcik, 1988), cortical (Middlemiss and Spedding, 1985), hippocampal (Gahwiler and Brown, 1987) and sympathetic ganglion neurons (Thayer et al., 1987;

Lipscombe et al., 1988), and brain synaptosomes (Dunn, 1988; Woodward et al., 1988). Antagonist DHPs are effective in some of these systems (Toll, 1982; Garcia et al., 1984; Enyeart et al., 1985; Turner and Goldin, 1985; Chang et al., 1986; Cohen and McCarthy, 1987; Gahwiler and Brown, 1987; Kunze et al., 1987; Carbone and Wojcik, 1988) but in others block only that portion of calcium influx or transmitter release that is potentiated by the DHP agonist (Middlemiss and Spedding, 1985; Middlemiss, 1985). Conflicting reports suggest that calcium influx and/or transmitter release in certain of these (and other) preparations is completely DHP resistant (reviewed by Miller, 1987). As with the work on the nerve-muscle junction described above, the negative results of some of these studies can be explained by the voltage dependence of antagonist DHP action. However, the antagonist DHP insensitivity of calcium influx and transmitter release, when stimulated under chronically depolarized conditions, is less easily explained. Two possibilities exist: (*a*) that the calcium-dependent processes are indeed regulated by DHP-insensitive calcium channels or (*b*) that the time-dependent actions of the DHP antagonists lag behind the calcium influx mechanism. For example, the fast phase of potassium-induced calcium influx into, and transmitter release from, synaptosomes is primarily limited to the first 1–2 s of sustained depolarization (see discussion, Holz et al., 1988). In such cases, no DHP inhibition would be expected, even if DHP-sensitive channels were involved. Transient kinetics are not limited to DHP-insensitive channels since, under conditions of normal intracellular calcium buffering, L-type calcium channels undergo relatively rapid inactivation (Eckert and Brehm, 1979; Chad, 1988).

Although a strong case can be made for the importance of DHP-sensitive channels in exocytosis, it is unlikely that all release mechanisms are controlled by the same type of calcium channel. In invertebrates, for example, high threshold calcium channels similar in many respects of those of L-type channels in dorsal root ganglion neurons are routinely observed and are thought to regulate synaptic transmission, yet neither the current nor the transmitter release is DHP sensitive (Augustine et al., 1987). In addition, transmitter release from sympathetic neurons stimulated by potassium depolarization is not inhibited by DHP antagonists, even when the cells are predepolarized for many seconds in the presence of nifedipine before the addition of calcium (Hirning et al., 1987). It thus seems likely that release mechanisms vary with the preparation.

What are the properties of DHP-insensitive channels that underlie exocytosis? It is too early to tell. Studies of calcium currents suggest tremendous heterogeneity among preparations. Although single-channel investigations offer the best evidence for multiple types of calcium channel, definitive correlations between single-channel and macroscopic currents have not been made. Furthermore, pharmacological means of identifying channel types has led to frustrations. For example, L- and N-type calcium currents in chick dorsal root ganglion (Feldman et al., 1987) and rat sympathetic (McCleskey et al., 1987) neurons are reported to be blocked by omega-conotoxin, while those in rat dorsal root ganglion neurons are not. Similarly, high threshold calcium currents in many central neurons are insensitive to the toxin. Does each of these pharmacological differences indicate the presence of a distinct calcium channel gene product, and if so, do they all regulate exocytosis? The development of more specific agents is necessary before definitive answers to these questions will be possible. Illuminating the rationale for the use of multiple types of calcium channel in the control of neurosecretion remains one of our most formidable future challenges.

References

Anderson, A. J., and A. L. Harvey. 1987. Omega-conotoxin does not block the verapamil-sensitive calcium channels at mouse motor nerve terminals. *Neuroscience Letters.* 82:177–180.

Atchison, W. D., and S. M. O'Leary. 1987. Bay K 8644 increases release of acetylcholine at the murine neuromuscular junction. *Brain Research.* 419:315–319.

Augustine, G., M. Charlton, and S. J. Smith. 1987. Calcium action in synaptic transmitter release. *Annual Review of Neuroscience.* 10:633–693.

Bean, B. P. 1984. Nitrendipine block of cardiac calcium channel binding to the inactivated state. *Proceedings of the National Academy of Sciences.* 81:6388–6392.

Carbone, E., and H. D. Lux. 1984*a*. A low voltage–activated, fully inactivating Ca channel in vertebrate sensory neurones. *Nature.* 310:501–502.

Carbone, E., and H. D. Lux. 1984*b*. A low voltage-activated calcium conductance in embryonic chick sensory neurons. *Biophysical Journal.* 46:413–418.

Carbone, E., and W. J. Wojcik. 1988. Dihydropyridine binding sites regulate calcium influx through specific voltage-sensitive calcium channels in cerebellar granule cells. *Journal of Neurochemistry.* 50:1279–1286.

Chad, J. 1988. Control of the generation and removal of calcium-mediated inactivation of the calcium current in *Helix aspersa.* In Calcium and Ion Channel Modulation. A. D. Grinnell, D. Armstrong, and M. B. Jackson, editors. Plenum Publishing, Corp., New York. 197–214.

Chang, J. P., E. E. McCoy, J. Graeter, K. Tasaka, and K. J. Catt. 1986. Participation of voltage-dependent calcium channels in the action of gonadotropin-releasing hormone. *Journal of Biological Chemistry.* 261:9105–9108.

Cohen, C. J., and R. T. McCarthy. 1987. Nimodipine block of calcium channels in rat anterior pituitary cells. *Journal of Physiology.* 387:195–225.

Dunn, S. M. J. 1988. Multiple calcium channels in synaptosomes: voltage dependence of 1,4-dihydropyridine binding and effects on function. *Biochemistry.* 27:5275–5281.

Eckert, R., and P. Brehm. 1979. Ionic mechanisms of excitation in *Paramecium. Annual Reviews of Biophysics and Bioengineering.* 8:353–383.

Enyeart, J. J., T. Aizawa, and P. M. Hinkle. 1985. Dihydropyridine calcium antagonists: potent inhibitors of secretion from normal and transformed pituitary cells. *American Journal of Physiology.* 248:C510–519.

Feldman, D. H., B. M. Olivera, and D. Yoshikami. 1987. Omega *Conus geographus* toxin: a peptide that blocks calcium channels. *FEBS Letters.* 214:295–300.

Fox, A. P., L. D. Hirning, D. V. Madison, E. W. McClesky, R. J. Miller, M. C. Nowycky, and R. W. Tsien. 1988. Physiology of multiple calcium channels. *In* Calcium and Ion Channel Modulation, A. D. Grinnell, D. Armstrong, and M. P. Jackson, editors. Plenum Press, New York. 63–74.

Fox, A. P., M. C. Nowycky, and R. W. Tsien. 1987*a*. Kinetic and pharmacological properties distinguishing three types of calcium currents in chick sensory neurones. *Journal of Physiology.* 394:149–172.

Fox, A. P., M. C. Nowycky, and R. W. Tsien. 1987*b*. Single-channel recordings of three types of calcium channels in chick neurones. *Journal of Physiology.* 394:173–200.

Gahwiler, B. H., and D. A. Brown. 1987. Effects of dihydropyridines on calcium in CA3 pyramidal cells in slice cultures of rat hippocampus. *Neuroscience.* 20:731–738.

Garcia, A. G., F. Sala, J. A. Reig, S. Viniegra, J. Frias, R. Ronteriz, and L. Gandia. 1984. Dihydropyridine Bay-K-8644 activates chromaffin cell calcium channels. *Nature.* 309:69–71.

Hirning, L. D., A. P. Fox, E. W. McCleskey, B. M. Olivera, S. A. Thayer, R. J. Miller, and R. W. Tsien. 1987. Dominant role of N-type calcium channels in evoked release of norepinephrine from sympathetic neurons. *Science.* 239:57–60.

Holz, G. G., K. Dunlap, and R. M. Kream. 1988. Characterization of the electrically evoked release of substance P from dorsal root ganglion neurons: methods and dihydropyridine sensitivity. *Journal of Neuroscience.* 8:463–471.

Hoshi, T., and S. J. Smith. 1986. Large depolarization induces long openings of voltage-dependent calcium channels in adrenal chromaffin cells. *Journal of Neuroscience.* 7:571–580.

Katz, B., and R. Miledi, 1967. The timing of calcium action during neuromuscular transmission. *Journal of Physiology.* 189:535–544.

Katz, B., and R. Miledi. 1973. Transmitter release induced by injection of calcium ions into nerve terminals. *Proceedings of the Royal Society of London.* 183:421–425.

Kerr, L. M., and D. Yoshikami. 1984. A venom peptide with a novel presynaptic blocking action. *Nature.* 308:282–284.

Kostyuk, P. G., Y. M. Shuba, and A. N. Savchenko. 1988. Three types of calcium channels in the membrane of mouse sensory neurons. *Pflügers Archiv.* 411:661–669.

Kream, R. M., T. A. Schoenfeld, R. Mancuso, A. Clancy, W. El-Bermani, and F. Macrides. 1985. Precursor forms of substance P (SP) in nervous tissue: detection with antisera to SP, SP-Gly, and SP-Gly-Lys. *Proceedings of the National Academy of Sciences.* 82:4832–4836.

Kunze, D. L., S. L. Hamilton, M. J. Hawkes, and A. M. Brown. 1987. Dihydropyridine binding and calcium channel function in clonal rat adrenal medullary tumor cells. *Molecular Pharmacology.* 31:401–409.

Lindgren, C. A., J. W. Moore, and A. H. Sostman. 1988. "L" type of calcium channel in nerve terminal is associated with transmitter release. *Society for Neuroscience Abstracts.* 14:507. (Abstr.)

Lipscombe, D., D. V. Madison, M. Poenie, H. Reuter, R. Y. Tsien, and R. W. Tsien. 1988. Spatial distribution of calcium channels and cytosolic calcium transients in growth cones and cell bodies of sympathetic neurons. *Proceedings of the National Academy Sciences.* 85:2398–2402.

McCleskey, E. W., A. P. Fox, D. H. Feldman, L. J. Cruz, B. M. Olivera, R. W. Tsien, and D. Yoshikami. 1987. Omega-conotoxin: direct and persistent block of specific types of calcium channels in neurons but not muscle. *Proceedings of the National Academy of Sciences.* 84:4327–4331.

Middlemiss, D. N. 1985. The calcium channel activator, Bay K 8644, enhances K-evoked efflux of acetylcholine and noradrenaline from rat brain slices. *Naunyn-Schmiedeberg's Archiv Pharmacology.* 331:114–116.

Middlemiss, D. N., and M. Spedding. 1985. A functional correlate for the dihydropyridine binding site in rat brain. *Nature.* 314:94–96.

Miller, R. J. 1987. Multiple calcium channels and neuronal function. *Science.* 235:46–52.

Mudge, A. W., S. E. Leeman, and G. D. Fischbach. 1979. Enkephalin inhibits release of substance P from sensory neurons in culture and decreases action potential duration. *Proceedings of the National Academy of Sciences.* 76:526–530.

Nowycky, M. C., A. P. Fox, and R. W. Tsien. 1985*a*. Three types of neuronal calcium channel with different calcium agonist sensitivity. *Nature*. 316:440–443.

Nowycky, M. C., A. P. Fox, and R. W. Tsien. 1985*b*. Long-opening mode of gating of neuronal calcium channels and its promotion by the dihydropyridine calcium agonist Bay K 8644. *Proceedings of the National Academy of Sciences*. 82:178–182.

Penner, R., and F. Dreyer. 1986. Two different presynaptic calcium currents in mouse motor nerve terminals. *Pflügers Archiv*. 406:190–197.

Perney, T. M., L. D. Hirning, S. E. Leeman, and R. J. Miller. 1986. Multiple calcium channels mediate neurotransmitter release from peripheral neurons. *Proceedings of the National Academy of Sciences*. 83:6656–6659.

Rane, S. G., G. G. Holz, and K. Dunlap. 1987. Dihydropyridine inhibition of neuronal calcium current and substance P release. *Pflügers Archiv*. 409:361–366.

Sanguinetti, M. C., and R. S. Kass. 1984. Voltage-dependent block of calcium channel current in the calf cardiac purkinje fiber by dihydropyridine calcium channel antagonists. *Circulation Research*. 55:336–348.

Sanguinetti, M. C., D. S. Krafte, and R. S. Kass. 1986. Voltage-dependent modulation of calcium channel current in heart cells by Bay K 8644. *Journal of General Physiology*. 88:369–392.

Sano, K., K. Enomoto, and T. Maeno. 1987. Effects of synthetic omega-conotoxin, a new type of Ca^{2+} antagonist, on frog and mouse neuromuscular transmission. *European Journal of Pharmacology*, 141:235–241.

Thayer, S. A., L. D. Hirning, and R. J. Miller, 1987. Distribution of multiple types of calcium channels in rat sympathetic neurons in vitro. *Molecular Pharmacology*. 32:579–586.

Toll, L. 1982. Calcium antagonists. *Journal of Biological Chemistry*. 257:13189–13192.

Turner, T. J., and S. M. Goldin. 1985. Calcium channels in rat brain synaptosomes: identification and pharmacological characterization. *Journal of Neuroscience*. 5:841–849.

Woodward, J. J., M. E. Cook, S. W. Leslie. 1988. Characterization of dihydropyridine-sensitive calcium channels in rat brain synaptosomes. *Proceedings of the National Academy of Sciences*. 85:7389–7393.

Chapter 16

Mechanisms Underlying the Inverse Control of Parathyroid Hormone Secretion by Calcium

Edward M. Brown, Chu J. Chen, Meryl S. LeBoff, Olga Kifor, Marguerite H. Oetting, and Ghada El-Hajj

Endocrine-Hypertension Unit, Department of Medicine, Brigham and Women's Hospital, and the Endocrinology Division, The Children's Hospital, Boston, Massachusetts 02115

Introduction

In mammalian species, the maintenance of a normal extracellular ionized calcium (Ca^{++}) concentration is critical for a variety of vital functions, such as cardiac contraction and neurotransmission. The extracellular Ca^{++} concentration is regulated by a complex homeostatic system (Stewart and Broadus, 1987), central within which are two Ca^{++}-elevating hormones, parathyroid hormone (PTH) and 1,25-dihydroxyvitamin D [1,25$(OH)_2$D]. In response to even slight decrements in the extracellular ionized Ca^{++} concentration, there are large increases in the secretion of PTH from the parathyroid glands. PTH then binds to specific, high-affinity receptors in target tissues, where it exerts its biological effects. In the kidney, PTH has several actions on Ca^{++} homeostasis: it increases distal tubular reabsorption of Ca^{++}, thereby effectively resetting the kidney to maintain a higher extracellular Ca^{++} concentration. It also promotes phosphaturia, which ensures excretion of phosphate mobilized into the extracellular fluid along with Ca^{++} from intestine and bone. Finally, PTH stimulates the synthesis of the active vitamin D metabolite, 1,25$(OH)_2$D. The latter increases the efficiency of the intestinal absorption of Ca^{++} and phosphate and also acts synergistically with PTH to enhance the net release of calcium and phosphate from bone. The resulting increase in the extracellular Ca^{++} concentration then inhibits PTH release, closing the negative feedback loop.

Proper functioning of the Ca^{++} homeostatic system, therefore, requires that low extracellular Ca^{++} concentrations stimulate while high levels of extracellular Ca^{++} inhibit PTH release. This inverse relationship between Ca^{++} and PTH release is, of course, just the opposite of the usual stimulatory effects of Ca^{++} on hormonal secretion in classical stimulus-secretion coupling (Douglas, 1968). In the present studies, we have used a variety of experimental approaches to study the mechanisms by which the parathyroid cell inverts the usual relationship between Ca^{++} and secretion. The results suggest that this cell type has modified the same basic mechanisms regulating exocytosis in other cell types to effect its own particular end.

Materials and Methods

Dispersed parathyroid cells were prepared by digestion of adult or neonatal bovine parathyroid glands with collegenase and DNase as described previously (LeBoff et al., 1985). Cells prepared in this fashion exclude trypan blue (90–95%), maintain high intracellular levels of ATP and ratios of cellular potassium (K^+) to sodium (Na^+), release PTH linearly for several hours, and show Ca^{++}-regulated PTH release similar to that present in vivo. After the cells were prepared, they were washed several times and then used immediately as acutely dispersed cells, maintained in culture for varying intervals, or permeabilized. Cells were cultured on fibronectin-coated cluster wells in media containing 2–15% fetal calf serum as outlined before (LeBoff et al., 1985). Parathyroid cells were permeabilized according to the method of Knight and Baker (1982), using five 2-kV discharges over a path length of 0.5 cm, after resuspension in a medium containing 120 mM K-glutamate, 5 mM ATP, 7 mM Mg acetate, 4.3 mM glucose, 17.2 mM piperazine-*N,N'*-bis[2-ethane-sulfonic acid]dipotassium salt, 1 mM EGTA, and 0.1% BSA, pH 6.9 (Oetting et al., 1986).

PTH release was measured using an antiserum (GW-1) raised in a sheep against bovine PTH (bPTH), which recognizes bPTH(1-84) as well as COOH-terminal fragments of hormone (LeBoff et al., 1985). bPTH(1-84) was used as standard and

^{125}I-bPTH(1-84) was used as a tracer. cAMP in cell pellets was extracted with 10% trichloroacetic acid. Samples and standards were then acetylated and cAMP was determined by radioimmunoassay (Brown et al., 1983).

For measurement of the cytosolic free Ca^{++} concentration, dispersed parathyroid cells were loaded with quin-2 or fura-2 by incubation with the respective acetoxymethyl esters of the dyes (10–15 or 1 μM, respectively) for 30 min at 37°C. The cells were then washed three times with a solution containing 0.025 M HEPES (adjusted to pH 7.47 with NaOH), 125 mM NaCl, 5 mM KCl, 1 g/liter dextrose, and 1 mM Na_2HPO_4. The washed pellet was then resuspended in a physiologic saline solution at a cellular concentration of 5–7 × 10^6 cells/ml for determination of cellular fluorescence. Fluorescence was monitored in thermostatted cuvettes (37°C) in a Perkin-Elmer 650-10S spectrofluorimeter (excitation, 337 nm and emission, 492 nm for quin-2 and excitation, 340 or 380 nm and excitation, 510 nm for fura-2). Cellular suspensions were mechanically stirred during all experiments except when recording was interrupted for 10–20 s during the addition of different reagents or resuspension of the cells. After each addition, fluorescence was monitored for 3–5 min or until the signal was stable. Calibration of fluorescence signals at the end of the study was achieved by cellular lysis with digitonin (96 μM) in 1 mM Ca^{++} (F_{max}) or after addition of 10 mM EGTA and alkalinization with 2.5 M Tris base to render the pH >8.3 (free Ca^{++} ≤1 nM, F_{min}). Cytosolic Ca^{++} concentration was then calculated according to the methods of Tsien et al. (1982) for quin-2 and Grynkiewicz et al. (1985) for fura-2.

Changes in the metabolism of inositol phosphates were assessed by examining the effects of metal ions on the levels of [^{3}H]inositol incorporation into IP, IP_2, IP_3, and IP_4. Cells were cultured for 20–24 h with [^{3}H]inositol, preincubated with 10 mM Li^+ for 30 min, washed, and then exposed to various polyvalent cations. The reaction was terminated with perchloric acid and the supernatant was neutralized with a mixture of 2 N KOH and 1 M Tris base. The neutralized supernatant was then applied to disposable polypropylene columns containing 0.8 ml of Dowex AG1X-8 in the formate form, washed with distilled water, and eluted sequentially with increasing concentrations of ammonium formate in 0.1 M formic acid (0.2, 0.4, 0.8, and 1.2 M to elute IP, IP_2, IP_3, and IP_4, respectively) (Downes et al., 1986). Diacylglycerol was measured in cells prelabeled with [^{3}H]arachidonic acid or [^{3}H]glycerol for 20–24 h, using thin layer chromatography (TLC) to separate diacylglycerol from other lipids. The TLC system employed used silica gel LK5 plates and was run two times with hexane/diethyl ether/acetic acid (70:30:10).

Results and Discussion

Relationship between Ca^{++} and PTH Release in Intact Cells

There is a remarkably steep, inverse sigmoidal relationship between the extracellular Ca^{++} concentration and PTH release in vivo and in vitro (Fig. 1). In vivo, most of the decrease in PTH occurs during an increase in the serum-ionized Ca^{++} concentration from 1.0 to 1.25 mM (e.g., from 8 to 10 mg/dl when expressed as total serum calcium concentration). The steepness of this relationship plays a key role in "setting" the extracellular Ca^{++} concentration at a level that is nearly invariant.

A similar relationship is apparent in vitro and is particularly striking when viewed within the context of extracellular Ca^{++} concentrations encompassing the range of 10^{-8} to 10^{-3} M (Fig. 1). In marked contrast to most other cell types, PTH secretion from parathyroid cells is near maximal at Ca^{++} concentrations ranging downward from just below 1 mM to 10^{-8} or less (Brown et al., 1983; Wallace and Scarpa, 1983). Over these levels of extracellular Ca^{++}, the cell appears to be essentially oblivious to a 5 order of magnitude change in Ca^{++} concentration. As Ca^{++} increases above 1 mM, however, and passes through the physiologic range of extracellular Ca^{++} concentrations, the secretory rate abruptly falls to low, although finite, levels.

The recent availability of intracellular Ca^{++}-sensitive dyes, such as quin-2 and fura-2, made it feasible to examine the relationships between the extracellular and cytosolic Ca^{++} concentrations, on the one hand, and PTH release, on the other. In quin-2–loaded parathyroid cells, raising the extracellular Ca^{++} concentration from 0.5 to 2 mM produced an increase in the cytosolic Ca^{++} concentration from ~200 nM to 600–700 nM, which correlated closely, in an inverse fashion, with the accompanying inhibition of PTH release (Fig. 2) (Shoback et al., 1983; Nemeth et al., 1986).

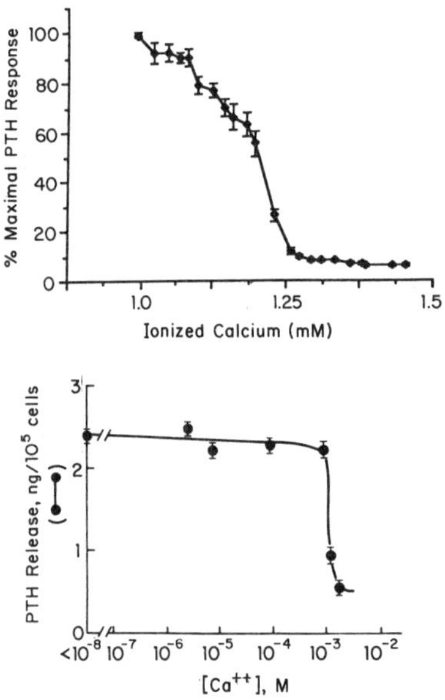

Figure 1. Relationship between PTH and the extracellular ionized Ca^{++} concentration in vivo (*top*) and in vitro (*bottom*). Normal human subjects were infused with EDTA to lower Ca^{++} or calcium gluconate to raise it, and serum ionized Ca^{++} and immunoreactive intact PTH were measured with a flow through electrode and immunoradiometric assay, respectively. Dispersed bovine parathyroid cells were incubated in vitro for 1 h with the extracellular ionized Ca^{++} concentrations shown. PTH in supernatant samples was measured by radioimmunoassay (COOH terminal).

Moreover, at intermediate calcium concentrations (e.g., 1 mM), addition of the divalent cation ionophore, ionomycin, at concentrations that raise the cytosolic Ca^{++} concentration to levels seen at high extracellular Ca^{++} (1.5–2 mM) inhibited PTH release to the extent expected with high extracellular Ca^{++} concentrations without ionophore (Fig. 3) (Shoback et al., 1984). Thus, under these conditions, there was an inverse relationship between not only extracellular but also intracellular Ca^{++} concentrations and PTH release. These observations suggested that the secretory apparatus of the parathyroid cell might differ in some fundamental way from cells exhibiting more classical stimulus-secretion coupling. Not only was secretion related to the cytosolic Ca^{++} concentration in an inverse fashion, but PTH release persisted at near maximal rates when the cytosolic Ca^{++} concentration was reduced to <100 nM by addition of chelating agents (i.e., EGTA) to the ambient medium (Shoback et al., 1984).

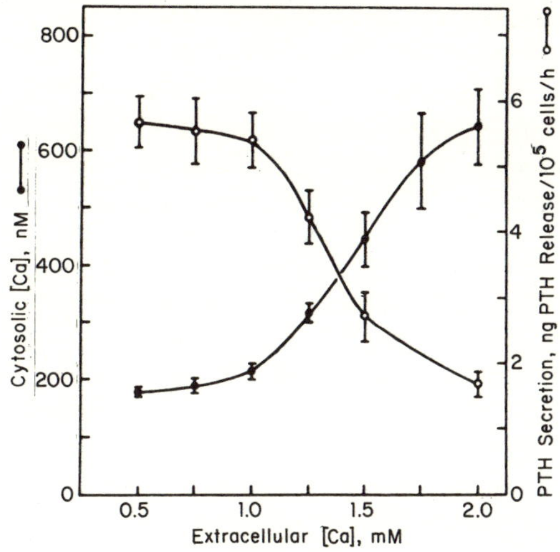

Figure 2. Relationship between cytosolic Ca^{++} concentration PTH release and the extracellular Ca^{++} concentration in quin-2–loaded bovine parathyroid cells. Cells were loaded with quin-2 and incubated with the indicated extracellular Ca^{++} concentrations. Cytosolic Ca^{++} concentrations and PTH release were determined as described in Materials and Methods. (From Shoback et al., 1983, with permission.)

It was apparent, however, that while the cytosolic Ca^{++} concentration was closely linked to changes in PTH release, the relationship was likely not a causal one. Addition of ionophore at high (2 mM) (Shoback et al., 1984) or low (0.5 mM) extracellular Ca^{++} concentrations (Nemeth et al., 1986), for example, had no effect on PTH release despite increasing the cytosolic Ca^{++} concentration 1.5–2-fold. Moreover, other secretagogues, such as dopamine (Shoback et al., 1984) or Li^{+} (Nemeth et al., 1986), enhanced PTH release without any apparent change in the cytosolic Ca^{++} concentration.

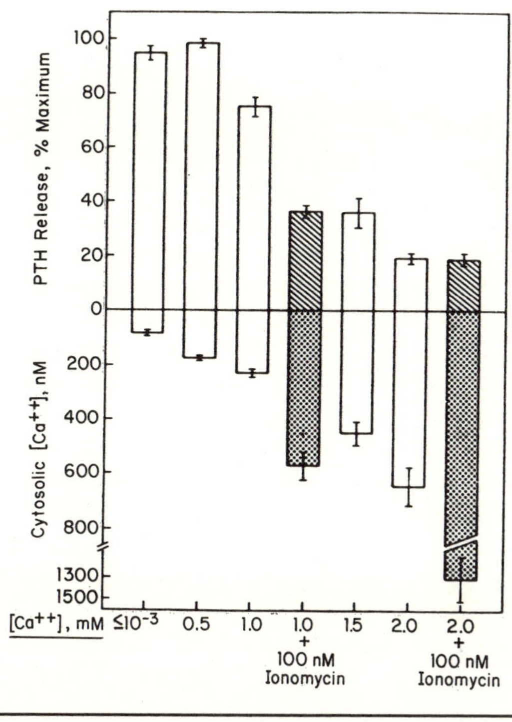

Figure 3. Relationhip between the extracellular and cytosolic Ca^{++} concentrations and PTH release in quin-2–loaded bovine parathyroid cells and their response to ionomycin. Addition of ionomycin at 1.0 mM extracellular Ca^{++} raises the cytosolic Ca^{++} concentration to a level similar to that seen with 1.5–2 mM Ca^{++} and inhibits PTH release to the extent observed with 1.5 mM Ca^{++}. Note, however, that addition of the ionophore at 2.0 mM Ca^{++} doubles cytosolic Ca^{++} without modifying PTH release, while removal of extracellular Ca^{++} lowers cytosolic Ca^{++} to <100 nM without inhibiting PTH release. (From Shoback et al., 1984, with permission.)

The Relationship between Ca^{++} and PTH Release in Permeabilized Parathyroid Cells

As an additional approach to exploring the interrelationships between Ca^{++} and PTH release, we used the model of electropermeabilized cells devised by Knight and Baker (1982). Exposure of dispersed parathyroid cells to high voltage discharge resulted in a cellular preparation that was freely permeable to small molecules, such as ^{86}Rb and trypan blue, but not to macromolecules such as lactate dehydrogenase (Oetting et al., 1986). The cells retained, however, the capacity for mediator-dependent hormonal secretion. Unexpectedly, exposure of permeabilized parathyroid cells to increases in ambient Ca^{++} concentration over the range of 10^{-7} to 10^{-5} M resulted in stimulation of PTH release (Fig. 4) (Oetting et al., 1987), a result similar to that seen in other exocytotic systems. Although others have reported no effect of Ca^{++} on PTH release from permeabilized cells (Muff and Fischer, 1986) these latter studies were carried out on cells shocked in a chamber of longer pathlength than we have used (1.0 vs. 0.5 cm). In our hands, cells shocked with 2 kV in a 1-cm chamber show either inconsistent or short-lived permeabilization.

The stimulation of PTH release from electropermeabilized cells by Ca^{++} suggested that, contrary to our expectations from studies in intact cells, the secretory apparatus of the parathyroid cell may not differ in a fundamental way from that of other secretory cells. That is, rather than possessing an entirely novel secretory "hardware," the parathyroid cells may have modified the "software" controlling exocytosis so as to achieve inverse control of secretion by Ca^{++} in the intact cell. On a phenomenologic level, "reconstitution" of the secretory behavior of the intact cell from that of the permeabilized cell would require at least two types of mechanisms: (*a*) those which enhance and maintain secretion at low extracellular Ca^{++} concentrations despite the presence of low or resting levels of cytosolic Ca^{++}, and (*b*) mechanisms that inhibit secretion over the narrow physiologic range of extracellular Ca^{++} concentrations. Since a major consequence of cellular permeabilization is the effective removal of the plasma membrane as a barrier between the extra- and intracellular spaces, one or another function related to the membrane could play an important role in regulating PTH release in intact parathyroid cells.

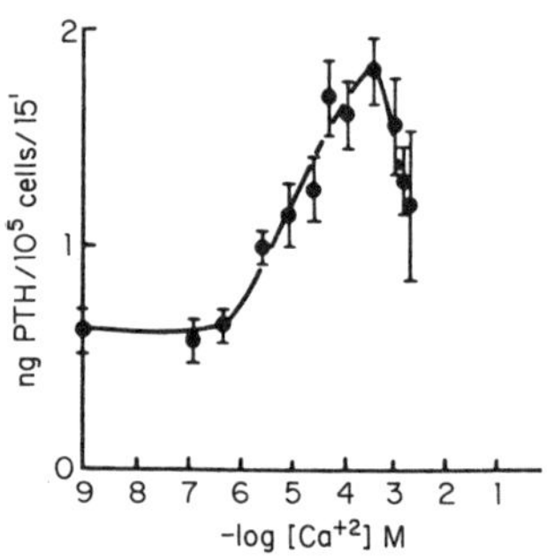

Figure 4. Ca^{++} stimulates PTH release from permeabilized parathyroid cells. Dispersed bovine parathyroid cells were permeabilized by electroshock and incubated for 15 min in the Ca^{++} concentrations shown in a medium mimicking the constituents of the cytosol, as described in Materials and Methods. PTH release was determined in supernatant samples by radioimmunoassay. (From Oetting et al., 1987, with permission.)

Factors Stimulating PTH Release at Low or Resting Cytosolic Ca^{++} Concentrations

At least three physiologically relevant cellular mediators stimulate PTH release at levels of Ca^{++} equivalent to or lower than those present in the intact cell at low extracellular Ca^{++} concentrations: (*a*) cAMP, (*b*) activators or protein kinase C, and (*c*) analogues of GTP. Agents increasing intracellular cAMP, such as dopamine or isoproterenol (Brown et al., 1978), stimulated PTH release in the absence of any measurable increase in cytosolic Ca^{++}. Moreover, dopamine also enhanced PTH secretion in the absence of extracellular Ca^{++}, when intracellular Ca^{++} was <100 nM (Shoback et al., 1984). Such secretagogues activate adenylate cyclase via G_s, the stimulatory guanine nucleotide regulatory (G) protein (Speigel et al., 1985).

The regulation of cAMP accumulation and PTH release by various hormonal factors can be bidirectional: prostaglandin $F_{2\alpha}$ ($PGF_{2\alpha}$) (Gardner et al., 1979) and alpha-adrenergic agonists, such as epinephrine and norepinephrine, lowered cAMP and PTH release in parallel (Brown et al., 1978). These agents inhibit adenylate cyclase via G_i, an effect which may be interrupted by pertussis toxin (Fitzpatrick et al., 1986*b*), which ADP-ribosylates and inactivates G_i (Speigel et al., 1985). Since cAMP rose in parathyroid cells at low extracellular Ca^{++}, this cyclic nucleotide could contribute to low Ca^{++}-stimulated PTH release. On a quantitative basis, however, it appeared unlikely that cAMP alone can account for the stimulatory effects of low extracellular Ca^{++} on PTH secretion (Brown et al., 1978). Activators of protein kinase C, such as phorbol esters and synthetic diacylglycerols, enhanced PTH secretion from

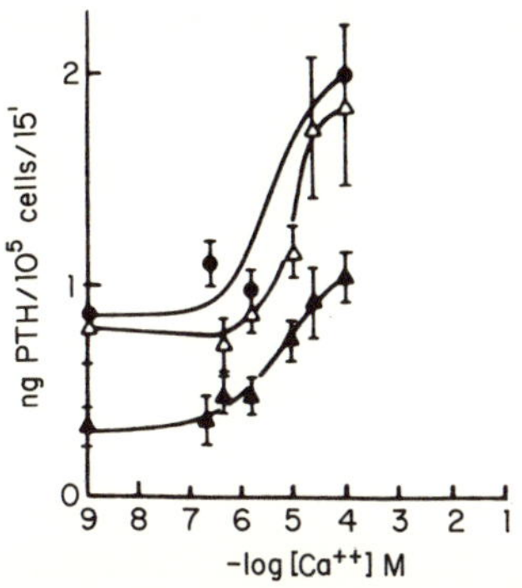

Figure 5. Stimulation of PTH release from permeabilized parathyroid cells by activators of protein kinase C. Cells were permeabilized, incubated for 15 min with the indicated additives and PTH release was determined as before. ▲, Ca^{++} alone; △, Ca^{++} + 10^{-6} M TPA; ●, Ca^{++} + 10^{-4} M dioctanoylglycerol. (From Oetting et al., 1987, with permission.)

both intact and permeabilized parathyroid cells (Brown et al., 1984*a*; Morissey, 1984; Nemeth et al., 1986). In intact cells, the stimulation of secretion occurred in association with either no change (Brown et al., 1984*a*) or a decrease (Nemeth et al., 1986) in the cytosolic Ca^{++} concentration, while in permeabilized cells, both tetradecanoyl-phorbol-13-acetate (TPA) and dioctanoylglycerol enhanced PTH secretion at Ca^{++} concentrations of 10^{-8} M or less (Oetting et al., 1987) (Fig. 5). The fraction of endogenous protein kinase C associated with particulate fractions in the intact parathyroid cell, an indirect measure of the activation of the enzyme, has been reported to be high at low extracellular Ca^{++} (Kobayashi et al., 1988). Curiously, while initial reports (Morrissey, 1985) indicated that diacylglycerol levels were higher at low than at high extracellular Ca^{++} concentrations in parathyroid cells, two subsequent studies studies found just the opposite (Bourdeau et al., 1988; Kifor and Brown, 1988). Possible explanations for these discrepancies will be considered subsequently.

Finally, nonhydrolyzable analogues of GTP, such as GppNHp and GTP-γ-S, were potent secretagogues in permeabilized bovine parathyroid cells, even at nominal Ca^{++} concentrations of 10^{-9} M or less (Oetting et al., 1986). This effect was not mimicked by other nucleotide triphosphates, by GMP or by GDP-β-S. The latter, in fact, blocked the stimulatory effects of GppNHp (Oetting et al., 1986). The specificity of this effect for analogues of GTP suggested that it was mediated by a G-protein, analogous to the G-protein–mediated stimulation of secretion from other cell types (Barrowman et al., 1985; Wollheim et al., 1987). Guanine nucleotides might enhance PTH secretion via several possible G-proteins: (*a*) activation of G_s, with attendant stimulation of cAMP accumulation, could stimulate PTH release. While GppNHp did promote a several fold increase in cellular levels of cAMP (LeBoff, Brown et al., unpublished observations), this effect occurred considerably more slowly than the increased rate of secretion. Moreover, in this permeabilized cell model, we found little or no effect of even high concentrations of cAMP or dibutyryl cAMP on PTH release, suggesting that this cyclic nucleotide could not account for the enhanced rate of PTH secretion observed with GppNHp. Guanine nucleotides could produce a G-protein–mediated activation of phospholipase C, with attendant generation of diacylglycerol and activation of protein kinase C. We have found that GppNHp does promote the accumulation of inositol phosphates, but addition of exogenous dioctanoylglycerol produced far less stimulation of PTH release than GppNHp (LeBoff, M. S., E. M. Brown, M. Oetting, unpublished observations). It seemed likely, therefore, that guanine nucleotides acted at some additional site to stimulate hormonal secretion in the parathyroid cell, perhaps at a "distal" G-protein, close to the site of exocytosis. It is not known whether the G-protein(s) presumed to mediate this effect is active in parathyroid cells at low extracellular Ca^{++}.

Evidence for Receptor-mediated Regulation of Parathyroid Function by Ca^{++} and Other Polyvalent Cations

An increasing body of data supports the concept that high extracellular concentrations of Ca^{++} and other polyvalent cations regulate parathyroid function by a receptor-like mechanism, which is coupled to intracellular effector systems by one or more G-proteins (Lopez-Barneo and Armstrong, 1983; Fitzpatrick et al., 1986*a*; Nemeth and Scarpa, 1986; Brown et al., 1987*b*; Chen et al., 1987). This concept was first suggested by the electrophysiological studies of Lopez-Barneo and Armstrong (1983), who hypothesized that the divalent cation-induced depolarization of rat parathyroid cells was most readily explained by a cell surface, polyvalent cation "receptor," which was likely coupled to inhibition of potassium channels. Fitzpatrick et al. (1986*a*) subsequently showed that pertussis toxin blocked the inhibitory effects of high extracellular Ca^{++} on PTH release, suggesting that a G-protein might be interposed between the recognition of changes in the extracellular Ca^{++} concentration and the subsequent control of PTH release.

Additional, recent data extended the analogy between Ca^{++}-regulated PTH secretion and regulation of other cells by more classical hormone-receptor systems. In contrast to parathyroid cells loaded with quin-2, in which high extracellular Ca^{++} promoted only a sustained rise in cytosolic Ca^{++} (Shoback et al., 1984; Nemeth et al., 1986), high concentrations of Ca^{++}, Mg^{++} (Fig. 6) (Nemeth and Scarpa, 1986; Chen et al., 1987), and other divalent cations, such as Ba^{++} and Sr^{++}, also evoked an initial, transient "spike" in cytosolic Ca^{++} in bovine parathyroid cells loaded with lower concentrations of the brighter dye, fura-2. This pattern of an "agonist"-induced spike in cytosolic Ca^{++}, resulting from release of intracellular stores, followed by a sustained increase, due to uptake of extracellular Ca^{++}, bore a close resemblance to the changes in intracellular Ca^{++} dynamics evoked by more classical Ca^{++}-mobilizing hormones, such as thyrotropin-releasing hormone (TRH) (Gershengorn et al., 1984). The latter act, at least in part, by promoting hydrolysis of phosphoinositides, with attendant generation of IP_3 and IP_4. The former releases intracellular Ca^{++} (Berridge and Irvine, 1984), while the latter has been postulated to enhance uptake of extracellular Ca^{++} (Irvine and Moor, 1986).

Effects of Polyvalent Cations on Inositol Phosphate Accumulation in Bovine Parathyroid Cells

To investigate whether Ca^{++} and other divalent cations might modulate cytosolic Ca^{++} in parathyroid cells via changes in phosphoinositide metabolism, we examined

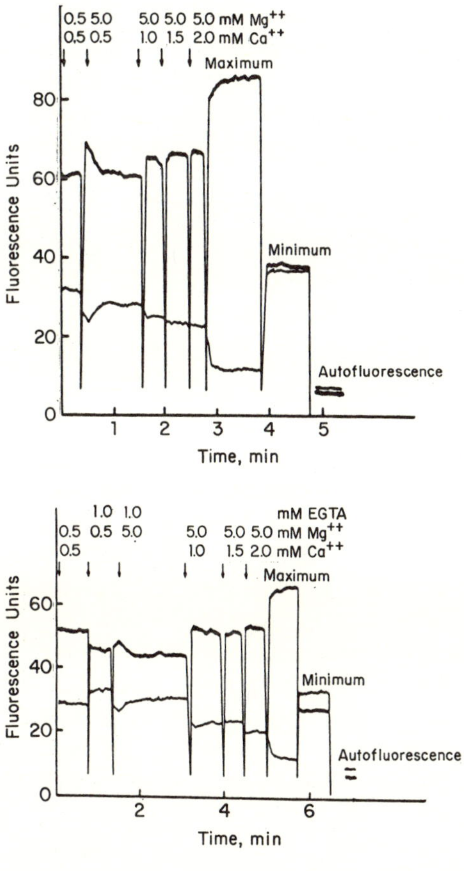

Figure 6. Mg^{++}-induced spikes in cytosolic Ca^{++} in fura-2–loaded bovine parathyroid cells. Cells were loaded with fura-2 and cytosolic Ca^{++} was determined as described in Materials and Methods. Note that raising extracellular Mg^{++} from 0.5 to 5 mM in the presence of 0.5 mM extracellular Ca^{++} (*top*) promotes a transient increase in fluorescence when the cells are excited at 340 nm (*upper tracing*) and decrease in fluorescence when they are excited at 380 nm (*lower tracing*), which is indicative of a spike in cytosolic Ca^{++}. When 1 mM EGTA is added prior to the high Mg^{++}, the spike is reduced but not abolished, showing that the transient arises, at least in part, from release of intracellular Ca^{++}. (From Chen et al., 1987; with permission.)

their effects on the accumulation of inositol phosphates (Brown et al., 1987*a*). A variety of divalent cations, including Ca^{++}, Mg^{++}, Ba^{++}, and Sr^{++}, enhanced the accumulation of inositol phosphates (Fig. 7). Moreover, the trivalent cation La^{3+} exerted similar effects. The order of potency of these polyvalent cations in modulating inositol phosphate metabolism was similar to their order of potency in modulating PTH release (Wallace and Scarpa, 1982) and cAMP accumulation (see below). In addition to stimulating the accumulation of IP, IP_2, and IP_3, Ca^{++} and Mg^{++} also raised the levels of IP_4 (Hawkins et al., 1989). It is possible that the increases in IP_3 mediated the divalent cation-induced spikes in cytosolic Ca^{++}, while the elevations in IP_4 accounted, in part, for the sustained increases in cytosolic Ca^{++} at high extracellular Ca^{++} and Mg^{++} concentrations.

It is also possible that changes in the metabolism of inositol phosphates were the result rather than the cause of polyvalent cation-induced changes in the cytosolic Ca^{++} concentration. For example, increases in cytosolic Ca^{++} could have directly activated phospholipase C, with attendant hydrolysis of phosphoinositides to yield inositol phosphates. Shoback et al. (1988), however, recently showed that concentrations of the divalent cation ionophore, ionomycin, which raised the cytosolic Ca^{++} concentration to levels higher than those observed at high extracellular Ca^{++} concentrations, had little or no effect on inositol phosphate accumulation. In addition, fluoride ion, which directly activates the G-protein–regulating phospholipase C in other cells types (Blackmore et al., 1986), promoted the accumulation of inositol phosphates in parathyroid cells. It is possible, therefore, that the effects of polyvalent cations on phosphoinositide metabolism in this cell type were mediated by a receptor that activated phospholipase C through a G-protein. If present, such a G-protein was not a substrate for pertussis toxin, however, since preincubation with the toxin had no effect

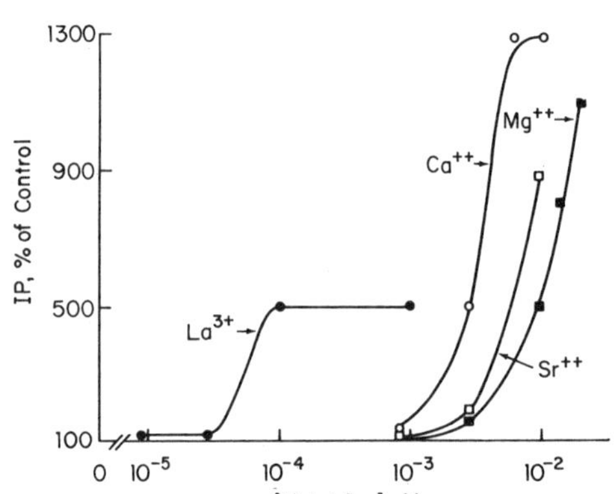

Figure 7. Di- and trivalent cations increase the accumulation of inositol phosphates in bovine parathyroid cells. Dispersed cells were cultured for 20–24 h with [^{3}H]inositol as described in Materials and Methods. The cells were then incubated with the indicated polyvalent cations for 30 min in the presence of 10 mM Mg^{++} and inositol phosphates were extracted and separated on Dowex AG1X-8 columns. Only results for IP are shown, but comparable increases in IP_2 and IP_3 were also noted.

on the stimulation of the accumulation of inositol phosphates by Ca^{++}(Hawkins et al., 1989).

Activation of phospholipase C would be expected to be accompanied by increased generation of diacylglycerol. While initial data suggested a high Ca^{++}-induced decrease in diacylglycerol (Morrissey, 1985), two subsequent groups have reported that levels of diacylglycerol increase at high Ca^{++} in parathyroid cells (Bourdeau et al., 1988; Kifor and Brown, 1988). Paradoxically, however, as assessed by its association with particulate fractions of the cell, protein kinase C activity was decreased at high Ca^{++} (Kobayashi, et al., 1988). While it is possible that additional data will resolve the apparent discrepancy between diacylglycerol levels and protein kinase C activity in parathyroid cells, perhaps, in addition to diacylglycerol one or more inhibitors or protein kinase C and/or secretion are produced at high extracellular Ca^{++}, which dominate in their effects on PTH release.

Effects of Polyvalent Cations on Dopamine-stimulated cAMP Accumulation in Bovine Parathyroid Cells

The regulation of cAMP accumulation by di- and trivalent cations in parathyroid cells provided more direct evidence for the regulation of parathyroid function by a receptor-like mechanism, coupled, in this case, to adenylate cyclase by G_i, the G-protein–inhibiting adenylate cyclase (Speigel et al., 1985). As shown in Fig. 8, Ca^{++} and a variety of divalent cations, as well as La^{3+}, potently inhibited dopamine-stimulated cAMP accumulation (Chen et al., 1989). Moreover, this inhibitory effect was totally blocked by preincubation with pertussis toxin, in association with ADP-ribosylation of G_i. The effects of Ca^{++} on cAMP accumulation and their reversal by pertussis toxin, therefore, were analogous to the regulation of cAMP metabolism by $PGF_{2\alpha}$ (Fitzpatrick et al., 1986*b*), which was also blocked by pertussis toxin and was known to regulate adenylate cyclase by G_i.

There are several possible ways, however, in which pertussis toxin could modulate the effects of polyvalent cations on cAMP metabolism in this cell type. Since pertussis toxin, if anything, potentiated the Ca^{++}-induced rise in cytosolic Ca^{++} (Chen et al., 1989), it appeared unlikely that the toxin blocked the effects of extracellular Ca^{++} on cAMP metabolism indirectly by preventing a rise in cytosolic Ca^{++} which secondarily

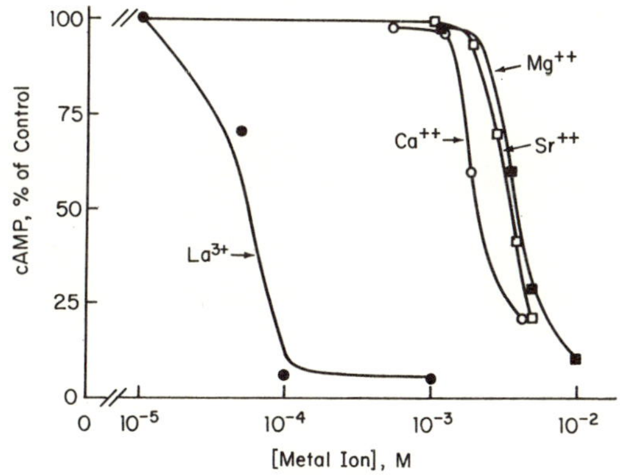

Figure 8. Inhibition of cAMP accumulation stimulated by dopamine in the presence of di- and trivalent cations. Parathyroid cells were incubated with 10^{-5} M dopamine and the indicated concentration of La^{3+}, Ca^{++}, Mg^{++}, or Sr^{++} for 5 min. cAMP was extracted with trichloroacetic acid and measured by radioimmunoassay as described in Materials and Methods.

regulated cAMP. In addition, pertussis toxin did not modify phosphodiesterase activity in parathyroid cells or cellular efflux of cAMP, two other processes that could modulate intracellular levels of cAMP.

Does the Polyvalent Cation Receptor Show Positive Cooperativity?

The data outlined above suggest that polyvalent cations might exert their effects on parathyroid function by a receptor-like mechanism, which is coupled to intracellular effector systems and ultimately to PTH release via one or more G-protein. There are several other properties of this regulatory mechanism that may shed additional light on the putative "receptor." One of the most striking characteristics of the control of parathyroid function by a variety of divalent cations is the steepness of these dose-response relationships. For example, nearly all of the effects of high Ca^{++} in elevating cytosolic Ca^{++} or inhibiting PTH release took place over a four to fivefold range of Ca^{++} concentrations or less (Shoback et al., 1983). Similarly, Ca^{++} or Mg^{++}-induced increases in inositol phosphates (Brown et al., 1987*a*) or lowering of agonist-stimulated cAMP occurred over a five to tenfold range of concentrations and showed Hill coefficients of 1.9–2. In contrast, the effects of other agents, either stimulatory or inhibitory, that modulate parathyroid function took place over about 2 orders of magnitude, which is characteristic of noncooperative one-on-one interactions of a ligand with its binding site. Examples of agents acting in this latter fashion include stimulatory ligands, such as isoproterenol and dopamine (Brown et al., 1977, 1978), and inhibitory ones, such as $PGF_{2\alpha}$ (Gardner et al., 1979). It is of interest that although both high extracellular Ca^{++} and $PGF_{2\alpha}$ lowered cAMP to a similar extent and the effects of both were blocked by pertussis toxin (presumably by inactivation of G_i), the slopes of their respective dose-response curves for effects on cAMP different markedly. It is possible, therefore, that this difference is the result of positive cooperativity at the level of the putative Ca^{++} receptor. Such cooperativity could be the result of multiple metal binding sites on one or more subunits and would be critical in ensuring that changes in parathyroid function take place over a narrow, physiologically relevant range of extracellular Ca^{++} concentrations in vivo.

There are also interesting interactions between the effects of Ca^{++} and Mg^{++} on parathyroid function which suggest additional complexities in this system. In the presence of physiologic concentrations of extracellular Ca^{++}, the effects of the two cations were additive in proportion to their relative potencies (Habener and Potts, 1976; Brown et al., 1984). With 0.1 mM extracellular Ca^{++}, on the other hand, Mg^{++} became markedly less potent in inhibiting PTH release and cAMP accumulation (Brown et al., 1984*b*), as well as in enhancing inositol phosphate generation (Brown, E. M., unpublished observations). This effect may have resulted from the requirement for a minimal level of intracellular Ca^{++} for proper functioning of the divalent cation–sensing mechanism of the cell. It is also conceivable, however, that not all metal ions are equivalent in promoting cooperative interactions at the level of the putative receptor. Perhaps Ca^{++} must be bound at one site for Mg^{++} to be able to modulate the function of the receptor.

Effects of Agents Binding to the Cell Surface on Ca^{++}-regulated Parathyroid Function

A variety of agents acting on the cell surface, such as enzymes, antibodies, and lectins, have been useful probes for defining the chemical nature, structural properties, and cell surface localization of classical receptors (Flier et al., 1972). The application of these techniques to parathyroid cells has provided additional, indirect evidence for the existence of a cell surface, polyvalent cation "receptor." Monoclonal antibodies, for instance, have either inhibited PTH release (Posillico et al., 1987) or blocked the suppressive effects of extracellular Ca^{++} on secretion (Juhlin et al., 1987). In both cases, the change in secretory function was accompanied by the expected alteration in cytosolic Ca^{++} if the antibody were interacting at or near the Ca^{++} "receptor" elevation in cytosolic Ca^{++} in the first case and prevention of the Ca^{++}-induced increase in intracellular Ca^{++} in the latter. In patients with autoimmune hypoparathyroidism, naturally occurring antiparathyroid antibodies can modulate parathyroid function and might have as their target antigen the Ca^{++} receptor or an associated structure (Posillico et al., 1986).

The lectin concanavalin A (conA) also blocked the effects of high extracellular Ca^{++} on PTH release and cytosolic Ca^{++}, presumably by binding to the cell surface (Posillico et al., 1986). In addition, conA reversed the inhibitory effect of Ca^{++} on dopamine-stimulated cAMP accumulation (Fig. 9), an effect which could be prevented by preincubation with α-methylglucoside. ConA had no effect, however, on dopamine-stimulated cAMP accumulation per se. By analogy with the modulatory effects of conA on other glycoprotein, cell surface receptors (Azhar and Menon, 1981),

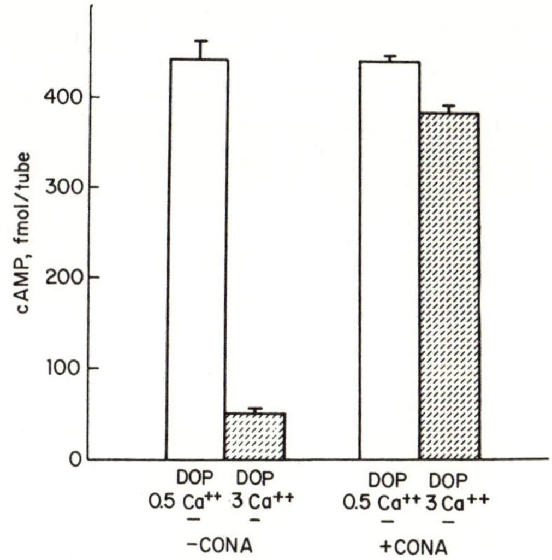

Figure 9. ConA reverses the inhibition of dopamine-stimulated cAMP accumulation by high Ca^{++}. Parathyroid cells were incubated for 5 min with 10^{-5} M dopamine and the indicated extracellular Ca^{++} concentrations, with or without 200 μg/ml of conA. cAMP was then extracted and measured as before.

therefore, it is possible that the polyvalent cation receptor or an associated membrane structure is a glycoprotein.

Summary

An increasing body of evidence suggests that the parathyroid cell recognizes changes in the extracellular calcium concentration via a cell surface receptor-like mechanism

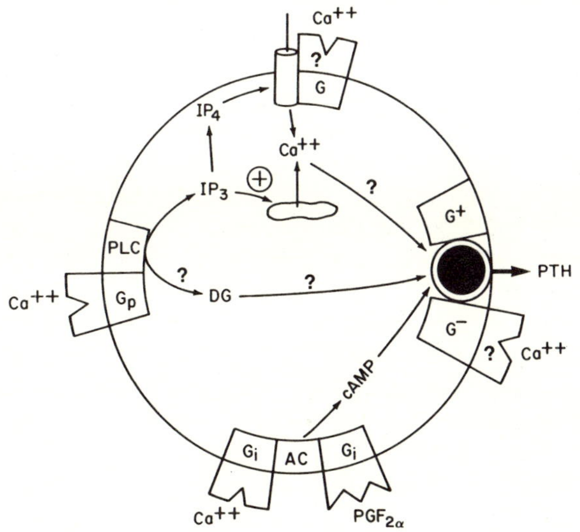

Figure 10. Possible mechanisms by which high extracellular Ca^{++} concentrations regulate PTH release. High extracellular Ca^{++} is visualized as binding to one or more distinct extracellular Ca^{++} receptors, each of which may interact with a separate intracellular effector system. In the case of the adenylate cyclase–cAMP system, the receptor is coupled via G_i in a manner entirely analogous to that by which $PGF_{2\alpha}$ receptors are linked in an inhibitory fashion to the enzyme.

Although a Ca^{++}-induced decrease in cAMP levels would be expected to inhibit PTH release, it is unlikely that this mediator can account quantitatively for the suppressive effects of high extracellular Ca^{++} on hormonal secretion. High extracellular Ca^{++} also activates phospholipase C, perhaps via G_p. The resultant increase in IP_3 may mediate the Ca^{++}-induced spike in cytosolic Ca^{++}, and the conversion of IP_3 to IP_4 may participate, along with extracellular Ca^{++}-gated channels, in promoting the uptake of extracellular Ca^{++}. Since high levels of Ca^{++} and diacylglycerol both stimulate PTH release from permeabilized cells, however, it is likely that extracellular Ca^{++} exerts some additional, "distal" inhibitory effect that reduces PTH release and, perhaps, protein kinase C activity.

(Fig. 10). One or more distinct polyvalent cation "receptors" may be linked to several intracellular effector systems through guanine nucleotide regulatory proteins. Thus, this signal recognition/transduction system may be analogous to the G-protein–coupled, cell surface receptors through which a variety of more classical hormones and neurotransmitters regulate cellular function (e.g., the beta-adrenergic or $PGF_{2\alpha}$ receptors). The precise coupling of this receptor-like mechanism to changes in hormonal secretion, however, remains uncertain. At extracellular Ca^{++} concentrations below 1 mM, the receptor appears to be inactive and hormonal secretion is tonically high. Activation of protein kinase C may play a role in maintaining high secretory rates at low Ca^{++} concentrations, but the factors regulating protein kinase C are unclear. Although diacylglycerol levels and inositol phosphates are increased at high extracellular Ca^{++}, perhaps ambient levels of diacylglycerol at low Ca^{++} are sufficiently elevated for activation of kinase C. Increased levels of cAMP, stimulatory G-proteins as well as reduced intracellular degradation of PTH (Habener et al., 1975; Morrissey et al., 1980) are also candidates for factors contributing to low Ca^{++}-stimulated PTH release.

As the extracellular Ca^{++} concentration increases above 1 mM, several changes in parathyroid cell function take place, which mimic receptor-mediated activation of intracellular effector systems in other cell types. High Ca^{++} concentrations promote spikes and sustained increases in cytosolic Ca^{++}, which are associated with increased levels of diacylglycerol, IP_3, and IP_4. High Ca^{++} also lowers cellular cAMP by a pertussis toxin–sensitive mechanism that is analogous to that for $PGF_{2\alpha}$, presumably the inhibition of adenylate cyclase by G_i. It appears unlikely, however, that the changes in cAMP can account on a quantitative basis for the inhibitory effects of Ca^{++} on PTH release. In addition, high Ca^{++} promotes the accumulation of two additional intracellular mediators (diacylglycerol and Ca^{++}), which are stimulatory to hormone release in permeabilized parathyroid cells. It is likely, therefore, the high Ca^{++} activates one or more additional mechanisms that dominate in their inhibitory effects on PTH release. These might include the generation of an inhibitor of protein kinase C, such as sphingosine (Hannun et al., 1986) or lysophosphatidylcholine (Oishi et al., 1988), or perhaps activation of a "distal" inhibitory mechanism (perhaps a G-protein) close to the site of exocytosis.

Acknowledgments

The authors gratefully acknowledge helpful discussions with Dr. Peter Baker as well as the expert secretarial help of Tom Boulet and Hilary Farmer.

This work was supported by National Institutes of Health grants DK-36801 and DK-35796.

References

Azhar S., and K. M. J. Menon. 1981. Receptor-mediated gonadotropin action in the ovary. Inhibitory actions of concanavalin A and wheat-germ agglutinin on gonadotropin-stimulated cAMP and progesterone responses in ovarian cells. *Biochemical Journal* 200:153–159.

Barrowman, M. M., S. Cockroft, and B. D. Gomperts. 1985. Two roles for guanine nucleotides in the stimulus-secretion sequence of neutrophils. Nature. 319:504–507.

Berridge, M. J., and R. F. Irvine. 1984. Inositol trisphosphate, a novel second messenger in cellular signal transduction. *Nature.* 312:315–321.

Blackmore, P. F., S. B. Bocckino, L. E. Waynick, and J. H. Exton. 1986. Role of a guanine nucletoide-binding regulatory protein in the hydrolysis of hepatocyte phosphatidylinositol 4,5-bisphosphate by calcium-mobilizing hormones and the control of cell calcium. *Journal of Biological Chemistry.* 260:14477–14483.

Bourdeau, A., B. Gross, and M. Lieberherr. 1988. Very early (2 min) in vitro effects of change in extracellular calcium on diacylglycerol formation in porcine parathyroid cells. *Journal of Bone and Mineral Research.* 3 (Suppl. I):35. (Abstr.)

Brown, E. M., R. J. Carroll, and G. D. Aurbach. 1977. Dopaminergic stimulation of cyclic AMP accumulation and parathyroid hormone release from bovine parathyroid cells. *Proceedings of the National Academy of Sciences.* 74:4210–4213.

Brown, E. M., P. Enyedi, M. LeBoff, J. Rotberg, J. Preston, and C. Chen. 1987*a*. High extracellular Ca^{++} and Mg^{++} stimulate accumulation of inositol phosphates in bovine parathyroid cells. *FEBS Letters.* 218:113–118.

Brown, E. M., M. S. LeBoff, M. Oetting, J. T. Posillico, and C. Chen. 1987*b*. Secretory control in normal and abnormal parathyroid tissue. *Recent Progress in Hormone Research.* 43:337–382.

Brown, E. M., D. G. Gardner, R. A. Windeck, and G. D. Aurbach. 1978. Relationship of intracellular 3′,5′-adenosine monophosphate to parathyroid hormone release from dispersed bovine parathyroid cells. *Endocrinology.* 103:2323–2333.

Brown, E. M., J. Redgrave, and J. Thatcher. 1984*a*. Effect of the phorbol ester TPA on PTH secretion. Evidence for a role for protein kinase C in the control of PTH release. *FEBS Letters.* 175:72–75.

Brown, E. M., J. G. Thatcher, E. J. Watson, and R. Leombruno. 1984b. Extracellular calcium potentiates the inhibitory effects of magnesium on parathyroid function in dispersed bovine parathyroid cells. *Metabolism.* 33:171–176.

Brown, E. M., E. J. Watson, R. Leombruno, and R. H. Underwood. 1983. Extracellular calcium is not necessary for acute, low calcium- or dopamine-stimulated PTH secretion in dispersed bovine parathyroid cells. *Metabolism.* 32:1038–1044.

Chen, C., C. S. Anast, J. T. Posillico, and E. M. Brown. 1987. Effects of extracellular calcium and magnesium on cytosolic calcium concentration in fura-2–loaded bovine parathyroid cells. *Journal of Bone and Mineral Research* 2:319–327.

Chen, C. J., J. Barnett, D. A. Congo and E. M. Brown. 1989. Divalent cations suppress cyclic adenosine monophosphate accumulation by stimulating a pertussis toxin-sensitive guanine nucleotide-binding protein in cultured bovine parathyroid cells. *Endocrinology.* 124:233–239.

Douglas, W. W. 1968. Stimulus-secretion coupling: the concept and clues from chromaffin and other cells. *British Journal of Pharmacology.* 34:451–474.

Downes, C. P., P. T. Hawkins, and R. F. Irvine. 1986. Inositol 1,3,4,5-tetrakisphosphate and not phosphatidylinositol 1,3,4-trisphosphate is the probable precursor of inositol 1,3,4-trisphosphate in agonist-stimulated parotid gland. *Biochemical Journal.* 238:501–506.

Fitzpatrick, L. A., M. L. Brandi, and G. D. Aurbach. 1986*a*. Calcium inhibition of parathyroid hormone secretion is mediated via a guanine nucleotide regulatory protein. *Endocrinology.* 119:2700-2703.

Fitzpatrick, L. A., M. L. Brandi, and G. D. Aurbach. 1986*b*. Prostaglandin $F_{2\alpha}$ and α-adrenergic agonists regulate parathyroid cell function via the inhibitory guanine nucleotide regulatory protein. *Endocrinology*. 118:2115–2119.

Flier J. S., C. R. Kahn, D. B. Jarrett, and J. Roth. 1972. Autoantibodies for the insulin receptor: effect on the insulin-receptor interaction in IM-9 lymphocytes. *Journal of Clinical Investigation*. 60:784–794.

Gardner, D. G., E. M. Brown, R. Windeck, and G. D. Aurbach. 1979. Prostaglandin $F_{2\alpha}$ inhibits 3′,5′-adenosine monophosphate accumulation and parathyroid hormone release from dispersed parathyroid cells. *Endocrinology*. 104:1–7.

Gershengorn, M. C., E. Geras, V. S. Purrello, and M. G. Rebecchi. 1984. Inositol trisphosphate mediates thyrotropin-releasing hormone mobilization of nonmitochondrial calcium in rat mammotropic pituitary cells. *Journal of Biological Chemistry*. 259:10675–10681.

Grynkiewicz, G., M. Poenie, and R. Y. Tsien. 1985. A new generation of Ca^{++}-indicators with greatly improved fluorescent properties. *Journal of Biological Chemistry*. 260:3440–3450.

Habener, J. F., B. Kemper, and J. T. Potts, Jr. 1975. Calcium-dependent intracellular degradation of parathyroid hormone: a possible mechanism for the regulation of hormone stores. *Endocrinology*. 97:431–437.

Habener, J. F., and J. T. Potts, Jr. 1976. Relative effectiveness of magnesium and calcium in the secretion and biosynthesis of parathyroid hormone *in vitro*. *Endocrinology*. 98:197–202.

Hannun, Y. A., C. R. Loomis, A. H. Merrill, Jr., and R. M. Bell. 1986. Sphingosine inhibition of protein kinase C activity and of phorbol dibutyrate *in vitro* and in platelets. *Journal of Biological Chemistry*. 261:12604–12609.

Hawkins, D., P. Enyedi, and E. Brown. 1989. The effects of high extracellular Ca^{2+} and Mg^{2+} concentrations on the levels of inositol 1,3,4,5-tetrakis-phosphate in bovine parathyroid cells. *Endocrinology*. 124:838–844.

Irvine, R. F., and R. M. Moor. 1986. Micro-injection of inositol 1,3,4,5-tetrakisphosphate activates sea urchin eggs by a mechanism dependent on external Ca^{++}. *Biochemical Journal*. 240:917–920.

Juhlin, C., R. Holmdahl, H. Johansson, J. Rasted, G. Akerstrom, and T. Klareskog. 1987. Monoclonal antibodies with exclusive reactivity against parathyroid cells and tubule cells of the kidney. *Proceedings of the National Academy of Sciences*. 84:2990–2994.

Kifor, O., and E. M. Brown. 1988. Relationship between diacylglycerol levels and extracellular Ca^{++} in dispersed bovine parathyroid cells. *Endocrinology*. 123:2723–2729.

Knight, D. E., and P. F. Baker. 1982. Calcium-dependence of catecholamine release from bovine adrenal medullary cells after exposure to intense electrical fields. *Journal of Membrane Biology*. 68:107–140.

Kobayashi, N., J. S. Russell, D. Lettieri, and L. M. Sherwood. 1988. Regulation of protein kinase C by extracellular calcium in bovine parathyroid cells. *Proceedings of the National Academy of Sciences*. 85:4857–4860.

LeBoff, M. S., D. Shoback, E. M. Brown, J. Thatcher, R. Leombruno, D. Beaudoin, M. Henry, R. Wilson, J. Pallotta, S. Marynick, J. Stock, and G. Leight. 1985. Regulation of parathyroid hormone release and cytosolic calcium by extracellular calcium in dispersed and cultured bovine and pathological human parathyroid cells. *Journal of Clinical Investigation*. 75:49–57.

Lopez-Barneo, J., and C. Armstrong. 1983. Depolarizing response of rat parathyroid cells to divalent cations. *Journal of General Physiology*. 82:209–249.

Morrissey, J. 1984. The influence of phorbol myristate acetate (PMA) on parathyroid hormone secretion. *Clinical Research.* 32:404A. (Abstr.)

Morrissey, J. J. 1985. Calcium-mediated changes in diglyceride levels in parathyroid cells. *Clinical Research.* 33:440. (Abstr.)

Morrissey, J. J., J. W. Hamilton, R. R. MacGregor, and D. V. Cohn. 1980. The secretion of parathormone fragments 34-84 and 37-84 by dispersed porcine parathyroid cells. *Endocrinology.* 107:164–171.

Muff, R., and J. Fischer. 1986. Parathyroid hormone secretion does not respond to changes of free calcium in electropermeabilized bovine parathyroid cells, but is stimulated with phorbol ester and cyclic AMP. *Biochemical and Biophysical Research Communications.* 139:1233–1237.

Nemeth, E. F., and A. Scarpa. 1986. Cytosolic Ca^{++} and the regulation of secretion in parathyroid cells. *FEBS Letters.* 213:15–19.

Nemeth, E. F., J. Wallace, and A. Scarpa. 1986. Stimulus-secretion coupling in bovine parathyroid cells. Dissociation between secretion and net changes in cytosolic Ca^{++}. *Journal of Biological Chemistry.* 261:2668–2674.

Oetting, M., M. S. LeBoff, S. Levy, L. Swiston, J. Preston, C. Chen, and E. M. Brown. 1987. Permeabilization reveals classical stimulus-secretion coupling in bovine parathyroid cells. *Endocrinology.* 121:1571–1576.

Oetting, M., M. S. LeBoff, L. Swiston, J. Preston, and E. Brown. 1986. Guanine nucleotides are potent secretagogues in permeabilized parathyroid cells. *FEBS Letters.* 208:99–104.

Oishi, K., R. L. Raynor, P. A. Charp, and J. F. Kuo. 1988. Regulation of protein kinase C by lysophospholipids. *Journal of Biological Chemistry.* 263:6865–6871.

Posillico, J. T., M. Burrowes, and E. M. Brown. 1986. Phospholipase C and concanavalin A stimulate PTH secretion from dispersed bovine parathyroid cells by different mechanisms. *8th Annual Science Meeting of the Society of Bone and Mineral Research.* Abstract 311.

Posillico, J. T., S. Srikanta, G. Eisenbarth, V. Quaranta, S. Kajiji, and E. M. Brown. 1987. Binding of monoclonal antibody (4F2) to its cell surface antigen in dispersed adenomatous parathyroid cells raises cytosolic Ca^{++} and inhibits parathyroid hormone secretion. *Journal of Clinical Endocrinology and Metabolism.* 64:43–50.

Posillico, J. T., J. Wortsman, S. Srikanta, G. S. Eisenbarth, L. E. Mallette, and E. M. Brown. 1986. Parathyroid cell surface autoantibodies that inhibit parathyroid hormone secretion from dispersed human parathyroid cells. *Journal of Bone and Mineral Research.* 1:475–483.

Shoback, D. M., L. A. Membreno, and J. G. McGhee. 1988. High calcium and other divalent cations increase inositol trisphosphate in bovine parathyroid cells. *Endocrinology.* 123:382–389.

Shoback, D. M., J. G. Thatcher, R. Leombruno, and E. M. Brown. 1983. Effects of extracellular Ca^{++} and Mg^{++} on cytosolic Ca^{++} and PTH release in dispersed bovine parathyroid cells. *Endocrinology.* 113:424–426.

Shoback, D. M., J. Thatcher, R. Leombruno, and E. M. Brown. 1984. The relationship between PTH secretion and cytosolic calcium concentration in dispersed bovine parathyroid cells. *Proceedings of the National Academy of Sciences.* 81:3113–3117.

Speigel, A. M., P. Gierschik, M. A. Levine, and R. W. Downs, Jr. 1985. Clinical implications of guanine nucleotide-binding proteins as receptor-effector couplers. *New England Journal of Medicine.* 312:26–33.

Stewart, A. F., and A. E. Broadus. 1987. Mineral metabolism. *In* Endocrinology and Metabolism. 2nd Ed. P. Felig, J. D. Baxter, A. E. Broadus and L. A. Frohman, editors. McGraw-Hill Book Co., New York. 1317–1453.

Tsien, R. Y., T. Pozzan, and T. J. Rink. 1982. Calcium homeostasis in intact lymphocytes: cytoplasmic free calcium monitored with a new intracellularly trapped fluorescent indicator. *Journal of Cell Biology*. 94:325–334.

Wallace, J., and A. Scarpa. 1982. Regulation of parathyroid hormone secretion *in vitro* by divalent cations and cell metabolism. *Journal of Biological Chemistry*. 257:10613–10616.

Wallace, J., and A. Scarpa. 1983. Parathyroid hormone secretion in the absence of extracellular free Ca^{++} and transmembrane Ca^{++} flux. *FEBS Letters*. 151:83–88.

Wollheim, C. B., S. Ullrich, P. Meda, and L. Vallar. 1987. Regulation of exocytosis in electrically permeabilized insulin-secreting cells. Evidence for Ca^{++} dependent and independent secretion. *Bioscience Reports*. 7:443–454.

Exocytosis

Chapter 17

The Mechanism of Exocytosis during Secretion in Mast Cells

W. Almers, L. J. Breckenridge, and A. E. Spruce

*Department of Physiology and Biophysics,
University of Washington, Seattle, Washington*

Introduction

All eukaryotic cells must continually engage in exocytosis in order to renew their plasma membrane. The supply of new membrane is maintained by cytoplasmic vesicles that derive from the Golgi and are transported to the cell surface. The membrane surrounding the vesicles fuses with, and becomes part of, the plasma membrane, along with the membrane proteins residing in the vesicle. At the same time, the contents of the vesicle escape into the extracellular space. This fundamental process is called constitutive exocytosis (Kelly, 1985; Moore, 1989). The rate at which it occurs is determined by the rate at which vesicles are produced and delivered to the plasma membrane.

In addition, some cells are specialized to perform stimulated exocytosis. They accumulate clusters of vesicles beneath the plasma membrane, sometimes, as in neurons, at specialized sites called active zones. The vesicles remain there dormant, until a stimulus (such as an action potential in a neuron, or a hormone in a secretory cell) results in the appearance of a cytosolic messenger substance. The messenger (Ca in a neuron) then causes conformational changes in intracellular receptor proteins, with the ultimate result that many or most of the vesicles are exocytosed in a burst.

How does the membrane of an exocytosing vesicle manage to fuse with the plasma membrane? This is not understood for constitutive nor for stimulated exocytosis. Biological membranes are notoriously disinclined to fuse. For example, many cells are connected by gap junctions and hence their membranes approach to within molecular dimensions, yet they never fuse. A resting mast cell is so tightly packed full of secretory vesicles (called granules) that many vesicles touch each other or the plasma membrane, yet fusion and exocytosis happen only when the cell is stimulated. The myelin sheath of vertebrate axons is another example with its indefinitely stable, extensive, and extremely close apposition of adjacent Schwann cell membranes. A major reason for the resistance of biological membranes to fusion is the stability of the lipid bilayer (see Rand, 1981). For the bilayers to fuse, lipid molecules must transiently leave the bilayer configuration, and this requires that a large energy barrier must be overcome. Probably, exocytotic membrane fusion is initiated by specific proteins associated with the plasma and/or the vesicle membranes. Stimulated exocytosis is probably triggered by the activation of such "fusogenic" proteins.

This article reviews our recent attempts to capture early events in membrane fusion during stimulated exocytosis in mast cells.

Electrical Recording of Exocytosis at the Level of Single Vesicles

Since all biological membranes have an electrical capacitance of 1 $\mu F/cm^2$ (or 10 $fF/\mu m^2$), the cell surface area may be assayed by measuring the cell membrane capacitance, C_m. Exocytosis inevitably leads to an increased cell surface area, and hence increases C_m (Jaffe et al., 1978; Gillespie, 1979). With sufficiently sensitive instrumentation (Neher and Marty, 1982), the increase in C_m may be seen to occur as a sequence of steps, each representing the fusion of one secretory vesicle with the cell membrane (Fernandez et al., 1984; Maruyama, 1986; Almers and Neher, 1987). In our work we chose mast cells from so-called "beige" mice. This strain of mice has a hereditary defect that prevents mast cells (and other granulocytes) from controlling the size of their secretory vesicles. Hence secretory vesicles are so large (1–5 μm

diameter) that the exocytosis of an individual vesicle can not only be detected as a stepwise increase in C_m, but also observed simultaneously under the light microscope.

Fig. 1 shows a recording of C_m vs. time, made on a peritoneal mast cell from a beige mouse. A fire-polished glass micropipette was sealed onto the mast cell. The micropipette was filled with a solution that was designed to mimic the cytosol, but also contained $GTP_{\gamma S}$, an irreversible activator of G proteins. $GTP_{\gamma S}$ is known to bypass the normal receptor-mediated signalling pathway of mast cells, and to stimulate secretion directly. The trace starts 20 s after the membrane patch beneath the pipette orifice was ruptured by a pulse of suction so that the pipette interior and the cytosol became continuous ("whole-cell" configuration, Hamill et al., 1981). The potential of the cell membrane was forced ("voltage clamped") to undergo sinusoidal oscillations, and the sinusoidal current necessary to do this was fed into a "lock-in amplifier" (Neher and Marty, 1982). This instrument decomposed the current into two components, one in-phase with the voltage and the other out-of-phase by 90°. The out-of-phase component is shown in Fig. 1; it is a measure of the cell membrane capacitance (more precisely, the imaginary part of the electrical admittance of the cell membrane). After $GTP_{\gamma S}$ had diffused into the cell and taken effect, C_m was observed to increase in steps, each representing exocytosis of one secretory vesicle. The in-phase component (not shown) was also recorded; it is proportional to the real part of the admittance or, in the simplest case, to the cell membrane conductance. It did not change significantly during the experiment, indicating that mast cell vesicles contain few open channels.

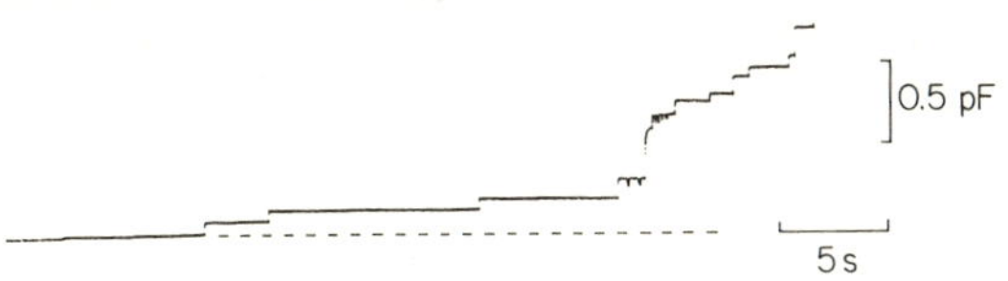

Figure 1. Capacitance plotted against time in a degranulating mast cell from a beige mouse. From Breckenridge and Almers, 1987*b*.

Membrane Fusion Is Not Initiated by Mechanical Stress

Experiments such as the one in Fig. 1 tell us when individual vesicles exocytose, but not how. A widely discussed hypothesis (reviewed by Finkelstein et al., 1986; Holz, 1986), holds that the fusion of vesicle and plasma membranes is driven by mechanical stress on the vesicle membrane, generated, perhaps, through osmotic imbalances caused by the opening of Ca-activated K channels (Stanley and Ehrenstein, 1985). The hypothesis seemed plausible, because osmotic stress can cause lipid vesicles to fuse to one another as well as to planar bilayers (Finkelstein et al., 1986), and because in several cell types, secretory vesicles swell during secretion. Mast cells of beige mice offered an opportunity to test this hypothesis. Cells were stimulated with $GTP_{\gamma S}$, C_m was recorded as in Fig. 1, and secretion was observed with a video camera. Capacitance steps were accompanied by the sudden swelling of individual vesicles. Fig. 2 shows a brief episode during the degranulation of a mast cell, during which one secretory vesicle underwent exotytosis. It plots C_m against time (continuous line) as well as the diameter of the exocytosing vesicle (dots), measured on polaroid photographs taken of successive video frames. In this particular experiment, the pipette solution had twice the tonicity of the external fluid and of the cytosol. This caused the vesicle to shrink during the first 30 s of whole-cell contact. Later, the granule swelled again, but the swelling was preceded by the C_m step. Clearly, C_m measurements indicate that fusion has occurred long before the first signs of vesicle swelling.

It could be argued that a very slight swelling may have occurred before the C_m step but was missed because of the limited optical resolution. If so, however, the swelling was insufficient to take up the slack introduced in the vesicle membrane by previous osmotic shrinkage. Evidently, the vesicle in Fig. 2 underwent exocytosis even though its membrane was slack. Hence, membrane fusion is not driven by mechanical stress (Breckenridge and Almers, 1987*b*). Similar observations were made independently by Zimmerberg et al. (1987). Osmotically shrunken vesicles are capable of exocytosis in adrenal chromaffin cells as well (Holz and Senter, 1986). There is not a single instance where a requirement of membrane stretch in exocytosis has been clearly established. Hence we are inclined to abandon theories that attribute a central role in membrane fusion to mechanical stresses or to the osmotic gradients that might lead to them.

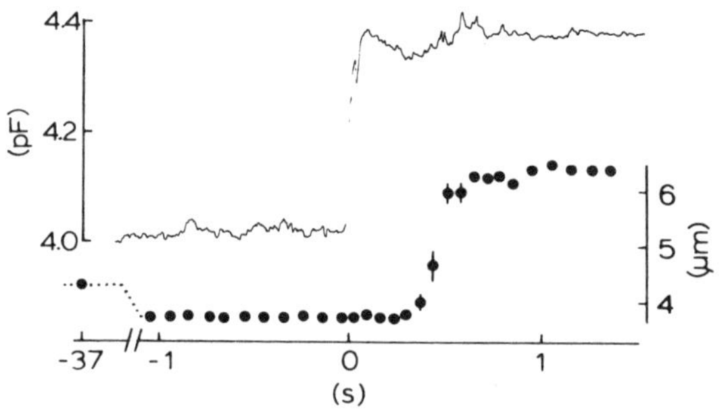

Figure 2. Vesicle fusion precedes swelling. 350 mM sucrose was added to the pipette solution to make it hypertonic and shrink the vesicle. The capacitance step occurred 37 s after pipette interior and cytosol became continuous. Details as in Breckenridge and Almers, 1987*a*. From Almers and Breckenridge, 1988.

Capacitance Flicker: Evidence for a Reversible Intermediate in Membrane Fusion

All secretory cells eventually reinternalize the membrane of the vesicles after exocytosis. The mechanisms for membrane retrieval may vary among different cells; in mast cells, membrane retrieval occurs approximately exponentially with a time constant of 3 h (Thilo, 1985). However, step decreases in C_m indicating membrane internalization may be recorded even during degranulation. Such "downward" steps have the same average size as the step increases in C_m, representing fusion of single vesicles, and occasionally, upward steps are followed by downward steps of identical size as if vesicles occasionally fused and unfused rapidly thereafter (Fernandez et al., 1984). This so-called capacitance flicker provides evidence for a reversible intermediate step in exocytosis, possibly representing a tenuous electrical connection between vesicle interior and cell exterior that can be made and broken repeatedly just as an ion channel may repeatedly open and close. Similar events are observed in mast cells of beige mice. Fig. 3 *A* shows a capacitance flicker that suggests a vesicle whose electrical connection with the cell exterior fluctuated for over a second, and once even broke completely, before the vesicle irreversibly committed itself to fusion.

To test how well vesicles can secrete while they "flicker," we took advantage of the fact that the lumen of most vesicles is acidic, and hence accumulates the fluorescent dye quinacrine (Johnson et al., 1980). When the vesicle exocytoses, the quinacrine is lost together with the remainder of its contents and washed away by the perfusion fluid. An example is shown in Fig. 3 *C*; this vesicle showed no evidence of flicker, and the loss of fluorescence began immediately after the step. In the vesicle of Fig. 3 *B*, however, there was no detectable fluorescence loss until "flickering" had ceased. Hence, vesicles secrete in earnest only after they have stopped flickering and have committed themselves irreversibly to fusion. The experiment supports the idea that the intermediate step, evident as a "flicker" in Fig. 3, represents an aqueous pore that connects the vesicle lumen with the cell exterior and is so narrow that quinacrine can only escape through it imperceptibly slowly. For lack of a better name, this aqueous connection was termed the "fusion pore" (Breckenridge and Almers, 1987*a*).

The Fusion Pore Dilates Rapidly after It Has Formed

The foregoing experiments show that exocytosis starts with the formation of a narrow pore that is usually short lived, but on rare occasions opens and closes repeatedly to cause capacitance flicker. If one could make capacitance measurements rapidly enough, it might be possible to demonstrate the existence of a transient, high-resistance pore even on vesicles that do not flicker. The rationale of an experiment to test this point is illustrated in Fig. 4 *C*, which shows a vesicle with capacitance C_v, connected to

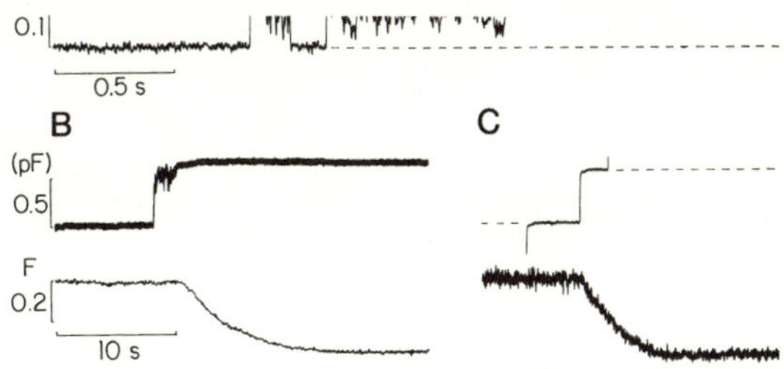

Figure 3. (*A* and *B*) Vesicles showing capacitance flicker. (*C*) A normally exocytosing vesicle. *B* and *C* were recorded in the presence of 8-μM external quinacrine; in *B* and *C* the upper trace shows the capacitance, the lower trace the fluorescence calibrated as a fraction of the starting value. The loss of fluorescence in *B* and *C* represents secretion of the quinacrine accumulated in the two vesicles. Loss of dye starts simultaneously with the capacitance step in *C*, but only after flicker has ceased in *B*. Assuming the vesicles to be spherical and their membrane capacitance to be 1 $\mu F/cm^2$, their diameters can be calculated from the capacitance as (*A*) 1.8 μm, (*B*) 4.5 μm, and (*C*) 4.2 μm. From Breckenridge and Almers, 1987*a*.

the extracellular space by a pore of conductance *g*. While the pore is narrow and *g* is small, the sinusoid applied to the cell membrane will fail to invade the exocytosing vesicle completely, so that the apparent capacitance contributed by the vesicle is less than C_v. Only when the conductance becomes large will the vesicle contribute its full share to the membrane capacitance. Hence, C_m steps are expected to appear gradual if observed at sufficient time resolution, and the gradual rise of the apparent capacitance to a final value C_v should reflect the gradual dilation of the pore. One also expects that the in-phase ("conductance") component of sinusoidal current should transiently increase as the pore dilates. When *g* is small, the sinusoidal voltage drops almost entirely across the pore. Hence the pore will pass current nearly in phase with the voltage, and the capacitance (out-of-phase component) of the vesicle membrane will contribute little: the vesicle will behave electrically like a resistor. Only when the pore has grown and *g* is larger will the voltage drop mostly across C_m; current through the pore will be increasingly out-of-phase and the resistor-capacitor combination will assume the appearance of a pure capacitance. The conductance of the pore, *g*, can be calculated either from the *C*- or the *G*-trace as follows (Breckenridge and Almers,

1987*b*; see also Zimmerberg et al., 1987):

$$g = 2\pi f C_v/(C_v/C - 1)^{1/2} \quad (1)$$

$$g = 2\,G/[1 + n\sqrt{1 - (G/2\pi f C_v)^2}] \quad (2)$$

where C is the apparent capacitance contributed by the vesicle and C_v its final value; $n = 1$ while $C < C_v/2$ and -1 otherwise; G is proportional to the in-phase component of sinusoidal current contributed by the vesicle and f is the frequency of the sinusoid.

To obtain rapid recordings, we designed a lock-in amplifier that sampled both the in-phase (G) and out-of-phase components (C) of sinusoidal current once every

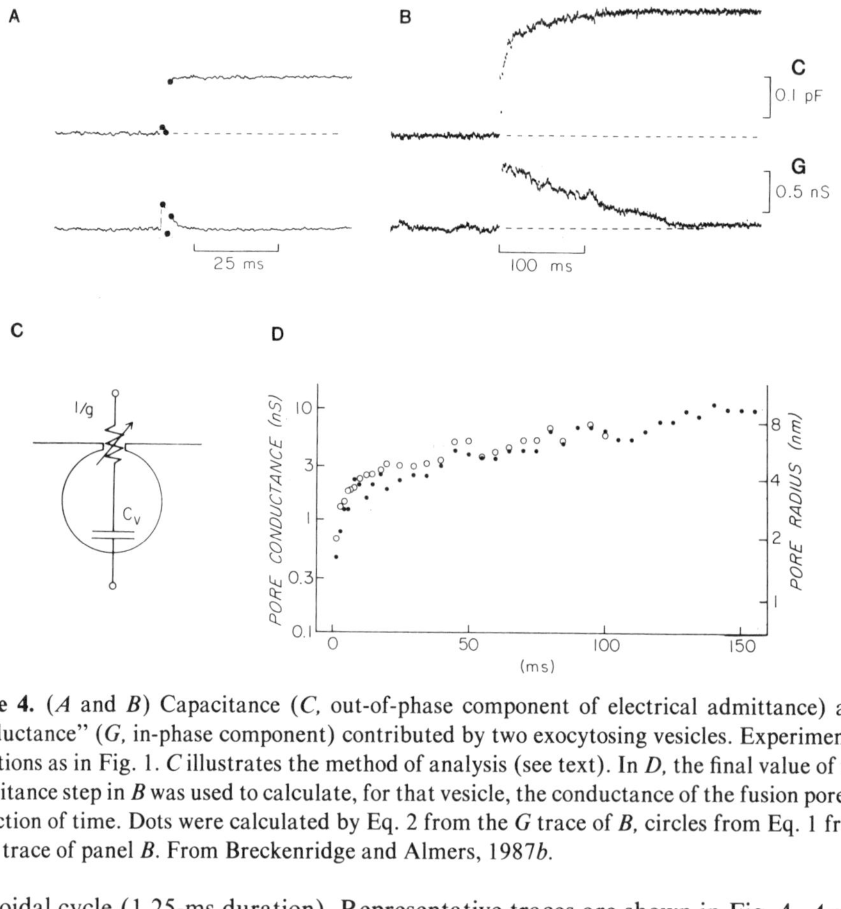

Figure 4. (*A* and *B*) Capacitance (*C*, out-of-phase component of electrical admittance) and "conductance" (*G*, in-phase component) contributed by two exocytosing vesicles. Experimental conditions as in Fig. 1. *C* illustrates the method of analysis (see text). In *D*, the final value of the capacitance step in *B* was used to calculate, for that vesicle, the conductance of the fusion pore as a function of time. Dots were calculated by Eq. 2 from the *G* trace of *B*, circles from Eq. 1 from the *C* trace of panel *B*. From Breckenridge and Almers, 1987*b*.

sinusoidal cycle (1.25-ms duration). Representative traces are shown in Fig. 4, *A* and *B*. The traces in Fig. 4 *B* showed the expected features, in that *C* grew gradually to a final value while *G* increased and then declined. In Fig. 4 *A*, however, *C* grew abruptly (within 3.75 ms) to its final value; evidently the dilation of the pore was much too rapid to be resolved. For Fig. 4 *B*, the conductance of the pore was calculated from Eqs. 1 and 2, and plotted in Fig. 4 *D*. *g* is seen to grow over a tenfold range, but even in the "slow" vesicle, *g* shows an early phase of growth that is too rapid to be resolved.

The radius of the pore may be estimated from the conductance (e.g., Eq. 8-1 in Hille, 1984), and is plotted on the right-hand side of Fig. 4 *D*. The length of the pore was assumed to be 15 nm as in a gap junction, an ion channel known to span two membranes; the resistivity of the pore lumen is taken as 100 Ω-cm, similar to the resistivity of the bathing medium. The earliest measurement that could be made in this experiment suggests a pore radius between 1.5 and 2 nm. The radius quickly increased to 4 nm, and then grew more slowly over the next 100 ms. The final values in Fig. 4 *D* approach the 12-nm radius observed in a quick-freeze electron micrograph of Chandler and Heuser (1980).

During flicker in Fig. 3 *A*, *C* grew reversibly to values nearly as large as C_v. Though the trace is noisy, one can estimate from Eq. 1 that *g* must have grown to ~2 nS when the pore first opened. The corresponding pore radius would be ~2.5 nm; even this dilated fusion pore could evidently close completely.

The experiment of Fig. 4 confirms that in the course of exocytosis, a fusion pore forms and then dilates. However, it is also clear that early steps in the formation of the pore are missed in such an experiment. To gain information on this point, one can take advantage of the small transient currents that flow through the pore at the moment it is first formed.

Currents through the Fusion Pore Generated by Electrical Discharge of the Vesicle Membrane Capacitance

Before exocytosis, mast cell vesicles have a lumen-positive membrane potential (Breckenridge and Almers, 1987*b*). As soon as the fusion pore opens, however, the potentials of vesicle and plasma membrane must begin to approach each other and ultimately become equal. Hence, a current must flow through the pore to adjust the charge stored on the vesicle membrane capacitance. Indeed, if the plasma membrane is held at zero or inside positive potentials, then each capacitance step is preceded by a transient outward current. Fig. 5 shows such a transient (upper trace) along with its time integral (lower trace). The following findings are consistent with the idea that the transients are generated when vesicles charge or discharge their membrane capacitance through fusion pores (Breckenridge and Almers, 1987*b*). (*a*) At a fixed plasma membrane potential, the charge carried by a transient (Q) is linearly related to the final amplitude of the capacitance step (C_v) that follows it. (*b*) The ratio Q/C_v is correlated with the potential at which the plasma membrane is held. The relationship is described by a straight line with a slope of one. This is exactly what is expected, since

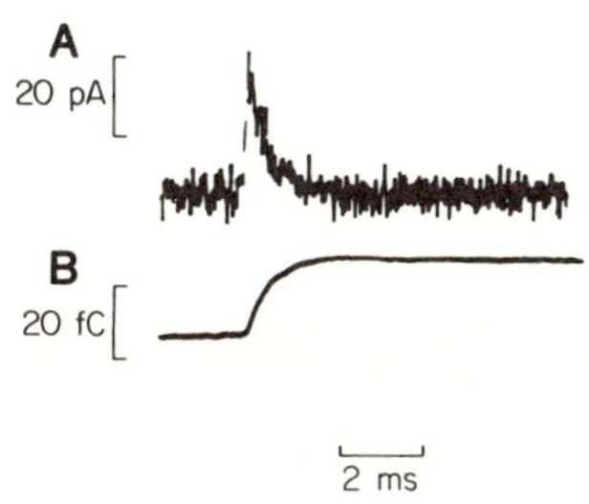

Figure 5. (*A*) Current transient observed as an exocytosing vesicle discharges its membrane capacitance through the fusion pore. (*B*) Time integral of *A*. From the plasma membrane holding potential (+30 mV), from the final amplitude of trace *B* (25 fC), and from the capacitance step that followed (170 fF, not shown), the vesicle membrane potential at the moment of fusion was calculated. It was (25 fC/170 fF −30 mV) equal to 117 mV, lumen positive. From Breckenridge and Almers, 1987*b*.

Q/C_v should equal the difference ($E_c - E_v$) between the potentials across the plasma (E_c) and vesicle membranes (E_v), that is, the potential difference that drives current through the pore. (*c*) Carbonyl cyanide 4-(trifluoromethoxy) phenylhydrazone (FCCP), a H^+-ionophore expected to collapse or reverse the vesicle potential, shifts the reversal potential of the current transients to 0 mV.

Since E_c is controlled experimentally, the ratio Q/C_v provides a direct measurement of the membrane potential of each vesicle (E_v) at the instant of exocytosis. Its average value is between +70 and +80 mV, lumen positive. The concentration gradients for several inorganic cations are known from X-ray microanalysis (Na^+, K^+, Cl^+, Ca^{++}; Kendall and Warley, 1986) or microfluorimetric measurements with the fluorescent dye quinacrine (H^+; Johnson et al., 1980; Breckenridge and Almers, 1987*a*) or with the Ca indicator dye fura-2 (Almers and Neher, 1985). Interestingly, none of these ions have concentration gradients of the right magnitude or direction to cause the observed lumen-positive potential of 70–80 mV. However, the positive potential may suggest the operation of an electrogenic H^+ pump. (*a*) An electrogenic H^+ pump is known to operate in the secretory vesicles of chromaffin cells, and H^+ pumps of probably similar type also operate in other secretory vesicles (reviewed by Johnson and Scarpa, 1984); (*b*) the vesicle lumen is known to be acidic (Johnson et al., 1980; Breckenridge and Almers, 1987*b*); (*c*) E_v depends on the pH of the intracellular medium, being increased at acid pH and diminished by alkaline pH (unpublished).

Interestingly, individual vesicles at the instant of exoctyosis had membrane potentials varying over a wide range (0–160 mV, lumen minus cytosol). Also, exocytosis is observed over a wide range of plasma membrane potentials (−120–+50 mV). Hence the potentials of neither vesicle nor plasma membrane are critically important for the ability of a vesicle to undergo exocytosis. This provides evidence against the otherwise attractive idea that ion channels in secretory vesicles might control fusogenic vesicle membrane proteins (and hence exocytosis) by controlling the membrane potential.

The Fusion Pore at the Instant of Its Formation

The abrupt appearance of the current in Fig. 5 indicates that the fusion pore forms suddenly. The potential driving this current is given by the ratio Q/C_v, and the conductance of the pore at the instant of its formation can be calculated by dividing the current at the beginning of the transient by Q/V_c. Values obtained were found to vary over a wide range, with a mean between 200 and 300 pS (Almers and Breckenridge, 1987*b*).

Transients as in Fig. 5 can be used to reconstruct the time course of the fusion pore conductance over the first 1 or 2 ms (Fig. 6 *A*). If the time integral is subtracted from its final value and divided by the final amplitude of the capacitance step, it represents the time course of the potential difference between vesicle and cell membrane ($E_c - E_v$), that is, the potential that drives the current through the pore. Dividing the current trace point by point by ($E_c - E_v$) yields the trace labeled *g*, that is, the conductance of the pore. It is seen that the conductance grows abruptly to a value, g_0 (~270 pS in Fig. 6, arrowhead) and then continues to increase. The later, comparatively gradual (but still rapid) increase corresponds to the rapid rise of conductance that was only poorly resolved in Fig. 4. The abrupt opening of the fusion pore is clearly reminiscent of the opening of an ion channel, and almost certainly represents a conformational change in a macromolecule. We imagine that the subsequent, more gradual growth of conductance may be due to the participation of many smaller molecules, e.g., lipids (Fig. 6 *B*).

Comparison of the Fusion Pore with Ion Channels

In comparing the fusion pore with ion channels, it is instructive to consider the gap junction channel, the only known ion channel that spans two membranes. Single gap

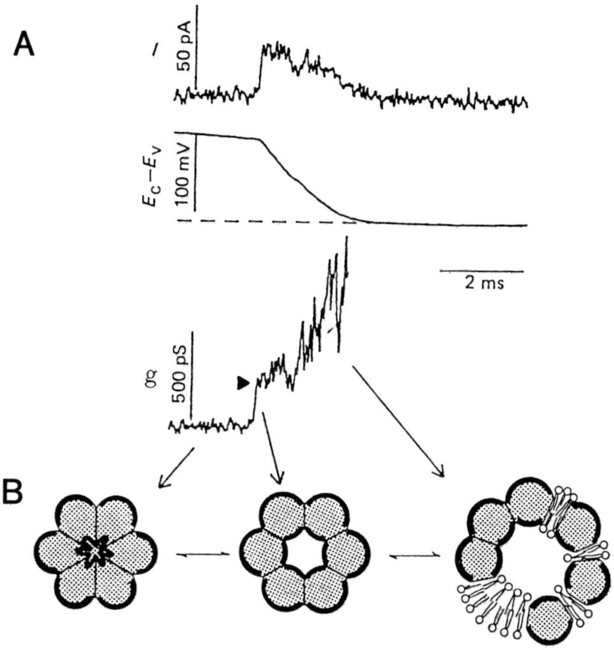

Figure 6. (*A*) From top to bottom: current transient similar to Fig. 5 *A*, the difference between vesicle and plasma membrane potentials as a function of time, and the conductance of the fusion pore. Experimental conditions were as in Fig. 5. (*B*) Hypothetical drawing showing a cross section through the pore made in the aqueous space between vesicle and plasma membranes. The macromolecule forming the fusion pore is shown as consisting of six monomers. Hydrophilic surfaces are drawn heavily, lipophilic ones lightly. Arrows relate the three different states of the pore to the time course of the conductance. From Almers et al., 1988.

junction channels in situ have been studied mainly by Neyton and Trautman (1985, 1986) and by Veenstra and DeHaan (1986, 1988). Interestingly, single gap junction channels in situ often do not open instantaneously, but instead rise to a stable level gradually, sometimes over milliseconds. The conductance of a single gap junction channel can vary even in the same type of cell (70–180 pS in lachrymal glands, Neyton and Trautman, 1985, 1986; 80–240 pS in cardiac muscle, Veenstra and DeHaan, 1988).

The conductance jump produced by the opening of the fusion pore is only poorly resolved in our recordings, but it, too, is probably often not instantaneous. However, the average rise time of conductance (102 ± 45 μs in 196 transients) is faster than in gap junction channels (280 μs, Veenstra and DeHaan, 1988). The histogram of g_o

values showed a skewed distribution, with a broad peak between 150 and 300 pS and a median value of ~270 pS. Hence g_0 is generally comparable to the conductance of single gap junction channels, and the initial diameter of the fusion pore must be comparable to that of a gap junction channel (<2 nm, Unwin and Zampighi, 1980).

However, neither gap junctions nor other ion channels are known to cause fusion. In our measurements, fusion pores have at least two features that are unusual for ordinary ion channels. (*a*) Unlike ion channels, fusion pores increase their conductance with time as they dilate. (*b*) The initial conductance of fusion pores varies over a much wider range than what is observed with other ion channels, including gap junction channels.

Structural Speculations on the Fusion Pore

Fig. 7 and Fig. 6 *B* illustrate the possible structure and function of the fusion pore. Its ability to open rapidly suggests to us that it is formed by a preexisting macromolecule

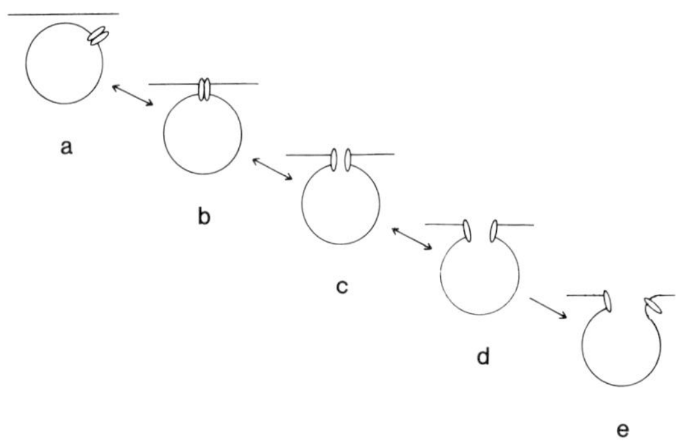

Figure 7. Hypothetical drawing of a secretory vescle that is free in the cytosol (*a*) or docked at the plasma membrane (*b*–*e*). In *b*, the fusion pore is preformed but closed; it has just opened in *c*, dilated in *d*, and partially disassembled in *e*.

that spans both the vesicle and the plasma membrane. How does the macromolecule reach this strategic position? The finding that vesicles may fuse with each other (compound exocytosis) suggests that the vesicle contains all the components necessary to initiate fusion, so we speculate that the fusion macromolecule originally resided in, and protruded from, the vesicle membrane (Fig. 7 *a*). We imagine that when the vesicle docks at the plasma membrane (perhaps by binding to a plasma membrane–bound receptor), the portion of the macromolecule protruding from the vesicle inserts itself into the plasma membrane (Fig. 7 *b*). Though the insertion of a macromolecule from an aqueous phase into a lipid bilayer is likely to be energetically unfavorable, it does occur; examples are the secreted proteins perforin and the ninth component of complement, both of which form ion channels in plasma membrane (Young et al., 1986), or the spike proteins of enveloped viruses that must insert themselves into the host cell membrane to cause fusion with the viral envelope (White et al., 1983).

We follow Satir et al. (1973, but see also Satir and Oberg, 1978) and speculate that the macromolecule consists of several monomers arranged in circular symmetry (Fig. 6 *B*), which is similar to the arrangement of connexin monomers in the gap junction channel. The variable size of the initial pore suggests that the number of monomers may be variable. When the appropriate cytosolic messenger (Ca^{++} in a neuron) binds to the macromolecule, it undergoes a rapid conformational change with two results. (*a*) The monomers realign to form a channel in their midst, causing the rapid jump in conductance (Figs. 6 *B* and 7 *c*). (*b*) The monomers lose their affinity for one another and drift apart, exposing between them amphipathic surfaces that may be invaded by lipid molecules (Fig. 6 *B*). As lipid molecules join the ring of monomers, the pore dilates (Fig. 7 *d*), and this explains the more gradual (but still rapid) subsequent rise in conductance. Given the fluidity of the lipid phase, we imagine that, while the monomers remain membrane attached on both their plasmalemmal and vesicle ends, the pore can fluctuate in size or even close completely (as in Fig. 3 *A*). But as more and more monomers detach themselves (Fig. 7 *e*), their constraint on the structure of the pore weakens, and the pore elongates to become a wide, seamless neck of lipid bilayer joining vesicle and plasma membrane. Now the vesicle has fused irreversibly.

The scheme in Figs. 7 and 6 *B* is speculative, but it helps to explain the enormous speed of neurotransmitter release. The delay between the upstroke of the presynaptic action potential and the excitatory postsynaptic potential may be as short as 0.5 ms at 20°C at the frog motor endplate (Katz and Miledi, 1965), or as short as 0.2 ms at 37°C in cat motorneurons (Munson and Sypert, 1979; Cope and Mendell, 1982). Within that time, several sequential events must take place in the presynaptic terminal: the opening of Ca^{++} channels, the influx of Ca^{++}, the binding of Ca^{++} to its intracellular receptor, the conformational changes that attend exocytosis, and finally the efflux of transmitter from the vesicle. If the vesicle is docked at the active zone via a preformed pore, then Ca^{++} could cause transmitter release by one single action, namely by opening the fusion pore. How rapidly could a vesicle discharge transmitter through the fusion pore? Consider that solute flux M is related to pore conductance by (compare, e.g., Eqs. 8-1 and 8-6 of Hille, 1984):

$$M = g \cdot \rho D \cdot c \tag{3}$$

where ρ is the resistivity of the pore lumen (e.g., 100 Ω-cm), D is the diffusion coefficient of the substance, and c is the concentration in the vesicle. If c is uniform in the vesicle (having the value C_o before the pore opens) and zero in the external medium, the substance is lost from the vesicle exponentially:

$$\begin{aligned} c &= C_o \cdot \exp(-t/\tau) \\ \tau &= V/\rho D g \end{aligned} \tag{4}$$

where V is the volume of the vesicle. If the diffusion coefficient of a typical transmitter (e.g., acetylcholine) is similar to that of tetraethylammonium ($8 \cdot 10^{-6}$ cm^2/s, Hille, 1984), Eq. 4 predicts that a typical synaptic vesicle of 50-nm diam would lose transmitter through a 300-pS pore with a time constant of only 0.24 ms. This time constant would lengthen in proportion to the volume of the vesicle, i.e., with the third power of its radius. For a 4.5-μm diam giant vesicle as in Fig. 3 *B*, for instance, the time constant would be ~3 min, imperceptibly slow on the time scale of that experiment. Clearly it is advantageous for a fast synapse to have small vesicles.

Two assumptions made in this calculation will not be valid in all cases: (*a*) all transmitter in the vesicle is in free solution, and (*b*) the fusion pore does not dilate. Nonetheless, the observed synaptic delay may give sufficient time for a vesicle to release much of its transmitter without disassembling its fusion pore, and hence without ever completely fusing with the plasma membrane. Avoidance of fusion would have advantages for the rapid and selective retrieval of vesicle membrane (Cecarelli and Hurlbut, 1980).

Acknowledgments

This work was supported by National Institutes of Health grants AR-17803 and GM-39520.

References

Almers, W., and L. J. Breckenridge. 1988. Early steps in the exocytosis of secretory vesicles in mast cells. *In* Molecular Mechanisms of Membrane Fusion (Proceedings). S. Ohki, D. Doyle, T. D. Flanagan, S. W. Hui, and E. Mayhew, editors. Plenum Publishing Corp., New York. 197–208.

Almers, W., L. Breckenridge, and A. Spruce. 1988. Time course of fusion pore conductance during exocytosis of secretory vesicles in beige mouse mast cells. *Journal of Physiology*. 407:96.

Almers, W., and E. Neher. 1985. The Ca signal from fura-2 loaded mast cells depends strongly on the method of dye-loading. *FEBS Letters*. 192:13–18.

Almers, W., and E. Neher. 1987. Gradual and stepwise changes in the membrane capacitance of rat peritoneal mast cells. *Journal of Physiology*. 386:205–217.

Breckenridge, L. J., and W. Almers. 1987*a*. Final steps in exocytosis in a cell with giant secretory granules. *Proceedings of the National Academy of Sciences*. 84:1945–1949.

Breckenridge, L. J., and W. Almers. 1987*b*. Currents through the fusion pore that forms during exocytosis of a secretory vesicle. *Nature*. 328:814-817.

Cecarelli, B., and W. P. Hurlbut. 1980. Vesicle hypothesis of the release of quanta of acetylcholine. *Physiological Reviews*. 60:396–441.

Chandler, D. E., and J. E. Heuser. 1980. Arrest of membrane fusion events in mast cells by quick-freezing. *Journal of Cell Biology*. 86:666–674.

Cope, T. C., and L. M. Mendell. 1982. Distributions of EPSP latency at different group Ia-fiber-alpha-motoneuron connections. *Journal of Neuroscience*. 47:469–478.

Fernandez, J. M., E. Neher, and B. D. Gomperts. 1984. Capacitance measurements reveal stepwise fusion events in degranulating mast cells. *Nature*. 312:453–455.

Finkelstein, A., J. Zimmerberg, and F. S. Cohen. 1986. Osmotic swelling of vesicles: its role in the fusion of vesicles with planar phospholipid bilayer membranes and its possible role in exocytosis. *Annual Reviews of Physiology*. 48:163–174.

Gillespie, J. I. 1979. The effect of repetitive stimulation on the passive electrical properties of the presynaptic terminal of the squid giant synapse. *Proceedings of the Royal Society of London Series B*. 206:293–306.

Hamill, O. P., A. Marty, E. Neher, B. Sakmann, and F. J. Sigworth. 1981. Improved patch-clamp techniques for high resolution current recording from cells and cell-free membrane patches. *Pflügers Archiv*. 391:85–100.

Hille, B. 1984. Ionic Channels in Excitable Membranes. Sinauer Associates, Sunderland, MA. 186–188.

Holz, R. W. 1986. The role of osmotic forces in exocytosis from adrenal chromaffin cells. *Annual Review of Physiology*. 48:175–192.

Holz, R. W., and R. A. Senter. 1986. The effects of osmolarity and ionic strength on secretion from adrenal chromaffin cells permeabilized with digitonin. *Journal of Neurochemistry*. 46:1835–1842.

Jaffe, L. A., S. Hagiwara, and R. T. Kado. 1978. The time course of cortical vesicle fusion in sea urchin eggs observed as membrane capacitance changes. *Developmental Biology*. 67:243–248.

Johnson, R. G., S. E. Carty, B. J. Fingerhood, and A. Scarpa. 1980. The internal pH of mast cell granules. *FEBS Letters*. 120:75–79.

Johnson, R. G., and A. Scarpa. 1984. Chemiosmotic coupling and its application to the accumulation of biological amines in secretory granules. *In* Electrogenic Transport: Fundamental Principles and Physiological Implications. M. P. Blaustein and M. Lieberman, editors. Raven Press, New York. 71–91.

Katz, B., and R. Miledi. 1965. The effect of temperature on the synaptic delay at the neuromuscular junction. *Journal of Physiology*. 181:656–670.

Kelly, R. B. 1985. Pathways of protein secretion in eukaryotes. *Science*. 230:25–31.

Kendall, M. D., and A. Warley. 1986. Elemental content of mast cell granules measured by x-ray microanalysis of rat thymic tissue sections. *Journal of Cell Science*. 83:77–87.

Maruyama, Y. 1986. Ca^{2+}-induced excess capacitance fluctuation studied by phase-sensitive detection method in exocrine pancreatic acinar cells. *Pflügers Archiv*. 407:561–563.

Moore, H.-P. H., C. Brion, K.-N. Chung, L. Lehmicke, R. Rivas, and D. Quinn. 1989. *In* Secretion and Its Control. G. S. Oxford and C. M. Armstrong, editors. 189–201.

Munson, J. B., and G. W. Sypert. 1979. Properties of single fibre excitatory post-synaptic potentials in triceps surae motoneurones. *Journal of Physiology*. 296:329–342.

Neher, E., and A. Marty. 1982. Discrete changes of cell membrane capacitance observed under conditions of enhanced secretion in bovine adrenal chromaffin cells. *Proceedings of the National Academy of Sciences*. 79:6712–6716.

Neyton, J., and A. Trautman. 1985. Single channel currents of an intercellular junction. *Nature*. 317:331–335.

Neyton, J., and A. Trautman. 1986. Physiological modulation of gap junction permeability. *Journal of Experimental Biology*. 124:93–114.

Rand, R. P. 1981. Interacting phospholipid bilayers: measured forces and induced structural changes. *Annual Reviews of Biophysics and Bioengineering*. 10:277–314.

Satir, B., and S. G. Oberg. 1978. *Paramecium* fusion rosettes: possible function as Ca^{2+} gates. *Science*. 199:536–538.

Satir, B., C. Schooley, and P. Satir. 1973. Membrane fusion in a model system. *Journal of Cell Biology*. 56:153–176.

Stanley, E. F., and G. Ehrenstein. 1985. A model for exocytosis based on the opening of calcium-activated potassium channels in vesicles. *Life Sciences*. 37:1985–1995.

Thilo, L. 1985. Selective internalization of granule membrane after secretion in mast cells. *Proceedings of the National Academy of Sciences*. 82:1711–1715.

Unwin, P. N. T., and G. Zampighi. 1980. Structure of the junction between communicating cells. *Nature*. 283:545–549.

Veenstra, R. D., and R. L. DeHaan. 1986. Measurement of single channel currents from cardiac gap junctions. *Science*. 233:972–974.

Veenstra, R. D., and R. L. DeHaan. 1988. Cardiac gap junction channel activity in embryonic chick ventricle cells. *American Journal of Physiology*. 254:H170–H180.

White, J., M. Kielian, and A. Helenius. 1983. Membrane fusion proteins of enveloped animal viruses. *Quarterly Review of Biophysics*. 16:151–195.

Young, J. D.-E., Z. A. Cohn, and E. R. Podack. 1986. The ninth component of complement and the pore-forming protein (Perforin 1) from cytotoxic T cells: structural, immunological, and functional similarities. *Science*. 233:184–190.

Zimmerberg, J., M. Curran, F. S. Cohen, and M. Brodwick. 1987. Simultaneous electrical and optical measurements show that membrane fusion preceds secretory granule swelling during exocytosis of beige mouse mast cells. *Proceedings of the National Academy of Sciences*. 84:1585–1589.

Chapter 18

Proton Pumps and Chloride Channels in Secretory Vesicles

Qais Al-Awqati, Jonathan Barasch, and Donald W. Landry

Departments of Medicine and Physiology, Columbia University, College of Physicians and Surgeons, New York, New York 10032

Regulation of Intravesicular pH by the Membrane Conductance

Secretory granules, such as chromaffin (Johnson et al., 1982), pituitary (Carty et al., 1980) and β cell granules (Orci et al., 1987) as well as lysosomes (Ohkuma et al., 1982) maintain an interior acid pH. This transmembrane pH gradient (Δph) is generated by a H^+ translocating ATPase located in the granule membrane. In addition to the ΔpH, a transmembrane electrical gradient ($\Delta\psi$), granule interior positive also develops as a result of proton pumping. The extent to which the H^+ ATPase generates either gradient is dependent on the conductance of the organelle membrane to counterions, which permit charge compensation. In fact, the fractional contribution that ΔpH and $\Delta\psi$ make to the proton gradient can be experimentally manipulated by altering the concentration of permeant ions in suspensions of isolated vesicles (Johnson and Scarpa, 1979; Glickman et al., 1983). Chromaffin granule (Johnson et al., 1979; Pazoles et al., 1980; Njus, 1983) and lysosomal membranes (Dell'Antone, 1979; Schneider, 1981) are sufficiently permeable to Cl^- that they rapidly acidify when treated with ATP in a Cl^--containing buffer. When Cl^- is absent, however, these vesicles fail to generate a ΔpH and instead form a $\Delta\psi$.

In intact cells, a variation has been found in the internal pH of endocytic and secretory vesicles. This has been demonstrated for organelles of the endocytic pathway by fluorescein-labeled macromolecules (Yamashiro and Maxfield, 1984) and for organelles of the secretory pathway by weak base trapping (Anderson et al., 1984). The cause of the heterogeneity in vesicular pH is unknown. Mechanisms accounting for variations in ΔpH may include differing conductance of vesicle membranes to ions, including H^+, or the presence of electrogenic ion translocating ATPases, in addition to the H^+ ATPase (Fuchs et al., 1988). Heterogeneity of organellar pH cannot result from the regulation of the number or activity of the vesicular H^+ ATPase, since a single H^+ ATPase molecule, with a turnover number of 50–100 ions/s, would maximally acidify a granule (radius, 0.2 μM) within 1 min. We propose that the heterogeneity of the granule pH results from physiological regulation of the conductance of individual granules for counterions.

While studying the characteristics of proton transport in the parafollicular cell secretory granule we recently found that the pH in these granules was not acidic, as in the chromaffin granule, but rather, was similar to that of the cytoplasm. However, when the parafollicular cells were stimulated by secretagogues they were able to acidify the interior of the granules. Secretagogues also increased the granule membrane conductance to Cl^- which would be either expected to collapse the membrane potential allowing granule acidification. In this way, these cells can maximize the granule ΔpH by reducing the granule $\Delta\psi$. Therefore, individual granules containing H^+ ATPases can vary in internal pH by varying the Cl^- conductance (Barasch et al., 1988).

Parafollicular Cell of the Thyroid

The parafollicular cell of the thyroid develops from emigres that migrate to the thyroid from the neural crest (Le Douarin et al., 1974; Polak et al., 1974). As a result of their embryological origin, these cells express characteristics typical of neurons, such as the elaboration of neuritic processes in response to nerve growth factor (Barasch et al., 1987*a*). In addition, parafollicular cells produce the neurotransmitter, 5-hydroxytryptamine (5-HT), which they co-store in granules with the peptide hormone, calcitonin

(Pearse, 1966; Bussolati and Pearse, 1967; Falck and Owman, 1968; Jaim-Etchevery and Zeiher, 1968; Gershon and Nunez, 1973; Nunez and Gershon, 1978*a*; Barasch et al., 1987*b*). The secretion of these hormones is stimulated by elevated extracellular Ca^{2+} (Hirsch and Munson, 1969).

Sheep parafollicular cells were isolated after enzymatic digestion of the thyroid gland, which was followed by treatment with thyroid-stimulating hormone. They were then separated on a thyroglobulin column. Follicular cells become phagocytic and are retained by the column while the parafollicular cells are eluted and purified further on Ficoll density gradients as described previously. Secretory granules from the cells were purified by a multistep cell fractionation procedure as described (Barasch et al., 1987*b*).

Intravesicular pH of the Parafollicular Cell Is Limited by a Low Membrane Conductance

We assayed ATP-driven acidification of parafollicular cell granules by measuring the accumulation of the weak base acridine orange in purified granules. Using a previously characterized technique (Barasch et al., 1987*b*) with two sequential discontinuous

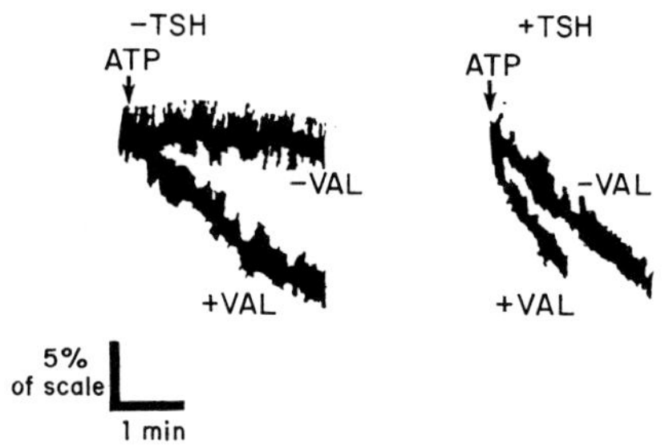

Figure 1. Uptake of acridine orange into parafollicular cell granules isolated from TSH-treated and from unstimulated thyroid cells was measured by fluorometry (490 excitation, 520 emission). A downward deflection reflects trapping of acridine orange, and thus H^+ transport. In the control vesicles there is little ATP-driven uptake of acridine orange; however, pretreatment with valinomycin stimulates the inward movement of acridine orange. In contrast, the addition of ATP to granules isolated from TSH-treated cells causes the rapid uptake of acridine orange that is not further enhanced by treatment with valinomycin. (From Barasch et al., 1988.)

metrizamide gradients, we isolated a granule fraction that was enriched 20-fold over the homogenate in the granule marker 5-HT, but was depleted in mitochondrial and lysosomal markers. In this fraction the granules, identified by electron microscopy, occupy ~63% of the particulate volume (Barasch et al., 1987*b*). When we added ATP (0.33 mM) to granules suspended in a KCl-containing "transport buffer" there was no trapping of acridine orange. However, when we preincubated the granules with valinomycin, a K^+-selective ionophore, the addition of ATP stimulated acridine orange trapping (from 0 to 8 ± 2.5, absorbance units 10^{-3}/min · mg protein; Fig. 1). Proton-conducting ionophores such as nigericin (a H/K exchanger) or CCCP (a H^+ carrier) released the trapped acridine orange, implying that the uptake of acridine orange was due the development of a transmembrane pH difference. Furthermore, *N*-ethylmaleimide (10 μM), which inhibits many nonmitochondrial ATPases, abolished the valinomycin/ATP–induced trapping of acridine orange.

Since the induction of a K^+ conductance in the granule membrane by valinomycin should collapse a membrane potential and thereby stimulate H^+ transport, these data suggest that H^+ transport, and thus acridine orange trapping in isolated granules is limited by a membrane potential generated by electrogenic H^+ translocation. Such a transmembrane potential difference could only arise if the granule membrane had little or no conductance to counterions such as K^+ or Cl^-.

To examine whether granule acidification in intact parafollicular cells is similarly limited we compared the ΔpH of individual granules with that in follicular cells and neutrophil lysosomes by visualizing the subcellular distribution of the weak base DAMP [3-(2,4-dinitroanilino)3′-amino-*N*-methyldipropyl-amine] with an electron microscopic immunogold technique (Anderson et al., 1984; Anderson and Pathak, 1985). When dissociated thyroid cells were incubated with DAMP for 15 min and fixed without pelleting, <1% of parafollicular cell granules showed a concentration of the tracer. In contrast, we found labeling of almost all lysosomes of follicular cells and of a subset of granules in neutrophils. When we incubated thyroid cells for 30 min with higher concentrations of DAMP (by washing the cells free of albumin, to which DAMP binds) we found a variable concentration of DAMP immunoreactivity in granules (Fig. 2 *A*). Follicular cell lysosomes, however, were uniformly and more

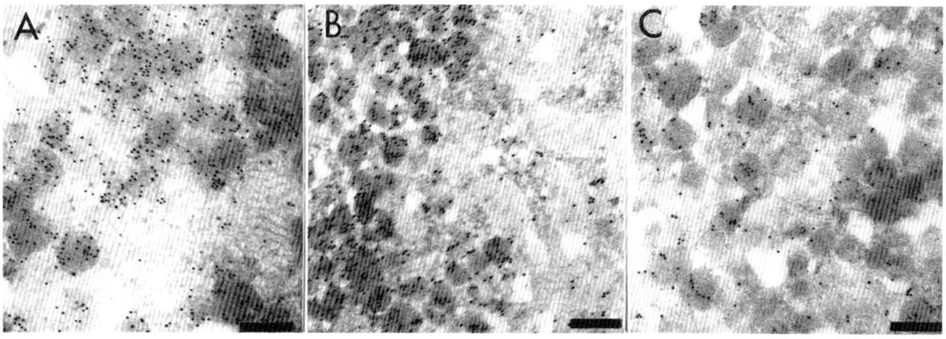

Figure 2. Electron microscopic immunocytochemical demonstration of subcellular sites of accumulation of DAMP in parafollicular cells from the sheep thyroid. Dissociated cells were incubated with DAMP in the absence (*A*) or presence of valinomycin (*B*) or nigericin (*C*). Valinomycin increases and nigericin abolishes labeling by DAMP of parafollicular cell granules. The markers equal 0.5 μm in *A* and *B*, and 0.25 μm in *C*. (From Barasch et al., 1988.)

intensely labeled than parafollicular granules. These labeling patterns suggest that ΔpH varies from granules to granule in parafollicular cells and is generally less than that across lysosomal membranes; however, when we treated thyroid cells with valinomycin (10 μM) for 1 min before fixation of the ΔpH probe, we found an increased percentage of parafollicular cell granules labeled per cell and increased numbers of immunogold particles over granules (Fig. 2 *B*). Valinomycin treatment therefore maximized the ΔpH across the membranes of parafollicular cell granules in intact cells just as it did in isolated granules. This observation suggests that in contrast to lysosomes, H^+ transport by parafollicular cell granules in situ is limited by a membrane potential. The addition of nigericin (7.5 μM) virtually eliminated all DAMP immunoreactivity from the granules (Fig. 2 *C*), indicating that accumulation of DAMP in granules reflects the ΔpH across the granular membrane. The effect of nigericin was apparent after a brief exposure of cells to the ionophore (0.5 min).

Secretagogues Increase the ΔpH of Secretory Granules

We found that TSH is a secretagogue for serotonin in these cells. Increasing extracellular Ca^{2+} is known to increase secretion of calcitonin and 5-HT from these cells. Using fura-2 measurements we also found that TSH increased intracellular Ca^{2+} slightly but significantly from 176 ± 6 to 211 ± 7 nM.

DAMP-loaded thyroid cells were treated with either TSH (30 mU/ml for 30 min), elevated $[Ca^{2+}]_e$ (15 mM for 5 min), or elevated $[K^+]_e$ (56 mM for 5 min). They were then examined by immunoelectron microscopy (as described above). With each of these treatments, the percentage of granules per cell labeled by DAMP increased, as did the number of immunogold particles over granules (Fig 3). In a similar manner, when we incubated isolated, acridine orange–loaded parafollicular

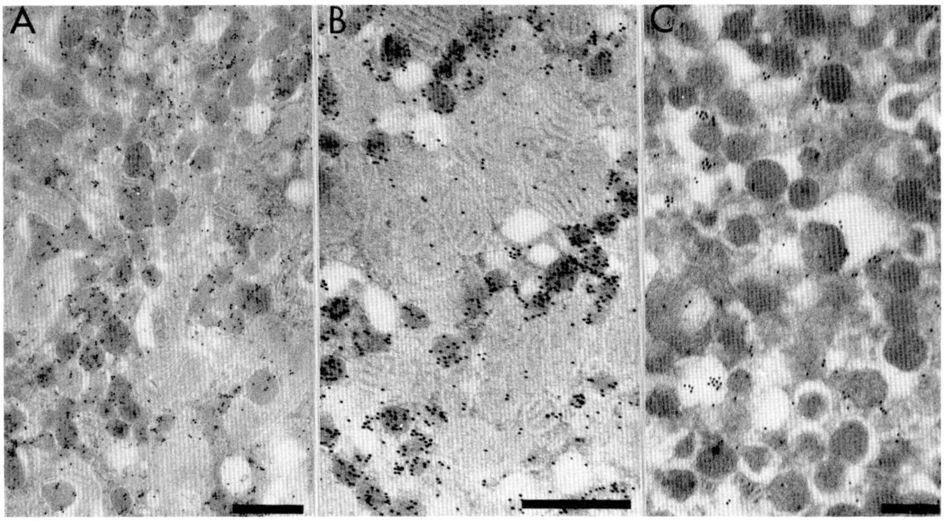

Figure 3. Electron microscopic immunocytochemical demonstration of subcellular sites of accumulation of DAMP in parafollicular cells from the sheep thyroid. Dissociated cells were incubated with DAMP (*A*) in the presence of TSH (30 mU/ml) or (*B*) elevated concentrations of extracellular Ca^{2+}. Both of these secretagogues increase DAMP labeling of parafollicular cell granules over the level (*C*) seen in control cells similarly incubated, but in the absence of TSH and in medium containing a physiological concentration of Ca^{2+} (1.8 mM). The markers equal 0.25 μm. (From Barasch et al., 1988.)

cells with these same agents and analyzed the fluorescence with a FACS (fluorescence-activated cell sorter), we found an increased intensity of fluorescence emission. Elevation of the $[Ca^{2+}]_e$ (>5 mM) was maximally effective when cells were incubated for ~1 min (113 ± 4% of control; $P < 0.005$). Finally, elevation of $[K^+]_e$ (56 mM, 1 min) rapidly increased the red fluorescence, approximately to the same extent as did TSH. The increased labeling of individual granules with DAMP or of populations of cells with acridine orange therefore demonstrates that secretagogues increase the transmembrane pH gradient of granules, as does valinomycin.

An increase in granule membrane ΔpH must result from either acidification of the parafollicular cell granule or alkalinization of its cytoplasm. To ascertain whether secretagogues and valinomycin stimulate alkalinization of the cell cytosol, we loaded purified cells with the pH-sensitive dye BCECF [(2′,7′)-bis(carboxyethyl)-(5,6)-carboxyfluorescein], and measured the pH of cells by excitation ratio fluorometry (490/460-nm excitation, 530-nm emission). The addition of NH_4Cl to BCECF-loaded cells increased the cell pH (fluorescence ratio), while nigericin caused a rapid acidification (probably by exchanging K^+/H^+ at the plasma membrane), which validates the use of BCECF as a probe of intracellular pH in parafollicular cells. Using the BCECF technique, we found that the initial cytoplasmic pH was 6.81 ± 0.26 ($n = 24$ experiments). The addition of TSH caused an immediate acidification by 0.13 ± 0.02 pH units ($P < 0.001$, $n = 13$ experiments) that persisted with further incubation. In contrast to TSH, neither the addition of Ca^{2+} ($n = 5$ experiments), nor valinomycin ($n = 5$ experiments) induced a change in the cytosolic pH. These results indicate that secretagogues, as well as valinomycin, stimulate the trapping of weak bases in granules by inducing granular acidification rather than cytoplasmic alkalinization.

Secretagogues Increase ΔpH by Opening Cl^- Channels

Many intracellular vesicles such as Golgi and clathrin-coated vesicles, whose contents are acidified by a H^+ translocating ATPase, have a Cl^- channel in parallel to the ATPase (Glickman et al., 1983). Removal of Cl^- (or closure of the Cl^- channel) reduces the pH gradient of these vesicles in vitro. Since valinomycin, an electrogenic ionophore, stimulated granule acidification, secretagogoues might similarly change the internal pH of granules by altering the granules' membrane conductance. We investigated the Cl^- conductance of granules purified from dissociated thyroid cells that had been treated with TSH by loading the isolated granules with $^{36}Cl^-$ in a KCl-buffered solution. We then measured tracer efflux from granules at 4°C upon dilution into 63 volumes of the same buffer by collecting aliquots of granules on filters over time after resuspension. We found that granules isolated from cells treated with TSH showed a rapid efflux of $^{36}Cl^-$ (reaching background in 3 min), while the $^{36}Cl^-$ efflux from granules of untreated cells was very slow (Fig. 4). To test whether $^{36}Cl^-$ efflux was due to a $^{36}Cl^-/Cl^-$ exchanger, we resuspended $^{36}Cl^-$ loaded granules in a buffered gluconate salt solution and collected aliquots over time. Again, we found that granules from TSH-stimulated cells showed $^{36}Cl^-$ efflux while control granules did not. This suggests that external Cl^- does not drive $^{36}Cl^-$ efflux and that the tracer diffuses from the granules through a channel rather than by Cl/Cl exchange. These results demonstrate that TSH induces the opening of a Cl^- channel, a modification that is stable during the period of granule isolation.

To investigate whether the increased Cl^- conductance of granules from TSH-treated cells results in increased granule acidification, we measured ATP-dependent proton transport in granules from TSH stimulated cells. Granules were rapidly prepared on the first metrizamide gradient. We found that the addition of ATP to these granules resulted in a rapid uptake of acridine orange. The addition of valinomycin did not enhance the rate of uptake. Granules isolated from untreated cells of the same preparation however, showed little trapping of acridine orange. Unlike the granules from TSH-stimulated cells, valinomycin increased the rate of H^+ transport by the control granules at least threefold (Fig. 1, right panel). The acidification of granules from TSH treated cells was due to the opening of a Cl^- channel since these granules failed to acidify in Cl^--free gluconate media (not shown). Granules from

TSH-treated cells, unlike granules from unstimulated cells, can therefore maximally acidify their contents in the absence of an electrogenic ionophore. This result implies that the granule membrane conductance no longer limits H^+ transport. TSH-induced Cl^- channel opening (as demonstrated above) would collapse a membrane potential and enhance granule acidification.

Secretagogues have been found to induce acidification of other organelles. For example, glucose acidifies β cell granules of pancreatic islets (Pace and Sachs, 1982). Moreover, intracellular canaliculi of parietal cells acidify in response to histamine (DiBona et al., 1979). In the latter system, Cuppoletti and Sachs found that microsomes from histamine-treated glands had increased membrane conductance for K^+ and Cl^- (Cuppoletti and Sachs, 1984). We demonstrated that TSH induces granule acidification by opening a vesicular Cl^- channel. Opening of a Cl^- channel would allow charge compensation for electrogenic H^+ pumping, stimulating the H^+ ATPase and inducing granule acidification. Furthermore, we found that substitution of K^+ gluconate for KCl reduced the rate of acidification of granules isolated from TSH-treated cells, suggesting that Cl^-, but not K^+, acts as the primary counterion.

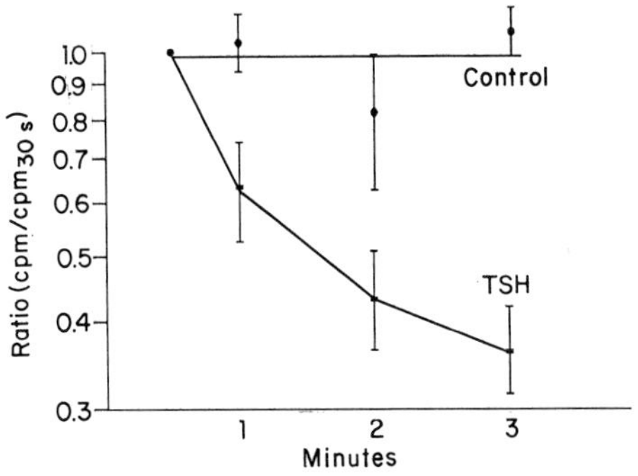

Figure 4. The effect of TSH on the granule membrane conductance for Cl^-, measured by ^{36}Cl-efflux from purified granules. Dissociated thyroid cells were treated with TSH and granules isolated by density gradient centrifugation. Isolated granules were loaded with $^{36}Cl^-$, then diluted and rapidly collected on filters. Data are expressed as a fraction of the initial content of $^{36}Cl^-$. Granules isolated from TSH-treated cells show the rapid efflux of the tracer, while control granules show little loss of the tracer over time. (From Barasch et al., 1988.)

Ion channels can be opened by the binding of specific ligands such as GABA (γ-aminobutyric acid) or glycine. Other channels can be regulated by the reversible binding of Ca^{2+}, as in the Ca^{2+}-activated K^+ channel. Alternatively, channels may be covalently modified, for example, by phosphorylation-dephosphorylation reactions. TSH increased the $[Ca^{2+}]_i$ of parafollicular cells; however, since increased membrane conductance for Cl^- was present in the isolated granules hours after the cells were stimulated by TSH, modification of the channel must have been covalent. TSH is known to increase the cyclic AMP concentration in follicular cells (Bastomysky and McKenzie, 1967) and thus a cyclic AMP–dependent kinase may be involved in the opening of the vesicular Cl^- channel. In fact, Schoumacher et al. (1987) has demonstrated that epithelial Cl^- channels are opened by cyclic AMP–dependent protein kinases. Alternatively, the increased intracellular Ca^{2+} that follows stimulation by TSH, may trigger Ca^{2+}-activated kinases to induce the covalent modification. Although we have not examined the Cl^- conductance of granules isolated from

KCl-depolarized cells or from cells stimulated by elevated extracellular Ca^{2+}, a similar mechanism might also occur in these cells.

Potential Role of Vesicular Cl^- Channels in Cell Function

Variation in Cl^- conductance can also explain the heterogeneity of organellar pH. Endosomes are less acidic than lysosomes (Murphy et al., 1984; Yamashiro et al., 1984) and differences in pH distinguish at least two type of endocytic vesicles involved in the recycling of receptors (Yamashiro et al., 1984; Yamashiro and Maxfield, 1984). Moreover, individual members of a single class of organelle can vary in pH, as occurs with maturing β cell granules (Orci et al., 1987). Vesicles that generate a large ΔpH, such as lysosomes (Dell'Antone, 1979; Schneider, 1981; Ohkuma et al., 1982) and chromaffin granules (Johnson et al., 1979; Njus, 1983), are highly permeable to Cl^-. In contrast, Golgi vesicles, which maintain a small transmembrane pH gradient, must conduct Cl^- poorly, since valinomycin increases proton pumping in these vesicles even in the presence of 150 mM Cl^- (Glickman et al., 1983). This suggests that Golgi and lysosomes maintain their Cl^- conductances at different levels. Variation in conductance may be regulated by membrane-associated kinases and phosphatases that can produce stable modification in Cl^- channels. Indeed, recent studies by Landry et al. (1987) demonstrated that the Cl^- conductance of vesicles enriched in Golgi membranes was reduced by treatment with ATP or ATP-γS, but not by nonhydrolyzable ATP analogues. This suggests that these vesicles contain a membrane-associated kinase that can close the channel.

Secretagogue-induced acidification may have multiple effects on storage granules. Acidification may regulate the proteolytic processing of procalcitonin to calcitonin. Acidification may also increase the uptake of 5-HT by granules from the cytosol by amine/H^+ exchange. In fact, granules appear to vary in 5-HT content (Barasch et al., 1987*b*) and this may reflect the heterogeneity of granule acidification.

Secretagogue-induced acidification may prepare granules for release of 5-HT. Like synaptic vesicles of serotonergic neurons (Tamir and Gershon, 1979), parafollicular cell granules contain a 5-HT–binding protein (SBP). SBP may bind many molecules of 5-HT within granules and reduce the osmotic pressure that would otherwise be generated by free amine; however, the intragranular stoichiometry is not yet known.

Purification of the Cl^- Channel

To study the mechanisms by which Cl^- channels are regulated it is necessary to be able to characterize the proteins responsible for this activity. Cl^- channels are present probably in every type of cell. They are located in plasma membranes as well as in intracellular vesicles. Their electrophysiological characteristics have been described using single-channel methods in a variety of cell types. In general, investigators have found three classes of anion channels. One has a large conductance in the range of 250–400 pS, which exhibits several substates. Another has conductances in the 10–80 pS range, some of these channels show outward rectification and some of them are regulated by the cyclic AMP–dependent protein kinase. A third class has been recently described that has a conductance of 0.5–1 pS and, at least in one instance, is also regulated by cyclic AMP. Whether these represent different molecular entities, a family of proteins, or minor variants of the same protein remains to be established.

To purify the Cl^- channel we used a strategy that depends on the identification of high affinity ligands. Initially one has to develop a simple but specific assay for the channel and identify a rich source of channel material, preferably from a solid organ. It is very likely that this channel is a rare protein in the cell. We used the kidney as our

Structure	Compound	^{36}Cl Transport Inhibition IC_{50}, μM	^3H-94 Binding Inhibition IC_{50}, μM
Indanyloxyacetic Acids			
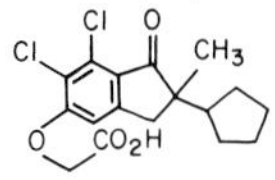	94(+) 95(−)	1.5 2	2 8
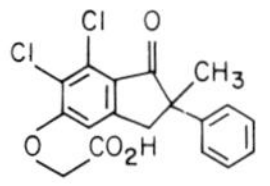	91(+) 92(−)	3 2	30 4
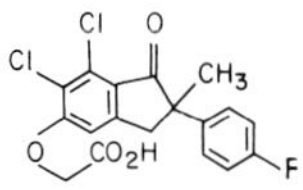	74(+) 75(−)	2 2	12 8
Ethacrynic Acid Analogues			
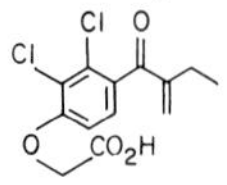	EA	20	45
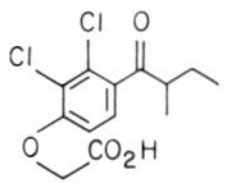	11	35	60
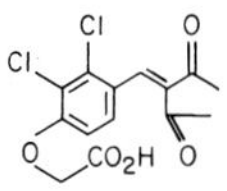	76	55	120
Other Inhibitors			
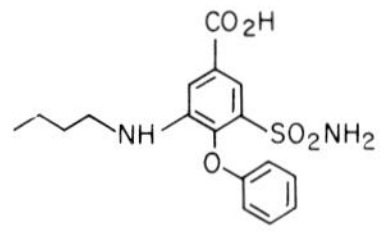	Bumetanide	30	200
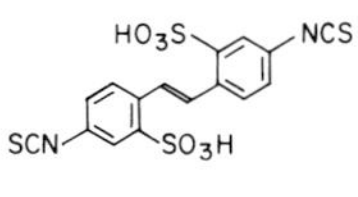	DIDS	10	120
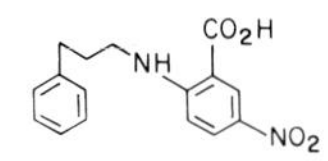	130B	5	5

Figure 5. Structure and potency of various inhibitors. (From Landry et al., 1987.)

source, and prepared light microsomal vesicles that were enriched in intracellular Golgi markers as well as plasma membrane markers such as the Na,K,ATPase. Using an assay based on the method devised by Garty et al. (1983) we demonstrated that these membranes contain a conductive Cl^- uptake. We screened a large number of drugs to identify potential structures that could be useful starting compounds.

Derivatives of ethacrynic acid and indanyl oxyacetic acids (IAA) were potent inhibitors, especially the compound termed IAA-94 (Fig. 5). This compound inhibited Cl^- uptake with a K_i of ~1–2 μM. We tritiated this drug and found that it bound to these vesicles with an K_d of 1 μM. Detailed correlation between inhibitory potency of a variety of drugs and their ability to displace [^{3}H]IAA-94 showed that the two functions correlated well, which suggests that the ligand indeed binds to the channel (Fig. 5). These results are described in detail in Landry et al., 1987.

Using this ligand we succeeded in solubilizing the binding activity with *N*-octyl glucoside. To demonstrate that the channel was indeed solubilized we were able to reconstitute the transport activity into liposomes. We now synthesized an affinity matrix based on this structure and were able to deplete the solubilized membranes of [^{3}H]IAA-94 binding activity. Elution of the column with an excess of IAA-94 released four proteins with molecular masses of 97, 64, 40, and 27 kD. We were able to incorporate these proteins into phospholipid vesicles and demonstrate transport of Cl^- in response to a membrane potential. Furthermore, these liposomes were incorporated into planar lipid bilayers and we found anion channels belonging to two of the classes mentioned above. A large conductance channel with several substates was seen as well as smaller channels with conductances of 20 and 60 pS. These results are described in Landry et al., 1989.

Conclusion

The pH of intracellular organelles is determined by a proton translocating ATPase acting in parallel with a Cl^- channel. Because the ATPase is electrogenic, it can generate an electrochemical potential ($\Delta\mu_H$) composed of a ΔpH and a $\Delta\psi$. The partition of the total $\Delta\mu_H$ among the two components depends on the electrical conductance of the membrane. We have shown that a Cl^- channel present in a secretory granule (of the thyroid parafollicular cell) can serve as this conductance. More significantly, we found that this conductance is regulated and hence, can control the pH of the granule. Because the Cl^- conductance is present in other intracellular vesicles such as endosomes, Golgi, and lysosomes, it is likely that its regulation can play a determining role in the control of the pH of these organelles. This could play a significant role in the traffic and sorting of membrane and secreted proteins.

References

Anderson, R. G. W., J. R. Falck, J. L. Goldstein, and M. S. Brown. 1984. Visualization of acid organelles in intact cells by electron microscopy. *Proceedings of the National Academy of Sciences*. 81:4838–4842.

Anderson, R. G. W., and R. K. Pathak. 1985. Vesicles and cisternae in the *trans* Golgi apparatus of human fibroblasts are acidic compartments. *Cell*. 40:635–643.

Barasch, J., M. D. Gershon, E. A. Nunez, H. Tamir, and Q. Al-Awqati. 1988. TSH induces the acidification of the secretory granules of parafollicular cells by increasing the choloride conductance of the granular membrane. *Journal of Cell Biology*. 107:2137–2147.

Barasch, J. M., H. Mackey, H. Tamir, E. A. Nunez, and M. D. Gershon. 1987*a*. Induction of a neural phenotype in a serotonergic endocrine cell derived from the neural crest. *Journal of Neuroscience*. 7:2874–2883.

Barasch, J. M., H. Tamir, E. A. Nunez, and M. D. Gershon. 1987*b*. Serotonin-storing secretory granules from thyroid parafollicular cells. *Journal of Neuroscience*. 7:4017–4033.

Bastomsky, C., and J. McKenzie. 1967. Cyclic AMP dependence of H^+ secretion. *Journal of Cell Biology*. 85:392–401.

Brown, M. S., R. G. W. Anderson, and J. L. Goldstein. 1983. Recycling receptors: the round-trip itinerary of migrant membrane proteins. *Cell*. 32:663–667.

Bussolati, G., and A. G. Pearse. 1967. Immunofluorescent localization of calcitonin in the c-cells of pig and dog thyroids. *Journal of Endocrinology*. 37:205–209.

Carty, S., R. G. Johnson, and A. Scarpa. 1980. The isolation of intact adrenal chromaffin granules using isotonic percoll density gradients. *Analytical Biochemistry*. 106:438–445.

Cuppoletti, J., and G. Sachs. 1984. Regulation of gastric acid secretion via modulation of a chloride conductance. *Journal of Biological Chemistry*. 259:14952–14959.

Dell'Antonne, P. 1979. Evidence for an ATP-driven "proton pump" in rat liver lysosomes by basic dye uptake. *Biochemical and Biophysical Research Communications*. 86:180–189.

DiBona, D. R., S. Ito, T. Berglindh, and G. Sachs. 1979. Cellular site of gastric acid secretion. *Proceedings of the National Academy of Sciences*. 76:6689–6692.

Falck, B., and C. Owman. 1968. 5-Hydroxytryptamine and related amines in endocrine cells systems. *Advances in Pharmacology*. 6A:211–231.

Fuchs, R., S. Schmid, and I. Mellman. 1988. A possible role for Na^+, K^+-ATPase in regulating ATP-dependent endosome acidification. *Proceedings of the National Academy of Sciences*. In press.

Garty, H., B. Rudy, and S. J. D. Karlish. 1983. A simple and sensitive procedure of measuring isotope fluxes through ion-specific channels in heterogeneous populations of membrane vesicles. *Journal of Biological Chemistry*. 258:13094–13099.

Gershon, M. D., and E. A. Nunez. 1973. Subcellular storage organelles for 5-hydroxytryptamine in parafollicular cells of the thyroid gland. *Journal of Cell Biology*. 56:676–689.

Glickman, J., K. Croen, S. Kelly, and Q. Al-Awqati. 1983. Golgi membranes contain an electrogenic H^+ pump in parallel to a chloride conductance. *Journal of Cell Biology*. 97:1303–1308.

Hirsch, P. R., and P. L. Munson, 1969. Thyrocalcitonin. *Physiological Reviews*. 49:548–622.

Jaim-Etcheverry, G., and L. M. Zeiher. 1968. Cytochemical localization of monoamine stores in sheep thyroid gland at the electron microscope level. *Experientia*. 24:593–596.

Johnson, R. G., and A. Scarpa. 1979. Proton motive force and catecholamine transport in isolated chromaffin granules. *Journal of Biological Chemistry*. 254:3750–3760.

Landry, D. W., M. H. Akabas, C. Redhead, A. Edelman, E. J. Cragoe, Jr., and Q. Al-Awqati. 1989. Purification of Cl channels from kidney and trachea. *Science*. In press.

Landry, D. W., M. Reitman, E. J. Cragoe, and Q. Al-Awquati. 1987. Epithelial chloride channel. *Journal of General Physiology*. 90:779–798.

LeDouarin, N. M., J. Fontaine, and C. Le Lievre. 1974. New studies of the neural crest origin of the avian ultimo-branchial granular cells. Interspecific combinations and cytochemical characterization of C cells based on the uptake of biogenic amine precursors. *Histochemistry*. 38:297–305.

Murphy, R. F., S. Powers, and C. R. Cantor. 1984. Endosome pH measured in single cells by dual fluorescence flow cytometry: rapid acidification of insulin to pH 6. *Journal of Cell Biology*. 98:1757–1962.

Njus, D. 1983. The chromaffin vesicle and the energetics of storage organelles. *Journal of the Autonomic Nervous System*. 7:35–40.

Nunez, E. A., and M. D. Gershon. 1978*b*. Formation of apical pseudopods by canine thyroid follicular cells: induction by thyrotropin and 5-hydroxytryptamine, antagonism by reserpine. *Anatomical Record*. 192:215.

Ohkuma, S., Y. Moriyama, and T. Takano. 1982. Identification and characterization of a proton pump on lysosomes by fluorescein isothiocyanate-dextran fluorescence. *Proceedings of the National Academy of Sciences*. 79:2758–2762.

Orci, L., M. Ravazzola, and R. G. W. Anderson. 1987. The condensing vacuole of exocrine cells is more acidic than the mature secretory vesicle. *Nature*. 326:77–79.

Pace, C. S., and G. Sachs. 1982. Glucose-induced proton uptake in secretory granules of β-cells in monolayer culture. *American Journal of Physiology*. 242:382–387.

Pazoles, C. J., C. E. Cruetz, A. Ramu, and H. B. Pollard. 1980. Permeant anion activation of MgATPase activity in chromaffin granules. *Journal of Biological Chemistry*. 255:7863–7869.

Pearse, A. G. E. 1966. Common cytochemical properties of cells producing polypeptide hormones, with particular reference to calcitonin and the thyroid C cells. *Veterinary Record*. 79:587–590.

Polak, J. M., A. G. E. Pearse, C. Le Lievre, J. Fontaine, and N. M. Le Douarin. 1974. Immunocytochemical confirmation of the neural crest origin of avian calcitonin-producing cells. *Histochemistry*. 40:209–214.

Schneider, D. L. 1981. ATP-dependent acidification of intact and disrupted lysosomes. *Journal of Biological Chemistry*. 256:3858–3864.

Schoumacher, R. A., R. L. Shoemaker, D. Halm, E. A. Tallant, R. W. Wallace, and R. A. Frizzell. 1987. Phosphorylation fails to activate chloride channels from cystic fibrosis airway cells. *Nature*. 330:752.

Tamir, H., and M. D. Gershon. 1979. Storage of serotonin and serotonin binding protein in synaptic vesicles. *Journal of Neurochemistry*. 33:35–44.

Tamir, H., and M. D. Gershon. 1981. Intracellular proteins that bind serotonin in neurons, pananeurons and platelets. *Journal of Physiology (Paris.)*. 77:283–286.

Yamashiro, D. J., and R. F. Maxfield. Acidification of endocytic compartments and the intracellular pathways of ligands and receptors. *Journal of Cellular Biochemistry*. 26:231–246.

Yamashiro, D. J., B. Tycko, S. R. Fluss, and F. R. Maxfield. 1984. Segregation of transferrin to a mildly acidic (pH 6.5) para-Golgi compartment in the recycling pathway. *Cell*. 37:789–800.

Chapter 19

Stimulus-Secretion Coupling in Mast Cells

Reinhold Penner and Erwin Neher

Max-Planck-Institut für Biophysikalische Chemie, Am Fassberg, D-3400 Göttingen, Federal Republic of Germany

Introduction

Secretion of vesicular contents via exocytosis is a cellular function common to cells as diverse as neurons, endocrine or exocrine gland cells, and cells of the immune system (Kelly, 1985). An elevation of intracellular calcium has been assigned the major controlling step in secretion (Douglas, 1968; Katz, 1969), although accumulating evidence indicates that the secretory event may be subject to complex regulation by a number of factors. Apart from the "traditional" signalling pathways (adenylate cyclase → cyclic AMP → protein kinase A, phospholipase A_2 → arachidonic acid → lipoxygenase/cyclooxygenase), the recently discovered second messenger system commonly referred to as the dual signal pathway (phospholipase C → inositol trisphosphate/diacylglycerol → protein kinase C) has particularly attracted much attention (Berridge and Irvine, 1984). It appears to be involved in a variety of cellular functions including exocytosis.

Mast cells are involved in early allergic and inflammatory reactions in that they release histamine and other chemical mediators (Ishizaka and Ishizaka, 1984). The secretory response of mast cells is one of the most fascinating "observable" biological processes since it is so vigorous that ~1,000 secretory granules stored in a single cell are exocytosed within a few seconds, resulting in complete degranulation of the cell (see Fig. 1 *A*). Because of this and because mast cells can be obtained in quantities large enough to carry out biochemical studies, they have become a model system for the study of secretion (reviewed by Ishizaka and Ishizaka, 1984; Metzger et al., 1986; Gomperts et al., 1988).

Physiologically, the secretory response can be induced by oligomerization of specific IgE receptors by the corresponding antigen (Metzger et al., 1986). Also, substance P is considered as a physiological stimulus (Foreman and Jordan, 1983), since mast cells are often clustered around substance P-containing nerve endings (Skofitsch et al., 1985) and the peptide is known as a potent mast cell secretagogue (Johnson and Erdös, 1973). In addition, mast cell degranulation can also be elicited by a number of substances including compound 48/80, polymyxin B, somatostatin, and mast cell–degranulating peptide (Lagunoff et al., 1983). The mechanisms by which these secretagogues bring about secretion are not understood. It is known, however, that a number of cellular events take place after stimulation of mast cells including (*a*) changes in phospholipid metabolism (Kennerly et al., 1979), (*b*) transient elevation of intracellular calcium (Beaven et al., 1984; White et al., 1984), and (*c*) rise of cyclic nucleotides (Winslow and Austen, 1984). It is not clear which of these stimulation pathways constitutes the trigger for secretion and whether they can be activated independently or, in case of acting synergistically, to what extent they contribute to the control of the exocytotic process.

Much progress in the understanding of secretion and the intracellular signalling pathways involved, has arisen from techniques that made it possible to effectively control the composition of the intracellular milieu (reviewed by Gomperts and Fernandez, 1985; Knight and Scrutton, 1986) and to monitor changes of intracellular calcium concentration ($[Ca]_i$) by fluorescent indicator dyes (reviewed by Tsien et al., 1984). We will discuss some of the major cellular events presently known to follow receptor-mediated stimulation of mast cells, primarily focussing on recent patch-clamp studies combined with fura-2 measurements obtained in our laboratory. We will emphasize the regulation of $[Ca]_i$, its role in secretion, and the interactive control of exocytosis by various second messengers.

Spotting the Secretory Event Measuring Membrane Capacitance, Membrane Currents, and Intracellular Calcium

The whole-cell configuration of the patch-clamp technique has proven to be a very powerful tool in the study of cellular physiology. The technique offers the possibility to control the intracellular milieu through the micropipette, which may be used to introduce second messengers or indicator dyes into the cytosol. It also provides the means to effectively control membrane voltage and to record transmembrane currents. Furthermore, changes in cell surface area resulting from incorporation of vesicle membranes into the plasma membrane during degranulation can be monitored by measuring membrane capacitance. Since all of these parameters can be recorded simultaneously in a single cell, one can obtain information about temporal and causal correlation of cellular events.

Fig. 1 illustrates the above-mentioned advantages of the patch-clamp technique. Traces *B–E* were obtained from a typical experiment carried out on a mast cell. After

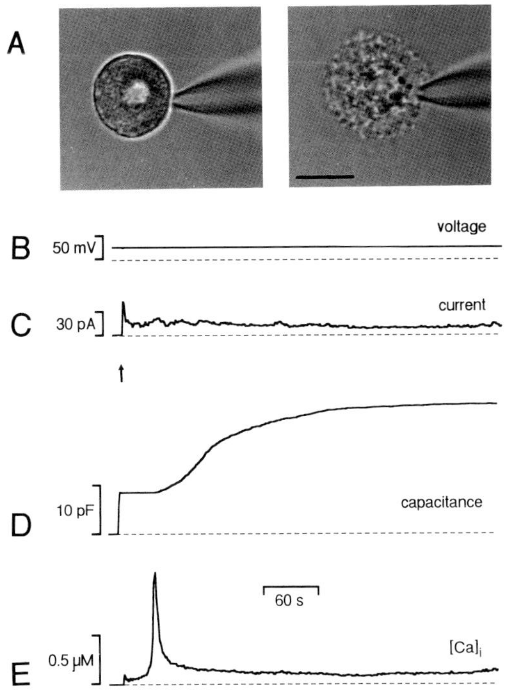

Figure 1. A typical patch-clamp experiment performed on a mast cell. (*A*) Morphological appearance before and after degranulation induced by intracellular perfusion with GTP-γ-S (calibration bar, 15 μm). (*B–E*) Some of the parameters being measured simultaneously on a single mast cell. Establishment of the whole-cell configuration is indicated by the arrow. The intracellular solution contained 100 μM GTP-γ-S (see text for details).

sealing the pipette to the surface of the cell (see Fig. 1 *A*), the membrane patch encircled by the rim of the tip is ruptured (indicated by arrow). Instantaneously, the membrane voltage of the whole cell is clamped (Fig. 1 *B*) and the current required for voltage control is measured (Fig. 1 *C*). At the moment of patch rupture, the membrane capacitance of the cell (~8 pF) becomes apparent and changes of this parameter can be measured continuously (Fig. 1 *D*). Subsequent to the establishment of the whole-cell configuration, the interior of the cell is dialyzed as the pipette's filling solution diffuses into the cytosol. In this experiment the standard pipette solution (mainly K-glutamate) was supplemented with GTP-γ-S and the Ca indicator dye fura-2 (Grynkiewicz et al.,

1985). The fluorescence ratio of fura-2 at two different excitation wavelengths is used to calculate $[Ca]_i$ (for details of fluorescence measurements see Neher, 1989). GTP-γ-S is a potent mast cell secretagogue (Fernandez et al., 1984) that acts by an irreversible activation of GTP-binding proteins within the cell. As a result, intracellular calcium undergoes a transient rise (Fig. 1 *E*) due to the release of Ca from intracellular stores (see Neher and Almers, 1986; Neher and Penner 1988). Thereafter, the initial cell membrane capacitance increases by more than threefold (Fig. 1 *D*) due to activation of the secretory process, eventually resulting in complete degranulation of the cell (see also Fig. 1 *A*). When looking at capacitance changes at higher

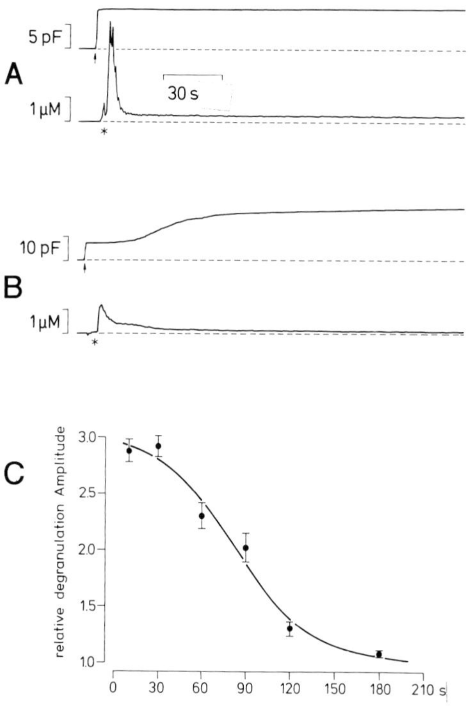

Figure 2. Washout phenomena in dialyzed mast cells. (*A*) Effect of compound 48/80 (5 μg/ml) on intracellular calcium concentration (*lower traces*) and cell membrane capacitance (*upper traces*) in the absence of GTP. Compound 48/80 was applied 5 s after the establishment of the whole-cell configuration. Time of patch rupture (↑) and drug application (∗) are indicated in the graphs. (*B*) Same as *A* except that GTP (300 μM) was present in the pipette filling solution. Shape and amplitude of Ca transients can be quite variable between cells (see Neher and Penner, 1988) (*C*) Reduction of the relative degranulation amplitude as a function of the time-lagged application of compound 48/80 in the presence of GTP (300 μM). Data points correspond to the ratio of the cell capacitance after completion of the secretory response and the initial capacitance value. Degranulation was considered terminated when capacitance values remained steady for 30–120 s. Each data point reflects mean values from 6–14 determinations ± SEM. Modified from Penner et al., 1987.

resolution, one can detect small stepwise increases in capacitance reflecting fusion events of single granules (Fernandez et al., 1984).

An apparent disadvantage pertinent to the whole-cell configuration comes from the effective dialysis of the cell against the intrapipette solution (Pusch and Neher, 1988). This may result in washout of cytosolic components that may be important for cellular signalling. In fact, while secretion can easily be induced by internally administered GTP-γ-S, mast cells rapidly lose their secretory response to external stimuli, such as compound 48/80, or antigen when perfusing with standard pipette solutions (Fig. 2 *A*). On the other hand, this washout also offers the experimental possibility to identify signal mediators by reconstituting the conditions required for cellular functions. Since secretion can be restored by including the nucleotide GTP in the pipette solution (Fig. 2 *B*), the loss of the exocytotic response in mast cells may be attributed mainly to the fast washout of internal GTP (see Penner et al., 1987), which indicates a major role for G proteins in secretion. The restoration of exocytosis by GTP is only transient and decays within 3 min (Fig. 2 *C*), suggesting diffusional escape of one or several other cytoplasmic constituents involved in stimulus-secretion coupling. Interestingly, Ca transients induced by external stimuli, which are also mediated by a G protein, are not affected by prolonged intracellular dialysis with GTP-free pipette solution.

The Role of Calcium in Stimulus-Secretion Coupling

Secretion in excitable cells, such as neurotransmitter release from nerve terminals (Katz, 1969), insulin release from beta cells (Wollheim and Sharp, 1981), or catecholamine release from chromaffin cells (Douglas, 1968) has been found to be

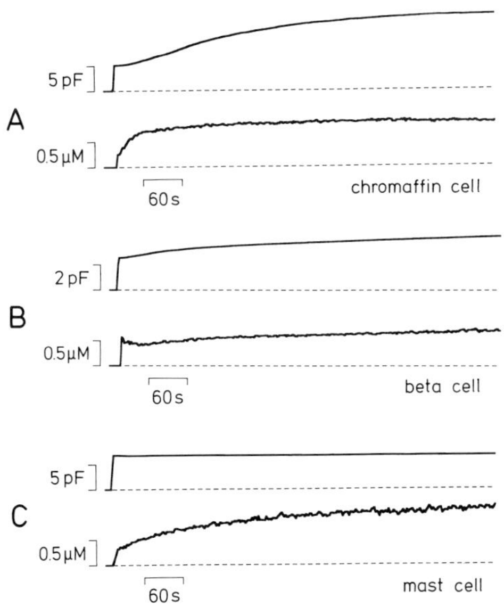

Figure 3. Secretory responses of different cell types stimulated by an injection of micromolar intracellular calcium concentrations. (*A*) Chromaffin cells: the cell was dialyzed with standard intracellular solution which also contained GTP (300 μM) and a combination of Ca-EGTA/EGTA at a ratio of 7:1 mM. (*B*) Pancreatic beta cells: conditions were the same as in *A* except that the EGTA buffer ratio in the internal solution was 6:1 mM. (*C*) Mast cells: conditions were the same as in *B*. Taken from Penner and Neher, 1988*b*.

tightly correlated with changes of intracellular levels of calcium. This view is confirmed by patch-clamp studies in which chromaffin cells and beta cells were perfused with intracellular solutions containing calcium buffered to defined levels (Penner and Neher, 1988*b*). As evidenced by Figs. 3, *A* and *B*, capacitance increases, indicative of secretion can be induced by $[Ca]_i$ clamped to ~ 1 μM (see also Neher and Marty, 1982; Penner et al., 1986). In mast cells, however, a similar rise in $[Ca]_i$ does not seem to be a sufficient stimulus for secretion (Fig. 3 *C*).

Unphysiologically high calcium can, however, induce secretion in mast cells (Penner and Neher, 1988*a*). This is illustrated in Fig. 4 *A*, where the calcium ionophore ionomycin was used to increase $[Ca]_i$ to different levels (by varying the application time) and measuring the resulting changes in capacitance. Calcium ionophores have long been known to induce secretion in mast cells (Cochrane and

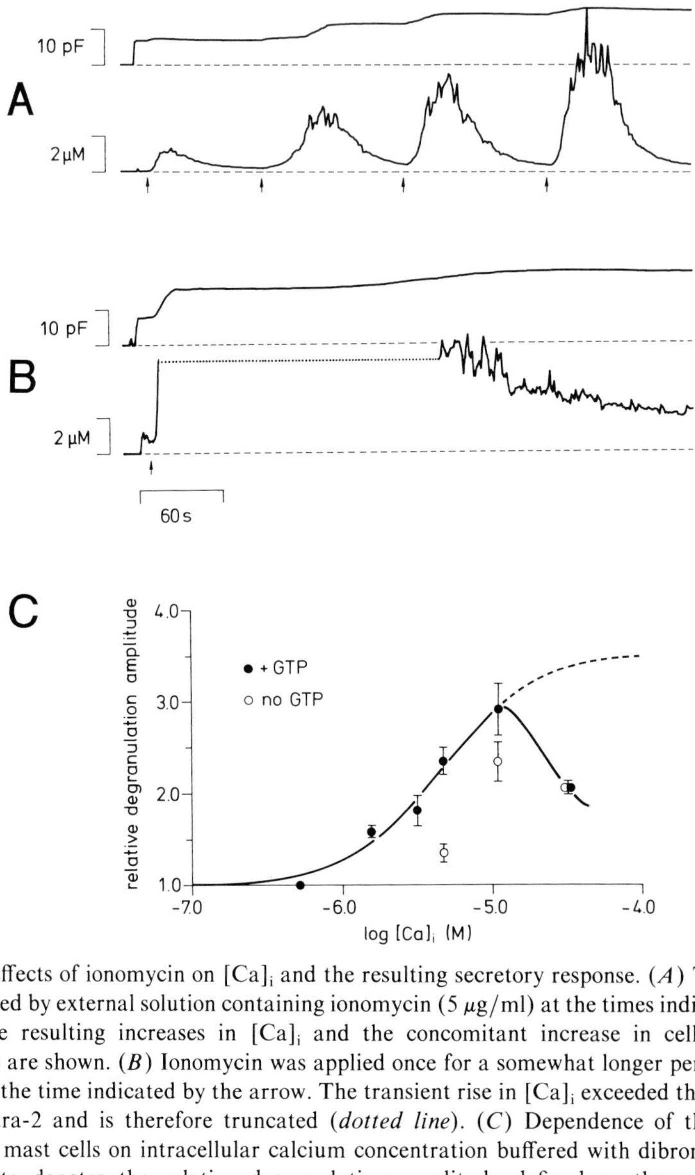

Figure 4. Effects of ionomycin on $[Ca]_i$ and the resulting secretory response. (*A*) The cell was briefly flushed by external solution containing ionomycin (5 μg/ml) at the times indicated by the arrows. The resulting increases in $[Ca]_i$ and the concomitant increase in cell membrane capacitance are shown. (*B*) Ionomycin was applied once for a somewhat longer period (several seconds) at the time indicated by the arrow. The transient rise in $[Ca]_i$ exceeded the measuring range of fura-2 and is therefore truncated (*dotted line*). (*C*) Dependence of the secretory response of mast cells on intracellular calcium concentration buffered with dibromo-BAPTA. The ordinate denotes the relative degranulation amplitude defined as the ratio of final capacitance to initial capacitance. The abscissa gives the $[Ca]_i$ as calculated from the ratio of Ca-free to Ca-bound dibromo-BAPTA. Data points reflect mean values ± SEM from 3–21 determinations. Experiments were conducted under identical conditions in the presence (●) and absence (○) of GTP (300 μM). The concentration-response curve was fitted to the data excluding values obtained in the absence of GTP, and $[Ca]_i$ of 32 μM, where secretion was inhibited. While the theoretical course of the fit is indicated by the dashed line, the data points for $[Ca]_i$ of 11 and 32 μM were connected by eye. Modified from Penner and Neher, 1988*a*.

Douglas, 1974; Bennett et al., 1979). On the single cell level it is seen (Fig. 4 *A*) that the first brief application of ionomycin increased $[Ca]_i$ to ~1.5 μM without causing significant secretion. Slightly longer application times of ionomycin resulted in transient rises of $[Ca]_i$ well above 3 μM that were paralleled by capacitance increases and ceased as $[Ca]_i$ returned to resting levels. In other experiments, where ionomycin was applied for several seconds, $[Ca]_i$ exceeded the measuring range of fura-2 (Fig. 4 *B*). In these cases, secretion proceeded as $[Ca]_i$ remained below ~10 μM but ceased as it rose to unmeasurable values, suggesting that high $[Ca]_i$ is inhibitory to secretion. Secretion resumed as $[Ca]_i$ returned to resting levels until it dropped below 1 μM. These findings were confirmed by quantitative experiments in which $[Ca]_i$ was buffered with dibromo-BAPTA to various levels between 0.5 and 32 μM (Penner and Neher, 1988*a*). The resulting concentration-response relationship between $[Ca]_i$ and secretory responses is illustrated in Fig. 4 *C*.

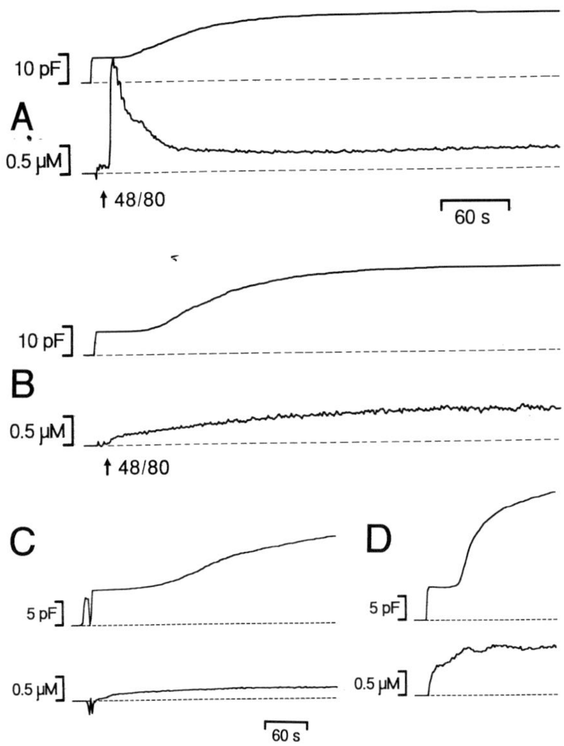

Figure 5. Secretory responses to compound 48/80 and GTP-γ-S stimulation under various intracellular buffering conditions as measured by the cell capacitance. (*A*) A standard pipette filling solution containing GTP (300 μM) was used, essentially corresponding to an "unbuffered" condition (5 μg/ml of compound 48/80 was applied at the time indicated). (*B*) The internal solution was strongly calcium buffered to slightly above basal levels by the addition of a Ca-EGTA/EGTA mixture of 5/5 mM. The cell was stimulated as in *A*. (*C*) $[Ca]_i$ was buffered as in *B* and stimulated by internal GTP-γ-S (40 μM). (*D*) $[Ca]_i$ was buffered to ~1 μM and stimulated by GTP-γ-S as in *C*. *A* and *B* taken from Penner and Neher, 1988*a*; *C* and *D* taken from Neher, 1988.

From the findings discussed above, a rise in $[Ca]_i$ to physiological levels is not sufficient by itself for inducing secretion in mast cells. Nevertheless, stimulation of mast cells by a variety of secretagogues does induce transient rises in intracellular calcium (Fig. 5 *A*). This raises the question whether Ca transients are necessary in some way for secretion. This was tested by buffering $[Ca]_i$ with EGTA close to resting levels and applying the secretagogue compound 48/80 (Fig. 5 *B*). As can be seen, the buffering prevented the secretagogue-induced Ca transient but, nevertheless, a normal degranulation response was induced (see also Neher and Penner, 1988). Thus, it may be concluded that physiologically induced Ca transients are neither sufficient nor necessary for secretion, which raises the question of what purpose intracellular calcium serves in mast cells.

While the extent of the secretory response to potent stimuli such as compound 48/80 or GTP-γ-S appears to be largely independent of the prevailing level of $[Ca]_i$ it is evident that the rate of secretion after secretagogue stimulation is markedly enhanced by high levels of $[Ca]_i$ as compared with low levels (Figs. 5, *C* and *D*). This "Ca sensitivity" of secretion rate develops gradually after the initial Ca transient (see Neher, 1988). The process that renders secretion sensitive to $[Ca]_i$ may be associated with activation of protein kinase C, which is known to reduce the Ca requirement for cellular functions (Nishizuka, 1984). Indeed, activation of this enzyme by phorbol esters has been found to lower the levels of $[Ca]_i$ required to induce secretion (Heiman and Crews, 1985; see Fig. 9 *A*).

Thus, an elevation of $[Ca]_i$ may be quite important for secretion, provided it can act in combination with an additional signal. However, the temporal correlation of Ca transients and secretion is not very good, since normally Ca transients happen too early to enhance secretion (see Figs. 1 and 5 *A*). To be effective in support of secretion, the increase in $[Ca]_i$ has to be more sustained. In a number of cells we have indeed obtained evidence for sustained plateaus of elevated $[Ca]_i$ after the initial transient release of Ca from internal stores (Neher and Penner, 1988; Penner et al., 1988). This plateau phase is barely noticeable in the examples shown so far, where membrane potential was clamped to +17 mV, but it can become quite prominent when holding the membrane potential at more negative values. We will show below that this sustained elevation of $[Ca]_i$ is due to Ca influx regulated by second messengers (Penner et al., 1988).

Second Messengers Regulate Calcium Influx

Calcium influx through voltage-activated Ca channels can be induced by depolarization in electrically excitable cells such as chromaffin cells. The resulting influx of calcium is determined by the number of open channels and the electrochemical driving force. The current-voltage relationship for Ca currents is typically bell-shaped with its maximum close to 0 mV (Fenwick et al., 1982). The changes in $[Ca]_i$ after depolarizing voltage steps are in agreement with this notion (Fig. 6 *A*): small depolarizations cause small changes in $[Ca]_i$ (activation of Ca channels is submaximal and driving force for Ca is high), depolarizations to ~0 mV cause large changes in $[Ca]_i$ (strong activation of Ca channels and moderate driving force), and with even stronger depolarizations changes in $[Ca]_i$ become smaller again (maximal activation of channels but small driving force). Quite a different dependence of $[Ca]_i$ and membrane voltage is observed in mast cells. Voltage changes in the range of ±50 mV normally do not influence $[Ca]_i$ in unstimulated mast cells (Fig. 6 *B*) to any significant extent. However, $[Ca]_i$ does change with voltage steps after agonist stimulation in a way that hyperpolarizing pulses increase $[Ca]_i$ and depolarizing pulses decrease it (Fig. 6 *C*), which is opposite to the case of cells that possess voltage-activated Ca channels. This "potential-dependent" behavior rather suggests the presence of Ca entry pathways across the plasma membrane that are permanently activated during the plateau phase and where Ca influx is mainly determined by the electrochemical driving force for Ca.

We found two mechanisms by which Ca influx can be accomplished in mast cells

(Penner et al., 1988). Agonist stimulation activates nonspecific cation channels (Fig. 7 *A*) with a reversal potential of 0 mV, and a single-channel conductance of ~50 pS in Ringer solution (Fig. 7 *B*) and 16 pS in isotonic Ba^{2+} (Fig. 7 *C*). This channel can be activated by basically all stimuli that induce phosphatidylinositol (PI) breakdown and secretion, indicating that it is second messenger regulated rather than directly agonist gated. High $[Ca]_i$ can block activity of the 50-pS channel. Although this channel resembles the IP_3-gated channel reported in lymphocytes (Kuno and Gardner, 1987), it is not gated by either IP_3 or IP_4, since high levels of IP_3 and IP_4 in the pipette solution do not activate the channel nor do they prevent its activation by agonists. Activation of the 50-pS channel can produce a whole-cell current of 1–15 pA at −50 mV and may

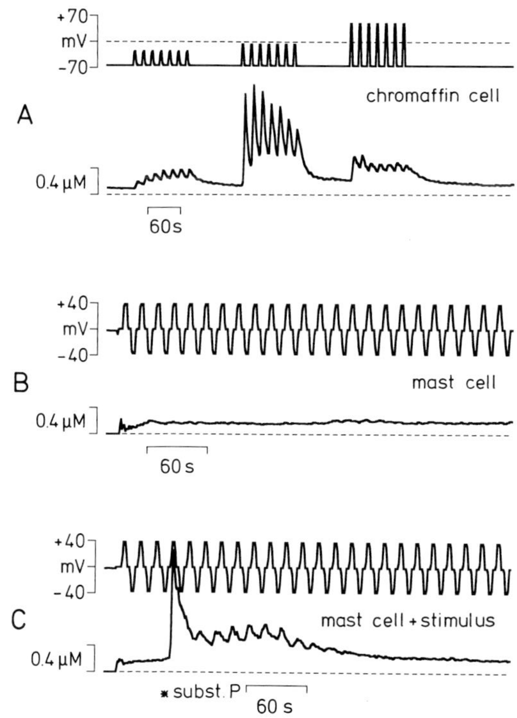

Figure 6. Voltage dependence of $[Ca]_i$ in excitable cells and nonexcitable cells. (*A*) The chromaffin cell was subjected to various depolarizing voltage pulses of 4 s duration as indicated by the voltage trace (see text for details). (*B*) The mast cell was stepped to various depolarized and hyperpolarized voltages as indicated in the voltage trace. No voltage-dependent changes in $[Ca]_i$ are associated with either depolarization or hyperpolarization. (*C*) Same conditions as in *B*, except that the mast cell was stimulated with substance P (50 µg/ml) at the time indicated. After the initial fast Ca-transient due to phospholipid turnover there is a phase in which $[Ca]_i$ becomes voltage dependent. Presumably, second messenger–activated channels are opened and $[Ca]_i$ increases as the driving force for Ca is high during the hyperpolarized episodes. Taken from Penner and Neher, 1988*b*.

contribute to Ca influx. However, in many instances, where the activation of 50-pS channels was small, the "potential-dependent" changes in $[Ca]_i$ were quite prominent, suggesting the presence of an additional pathway for Ca entry.

This alternative pathway appears to be controlled by IP_3, since perfusion of IP_3 into the cell could mimic not only the initial transient rise of $[Ca]_i$ but also the potential-dependent plateau phase of $[Ca]_i$ (Fig. 8 *A*). This was not accompanied by an increase of the activity of 50-pS channels, providing additional evidence for a Ca entry mechanism besides that of 50-pS channels. In favorable experiments, a close inspection of current records revealed a "smooth" inward current closely related to the phase of Ca entry (see the bar graph in Fig. 8). At present, this current is not resolvable into single channels and amounts to not more than 1–2 pA in the whole cell (at −40 mV). This current must therefore be very Ca specific, since nonspecific currents through leaks (which may compare with currents through nonspecific cation channels) have to be in excess of 10–20 pA to produce comparable changes in $[Ca]_i$. We therefore infer that the IP_3-mediated Ca-specific entry mechanism is the main determinant of intracellular calcium during the plateau phase.

Since the two pathways for Ca entry described above are gated by second

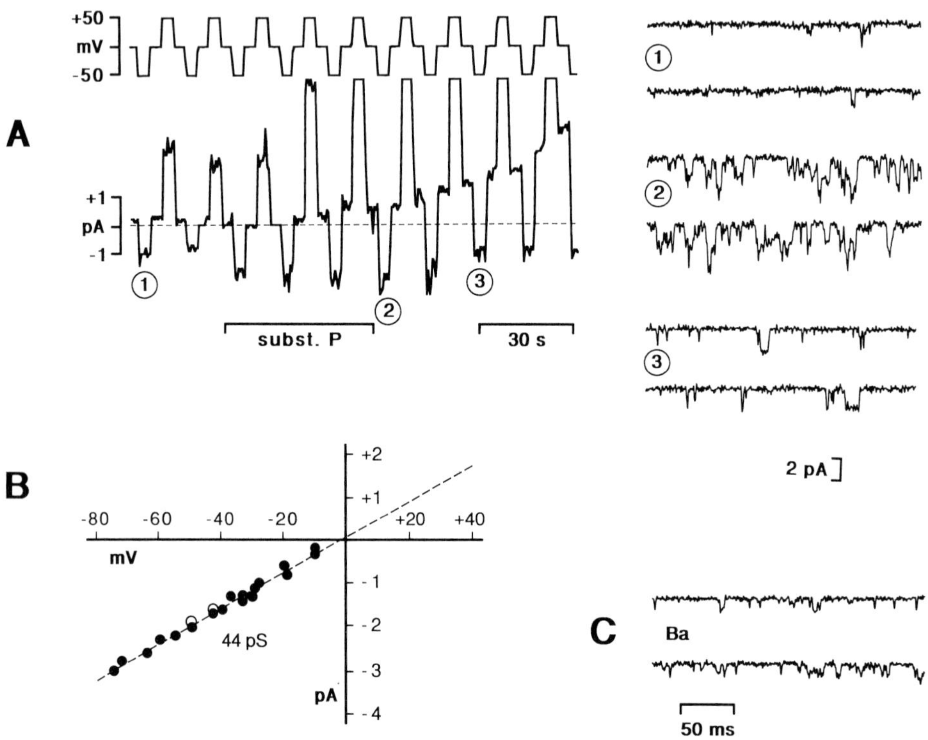

Figure 7. The characteristics of single-channel events after agonist stimulation. (*A*) The cell was stimulated with substance P (50 μg/ml) for the time indicated, and subjected to voltages steps. Note the transient increase in inward current after agonist stimulation. The delayed Cl current, which also activated in this experiment, is truncated at depolarized voltage pulses as it exceeded 5 pA. Representative traces were taken before (*1*), during (*2*), and after (*3*) the inward current change, showing single-channel events detected in the whole-cell mode during hyperpolarizing episodes at −50 mV. (*B*) Current-voltage relationship for agonist-induced single channels from different cells (●, compound 48/80-induced; ○, substance P-induced). (*C*) Single channels as in *A*, but activated by compound 48/80 and recorded at −60 mV in external saline of the following composition (in millimolar): 95 $BaCl_2$, 10 HEPES-NaOH, 11 glucose, pH 7.2. Taken from Penner et al., 1988.

messengers and do not show up as large currents, they have gone unnoticed in previous patch-clamp studies. The predominant conductance change after the application of secretagogues is a large Cl current (Fig. 8 *B*) with a small single-channel conductance of ~1 pS. This current can amount to several 100 pA in a whole cell (at +40 mV). It can also be activated by internally applied cAMP (Fig. 8 *C*) and high $[Ca]_i$ in the absence of agonist stimulation (Penner et al., 1988). The physiological role of this Cl

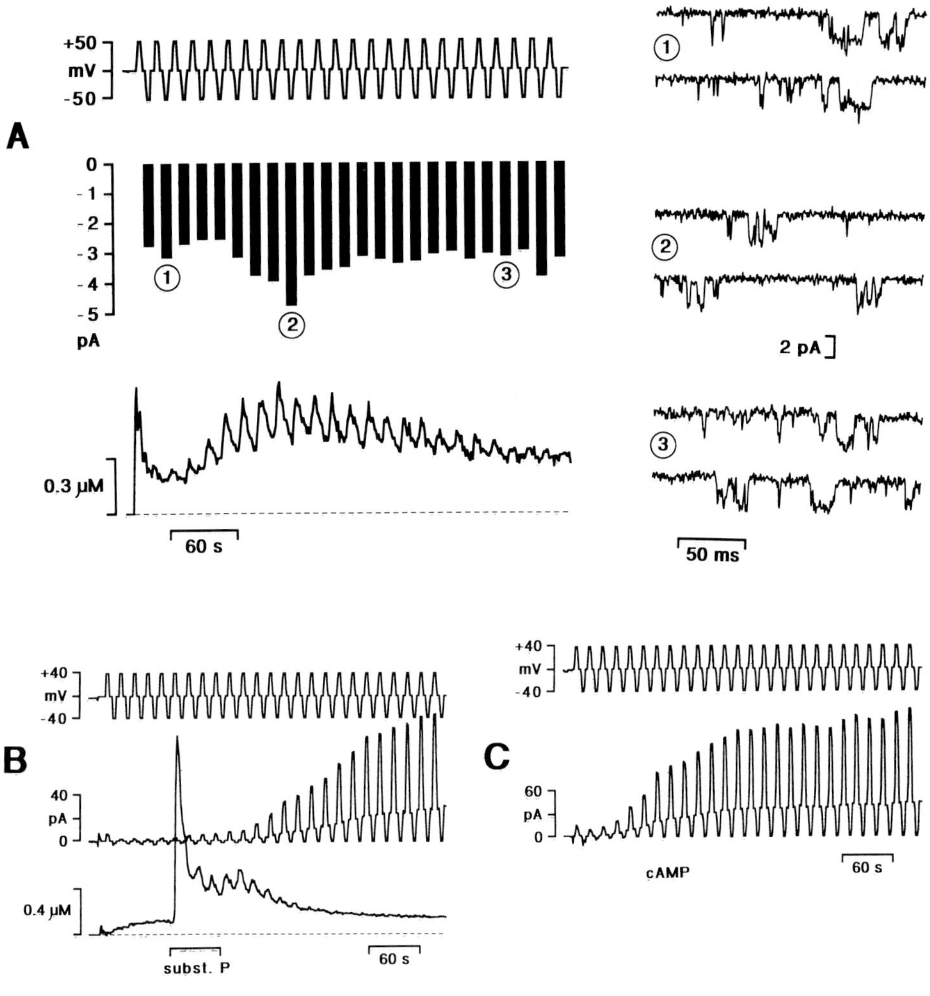

Figure 8. Effects of IP_3 on inward current and single-channel activity. (*A*) The cell was perfused with standard internal solution containing 4 μM IP_3 and voltage clamped to three different potentials. Inward currents for hyperpolarizing voltage pulses are illustrated as a bar graph. Note the temporal correlation between inward current amplitude and the resulting voltage-dependent changes in $[Ca]_i$. In addition, representative traces showing single-channel events before (*1*), during (*2*), and after (*3*) the voltage-dependent plateau phase of $[Ca]_i$ were taken during hyperpolarizing episodes. (*B*) Substance P (50 μg/ml) was applied for the time indicated. The cell was voltage clamped to different potentials (top trace) and the resulting currents (middle trace) and changes in $[Ca]_i$ (lower trace) are shown. (*C*) The cell was perfused with standard internal solution supplemented with 50 μM cAMP and subjected to the same voltage protocol. Modified from Penner et al., 1988.

current may consist of providing the driving force for Ca influx by keeping membrane voltage at negative potentials.

In summary, PI breakdown induces both release of Ca from intracellular stores and sustained Ca influx, both of which appear to be mediated by intracellular IP_3. Since this pathway also generates diacylglycerol (DAG) with concomitant activation of protein kinase C (PKC), an enzyme known to reduce the Ca requirement for cellular functions, the stage appears to be set for the control of secretion via activation of the dual signal pathway.

Multiple Signalling Pathways Control Stimulus-Secretion Coupling

An experiment that gives support to a potential role of the dual signal pathway in secretion is demonstrated in Fig. 9 *A*, where physiological levels of $[Ca]_i$ can indeed

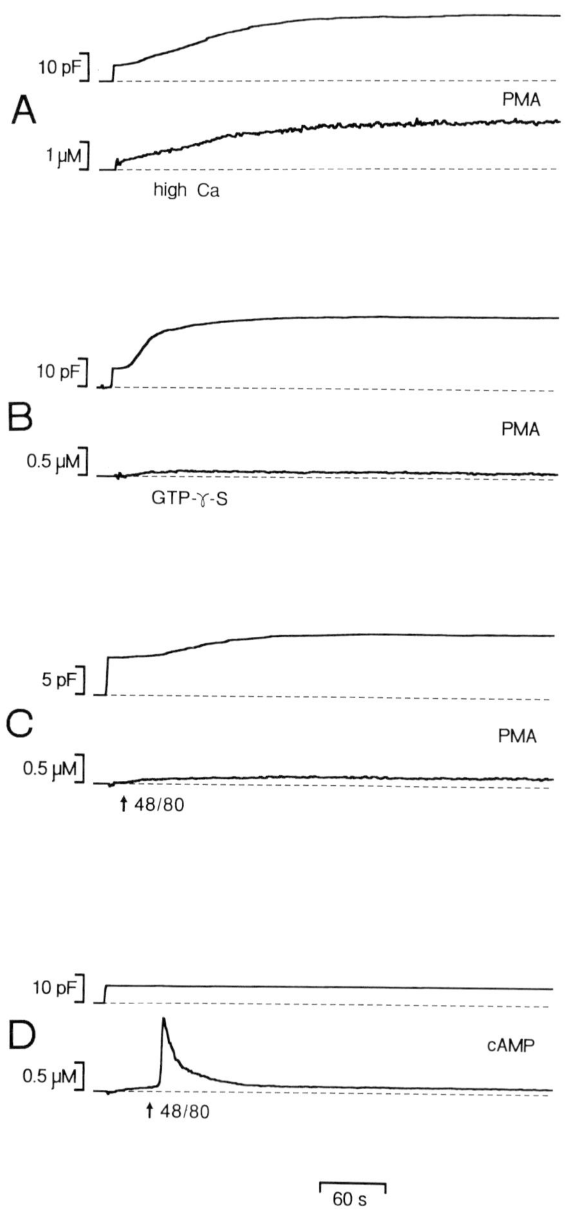

Figure 9. Effects of phorbol ester on secretory responses and $[Ca]_i$. Cells were pretreated with phorbol 12-myristate 13-acetate (PMA) for 30 min at 37°C. (*A*) The cell was perfused with intracellular solution containing Ca-EGTA/EGTA at a ratio of 7:1 mM. (*B*) The cell was challenged by internal perfusion with a pipette solution containing GTP-γ-S (100 μM). (*C*) In this experiment compound 48/80 was used to externally stimulate the cell at the time indicated. (*D*) The cell was perfused with an internal solution containing cAMP (50 μM) and subsequently stimulated with compound 48/80 at the time indicated. *B*, *C*, and *D* were taken from Penner, 1988.

cause degranulation of a mast cell pretreated with the phorbol ester PMA (phorbol esters can substitute for endogenous DAG in activating PKC). It is clear, however, that activation of PKC by PMA is not sufficient to induce secretion at basal $[Ca]_i$, as is the case in platelets or neutrophils (Rink et al., 1983; Sha'afi et al., 1983). Since mast cell degranulation can proceed at basal $[Ca]_i$ when stimulating with GTP-γ-S (see Fig. 5 *C*; Neher, 1988), there may be an alternative signalling pathway for secretion that is mediated by a GTP-binding protein and that is largely independent of the dual signal pathway (see also Beaven et al., 1987; Cockcroft et al., 1987; Penner et al., 1987).

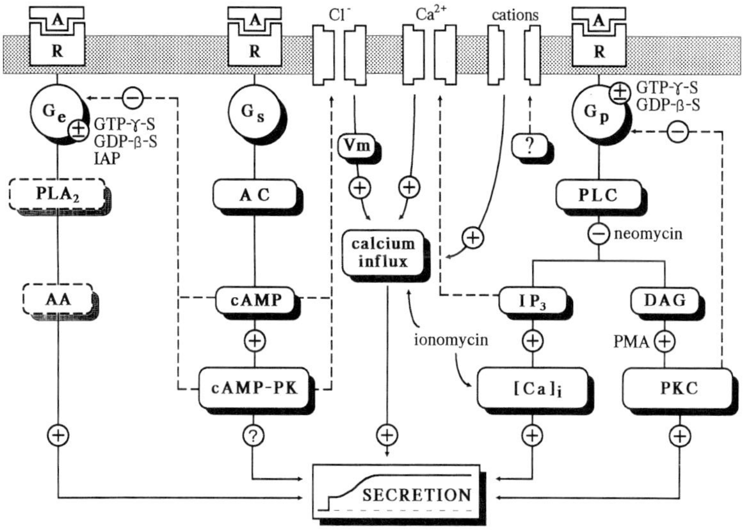

Figure 10. Hypothetical model of the signalling pathways involved in stimulus-secretion coupling in mast cells and the presumed sites of action of various pharmacological manipulations. The model assumes three different G proteins (Ge, Gs, and Gp) activated by receptor-agonist interaction. Each G protein mediates the activation of membrane-associated enzymes (phospholipase A_2, adenylate cyclase, and phospholipase C), which in turn liberate second messengers (arachidonic acid, cyclic AMP, DAG, and inositol trisphosphate) that act to activate another set of enzymes (cAMP-dependent protein kinase and protein kinase C) or to induce release of Ca from internal stores. In addition, these second messengers act to facilitate Ca influx into the cytosol. Inhibitory effects on the "Ge pathway" are provided by pertussis toxin (islet-activating protein) and cAMP (possibly via the cAMP-dependent protein kinase), while the "Gp pathway" is blocked by neomycin and recurrently inhibited by protein kinase C. The coupling between Ge and PLA_2 is only tentative, as indicated by dashed boxes, and awaits investigations with pharmacological tools more selective than the ones currently available. Modified from Penner, 1988.

Additional evidence that gives credence to this view comes from a number of observations: (*a*) Secretion in neutrophils and mast cells (Barrowman et al., 1986; Cockcroft et al., 1987) can be induced by GTP-γ-S in the presence of neomycin (which blocks PI breakdown) and this has led to the postulation of a G protein involved in exocytosis (Ge). (*b*) Ca transients (which may be taken as evidence for PI breakdown and activation of the dual signal pathway) and secretion are differentially affected by "washout," i.e., the generation of Ca transients is not impeded by prolonged intracellular dialysis of cells, whereas secretory responses are rapidly lost due to washout

(Penner et al., 1987). (*c*) Ca transients and degranulation do not always show a strict temporal correlation, i.e., in some cells secretion can start before Ca transients are manifest (Neher, 1988). (*d*) Secretory responses after receptor-stimulation can selectively be blocked by intracellular cAMP without blocking PI breakdown, as evidenced by the presence of Ca transients (Penner, 1988; see Fig. 9 *D*). (*e*) Conversely, GTP-γ-S or compound 48/80 can induce secretion without changes of $[Ca]_i$ in the presence of PMA (Penner, 1988; see Fig. 9, *B* and *C*) and this effect is probably not caused by the activation of PKC via the dual signal pathway since the enzyme is presumably activated maximally already. Moreover, maximal stimulation of PKC imposes negative feedback on PI breakdown by both receptor-stimulation and GTP-γ-S (see also Orellana et al., 1985; Watson and Lapetina, 1985).

Taken together, these findings clearly point towards an as yet unidentified signalling pathway for exocytosis that is under control of a G protein, tentatively termed Ge, and is distinct from the G protein that transduces PI breakdown (Gp). Whether this G protein mediates activation of a signalling cascade or controls exocytotis directly is not yet known. A hypothetical model that shows the possible arrangement of the various signalling pathways in mast cells is depicted in Fig. 10.

References

Barrowman, M. M., S. Cockcroft, and B. D. Gomperts. 1986. Two roles for guanine nucleotides in the stimulus-secretion sequence of neutrophils. *Nature*. 319:504–507.

Beaven, M. A., D. F. Guthrie, J. P. Moore, G. A. Smith, T. R. Hesketh, and J. C. Metcalfe. 1987. Synergistic signals in the mechanism of antigen-induced exocytosis in 2H3 cells: evidence for an unidentified signal required for histamine release. *Journal of Cell Biology*. 105:1129–1136.

Beaven, M. A., J. P. Moore, G. A. Smith, T. R. Hesketh, and J. C. Metcalfe. 1984. The calcium signal and phosphatidylinositol breakdown in 2H3 cells. *Journal of Biological Chemistry*. 259:7137–7142.

Bennett, J. P., S. Cockcroft, and B. D. Gomperts. 1979. Ionomycin stimulates mast cell histamine secretion by forming a lipid soluble calcium complex. *Nature*. 282:851–853.

Berridge, M. J., and R. F. Irvine. 1984. Inositol trisphosphate, a novel second messenger in cellular signal transduction. *Nature*. 312:315–321.

Cochrane, D. E., and W. W. Douglas. 1974. Calcium-induced extrusion of secretory granules (exocytosis) in mast cells exposed to 48/80 or the ionophores A-23187 and X-537A. *Proceedings of the National Academy of Sciences*. 71:408–412.

Cockcroft, S., T. H. Howell, and B. D. Gomperts. 1987. Two G-proteins act in series to control stimulus-secretion coupling in mast cells: use of neomycin to distinguish between G-proteins controlling polyphosphoinositide phosphodiesterase and exocytosis. *Journal of Cell Biology*. 105:2745–2750.

Douglas, W. W. 1968. Stimulus-secretion coupling: the concept and clues from chromaffin and other cells. *British Journal of Pharmacology*. 34:451–474.

Fenwick, E. M., A. Marty, and E. Neher. 1982. Sodium and calcium channels in bovine chromaffin cells. *Journal of Physiology*. 331:599–635.

Fernandez, J. M., E. Neher, and B. D. Gomperts. 1984. Capacitance measurements reveal stepwise fusion events in degranulating mast cells. *Nature*. 312:453–455.

Foreman, J., and C. Jordan. 1983. Histamine release and vascular changes induced by neuropeptides. *Agents and Actions*. 13:105–116.

Gomperts, B. D., S. Cockcroft, T. W. Howell, and P. E. R. Tatham. 1988. Intracellular Ca^{2+}, GTP and ATP as effectors and modulators of exocytotic secretion from rat mast cells. *In* Molecular Mechanisms in Secretion. N. A. Thorn, M. Treiman, and O. H. Petersen, editors. Munksgaard, Copenhagen. 248–258.

Gomperts, B. D., and J. M. Fernandez. 1985. Techniques for membrane permeabilization. *Trends in Biochemical Sciences*. 10:414–417.

Grynkiewicz, G., M. Poenie, and R. Y. Tsien. 1985. A new generation of Ca^{2+} indicators with greatly improved fluorescence properties. *Journal of Biological Chemistry*. 260:3440–3450.

Heiman, A. S., and F. T. Crews. 1985. Characterization of the effects of phorbol esters on rat mast cell secretion. *Journal of Immunology*. 134:548–554.

Ishizaka, T., and K. Ishizaka. 1984. Activation of mast cells for mediator release through IgE receptors. *In* Progress in Allergy. K. Ishizaka, editor. Karger, Basel. 188–235.

Johnson, S. R., and E. G. Erdös. 1973. Release of histamine from mast cells by vasoactive peptides. *Proceedings of the Society of Experimental Biology and Medicine*. 142:1253–1256.

Katz, B. 1969. The release of neural transmitter substances. Liverpool University Press, Liverpool.

Kelly, R. B. 1985. Pathways of protein secretion. *Science*. 230:25–32.

Kennerly, D. A., T. J. Sullivan, and C. W. Parker. 1979. Activation of phospholipid metabolism during mediator release from stimulated rat mast cells. *Journal of Immunology*. 122:152–159.

Knight, D. E., and M. C. Scrutton. 1986. Gaining access to the cytosol: the technique and some applications of electropermeabilization. *Biochemical Journal*. 234:497–506.

Kuno, M., and P. Gardner. 1987. Ion channels activated by inositol 1,4,5-tris-phosphate in plasma membrane of human T-lymphocytes. *Nature*. 326:301–304.

Lagunoff, D., T. W. Martin, and G. Read. 1983. Agents that release histamine from mast cells. *Annual Review of Pharmacology and Toxicology*. 23:331–351.

Metzger, H., G. Alcaraz, R. Hohman, J.-P. Kinet, V. Pribluda, and R. Quarto. 1986. The receptor with high affinity for immunoglobulin E. *Annual Review of Immunology*. 4:419–470.

Neher, E. 1988. The influence of intracellular calcium concentration of degranulation of dialysed mast cells from rat peritoneum. *Journal of Physiology*. 395:193–214.

Neher, E. 1989. Combined fura-2 and patch clamp measurements in rat peritoneal mast cells. *In* Proceedings of the Eric K. Fernström Symposium. Vol. 13. Neuromuscular Junction. Elsevier, Amsterdam. In press.

Neher, E., and W. Almers. 1986. Fast calcium transients in rat peritoneal mast cells are not sufficient to trigger exocytosis. *EMBO (European Molecular Biology Organization) Journal* 5:51–53.

Neher, E., and A. Marty. 1982. Discrete changes of cell membrane capacitance observed under conditions of enhanced secretion in bovine adrenal chromaffin cells. *Proceedings of the National Academy of Sciences*. 79:6712–6716.

Neher, E., and R. Penner. 1988. The influence of intracellular calcium concentration on the secretory response of mast cells. *In* Molecular Mechanisms in Secretion. N. A. Thorn, M. Treiman, and O. H. Petersen, editors. Munksgaard, Copenhagen. 262–270.

Nishizuka, Y. 1984. The role of protein kinase C in cell surface signal transduction and tumour promotion. *Nature*. 308:693–698.

Orellana, S. A., P. A. Solski, and J. H. Brown. 1985. Phorbol ester inhibits phosphoinositide hydrolysis and calcium mobilization in cultured astrocytoma cells. *Journal of Biological Chemistry*. 260:5236–5239.

Penner, R. 1988. Multiple signalling pathways control stimulus-secretion coupling in rat peritoneal mast cells. *Proceedings of the National Academy of Sciences*. 85:9856–9860.

Penner, R., E. Neher, and F. Dreyer. 1986. Intracellularly injected tetanus toxin inhibits exocytosis in bovine adrenal chromaffin cells. *Nature*. 324:76–78.

Penner, R., and E. Neher. 1988*a*. Secretory responses of rat peritoneal mast cells to high intracellular calcium. *FEBS Letters*. 226:307–313.

Penner, R., and E. Neher. 1988*b*. The role of calcium in stimulus-secretion coupling in excitable and non-excitable cells. *Journal of Experimental Biology*. 139:329–345.

Penner, R., G. Matthews, and E. Neher. 1988. Regulation of calcium influx by second messengers in rat mast cells. *Nature*. 334:499–504.

Penner, R., M. Pusch, and E. Neher. 1987. Washout phenomena in dialyzed mast cells allow discrimination of different steps in stimulus-secretion coupling. *Bioscience Reports*. 7:313–321.

Pusch, M., and E. Neher. 1988. Rates of diffusional exchange between small cells and a measuring patch pipette. *Pflügers Archiv*. 411:204–211.

Rink, T. J., A. Sanchez, and T. J. Hallam. 1983. Diacylglycerol and phorbol ester stimulate secretion without raising cytoplasmic free calcium in human platelets. *Nature*. 305:317–319.

Sha'afi, R. I., J. R. White, T. F. P. Molski, J. Shefcyk, M. Volpi, P. H. Naccache, and M. B. Feinstein. 1983. Phorbol 12-myristate 13-acetate activates rabbit neutrophils without an apparent rise in the level of intracellular free calcium. *Biochemical and Biophysical Research Communications*. 114:638–645.

Skofitsch, G., J. M. Savitt, and D. M. Jacobowitz. 1985. Suggestive evidence for a functional unit between mast cells and substance P fibres in the rat diaphragm and mesentery. *Histochemistry*. 2:5–8.

Tsien, R. Y., T. Pozzan, and T. J. Rink. 1984. Measuring and manipulating cytosolic Ca^{2+} with trapped indicators. *Trends in Biochemical Sciences*. 9:263–266.

Watson, S. P., and E. L. Lapetina. 1985. 1,2-Diacylglycerol and phorbol ester inhibit agonist-induced formation of inositol phosphates in human platelets. Possible implications for negative feedback regulation of inositol phospholipid hydrolysis. *Proceedings of the National Academy of Sciences*. 82:2623–2626.

White, J. R., T. Ishizaka, K. Ishizaka, and R. I. Sha'afi. 1984. Direct demonstration of increased intracellular concentration of free calcium as measured by quin-2 in stimulated rat peritoneal mast cells. *Proceedings of the National Academy of Sciences*. 81:3978–3982.

Winslow, C. M., and K. F. Austen. 1984. Role of cyclic nucleotides in the activation-secretion response. *In* Progress in Allergy. K. Ishizaka, editor. Karger, Basel. 236–270.

Wollheim, C. B., and G. W. G. Sharp. 1981. Regulation of insulin release by calcium. *Physiological Reviews*. 61:914–973.

Chapter 20

Excitation and Secretion at the Nerve Terminals of Vertebrates: Optical Measurements with and without Voltage-sensitive Dyes

Brian M. Salzberg and Ana Lia Obaid

Department of Physiology, University of Pennsylvania, School of Medicine and The David Mahoney Institute of Neurological Sciences, Philadelphia, Pennsylvania 19104-6085; and the Marine Biological Laboratory, Woods Hole, Massachusetts 02543

Introduction

For the past six years, we have been using optical measurement of membrane potential to study the secretory event at the very fine nerve terminals of vertebrates by monitoring directly the voltage changes that occur in this relatively inaccessible region of the nervous system. The study of the secretory event per se, as well as the excitation that provokes it, has proven especially difficult at these terminals, largely because of the small size of the structures of interest. Consequently, much of our mental imagery, if not our understanding, related to the release of transmitter substances, comes from the electron microscope (Heuser et al., 1979; Heuser and Reese, 1981). So far as electrophysiology is concerned, we have depended for our understanding largely upon a series of elegant studies of synaptic transmission at the frog neuromuscular junction (Katz, 1969), where most measurements are actually made on the postsynaptic element, upon the observations of Martin and Pilar (1963*a*, *b*; 1964*a*, *b*) in the caliciform nerve terminals of the avian ciliary ganglion, and upon analogies drawn from that most useful of invertebrates, the Atlantic squid, *Loligo*.

Patch-clamp methods have begun to provide important tools for the study of secretion (Fernandez et al., 1984; Lindau and Fernandez, 1986) in some systems (e.g., mast cells), but their reliable application to vertebrate nerve terminals is only, perhaps, beginning (Mason and Dyball, 1986; Lemos and Nowycky, 1987). Optical techniques that use potentiometric probes for the measurement and analysis of transmembrane electrical events have also found a wide range of applications over the past decade because, while they often provide less detailed information than does the voltage clamp, they can offer certain advantages over more conventional measurements. These probes are small (300–500 mol wt) and bind to, but do not cross, cell membranes, and they behave as fast linear transducers of membrane voltage. Their optical properties, particularly absorbance and fluorescence, respond to changes in potential in <2 μs, and they may be used to follow electrical events in membranes that are inaccessible to microelectrodes. Because the membranes of interest are not mechanically violated, the methods may be comparatively noninvasive. Spatial resolution is limited only by microscope optics and noise considerations; it is possible to measure changes in membrane potential from regions of a cell having linear dimensions on the order of 1 μm, and optical sectioning, with elimination of scattering, by means of confocal imaging may soon permit optical recording from small regions of cells in intact ganglia or brains. Temporal resolution is limited by the physical response of the probes and by the bandwidth imposed upon the measurement, again usually by noise considerations, and response times faster than typical membrane time constants may be achieved. Because mechanical access is not required, unusual latitude is possible in the choice of preparation, and voltage changes may be monitored in membranes that are otherwise inaccessible. Also, since no recording electrodes are used, and the measurements are actually made at a distance from the preparation—in the image plane of an optical apparatus—it is possible to record changes in potential simultaneously from a large number of spatially separated sites. Several recent reviews had surveyed the literature on optical measurement of membrane potential (Cohen and Salzberg, 1978; Waggoner, 1979; Freedman and Laris, 1981; Salzberg, 1983; Grinvald, 1985) and they should be consulted by the reader interested in the details of these techniques.

The evidence that optical methods are equivalent, at least in a limited sense, to electrode measurements, has been detailed elsewhere (Cohen et al., 1974; Ross et al.,

1977; Salzberg, et al., 1977; Cohen and Salzberg, 1978; Gupta et al., 1981). Indeed, under certain conditions (Salzberg and Bezanilla, 1983), an optical measurement of membrane potential is superior to an electrode measurement. Our purpose here, however, is to illustrate the application of optical techniques to the coupled problems of excitation and secretion at vertebrate nerve terminals, problems for which these methods are particularly well suited.

The direct optical recording of the action potential from the nerve terminals of a vertebrate, and the study of its ionic basis (Salzberg et al., 1983; Obaid et al., 1985) provides information that is presently unobtainable by other means. A variety of linear potentiometric probes can now be used to make this kind of measurement, by exploiting the extrinsic absorption or fluorescence signals that result when the potential varies across membrane stained with these dyes. We shall begin by describing some of the properties of the nerve terminal action potential, as revealed by measurements of the extrinsic absorption changes exhibited by linear potentiometric probes.

Analysis of optical recordings of the action potential in the nerve terminals of a mammal disclosed the presence of at least two large and rapid changes in the intrinsic optical properties of the neural lobe, in addition to the extrinsic absorption changes that report changes in membrane potential. One of these, a change in light scattering, has now been shown (Salzberg et al., 1985) to be intimately associated with the physiological events that attend the release of the neuropeptides arginine vasopressin (AVP) and oxytocin. After describing very briefly the properties of these intrinsic signals, and summarizing some of the evidence that this optical signal is related to the secretory event, we will conclude this chapter by discussing some of the pharmacology of the calcium channels that are required for the release of neuropeptides in the neurohypophysis.

Since the classic early work of Douglas and his co-workers, the vertebrate neurohypophysis has provided a very useful model system for understanding the release of neurotransmitters and neuropeptides. Magnocellular neurons located in the hypothalamus (supraoptic and paraventricular nuclei in mammals, preoptic nucleus of lower vertebrates) project their axons as bundles of fibers through the median eminence and infundibular stalk to terminate in the neurohypophysis, where the neurohypophysial peptides (oxytocin and AVP, or their homologues) and associated proteins (neurophysins) are secreted into the circulation. Indeed, the neurohypophysis has been a classical in vitro preparation for measuring the calcium-dependent release of peptide hormones under various conditions of stimulation (Douglas, 1963; Douglas and Poisner, 1964), and extracellularly recorded compound action potentials are readily measured in this organ (Dreifuss et al., 1971). The analysis of excitation-secretion coupling in the neurohypophysis has been compromised, however, by the circumstance that these nerve terminals are also too small for intracellular measurement of the electrophysiological events that affect release. However, the intact terminals of the frog neurohypophysis provide a nearly ideal preparation for the use of fast voltage-sensitive dyes (Cohen and Salzberg, 1978). This is because of the relative homogeneity of the region; the enormous proliferation of magnocellular neuron terminals, which provide a very large surface-to-volume ratio for the binding of impermeant probe molecules; the precise control of the timing of the action potential in the terminals; and because of the absence of any postsynaptic structure to confound the interpretation of the extrinsic optical signals.

Properties of the Action Potential in the Intact Nerve Terminals of the Frog (*Xenopus laevis*) Neurohypophysis

Resting and slowly changing transmembrane voltages in synaptosomes prepared from rat brain homogenates had already been measured using cyanine (Blaustein and Goldring, 1975) and merocyanine (Kamino and Inouye, 1978) dyes a decade ago, and action potentials have been recorded from the growth cones of tumor cells in tissue culture using an oxonol dye (Grinvald and Farber, 1981). In 1983, we took advantage of the extraordinary qualities of the vertebrate neurohyophysis and obtained optical recordings of the action potential with good fidelity to the time course of the transmembrane potential change (Salzberg et al., 1983). Because these recordings exhibited very large signal-to-noise ratios, we could manipulate the shape of the action potential by extracellular calcium and other agents known to alter the release of neurohormones and neurotransmitters (Salzberg et al., 1983). Organic, as well as inorganic calcium channel modifiers could be studied (Salzberg et al., 1983; Obaid et al., 1989), and we could also show that when voltage-sensitive Na^+ channels are blocked by tetrodotoxin (TTX) and K^+ channels are blocked by tetraethylammonium (TEA) (Katz and Miledi, 1969), direct electric field stimulation of the nerve terminals evokes large active responses that reflect the same inward Ca^{++} current that is associated with hormone release from these nerve terminals (Obaid et al., 1985).

Fig. 1 *B* shows an optical recording that represents the intracellular potential changes during the action potentials in a population of synchronously activated nerve terminals in the neurohypophysis of the frog, *Xenopus laevis*. This recording was photographed from an oscilloscope screen without signal averaging. The technique used here, multiple site optical recording of transmembrane voltage (MSORTV), is an extension of the method used to record extrinsic absorption changes in the squid giant axon and in many other preparations (Salzberg et al., 1977; Grinvald et al., 1981; Salzberg et al., 1983; Salzberg, 1983; Senseman et al., 1983) and has been described in detail in these reports and in the article by Cohen and Lesher in this series of volumes (1986). The MSORTV system is illustrated schematically to the left of the optical spike (Fig. 1 *A*). Briefly, light from a tungsten-halogen lamp is collimated, made quasimonochromatic with a heat filter (KG-1; Schott Optical Co., Duryea, PA) and an interference filter (700 nm; 70-nm full width at half maximum), and focused by means of a bright field condenser onto the neurohypophysis (pars nervosa). The neurohypophysis had been excised together with the hypothalamus and the infundibulum, and the entire preparation, vitally stained by incubating it for 25 min in a 100 μg/ml solution of the merocyanine-rhodanine dye, NK 2761 (Nippon Kankoh Shikiso Kenkyusho Inc., Okayama, Japan) (Gupta et al., 1981; Kamino et al., 1981) in *Xenopus* Ringer's solution (112 mM NaCl, 2 mM KCl, 2 mM $CaCl_2$, 15 mM HEPES, 33 mM glucose, pH adjusted to 7.35). Light transmitted by the stained preparation was collected by a high numerical aperture water immersion objective, which formed a real image on a 12 × 12 element silicon photodiode matrix array (MD 144-0; Centronic, Inc., Newbury Park, CA) located in the image plane of a compound microscope (UEM; Carl Zeiss, Inc., Oberkochen, FRG). The photocurrents generated by the central 124 array elements were separately converted to voltages and amplified as described earlier (Salzberg et al., 1977; Grinvald et al., 1981). In most of the experiments, all the amplifier outputs were passed to a data acquisition system based on a PDP 11/34A computer (Digital Equipment Corp., Maynard, MA) capable of acquiring a complete

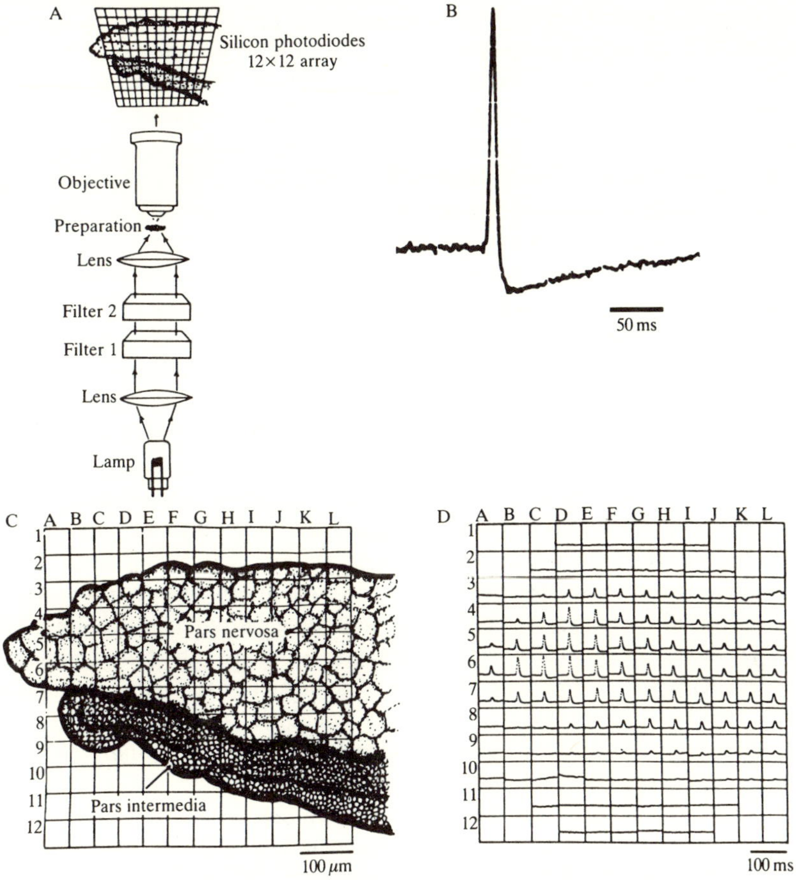

Figure 1. MSORTV from nerve terminals in the neurohypophysis of *Xenopus*. (*A*) Schematic diagram of the optical portion of the MSORTV system. Collimated light from an incandescent source is made quasimonochromatic with interference and heat filters and focused on the preparation by means of a bright-field condenser with numerical aperture matched to that of the objective. A high numerical aperture water immersion objective projects a real image of a portion of the preparation onto the central 124 elements of a 144-element photodiode array, whose photocurrent outputs are converted to voltages, AC coupled and amplified, multiplexed, digitized, and stored in a PDP 11/34A computer under direct memory access (DMA). A full frame is recorded, with an effective resolution of 18 bits, every 800 μs. (*B*) A photograph of an oscilloscope recording of the change in the transmitted intensity monitored by one channel (element E5 of the array) of the MSORTV system, after a brief shock to the hypothalamus. This element monitored intensity changes from a region of the posterior pituitary entirely within the pars nervosa. The fractional change in intensity during the action potential was ~0.25%. Single sweep, 722 ± 21 nm. Rise time of the optical system (10–90%), 1.1 ms. AC coupling time constant, 1.0 s. (*C*) Drawing of the region of the posterior pituitary imaged on the photodiode matrix array, showing the positions of the individual detector elements with respect to the tissue. Drawn from a photomicrograph of the preparation and a transparent overlay representing the detector array. Objective 20×, 0.33 n.a. (*D*) Extrinsic absorption changes obtained in a single trial, after stimulation of the hypothalamus, superimposed on corresponding elements of the photodiode matrix array. The five elements in each corner of the array were not connected. The largest signals represent fractional changes in transmitted intensity of ~0.3%. (After Salzberg et al., 1983.)

frame every 0.8 ms with an effective resolution of 18 bits. This data acquisition system is similar to that described previously by Grinvald et al. (1981) and used by Grinvald et al. (1982), Senseman et al. (1983), and Salzberg et al. (1983). In some experiments, the temporal resolution was markedly improved (17 μs per point for a single detector element) when selected outputs of the current-to-voltage converters were also passed in parallel to a 16-channel signal averager (TN 1500; Tracor Northern Inc., Middleton, WI). These digitized output signals could be stored on magnetic tape for later display and analysis. A map of the preparation was obtained by superimposing a transparent overlay representing the photodiode array elements on a photograph taken through the trinocular tube of the microscope after removing the photodiode array housing. The experiment shown in Fig. 1 *B* used a 20× 0.33-numerical aperture water immersion objective (Nikon Inc., Garden City, NY) to image part of the neurointermediate lobe (neurohypophysis plus pars intermedia) onto the photodetector array. Fig. 1 *C* is a drawing of the projected region of the gland, prepared from the photomicrograph. Fig. 1 *D* shows the MSORTV display of the action potentials recorded optically, in a single sweep, from different areas of the pars nervosa, after a single brief stimulus had been applied to the hypothalamus. The optical signals are invariably confined to elements of the grid that correspond to the neurohypophysis; note that in this instance, a small portion of the neurohypophysis lies under the pars intermedia (row *9*). The optical signal illustrated in Fig. 1 *B* represents the analog output from channel E5. This record, which is typical, illustrates the large signal-to-noise ratio that may be achieved in these experiments. A variety of control experiments showed that these signals represent voltage changes, e.g., no signals could be recorded in the absence of illumination or in white light; the signals could be inverted at shorter wavelengths (Senseman and Salzberg, 1980), etc.

Evidence that the optical signals shown in Fig. 1 do, in fact, represent the membrane potential changes associated with the invasion of the nerve terminals of the neurohypophysis by the action potential, was provided indirectly by some morphometric observations. In the rat, it is estimated that the 18,000 magnocellular neurons in the hypothalamus give rise to ~40,000,000 terminal endings in the neurohypophysis, and morphometric data suggest that here, secretory endings and swellings comprise 99% of the excitable membrane area with axons making up the remaining 1% (Nordmann, 1977). Comparably detailed morphometric studies do not yet exist for the neurohypophysis of *Xenopus,* but the structural resemblance of the anuran neurohypophysis to that of the rat (Gerschenfeld et al., 1960; Rodriguez and Dellman, 1970; Dellman, 1973) suggests that, in *Xenopus,* it is also very heavily enriched with nerve terminals. Thus, we concluded that the optical signals shown in Fig. 1, originating in the lateral regions of the neurohypophysis, are, at the least, dominated by the membrane potential changes in the terminals of the magnocellular neurons. (For additional arguments, see Salzberg et al., 1983.)

The nerve terminal action potentials shown in Fig. 1 exhibit a rapid upstroke that is primarily the result of a fast inward sodium current, a rapid repolarizing phase that results from potassium efflux, and a prominent afterhyperpolarization that depends on a calcium-mediated potassium conductance (Salzberg et al., 1983; Obaid et al., 1985) that is reminiscent of other manifestations of this channel, e.g., snail neurons (Meech and Strumwasser, 1970), rat myotube (Barrett et al., 1982), and guinea pig Purkinje cells (Llinas and Sugimori, 1980*a*, *b*).

When voltage-sensitive Na channels are blocked by TTX, direct electric field stimulation of the intact terminals reveals a small regenerative Ca response, and when the terminal depolarization is prolonged by blocking voltage-sensitive potassium channels with TEA, the large active response that one sees in Fig. 2 *B* is observed. This is the optical response recorded from the same population of terminals represented in Fig. 2 *A*, after 20 min of exposure to 5 mM TEA and 2 μM TTX. The regenerative response illustrated is smaller and slower than the normal action potential (Fig. 2 *A*), and appears to depend entirely upon the influx of calcium and the subsequent rise in a TEA-insensitive K^+ conductance. These responses are blocked reversibly by 100–500 μM Cd^{++} (Standen, 1981), and are sensitive to Ca^{++} concentration in the range of 0.1–10 mM. Thus, under these conditions, sufficient inward calcium current may develop to give rise to a relatively large calcium action potential. The reader is reminded that these recordings were obtained without signal averaging; they represent voltage-induced changes in the absorbance of a membrane-bound dye. The residual signal seen in Fig. 2 *C* is the passive depolarization produced by the field stimulation; it

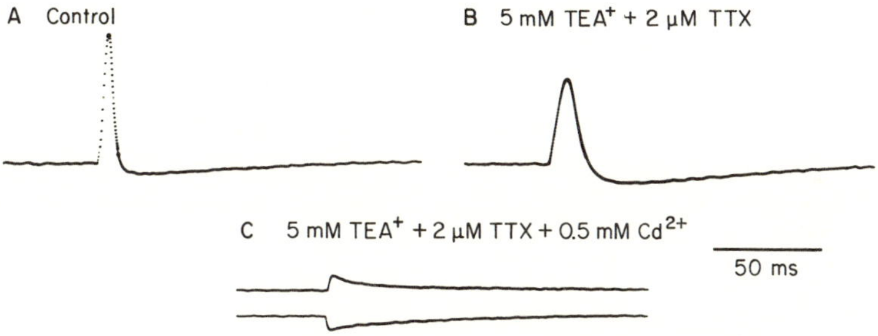

Figure 2. Cadmium sensitivity of the active responses (calcium spikes) elicited by direct field stimulation of the neurohypophysis in the presence of TTX and TEA. Optical recording of action potentials from nerve terminals of the *Xenopus* neurohypophysis after staining for 25 min in 0.1 mg/ml NK 2761. Single sweep outputs of a single channel from the photodiode matrix array in the MSORTV system. (*A*) Action potential recorded in normal Ringer's solution. (*B*) Active response of the nerve terminals after 20 min exposure to 5 mM TEA plus 2 μM TTX in normal Ringer's solution. (*C*) Passive response (electrotonus) remaining 15 min after the addition of 0.5 mM Cd^{++} to the TEA-TTX Ringer's solution bathing the preparation, upon stimulation with normal and reversed polarity. 700 nm. Rise time of the light measuring system (10–90%) was 1.1 ms. Temperature 18–22°C. (After Obaid et al., 1985.)

was reversed with reversal of stimulus polarity, and it exhibited the same wavelength dependence as that of the optical recording of the normal action potential.

The calcium-mediated active responses, recorded optically and described here, appear to share some properties with the local subthreshold action potentials described by Hodgkin (1938) in unmyelinated crab nerve; with TTX blocking any propagated action potential in the axons of the infundibulum, the small calcium currents cannot excite a sufficient length of nerve to produce a propagated action potential. Nonetheless, their dependence on $[Ca^{++}]_o$ and sensitivity to Cd^{++} block indicates that they probably result from a voltage-sensitive calcium influx into the terminals (Kostyuk and Krishtal, 1977; Hagiwara and Byerly, 1981; Obaid et al., 1985), which mediates hormone release. Thus, our data demonstrate that although the rising phase of the nerve terminal action potential in the neurohypophysis of *Xenopus* is mediated predominantly by sodium, at least a small calcium component is present. The important role for calcium entry in excitation-secretion coupling is fully consistent with its apparently small contribution to the terminal action potential (Katz and Miledi, 1969). In adrenal chromaffin cells, for example, where calcium influx is the critical event in excitation-secretion coupling (Douglas, 1978), the Ca^{++} component of the action potential is also small (Brandt et al., 1976).

The afterhyperpolarization that is such a prominent feature of both the normal action potential and the calcium response in the terminals of the frog neurohypophysis is blocked by nanomolar concentrations of charybdotoxin (CTX) (Miller et al., 1985), a protein toxin derived from scorpion venom. This toxin blocks calcium-sensitive potassium channels from skeletal muscle in planar lipid bilayers (Miller et al., 1985) with a K_i of 10 nM. We have observed (Obaid and Salzberg, 1985; Obaid et al., 1989) that CTX specifically blocks the afterhyperpolarization of the nerve terminal action potential in *Xenopus*. In addition, both the width and height of the action potential are increased in the presence of the toxin. Fig. 3 shows the effect of 50 nM CTX on the nerve terminal action potential recorded optically. *A* depicts the action potential

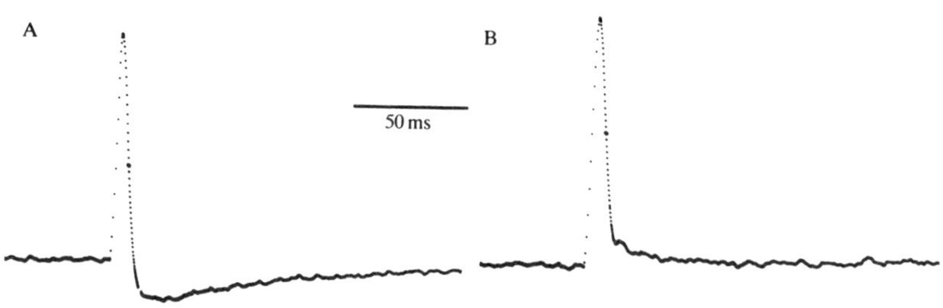

Figure 3. Effect of CTX on the afterhyperpolarization of the nerve terminal action potential in *Xenopus*. Optical recordings of the action potential resulting from direct field stimulation of the nerve terminals of the *Xenopus* neurohypophysis after staining for 25 min in 0.1 mg/ml NK 2761. Outputs of a single channel of the MSORTV system. (*A*) The action potential recorded in normal Ringer's solution. (*B*) The action potential recorded 8 min after the addition of 50 nM CTX to the Ringer's solution bathing the neurohypophysis. The rise time (10–90%) of the light measuring system was 1.1 ms; single sweeps; 700 nm; AC coupling time constant was 3 s. (After Obaid et al., 1989.)

elicited in normal Ringer's solution by direct field stimulation, and *B* shows the complete elimination of the afterhyperpolarization after an 8-min exposure to CTX. (Note that there is a 10% increase in the height of the action potential, in addition to the loss of the undershoot.) Evidence that CTX acts directly on a calcium-mediated potassium conductance, rather than indirectly by blocking calcium entry is provided in Fig. 4. Again, *A* shows the action potential in normal Ringer's solution. *B* illustrates the calcium response obtained 8 min after the addition of 5 mM TEA and 1 μM TTX to the Ringer's solution. This response also exhibits an afterhyperpolarization similar to that seen in Fig. 2. *C* demonstrates the effect of a 23-min exposure to 50 nM CTX (in TEA-TTX Ringer's solution). The afterhyperpolarization is again eliminated, while calcium entry is actually enhanced. At the same time, the loss of an outward current prolongs the duration of the response. Evidence that the active response is mediated by calcium entry is given in *D*, where a 2-min exposure to 0.5 mM Cd^{++} entirely abolishes the active response.

Because the $g_{K(Ca)}$-dependent afterpotential of the spike is such a sensitive indicator of Ca^{++} entry into the intact terminals of the neurohypophysis, we were able to use this component of the optical signal to study some of the properties of the calcium channels that are present in a population of intact vertebrate nerve terminals. Inorganic blockers of calcium channels such as nickel (Kaufmann and Fleckenstein, 1965), cobalt, manganese, gadolinium, and lanthanum blocked calcium currents in these terminals, although with a potency less than that of cadmium ion (Salzberg et al., 1983; Obaid, A. L., and B. M. Salzberg, unpublished observations).

Aminoglycoside antibiotics, including neomycin, streptomycin, and gentamicin, depress evoked transmitter release at the neuromuscular junction by competing with extracellular calcium (Fiekers, 1983). We were able to show (Parsons et al., 1986) that neomycin (220 μM) reduced the height of the upstroke of the nerve terminal action

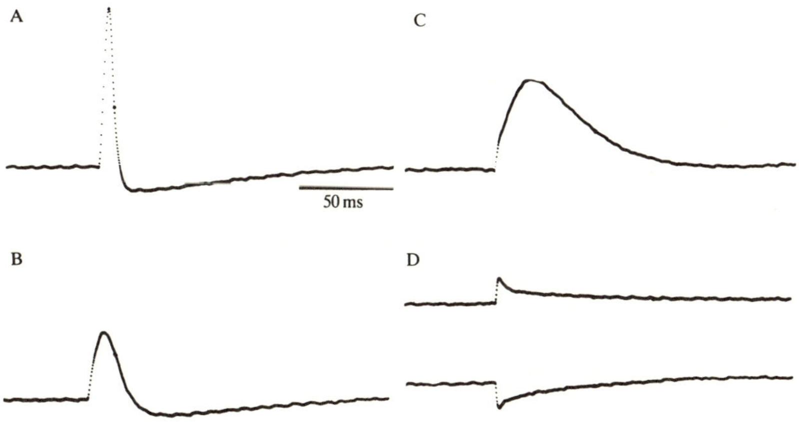

Figure 4. CTX eliminates the nerve terminal afterhyperpolarization without blocking calcium entry. Optical recordings of the action potential resulting from direct field stimulation of the nerve terminals of the *Xenopus* neurohypophysis after staining for 25 min in 0.1 mg/ml NK 2761. Outputs of a single channel of the MSORTV system. (*A*) The action potential recorded in normal Ringer's solution. (*B*) The action potential recorded 8 min after the addition of 5 mM TEA and 1 μM TTX to the Ringer's solution bathing the neurohypophysis. (*C*) The action potential recorded 23 min after the addition of 50 nM CTX to the Ringer's solution already containing 5 mM TEA and 1 μM TTX. The afterhyperpolarization is eliminated. (*D*) Purely passive response (electrotonus) 1 min after the addition of 500 μM Cd^{++} to the Ringer's solution containing 50 nM CTX, 1 μM TTX, and 5 mM TEA. The rise time (10–90%) of the light measuring system was 1.1 ms; single sweeps; 700 nm; AC coupling time constant was 400 ms. (After Obaid et al., 1989.)

potential, increased the spike duration, and abolished the afterhyperpolarization in the neurohypophysis of *Xenopus*. These effects were reversible upon washing, and could be antagonized by raising $[Ca^{++}]_o$ from 2 to 5 mM (Parsons et al., 1986). When the effects of these agents on the regenerative calcium response were examined, we found that both neomycin (220 μM) and gentamicin (190 μM) decreased the height of the calcium spike, and practically eliminated the afterhyperpolarization. These observations all supported the idea that aminoglycoside antibiotics antagonize the entry of calcium into vertebrate nerve terminals during excitation.

Other organic modifiers of calcium channel behavior have been used by several writers to distinguish different calcium channel types in a variety of preparations (Miller, 1987). Previously, however, it has not been possible to use these pharmacolog-

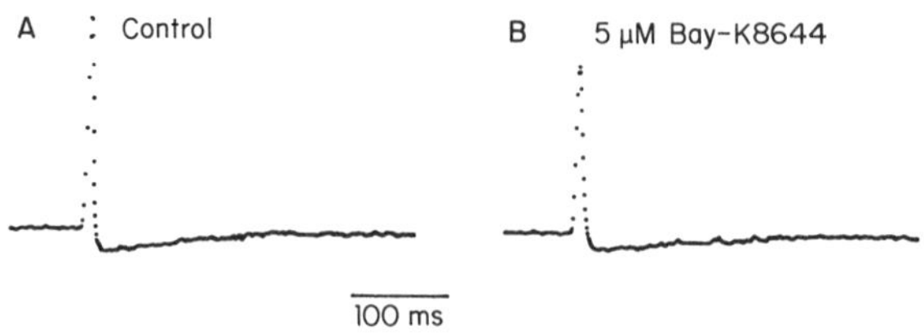

Figure 5. Bay-K 8644 fails to enhance the calcium-dependent components of the nerve terminal action potential in the neurohypophysis of *Xenopus*. (*A*) Optical recording of the action potential in control Ringer's solution after staining with NK 2761. (*B*) Optical recording of the action potential after 32 min exposure to 5 μM Bay-K 8644 (0.25% ethanol). Spatial average of the digitized outputs of three contiguous elements of the photodiode array; single sweep; 20×; 0.33 n.a.; 700 ± 35 nm; infundibular stimulation; AC coupling time constant, 400 ms; response time constant (10–90%), 1.1 ms. (After Obaid et al., 1989.)

ical tools to study calcium channels in the intact nerve terminals of a vertebrate. Using the extrinsic absorption changes exhibited by the terminals of the frog neurohypophysis stained with the merocyanine-rhodanine dye NK 2761, we carried out a series of experiments (Salzberg et al., 1987; Obaid et al., 1989) that help to define the pharmacological profile of the calcium channels present in intact neurosecretory terminals of vertebrates. Again, we used the $g_{K(Ca)}$-dependent afterpotential of the optical spike to examine the sensitivities of the calcium channels that are present in this population of vertebrate nerve terminals. We used the dihydropyridine compounds nifedipine and Bay-K 8644, at concentrations ranging from 2 to 5 μM, and the peptide toxin ω-conotoxin GVIA (Olivera et al., 1985). In the frog neurohypophysis, neither dihydropyridine compound had any effect on the calcium-dependent components of the action potential (Figs. 5 and 6); the spikes remained absolutely unchanged for up to

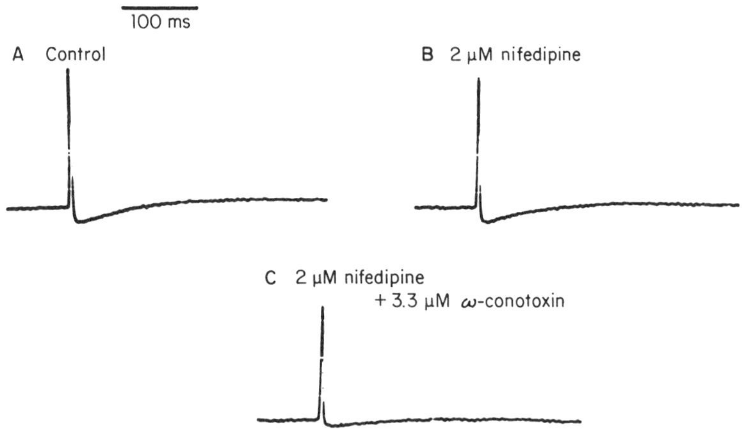

Figure 6. ω-Conotoxin reduces the amplitude of the calcium-dependent components of the nerve terminal action potential in a preparation in which 2 μM nifedipine had no effect. (*A*) Optical recording of the action potential in control Ringer's solution after staining with NK 2761. (*B*) Optical recording of the action potential after 30 min exposure to 2 μM nifedipine (0.1% ethanol). (*C*) Optical recording of the action potential 10 min after the addition of 3.3 μM ω-conotoxin GVIA to the 2 μM nifedipine Ringer's solution. Analogue output of a single representative element of the 124-element photodiode array; single sweep; 10×; 0.4 n.a.; 700 ± 35 nm; infundibular stimulation; AC coupling time constant, 400 ms; response time constant (10–90%), 1.1 ms. (After Obaid et al., 1989.)

1 h in the presence of either nifedipine or Bay-K 8644. ω-conotoxin GVIA (1 μM), however, rapidly abolished the afterhyperpolarization of the normal action potential (Fig. 6 *C*) by blocking calcium entry (and not by reducing $g_{K(Ca)}$). These results demonstrated clearly the effect of ω-conotoxin on the calcium channels involved in the release of neuropeptides (see below) from the intact nerve terminals of vertebrates, and, although this is the first direct evidence from measurements on intact nerve terminals, it is consistent with the findings of others using patch clamp of neuronal preparations (Fox et al., 1987; McClesky et al., 1987). The insensitivity to dihydropyridines exhibited by these same calcium channels is apparently at variance with reports of dihydropyridine block of neuronal calcium channels (for review, see Tsien, 1986). However, Rane et al. (1987) concluded that chronic depolarization is required to achieve nifedipine block of calcium channels in chick DRG neurons, but is not required for Bay-K 8644 action (Holz et al. 1988). In our hands, $[K^+]_o$ depolarization estimated at 25 mV failed to increase substantially the effects of the dihydropyridines (Obaid et al., 1989), but it may be that larger depolarizations, inconsistent with excitability, are required to exhibit significant nifedipine binding. Our results are consistent with this interpretation, although our failure to observe any enhancement of calcium channel activity in the presence of Bay-K 8644 is not.

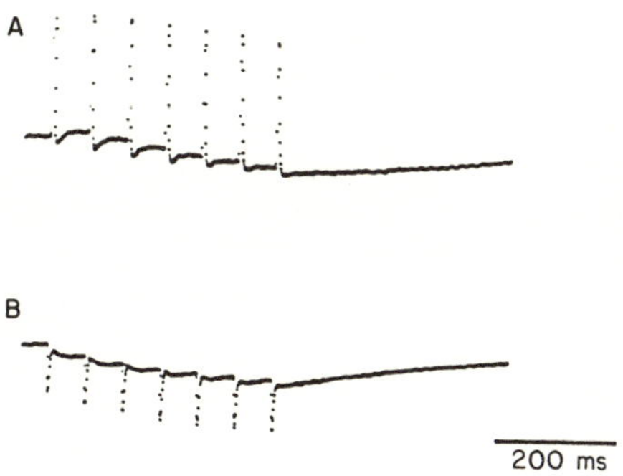

Figure 7. Extrinsic and intrinsic optical changes recorded from nerve terminals in the neurohypophysis of the mouse. (*A*) Changes in the extrinsic absorption at 675 nm (52 nm full width at half maximum) of nerve terminals after staining for 25 min in 0.1 mg/ml NK 2367. A single train of 500 μs (60 V) stimuli at 16 Hz was delivered to the posterior pituitary through the axons of the infundibulum for 400 ms. The resulting changes in transmitted light intensity are shown, recorded by a single element of the photodiode matrix array. Here, a downward deflection of the trace represents an increase in transmitted light intensity. The AC coupling time constant for the light measuring system was 400 ms. The fractional change in light intensity during the action potentials was ~0.3%. Single sweep; single array element. (*B*) Same experiment as in *A*, except that the transmitted intensity was monitored at 540 nm (60 nm full width at half maximum). At this wavelength, the voltage-dependent extrinsic absorption change exhibited by the merocyanine-oxazolone probe NK 2367 is known to be opposite in sign to that observed at 675 nm. Single sweep; single array element. All trials were carried out in normal Ringer's solution. The rise time of the light measuring system (10–90%) was 1.1 ms. (After Salzberg et al., 1985.)

Intrinsic Optical Changes That Accompany Secretion from Mammalian Nerve Terminals

The anatomy of the amphibian posterior pituitary provides a unique opportunity for the study in vitro of excitation-secretion coupling in the vertebrate; as we pointed out earlier, there is no postsynaptic excitable membrane to confound the interpretation of the optical signals (Salzberg et al., 1983; Obaid et al., 1985) and, more important for biochemical studies, no barrier to the collection of its secretory products. However, relatively little is known about the secretion of amphibian neuropeptides and neurohormones, while an extensive literature has accumulated on the biochemistry of secretion in the mammalian, and particularly the rodent, neurohypophysis. Together with Hal Gainer of the National Institutes of Health, we felt that it would be important to be able to correlate the electrical events in the terminals with the kinetics of release of the secretory products, and we thought that this might be accomplished by means of optical recording of membrane potential combined with sensitive radioimmunoassays using the CD-1 mouse.

Fig. 7 shows an optical recording of a train of seven action potentials, stimulated at 16 Hz, in a population of nerve terminals in the neurohypophysis of the CD-1 (Charles River Breeding Laboratories, Inc., Wilmington, MA) mouse. It represents the change in transmitted light intensity at 675 nm, monitored in a single sweep by one

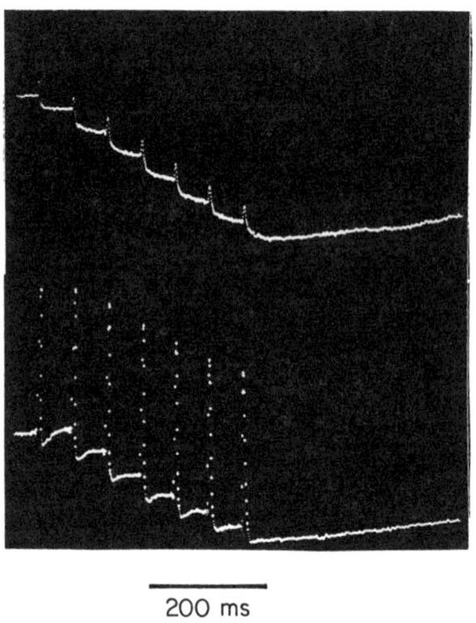

Figure 8. Intrinsic and extrinsic optical signals recorded from the same population of intact nerve terminals. The top panel shows the changes in light scattering recorded at 675 nm from a mouse neurohypophysis during a train of stimuli delivered at 16 Hz for 400 ms. The rapid upward deflection of the trace (decrease in transparency), which we call the "E-wave," appears to coincide with the arrival of the action potential at the terminals, while the subsequent increase in transparency (downward deflection), "S-wave," is intimately related to neurosecretion. Single sweep. The bottom panel shows the change in transmitted light intensity at 675 nm during identical stimulation of the same neurohypophysis after it had been stained for 25 min in a Ringer's solution containing 0.1 mg/ml of the voltage-sensitive merocyanine-oxazolone dye NK 2367. Single sweep.

channel of the 124-channel MSORTV system. The preparation had previously been incubated in a Ringer's solution (in millimolar: 154 NaCl, 5.6 KCl, 1 $MgCl_2$, 2.2 $CaCl_2$, 10 glucose, 20 HEPES, adjusted to pH 7.4 with NaOH) containing 0.1 mg/ml of the merocyanine-oxazolone dye NK 2367 (Salzberg et al., 1977). The method used here was virtually identical to that used to record electrical activity from the neurosecretory terminals of the frog neurohypophysis. Aside from the pleasing fact that it is possible to monitor the action potential in the intact terminals of a furry animal, the interesting feature to notice here is the progressive decline in the baseline of the optical signal that is obvious in Fig. 7 *A*. At first glance, it appears to be a hyperpolarization, perhaps the turn-on of a pump, but it's clear from the constancy in the heights of the action potentials, as their peaks are shifted downwards, that something else is going on. Fig. 7 *B* indicates that the wavelength dependence of the slow signal is entirely different from the action spectrum of the dye. This record was obtained at 540 nm, where the extrinsic absorption signal is inverted, and, in short, the slow component of the optical signal represents a change in light scattering rather than dye absorption (Cohen and Landowne, 1970; Cohen et al., 1972*a*, *b*). This interpretation is consistent with experiments that explore, in more detail, the wavelength dependence of the slow signal, in unstained preparations (Salzberg et al., 1985). Fig. 8 illustrates the properties of the intrinsic signal more clearly. The top and bottom panels show records from the same population of terminals before and after staining. In the record from the unstained preparation (top panel), the decline in the baseline resulting from accumulation of slow changes in transparency is seen clearly. (Note that in these records, a downward deflection in the trace represents an increase in transmitted light intensity, and a decrease in large angle scattering). This rather large optical signal was also recorded in a single sweep, and represents a change in light scattering consisting of

TABLE I
Effects of Calcium Ions and Selected Experimental Treatments on the Secretion of AVP from Mouse Neural Lobes

Experimental treatment	*n*	ng AVP released/neural lobe*	$P^{\ddagger}$
Effect of external calcium			
0.1 mM Ca^{2+}	6	0.2 ± 0.1	$P < 0.01$
1.0 mM Ca^{2+}	6	2.4 ± 0.3	$P < 0.01$
2.0 mM Ca^{2+}	6	3.7 ± 0.5	$P < 0.01$
5.0 mM Ca^{2+}	6	6.7 ± 0.8	—
10.0 mM Ca^{2+}	6	6.2 ± 1.2	$P > 0.5$ (NS)
Effect of $D_2O^{\S}$			
Control	6	1.0 ± 0.2	—
D_2O	6	0.4 ± 0.1	$0.05 > P > 0.01$

*Radioimmunoassay measurement of AVP released (after subtracting 30 min basal AVP release) over a 30-min sampling period, after 600 stimulation pulses at 20 Hz. Data expressed as average values ± SEM for *n* samples.

‡*P* values determined using the Student's *t* test. NS, not significantly different. In *A*, each measurement was independently compared to the maximum release at 5 mM Ca^{2+}.

§Experiments were performed as shown in Fig. 12, except that the first 20-Hz stimulation was the control, and the second 20-Hz stimulation was performed under the experimental condition, i.e., where the H_2O in the medium was completely replaced by D_2O. (After Gainer et al., 1986.)

several components. The small upward deflection marks the arrival of the impulse, or excitation, in the terminals and we have called it, therefore, the E-wave (Salzberg et al., 1985). The larger, downward, slow signal appears to reflect secretion from the terminals, so we have called it the S-wave. If this association were more than accidental, we expected that the changes in transparency would have certain properties that are characteristic of secretory systems in general, and the release of AVP in particular. These included dependence upon frequency, with facilitation, dependence upon extracellular calcium concentration, and sensitivity to calcium antagonists and known secretagogues. All of these expectations were fulfilled, including a dramatic effect of D_2O substitution for water in the Ringer's solution bathing the preparation

(Salzberg et al., 1985), and, indeed, all of the manipulations that we've found to alter the light-scattering signal; changes in $[Ca^{++}]_o$, stimulation frequency (Fig. 11), various secretagogues, and blockers, have now been shown to produce nearly identical effects on AVP release, as determined by radioimmunoassay carried out in collaboration with Hal Gainer's laboratory at the National Institutes of Health (Gainer et al., 1986) (cf. Table I).

Thus, the light-scattering changes illustrated in Figs. 7 and 8 are clearly related intimately to secretion of peptides from the intact nerve terminals of the mouse neurohypophysis, and this observation allowed us (Obaid et al., 1989) to ask an interesting question, viz., whether the calcium channels that are affected by ω-conotoxin are those required for normal secretion from these terminals. In Fig. 9, the left-hand record (*A*) shows the light-scattering signal during seven stimuli delivered at 16 Hz. The right-hand record (*B*) shows that after 18 min, 5 μM ω-conotoxin blocks ~50% of the intrinsic optical signal. This strongly suggests that conotoxin blocks at least some of the calcium channels required for neuropeptide release in the intact nerve terminals of at least this vertebrate neurohypophysis. Fig. 10 shows, however, that 22 min in 5 μM nifedipine has no effect on the light-scattering signal, and this result is consistent with the absence of any effect of 5 μM nifedipine on the action potential in

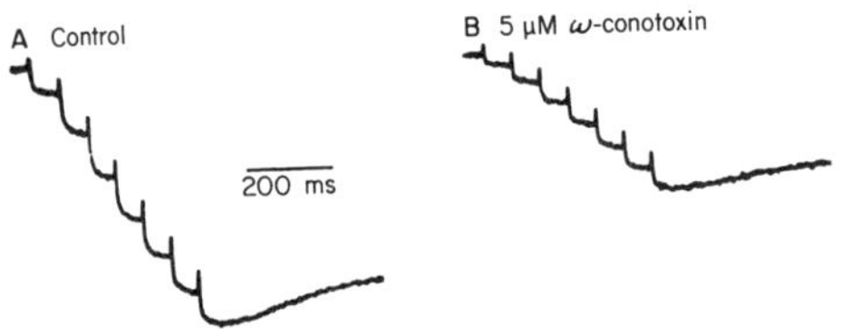

Figure 9. ω-Conotoxin GVIA (5 μM) blocks calcium channels required for neuropeptide release in the nerve terminals of the mouse neurohypophysis. (*A*) Light-scattering signal resulting from stimulation of the infundibulum of an unstained mouse (CD-1) neurohypophysis at 16 Hz for 400 ms in control Ringer's solution. A downward deflection of the trace represents an increase in the transparency of the tissue. (*B*) The effect of an 18-min exposure to a Ringer's solution containing 5 μM ω-conotoxin GVIA. Spatial average of the digitized outputs of twelve contiguous elements of the photodiode array; single sweep. 10×; 0.5 n.a.; 675 ± 26 nm; infundibular stimulation; AC coupling time constant, 3 s; response time constant (10–90%), 1.1 ms. (After Obaid et al., 1989.)

the neurohypophysial terminals of the frog. A similar null result was obtained (Obaid et al., 1989) with 5 μM Bay-K 8644 applied for at least 40 min.

Effect of Frequency on Nerve Terminal Spike Duration

Before concluding, we would like to turn to a slightly different aspect of the problem of secretion from the intact nerve terminals of vertebrates. This has to do with the effect of stimulation frequency on facilitation of neuropeptide release. Fig. 11 shows the release of AVP from the CD-1 mouse neurohypophysis, monitored by radioimmunoassay, under two conditions of stimulation. We had determined previously (Salzberg et al., 1985) that the size of the S-wave increases with an increase in stimulation frequency, and the radioimmunoassay data demonstrate that the effect is paralleled by the release of the neuropeptide itself. Here, release at 1 Hz is seen to be indistinguishable from background levels by radioimmunoassay, while at 20 Hz the release of AVP is enhanced (or facilitated) about 20-fold. (The AVP released was ~0.4% of total lobe content.) Notice the high degree of reproducibility of the data (Gainer et al., 1986).

An important hypothesis to explain the observed increase in the release of

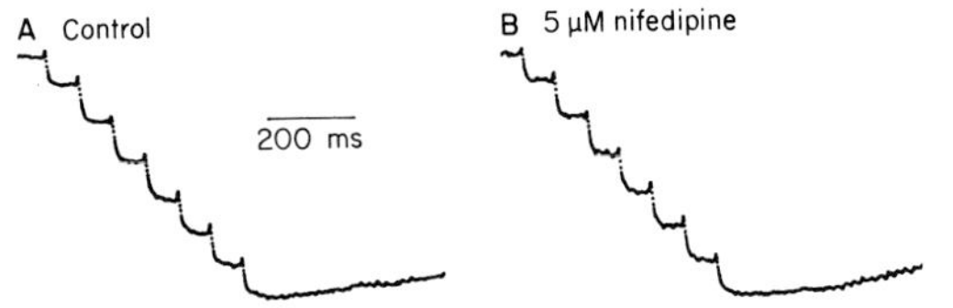

Figure 10. Nifedipine has no effect on the light-scattering change associated with the release of hormones from the nerve terminals of the mouse neurohypophysis. (*A*) Light-scattering signal resulting from stimulation of the infundibulum of an unstained mouse (CD-1) neurohypophysis at 16 Hz for 400 ms in control Ringer's solution. (*B*) Light-scattering signal after a 22-min exposure to a Ringer's solution containing 5 μM nifedipine (0.05% ethanol). Spatial average of the digitized outputs of four contiguous elements of the photodiode array; single sweep. 10×; 0.4 n.a; 675 ± 26 nm; infundibular stimulation; AC coupling time constant, 3 s; response time constant (10–90%), 1.1 ms. (After Obaid et al., 1989.)

oxytocin and vasopressin with frequency of stimulation focuses on the possible increase in the duration of the action potential at the terminal itself. Fig. 12 *A* shows two trains of action potentials recorded optically from the mouse neurohypophysis using the potentiometric merocyanine-oxazolone dye NK 2367. Stimulation was at 10 (top) and 16 Hz (bottom), respectively. Here, one sees the sum of the voltage change in the terminals and the intrinsic optical signal, the light-scattering change. Our purpose here was to try to evaluate any changes in spike duration during facilitated secretion. More precisely, the issue was whether the increase in AVP release at increased frequencies could be related to prolongation of the spike. Fig. 12 *B* shows the analysis of this sort of experiment. Traces *a–c* show the 10-Hz train, and *d–f* show the 16-Hz train. Traces *a* and *d* represent the first spike in each of the trains, and, naturally, they are virtually

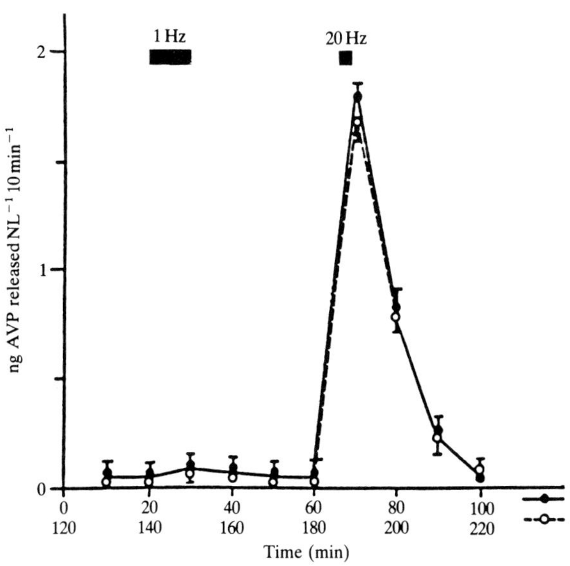

Figure 11. Radioimmunoassay measurements of AVP release from mouse neural lobes stimulated electrically in vitro. Each point on the graph represents the average (±SEM) value of AVP released in six independent experiments. Electrical stimulation consisted of 600 pulses (0.5 ms duration) at either 1 Hz (600 s) or 20 Hz (30 s). The neural lobes were stimulated in two cycles of 1 and 20 Hz. The first cycle is shown by filled circles (0–100 min, followed by wash-out, data not shown), and a repeat second cycle shown by open circles (120–220 min). Calcium concentration in the medium was 1 mM in these experiments. After the experiments, the total residual AVP in the lobes was determined, and AVP release at 20 Hz was found to be 0.4% of total lobe content. (After Gainer et al., 1986.)

identical. In traces *b* and *e*, we (Gainer et al., 1986) have grouped the individual spikes in each train, displacing them from one another in time by 0.5 ms. In this way, it is easy to see the decrease in spike amplitude for each subsequent spike in the train. This is consistent with the expected accumulation of K^+ in the restricted extracellular space. In *c* and *f*, the spike heights were normalized to the height of the first spike in the train and superimposed. This shows that there was no significant change in the rates of rise of the spikes, but, rather, a measurable broadening of each successive spike in the train. The increase between the first and the fifth spike in *c* was ~18% and, between the first and seventh spike in *f*, ~32%. These values are actually similar to those found in other systems, for example, in the cell body of the parabolic burster, R15, in *Aplysia.* One must be very cautious, however, in evaluating this data. Although spike broadening clearly occurs in the terminals of the mouse neurohypophysis, and could contribute to

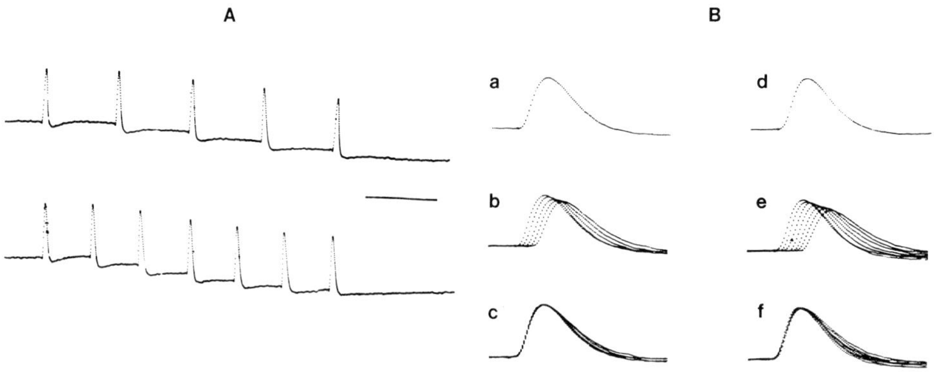

Figure 12. Analysis of spike broadening with frequency in terminals of the mouse neurohypophysis. (*A*) Optical recordings of action potentials from the mouse neurohypophysis in vitro at 10 Hz (*top*) and 16 Hz (*bottom*). Calibration bar equals 100 ms. (*B*) Analysis of the optical recordings from the mouse neurohypophysis evoked by 10 Hz (*a–c*) and 16 Hz (*d–f*) stimulation. The first spike in the trains are illustrated in *a* (for 10 Hz) and *d* (for 16 Hz). In *b* and *e*, the spikes were grouped in the order of stimulus, but separated by a 0.5-ms interval. This sort of display permitted easy visualization of the decrease in spike amplitude for each subsequent spike in the train. In *c* and *f*, the spike heights were normalized to the height of the first spike in each train, and superimposed. There are no significant changes in the rates of rise of the spikes, but a measurable broadening of each subsequent spike in the train is apparent. (After Gainer et al., 1986.)

enhanced release, possibly though increased calcium entry, this is by no means demonstrated.

Conclusion

Although the intrinsic optical signals discussed here are undoubtedly related to secretion, we have, as yet, no evidence to implicate any particular step among the sequence of events that couples excitation to secretion in the generation of these large intrinsic optical signals, and the identity of the physiological event or events responsible remains clouded. We are hoping that during the next few years we will better understand the origin of these very large and, to us, interesting light-scattering changes. To this end, we are planning to use quasielastic laser light scattering and photon correlation spectroscopy on model systems derived from the neurohypophysis, and, perhaps, to reconstruct all of the optical signals, including the action potential, in cuvette. We will then be in a better position to try to understand how excitation is coupled to secretion in vertebrate nerve terminals.

Acknowledgments

We are grateful to Dr. H. Gainer for introducing us to the neurohypophysis as an experimental system and for his participation in many of the experiments described here; to Drs. C. M. Armstrong, R. K. Orkand, D. B. Sattelle, and R. W. Tsien for valuable discussions; and to Drs. Orkand, R. Flores, T. D. Parsons, and D. M. Senseman for their participation in some of the experiments described. We are grateful, as well, to L. B. Cohen and S. Lesher for providing the software used for much of the data acquisition and display.

This work was supported by United States Public Health Services grant NS-16824.

References

Barrett, J. N., K. L. Magleby, and B. S. Palotta. 1982. Properties of single calcium activated potassium channels in cultured rat muscle. *Journal of Physiology*. 331:221–230.

Blaustein, M. P., and J. M. Goldring. 1975. Membrane potentials in pinched-off presynaptic nerve terminals monitored with a fluorescence probe: evidence that synaptosomes have potassium diffusion potentials. *Journal of Physiology*. 247:589–615.

Brandt, B. L., S. Hagiwara, Y. Kidokoro, and S. Miyazaki. 1976. Action potentials in the rat chromaffin cell and effects of acetylcholine. *Journal of Physiology*. 263:417–439.

Cohen, L. B., R. D. Keynes, and D. Landowne. 1972*a*. Changes in light scattering that accompany the action potential in squid giant axons: potential-dependent components. *Journal of Physiology*. 224:701–725.

Cohen, L. B., R. D. Keynes, and D. Landowne. 1972*b*. Changes in light scattering that accompany the action potential in squid giant axons: current-dependent components. *Journal of Physiology*. 224:727–752.

Cohen, L. B., and D. Landowne. 1970. Changes in light scattering during synaptic activity in the electric organ of the skate, *Raia erinacea*. *Biophysical Journal*. 10:16*a*. (Abstr.)

Cohen, L. B., and S. Lesher. 1986. Optical monitoring of membrane potential: methods of multisite optical measurement. *In* Optical Methods in Cell Physiology. P. DeWeer and B. M. Salzberg, editors. John Wiley and Sons, New York. 71–99.

Cohen, L. B., and B. M. Salzberg. 1978. Optical measurement of membrane potential. *Reviews of Physiology, Biochemistry, and Pharmacology*.83:33–88.

Cohen, L. B., B. M. Salzberg, H. V. Davila, W. N. Ross, D. Landowne, A. S. Waggoner, and C.-H. Wang. 1974. Changes in axon fluorescence during activity: molecular probes of membrane potential. *Journal of Membrane Biology*. 19:1–36.

Dellman, H.-D. 1973. Degeneration and regeneration of neurosecretory systems. *International Review of Cytology*. 36:215–315.

Douglas, W. W. 1963. A possible mechanism of neurosecretion-release of vasopressin by depolarization and its dependence on calcium. *Nature*. 197:81–82.

Douglas, W. W. 1978. Stimulus secretion coupling: variations on the theme of calcium activated exocytosis involving cellular and extracellular sources of calcium. *Ciba Foundation Symposia*. 54:61–90.

Douglas, W. W., and A. M. Poisner. 1964. Stimulus secretion coupling in a neurosecretory organ and the role of calcium in the release of vasopressin from the neurohypophysis. *Journal of Physiology.*172:1–18.

Dreifuss, J. J., I. Kalnins, J. S. Kelley, and K. B. Ruf. 1971. Action potentials and the release of neurohypophysial hormones in vitro. *Journal of Physiology.* 215:805–817.

Fernandez, J. M., E. Neher, and B. D. Gomperts. 1984. Capacitance measurements reveal stepwise fusion events in degranulating mast cells. *Nature.* 312:453–455.

Fiekers, J. F. 1983. Effects of the aminoglycoside antibiotics, streptomycin and neomycin, on neuromuscular transmission. I. Presynaptic considerations. *Journal of Pharmacology and Experimental Therapeutics.* 225:487–495.

Fox, A. P., M. C. Nowycky, and R. W. Tsien. 1987. Kinetic and pharmacological properties distinguishing three types of calcium currents in chick sensory neurons. *Journal of Physiology.* 394:149–172.

Freedman, J. C., and P. C. Laris. 1981. Electrophysiology of cells and organelles: studies with optical potentiometric indicators. *International Review of Cytology (Supplement).* 12:177–246.

Gainer, H., S. A. Wolfe, Jr., A. L. Obaid, and B. M. Salzberg. 1986. Action potentials and frequency-dependent secretion in the mouse neurohypophysis. *Neuroendocrinology.*43:557–563.

Gerschenfeld, H. M., J. H. Tramezzani, and E. De Robertis. 1960. Ultrastructure and function in the neurohypophysis of the toad. *Endocrinology.* 66:741–762.

Grinvald, A. 1985. Real-time optical mapping of neuronal activity: from single growth cones to the intact mammalian brain. *Annual Review of Neuroscience.* 8:263–305.

Grinvald, A., L. B. Cohen, S. Lesher, and M. B. Boyle. 1981. Simultaneous optical monitoring of activity of many neurons in invertebrate ganglia, using a 124 element photodiode array. *Journal of Neurophysiology.* 45:829–840.

Grinvald, A., and I. C. Farber. 1981. Optical recording of Ca^{2+} action potentials from growth cones of cultured neurons using a laser microbeam. *Science.* 212:1164–1169.

Grinvald, A., A. Manker, and M. Segal. 1982. Visualization of the spread of electrical activity in rat hippocampal slices by voltage-sensitive optical probes. *Journal of Physiology.* 333:269–291.

Gupta, R. K., B. M. Salzberg, A. Grinvald, L. B. Cohen, K. Kamino, S. Lesher, M. B. Boyle, A. S. Waggoner, and C.-H. Wang. 1981. Improvements in optical methods for measuring rapid changes in membrane potential. *Journal of Membrane Biology.* 58:123–137.

Hagiwara, S., and L. Byerly. 1981. Calcium channel. *Annual Review of Neuroscience.*4:69–125.

Heuser, J. E., and T. S. Reese. 1981. Structural changes after transmitter release at the frog neuromuscular junction. *Journal of Cell Biology.* 88:564–580.

Heuser, J. E., T. S. Reese, M. J. Dennis, Y. Jan, L. Jan, and L. Evans. 1979. Synaptic vesicle exocytosis captured by quick freezing and correlated with quantal transmitter release. *Journal of Cell Biology.* 81:275–300.

Hodgkin, A. L. 1938. The sub-threshold potentials in a crustacean nerve fibre. *Proceedings of the Royal Society of London, Biological Sciences.* 126:87–121.

Holz, G. G., K. Dunlap, and R. M. Kream. 1988. Characterization of the electrically-evoked release of substance P from dorsal root ganglion neurons: methods and dihydropyridine sensitivity. *Journal of Neuroscience*. 8:463–471.

Kamino, K., A. Hirota, and S. Fujii. 1981. Localization of pacemaking activity in early embryonic heart monitored using a voltage sensitive dye. *Nature*. 290:595–597.

Kamino, K., and A. Inouye. 1978. Evidence for membrane potential changes in isolated synaptic membrane ghosts monitored with a merocyanine dye. *Japanese Journal of Physiology*. 28:225–237.

Katz, B. 1969. The release of neural transmitter substances. Charles C. Thomas, Publisher, Springfield, IL. 193 pp.

Katz, B., and R. Miledi. 1969. Tetrodotoxin-resistant electrical activity in presynaptic terminals. *Journal of Physiology*. 203:459–487.

Kaufmann, R. and A. Fleckenstein. 1965. Ca^{++}-kompetitive electromechanische Entkoppelung durch Ni^{++}-und Co^{++}-Ionen am Warmblutermyokard. *Pflügers Archiv für Gesamte Physiologie*. 282:290–297.

Kostyuk, P. G., and O. A. Krishtal. 1977. Separation of sodium and calcium currents in the somatic membrane of mollusc neurons. *Journal of Physiology*.270:545–568.

Lemos, J. R., and M. C. Nowycky. 1987. One type of calcium channel in nerve terminals of the rat neurohypophysis is sensitive to dihydropyridines. *Society for Neuroscience Abstracts*. 13:793. (Abstr.)

Lindau, M., and J. M. Fernandez. 1986. IgE-mediated degranulation of mast cells does not require opening of ion channels. *Nature*. 319:150–153.

Llinas, R., and M. Sugimori. 1980*a*. Electrophysiological properties of in vitro Purkinje cell somata in mammalian cerebellar slices. *Journal of Physiology*. 305:171–195.

Llinas, R., and M. Sugimori. 1980*b*. Electrophysiological properties of in vitro Purkinje cell dendrites in mammalian cerebellar slices. *Journal of Physiology*. 305:197–213.

Martin, A. R., and G. Pilar. 1963*a*. Dual mode of synaptic transmission in the avian ciliary ganglion. *Journal of Physiology*. 168:443–463.

Martin, A. R., and G. Pilar. 1963*b*. Transmission through the ciliary ganglion of the chick. *Journal of Physiology*. 168:464–475.

Martin, A. R., and G. Pilar. 1964*a*. An analysis of electrical coupling at synapses in the avian ciliary ganglion. *Journal of Physiology*. 171:454–475.

Martin, A. R., and G. Pilar. 1964*b*. Presynaptic and post-synaptic events during post-tetanic potentiation and facilitation in the avian ciliary ganglion. *Journal of Physiology*. 175:17–30.

Mason, W. T., and R. E. Dyball. 1986. Single ion channel activity in peptidergic nerve terminals of the isolated rat neurohypophysis related to stimulation of neural stalk axons. *Brain Research*. 383:279–286.

McCleskey, E. W., A. P. Fox, D. H. Feldman, L. J. Cruz, B. M. Olivera, R. W. Tsien, and D. Yoshikami. 1987. ω-Conotoxin: direct and persistent block of specific types of calcium channels in neurons but not muscle. *Proceedings of the National Academy of Sciences*. 84:4327–4331.

Meech, R. W., and F. Strumwasser. 1970. Intracellular calcium injection activates potassium conductance in *Aplysia* nerve cells. *Federation Proceedings*. 29:834. (Abstr.)

Miller, C., E. Moczydlowski, R. Latorre, and M. Phillips. 1985. Charybdotoxin, a protein inhibitor of single Ca^{2+}-activated K^{+} channels from mammalian skeletal muscle. *Nature*. 313:316–318.

Miller, R. J. 1987. Multiple calcium channels and neuronal function. *Science.* 235:46–52.

Nordmann, J. J. 1977. Ultrastructural morphometry of the rat neurohypophysis. *Journal of Anatomy.* 123:213–218.

Obaid, A. L., R. Flores, and B. M. Salzberg. 1989. Calcium channels that are required for secretion from intact nerve terminals of vertebrates are sensitive to ω-conotoxin and relatively insensitive to dihydropyridines: optical studies with and without voltage-sensitive dyes. *Journal of General Physiology.* 93:715–729.

Obaid, A. L., R. K. Orkand, H. Gainer, and B. M. Salzberg. 1985. Active calcium responses recorded optically from nerve terminals of the frog neurohypophysis. *Journal of General Physiology.* 85:481–489.

Obaid, A. L., and B. M. Salzberg. 1985. Selective block of the calcium-mediated potassium conductance [$g_{K(Ca)}$] in vertebrate nerve terminals by charybdotoxin (CTX): an optical demonstration using the neurohypophysis of *Xenopus. Journal of General Physiology.* 86:19*a*. (Abstr.)

Olivera, B. M., W. R. Gray, R. Zeikus, J. M. McIntosh, J. Varga, J. Rivier, V. de Santos, and L. J. Cruz. 1985. Peptide neurotoxins from fish-hunting cone snails. *Science.* 230:1338–43.

Parsons, T. D., A. L. Obaid, and B. M. Salzberg. 1986. Aminoglycoside antibiotics block calcium currents in nerve terminals of the frog neurohypophysis: an optical demonstration using potentiometric probes. *Biological Bulletin.* 171:497. (Abstr.)

Rane, S. G., G. G. Holz, and K. Dunlap. 1987. Dihydropyridine inhibition of neuronal calcium current and substance P release. *Pflügers Archiv.* 409:361–366.

Rodriguez, E. M., and H.-D. Dellman. 1970. Hormonal content and the ultrastructure of the disconnected neural lobe of the grass frog (*Rana pipiens*). *General and Comparative Endocrinology.* 15:272–288.

Ross, W. N., B. M. Salzberg, L. B. Cohen, A. Grinvald, H. V. Davila, A. S. Waggoner, and C.-H. Wang. 1977. Changes in absorption, fluorescence, dichroism, and birefringence in stained giant axons: optical measurement of membrane potential. *Journal of Membrane Biology.* 33:141–183.

Salzberg, B. M. 1983. Optical recording of electrical activity in neurons using molecular probes. *In* Current Methods in Cellular Neurobiology, Vol. 3: Electrophysiological Techniques. J. L. Barker and J. E. McKelvey, editors. John Wiley and Sons, Inc., New York.

Salzberg, B. M., and F. Bezanilla. 1983. An optical determination of the series resistance in *Loligo. Journal of General Physiology.*82:807–818.

Salzberg, B. M., A. Grinvald, L. B. Cohen, H. V. Davila, and W. N. Ross. 1977. Optical recording of neuronal activity in an invertebrate central nervous system: simultaneous monitoring of several neurons. *Journal of Neurophysiology.* 40:1281–1291.

Salzberg, B. M., A. L. Obaid, and R. Flores. 1987. Calcium channels required for neuropeptide release in the intact nerve terminals of vertebrate neurohypophyses are sensitive to omega-conotoxin and insensitive to dihydropyridines: optical studies with and without voltage-sensitive dyes. *Biological Bulletin.* 173:446. (Abstr.)

Salzberg, B. M., A. L. Obaid, and H. Gainer. 1985. Large and rapid changes in light scattering accompany secretion by nerve terminals in the mammalian neurohypophysis. *Journal of General Physiology.* 86:395–411.

Salzberg, B. M., A. L. Obaid, D. M. Senseman, and H. Gainer. 1983. Optical recording of action potentials from vertebrate nerve terminals using potentiometric probes provides evidence for sodium and calcium components. *Nature.* 306:36–40.

Senseman, D. M., and B. M. Salzberg. 1980. Electrical activity in an exocrine gland: optical recording using a potentiometric dye. *Science.* 208:1269–1271.

Senseman, D. M., H. Shimizu, I. S. Horwitz, and B. M. Salzberg. 1983. Multiple site optical recording of membrane potential from a salivary gland: interaction of synaptic and electrotonic excitation. *Journal of General Physiology.* 81:887–908.

Standen, N. B. 1981. Ca channel inactivation by intracellular Ca injection into *Helix* neurones. *Nature.* 293:158–159.

Tsien, R. W. 1986. Modulation of calcium currents in heart cells and neurons. *In* Neuromodulation. I. B. Levitan and L. K. Kaczmarek, editors. Oxford University Press, Oxford. 206–242.

Waggoner, A. S. 1979. Dye indicators of membrane potential. *Annual Review of Biophysics and Bioengineering.* 8:47–68.

Chapter 21

Possible Role for Ionic Channels in Neurosecretory Granules of the Rat Neurohypophysis

José R. Lemos, Karen A. Ocorr, and Jean J. Nordmann

Worcester Foundation for Experimental Biology, Shrewsbury, Massachusetts 01545

Introduction

The pioneering work of Douglas and co-workers, on catecholamine release from the adrenal medulla (Douglas and Rubin, 1963) and peptide release from the neurohypophysis (Douglas and Poisner, 1964*a*, *b*), has established the basic hypothesis for the mechanism by which a molecule, enclosed in a granule (or vesicle), is released into the external medium.

Depolarization-Secretion Coupling

The following steps in the above process of depolarization-secretion coupling have been demonstrated in the neurohypophysis (Nordmann, 1983): the arrival of action potentials (Salzberg et al., 1983) induces the depolarization of these nerve endings which have been shown to have in their plasma membrane both Na (Nordmann and Dyball, 1978) and two types of Ca channels (Lemos and Nowycky, 1987, 1989; Nowycky and Lemos, 1988). The electrically induced depolarization promotes the entry of calcium into the nerve terminals (Brethes et al., 1987), which then triggers the release of the neurosecretory granule (NSG) contents via exocytosis (Nordmann et al., 1987). The key question, however, remains unanswered: by what process(es) does calcium entry lead to exocytosis and the release of hormones?

Release Mechanisms

In order for exocytosis to occur the NSG must be brought into close proximity or, perhaps, even apposition with the plasma membrane. Most observers have invoked either cytoskeletal or enzymatic components as agents involved in "priming" the release process (Augustine et al., 1987). For example, proteins, such as synapsin I, have been found to be associated with secretory vesicles and it is hypothesized that their phosphorylation by Ca-activated kinases enables the movement of the vesicles toward the plasma membrane and "active release sites" (Llinas et al., 1985). Ca-dependent processes such as actin or microtubule polymerization/depolymerization could also be responsible for NSG movement within neuroendocrine terminal structures (not to mention, of course, their transport from sites of synthesis). For example, fodrin, which may interact between actin filaments and NSG, has been implicated in the process of release from chromaffin cells (Perrin et al., 1987).

It has also been hypothesized that entry of Ca^{2+} into neurosecretory cells activates a channel in the NSG membrane and an elevation of cations and anions inside the vesicle. The resulting increase in osmolarity would then lead to swelling of the NSG and thus promote fusion with the plasma membrane and the release of their contents (Pollard et al., 1979; Cohen et al., 1982; Stanley and Ehrenstein, 1985).

One way to test this last hypothesis is to examine the NSG membranes for the presence of ion channels. Functional reconstitution of channel-forming proteins originating from sarcoplasmic reticulum, transverse tubules, axon membrane, *Escherichia coli,* mitochondria, brain, and epithelial membranes has already been successful (Latorre et al., 1985). It is now possible to isolate, on isoosmotic gradients, relatively pure NSG (Nordmann et al., 1979). The ability to obtain such NSG membranes (Fig. 1) has made it possible for us to analyze NSG ion fluxes (Lemos and Nordmann, 1986; Lemos et al., 1987) by using "tip-dip" methodology (Coronado and Latorre, 1983). We now report the characterization of two ionic channels in NSG membranes and speculate on their role in depolarization-secretion coupling.

Materials and Methods

Purified brain phosphatidyl ethanolamine (PE) and brain phosphatidyl serine (PS) were obtained from Avanti Polar Lipids, Inc. (Birmingham, AL). Lipid purity was verified using thin layer chromatography prior to use (K6 TLC plates, solvent system of chloroform, methanol, distilled water, 65:25:4; Whatman, Inc., Clifton, NJ). The PE and PS were mixed at a ratio of 3:1 (wt:wt) in choloroform-washed test tubes, and 5.0 mg aliquots were dried under a stream of filtered nitrogen. Dried lipids were resuspended (10 mg lipids/ml) in sterile filtered HEPES (20 mM) at pH 7.0 with sonication for 30 min. Purified membranes were mixed with the lipid mixture at an

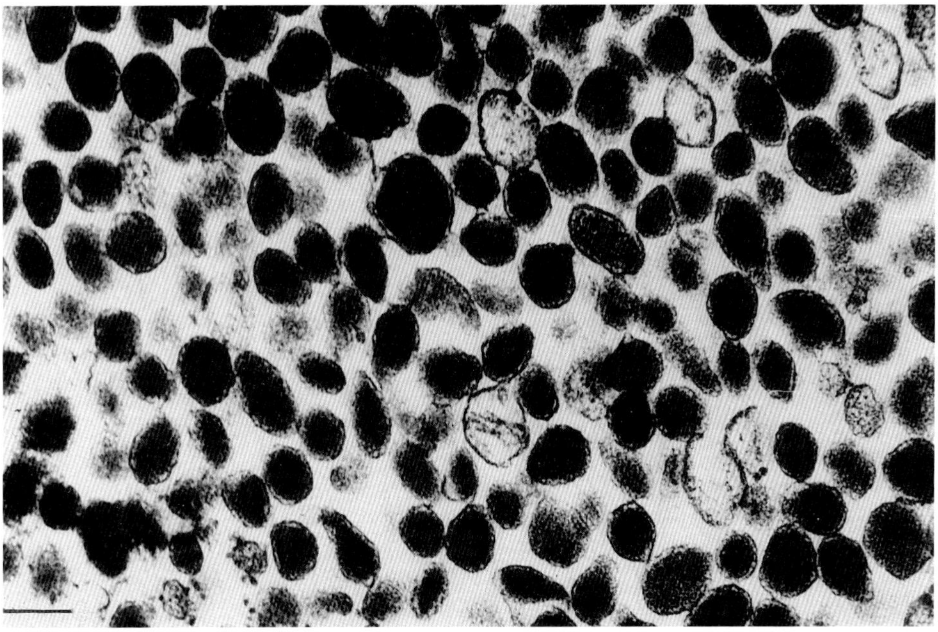

Figure 1. Electron micrograph of purified preparation of neurohypophysial NSGs that exhibited vasopressin, oxytocin, or neurophysin reactivity. The granules were prepared as described in Materials and Methods. This micrograph is part of a section through the 27,000 *g* pellet. The NSG were fixed for 90 min with 2.5% gluteraldehyde in 0.1 M Na-cacodylate buffer and postfixed with 1% OsO_4. After dehydration with alcohol they were embedded in Epon resin. Note the relatively small contamination by mitochondrial fragments. Scale bar, 200 nm.

approximate ratio of 10:1 (wt:wt) and the mixture was sonicated for ~30 min until it was clear. All steps were performed at 4°C under filtered nitrogen. Aliquots of the membrane/lipid mixture were stored at −20°C. 4-acetamido-4′-isothiocyanostilbene 2-2′-disulfonic acid (SITS) and all other compounds were obtained from Sigma Chemical Co. (St. Louis, MO).

NSG Membrane Preparation

NSG membranes were prepared from the posterior pituitary gland, or neural lobe, as described by Nordmann et al. (1979). Neural lobes from 200–300 g rats were removed and dipped first in the antibiotic Listerene and then in Locke's solution. The lobes were then homogenized on ice in 0.3 M sucrose containing 10 μM EGTA and 10 mM HEPES (pH 7.4). The homogenate was centrifuged for 4 min at 800 *g*. The supernatant fraction was centrifuged three more times, for 15 min each, at 3,400, 8,000, and 27,000 *g*. After the spin at 27,000 *g* the supernatant was discarded and the pellet (see Fig. 1) was resuspended in 10 mM HEPES, 10 μM EGTA (pH 7.4); some preparations were further purified using a 20% isoosmotic Percoll gradient, and then recentrifuged at 100,000 *g* for 60 min. Both methods of purification gave similar electrophysiological results. Purification using isoosmotic sucrose gradients, however, tended to inhibit later fusion of NSG with lipids. The final membrane pellet was resuspended in 200 mM KCl, 20 mM HEPES (final concentrations, pH 7.4) and stored frozen at −70°C in 100 μl aliquots. All steps were performed at 4°C.

Preparation of Artificial Bilayers

Phospholipid bilayers were made using a modification of the procedure described by Coronado and Latorre (1983). At the start of each experiment an aliquot of the membrane/lipid mixture was thawed and refrozen three times in an acetone dry-ice bath to promote vesicle fusion. Approximately 20 μl of the mixture was layered over a central well of a cell well plate (Corning Glass Works, Corning, NY) containing a sterile filtered solution of 100 mM $BaCl_2$ (unless otherwise indicated). A stream of filtered nitrogen was then directed over the well containing the membranes. Adjacent wells were filled with the sterile filtered solutions of various ionic compositions. All solutions were buffered to pH 7.4 with 20 mM HEPES. All wells were connected to ground via a silver-chlorided wire and the cell well plate, ground wires, and electrodes were washed in methanol before use. Electrodes were pulled from 2-mm capillary glass (Jencon Scientific Ltd., Leighton Buzzard, UK) which had been washed in distilled water for 30 min, boiled for 30 min in acetone, then boiled in distilled water for an additional 30 min, and allowed to dry overnight in an oven. Electrodes were then fire polished with a Narashige Scientific Instruments (Tokyo, Japan) electrode polisher (model MF-83) giving final resistances of 5–10 MΩ. Electrodes were normally filled with sterile filtered 50 mM $BaCl_2$ in order to induce an osmotic gradient between the pipette, bath, and NSG interior. An osmotic gradient across the pipette tip (vesicle-containing side, hyperosomotic) promotes channel incorporation into the bilayer from channel-containing membrane fragments, NSG, or liposomes (Woodburry and Hall, 1988). The electrode was lowered into the well containing the membranes while constant, positive pressure was applied. The pressure was then removed and the electrode was moved to a position close to the membrane/lipid layer. The electrode was rapidly raised out of the saline bath and immediately reinserted. Unless bilayers were formed with seals of > 50 GΩ the electrode was discarded. Occasionally, suction or hyperpolarizing current pulses were used to improve the quality of the seal.

Electrophysiology

Once the bilayer had been formed on the tip of the patch pipette, conventional patch-clamp techniques (Hamill et al., 1981; Lemos et al., 1986) were applied to study the channel proteins that may have been incorporated into this bilayer. Electrodes were placed in a head stage amplifier and connected to a patch clamp (model EPC-7; Medical Systems Corp., Great Neck, NY). Signals were filtered with an 8-pole Bessel filter (model 902LPF; Frequency Devices, Haverhill, MA) and digitized using a Sony PCM digitizer (model 501ES; Tokyo, Japan). Data were recorded on a Sony Beta HiFi recorder (model SL-HF900) and analyzed using P-clamp software (Axon Instruments, Inc., Burlingame, CA).

Results

NSG Membranes

The anatomy of the neurohypophysis and biochemical separation techniques made possible the reconstitution of ionic channel proteins from specific intracellular organelles. In particular, ion channels in NSGs, from which neurohormones are released, were studied to analyze their possible roles in depolarization-secretion coupling.

Electron micrographs of the purified NSG preparation (Fig. 1; see also Nordmann et al., 1979) show its relative purity. Radioimmunoassays for vasopressin, oxytocin, and associated neurophysins were used to follow NSG purification. Such assays and specific membrane markers allowed us to calculate the relative purity of the NSG at 90–95%, with mitochondria as the major contaminant (Nordmann et al., 1979), as can be seen in Fig. 1. The purified NSG membranes were then examined for the presence of ion channels using tip-dip methodology. Our evidence indicates that these NSG membranes contain at least two ionic channels.

Anion Channel

We have observed a small amplitude channel (Fig. 2) in membranes prepared from isolated rat neurohypophysial NSGs. Increasing depolarizations (of the pipette) tend to increase the activity of this channel slightly (compare Fig. 2, *A–D*), but this voltage dependence is relatively weak. The channel does not seem to be affected by Ca^{2+} in the bath as is evident by its normal activity even in concentrations as low as 10 nM of this divalent cation (Figs. 2 and 3 *A*).

To investigate whether this channel could be important in anion transport we tested the effects of the anion transport blocker SITS. Fig. 3 shows normal channel activity in 100 mM KCl and 10 nM free Ca^{2+} at a holding potential of +90 mV. Moving the patch into the same solution but with 30 μM SITS (Fig. 3 *B*) reduced the activity of the channels (minimum of two in this patch). The effect of SITS appears to be mainly a reduction in the mean open time of the channel. Higher concentrations (100 μM) of SITS (Fig. 3 *C*) further reduced the channels' activity. This result is important since 100 μM SITS also inhibits release, by 72%, from digitonin-permeabilized neurohypophysial terminals (Lemos and Nordmann, 1986; Cazalis et al., 1987*b*).

The current-voltage relationship for the open channel, in symmetrical 150 mM KCl (Fig. 4 *A*), is linear with a slope conductance of 30 pS. The single-channel conductance increases to >40 pS in higher salt concentrations (e.g., Fig. 2). In asymmetrical salt gradients the open channel reversal potential indicates that it is permeable to anions, such as Cl^-, and not to cations (see Fig. 2). Furthermore, this NSG anion channel is selective between anions, since the channel I-V shows rectification if SO_4^{2-} is on one side of the bilayer.

Cation Channel

We have observed another, larger amplitude channel in membranes prepared from isolated neurohypophysial neurosecretory granules. Fig. 4 *B* shows that this NSG channel, in symmetrical 200 mM KCl, has a linear slope conductance of 130 pS. In

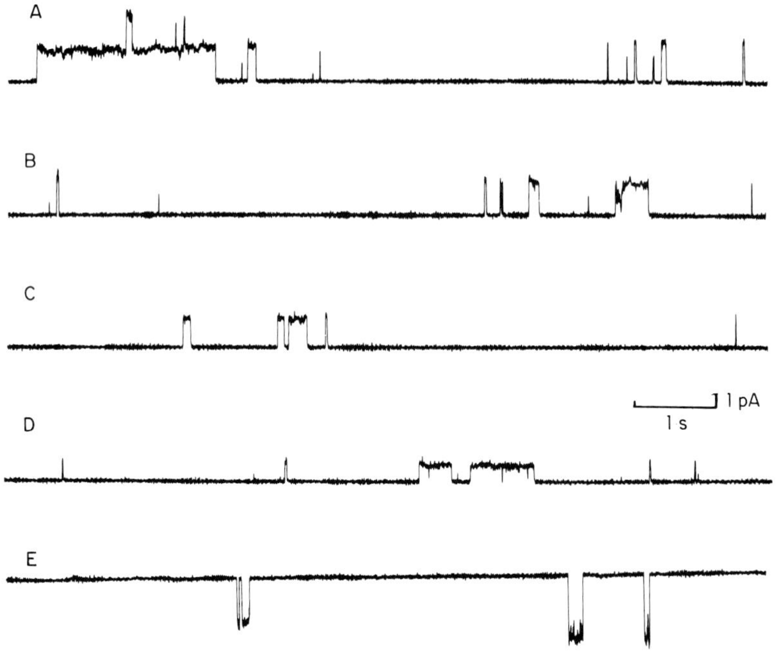

Figure 2. Anion channel from rat neurohypophysial secretory granules. Current records from rat posterior pituitary NSG membranes incorporated into an exogenous lipid (PE/PS; 3/1) bilayer formed on the tip of a pipette. 200 mM $BaCl_2$ in the pipette and 100 mM KCl with 10 nM free Ca^{2+} in the bath. The patch was held at the following pipette potentials (V_p): (*A*) +90 mV, (*B*) +75 mV, (*C*) +60 mV, (*D*) +45 mV, (*E*) −45 mV. Recorded at 3 kHz bandwidth, displayed on chart recorder.

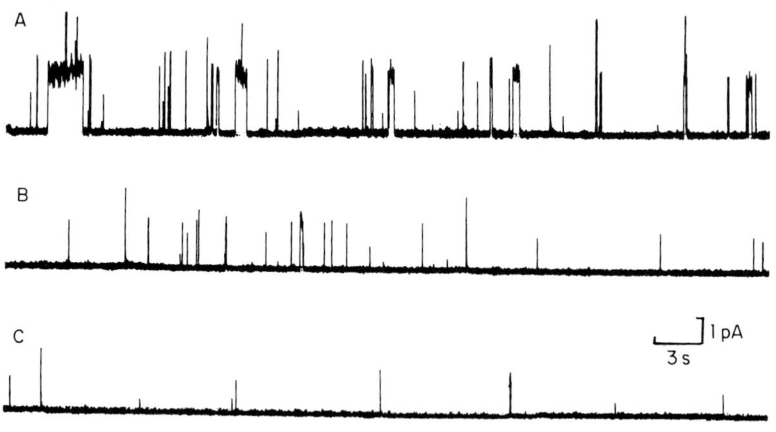

Figure 3. Effect of SITS on NSG anion channel. (*A*) Control recording in same solutions as in Fig. 2. Patch held at V_p = +90 mV. (*B*) Same patch after addition of 30 μM SITS to the bath. (*C*) In the presence of 100 μM SITS. Recorded at 3 kHz bandwidth, displayed on chart recorder.

asymmetrical salt gradients (e.g., Fig. 5) the reversal potentials indicate that this channel appears to be permeable primarily to cations, with K^+ permeating twice as well as Na^+ or Cs^+.

Fig. 5 indicates that the channel can also allow divalent cations, such as Ba^{2+} or Ca^{2+}, to permeate. The gating kinetics for this channel are much more complicated than for the anion channel. For example, the cation channel can open to more than one conductance level at a variety of pipette potentials (Fig. 5, *A–C*). The two levels of openings seen here with 50 mM $BaCl_2$ in the pipette and 100 mM KCl and 0.1 μM

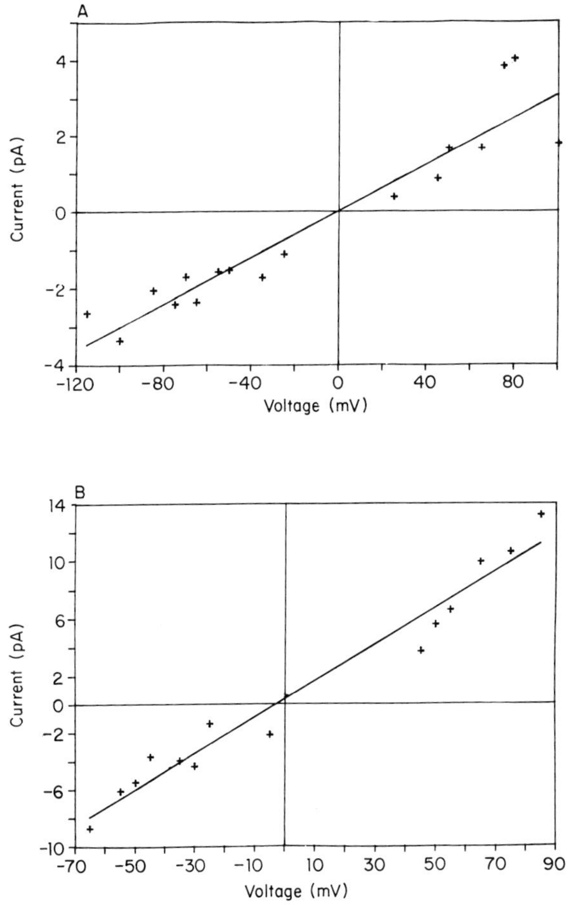

Figure 4. Conductance of the two ionic channels of neurohypophysial NSG. (*A*) I-V relationship of the smaller NSG ionic channel in symmetrical 150 mM KCl. The amplitude (current) of small channel openings (see Fig. 2) were plotted against the pipette potential (voltage). Calculated regression shows a slope conductance of 30 pS. (*B*) I-V relationship of the larger NSG ionic channel in symmetrical 200 mM KCl. The primary amplitude (see Fig. 6) of large channel openings were plotted against the membrane potential. Calculated regression gives a slope conductance of 130 pS.

Ca^{2+} in the bath have conductances of 89 and 177 pS, respectively. If an I-V curve is constructed in such asymmetrical solutions, both conductances reverse at the same potential. As many as four different conductance levels have been observed in a single patch, with openings occurring, as in Fig. 5, between any and all levels. As is evident here, its gating shows little voltage dependence at pH 7.4.

Fig. 6 shows an unusual instance of a patch with a single conductance level. Under these conditions it is clear that this cation channel is bimodally regulated by Ca^{2+}. It opens only in the presence of >30 nM free Ca^{2+}, with maximal effects at ~3 μM, where

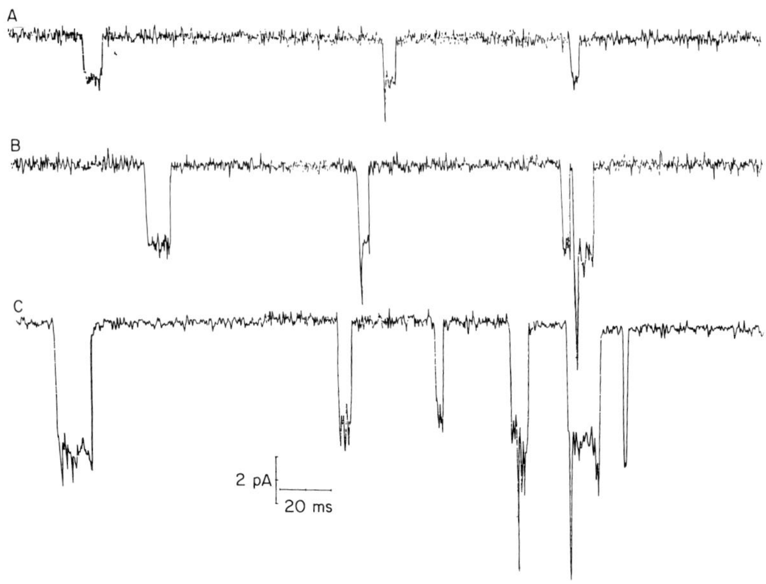

Figure 5. Ionic channel currents from tip-dip patches showing large channel type observed in isolated rat neurohypophysial NSG. The patch was held at V_p = (*A*) −45, (*B*) −65, and (*C*) −75 mV, with 100 mM KCl and 0.1 μM free Ca^{2+} in bath and 50 mM $BaCl_2$ in pipette. 1 kHz bandwidth. Note multiple levels of openings at each V_p.

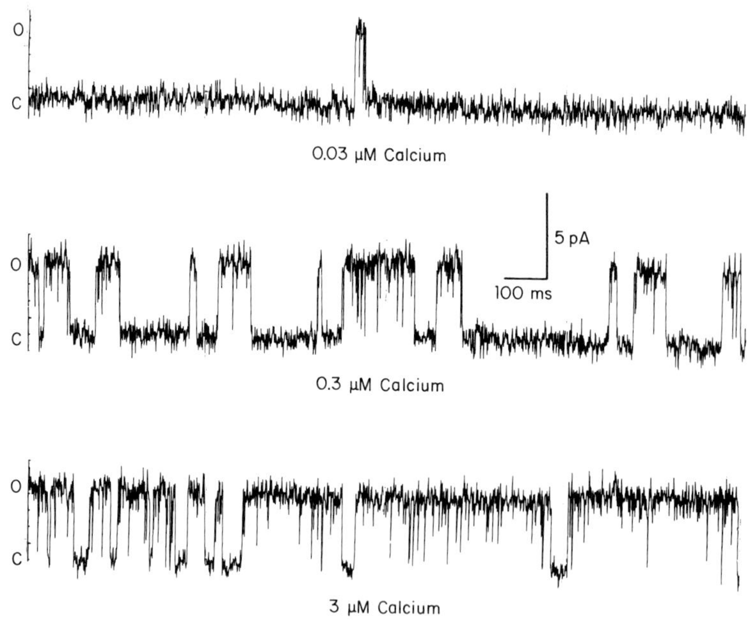

Figure 6. Ca^{2+} dependence of NSG cation channel openings. The patch showing a single conductance level was held at V_p = +40 mV. The pipette contained 200 mM NaCl and the bath consisted of 200 mM KCl and the indicated free Ca^{2+} concentrations. Open (o) and closed (c) levels for this channel are indicated at left of each trace. 3 kHz bandwidth.

the channel is almost always open. In contrast, its gating is inhibited at concentrations above 10 μM (data not shown).

Discussion

Is the Source of Channels the NSG Membranes?

Neurohypophysial nerve terminals are especially well suited for purification techniques because of the near absence of non–terminal structures (Dayanithi and Nordmann, 1989). Another point in their favor, at least in comparison to brain "synaptosomes," is that these preparations consist of only "presynaptic" nerve endings with no contamination by "postsynaptic" elements (see Lemos and Nordmann, 1986). Furthermore, the released material consists only of the two neuropeptides, oxytocin and vasopressin, and their associated neurophysins (Nordmann, 1983). Thus the starting material is already highly enriched in relatively homogenous NSG membranes compared with other membranes (Nordmann and Cazalis, 1986).

The above argues that it should be feasible to obtain relatively pure NSG membranes from the neurohypophysis, and, indeed, they appear to be as much as 95% pure as judged by electronmicroscopy (Fig. 1) and the use of specific markers (Nordmann et al., 1979). The major contaminant, however, appears to be mitochondrial membranes. We do not feel that this is the source for the activity described here since we have found neither the small anion nor larger cation channel in fractions from the neurohypophysis enriched in mitochondria ($n = 21$), although other channels are evident. In control recordings, where no NSG membranes were added to the exogenous lipids during sonication, no openings showing the characteristics of the two channels described here have been seen ($n = 77$). Care must be taken, however, not to use high lipid concentrations or to allow the phospholipids to oxidize, otherwise spurious "unitary events" are possible (Ocorr, K. A. and J. R. Lemos, unpublished; Woodbury, 1989). Furthermore, in hundreds of cell-attached and inside-out recordings from the nerve terminal plasma membrane, no channel with the characteristics of the NSG anion channel has been observed (Lemos, J. R., and P. Thorn, unpublished). The cation channel, however, may have a counterpart in the plasma membrane (Lemos and Nordmann, 1986; Thorn and Lemos, 1988). Our evidence indicates that both an anion and a Ca^{2+}-regulated cation channel exist in neurohypophysial NSG. However, because of our reconstitution methods, we cannot make any estimates as to the numbers of such channels per NSG or prove that they are only found in peptide-hormone–containing structures.

Comparison of NSG Channels with Other Ion Channels

Similar channels have been recently reported in fused *Torpedo* synaptic vesicles and bovine neurohypophysial NSG. The anion channel described here and in Lemos and Nordmann (1986) appears to be similar to a channel reported in NSG membranes from bovine pituitary glands studied in planar bilayers (Stanley et al., 1988). Both preparations exhibit channels with similar conductances that become larger with increasing Cl^- concentrations and have a similar weak voltage dependence. Furthermore, each channel is blocked by anion transport blockers. The opening probabilities for the two channels, however, are quite different, but this could be due to the different recording techniques utilized. A less completely characterized anion channel has also been found in *Torpedo* vesicles (DeRiemer et al., 1988). In contrast, a much larger, strongly voltage-dependent anion channel from chromaffin cell membranes has been recently localized to the mitochondrial fraction (Thieffry et al., 1988).

The NSG cation channel is similar in size (Fig. 4 *B*) to one seen in fused *Torpedo* synaptic vesicles (Rahamimoff et al., 1988). Both channels are more selective for K^+ than Na^+, although it is not known if divalent cations, such as Ca^{2+} or Ba^{2+}, can also permeate through or activate the vesicular channel. In addition, these cation channels from both preparations show complex gating kinetics. Records from each preparation exhibit multiple conductance levels of approximately similar amplitudes (Fig. 5). This behavior could be explained in a number of ways (see Rahamimoff et al., 1988). Since the reversal potential for each conductance level is identical and openings occur without intervening levels being evident (Fig. 5), the most probable explanations are the cooperative opening of similar channels or the opening of a single channel with multiple conductance levels.

The ryanodine receptor of sarcoplasmic reticulum (SR) also exhibits this type of gating kinetics, again with similar conductances (Ma et al., 1988). The NSG and SR channel are both activated at similar $[Ca^{2+}]$ and inhibited at higher concentrations. These two channels are permeable to both monovalent and divalent cations (see Figs. 4 *B* and 5), but less so to anions. Neither channel appears to be voltage dependent in its gating at neutral pH. Furthermore, both channels resemble gap junction channels (Spray et al., 1982, 1985; Lowenstein, 1987) in these respects (Ma et al., 1988). The NSG channel, however, has not yet been tested for pH regulation and specific block by heptanol and octanol so that we cannot conclude that it is a gap junction–type channel (Spray et al., 1986; Lowenstein, 1987). We would not expect the channel to be identical to gap junctions, in any case.

Possible Functions of NSG Channels

The two channels could simply function as anion and cation transporters (Pollard et al., 1979; Stanley et al., 1988) in order to establish concentrations of these ions inside the NSG. Another possibility is that the insertion of these channels into the plasma membrane during fusion of NSGs would be a form of electrical feedback that could alter transmitter release in response to subsequent action potentials, or it might act as a signal for further synthesis at somata via Ca-dependent activation of protein synthesis (Kley et al., 1987).

Perhaps most interestingly, these NSG ion channels could be directly involved in the molecular mechanism underlying release. Since the cation NSG channel shows some similarities to gap junctions, this Ca-activated channel could be the analogous "fusion pore" for nerve cells (Rahamimoff et al., 1988) reported (Breckenridge and Almers, 1987*b*) to occur during mast cell exocytosis. In the case of gap junctions, the two hemi-channels or "connexons" are coupled together in their gating when the two bilayers are apposed. An analogous situation could exist for exocytosis where the NSG bilayer has to come into close apposition with the plasma membrane of the terminal. A similar hypothesis has recently been tendered for excitation-contraction coupling by Ma et al. (1988).

The mechanism of electrical signal transmission across gap junctions by means of two ion-permeable pores in series seems applicable to depolarization-secretion coupling (see Fig. 7). In analogy with gap junctions, the NSG cation channel would be coupled to a similar channel in the terminal plasma membrane and this combination would then span the two bilayers. In contrast to conventional gap junctions, however, the "fusion pore" channel would have its "gap" in the terminal cytoplasm rather than in the extracellular milieu. Our hypothesis (Fig. 7) is that, once the vesicle is near the plasma membrane, elevated intracellular Ca^{2+} would cause the simultaneous opening of the two hemi-channels and/or would increase their probability of forming a complete "fusion pore." Unfortunately it is not clear if gap junctions are coupled or activated by low Ca^{2+} acting between the connexons (Spray et al., 1985), but in

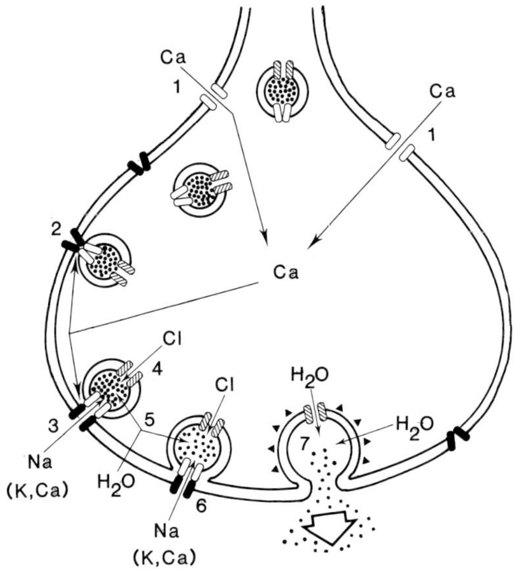

Figure 7. Model of depolarization-secretion coupling showing hypothesized role(s) of NSG channels and calcium. It is postulated that channels in both plasma and NSG membranes, when apposed, could form a "fusion pore" analogous to a gap junction–like connection between the inside of the NSG and the extracellular milieu. Representational drawing of neurohypophysial nerve terminal. NSG are dense-core vesicles entering from axon (at top). (*1*) Ca enters through two types of channels and elevates its intracellular levels. (*2*) In this hypothesis Ca could stabilize the interaction between the NSG cation channel (*clear*) and a similar channel in plasma membrane (*dark*). In this context the plasma membrane hemichannel would act as a "docking protein" localizing release to specific sites on the plasma membrane. (*3*) Ca would also lead to the opening of the two channels, and, when apposed, they could form a gap junction–like "fusion pore." When the fusion pore is open, extracellular cations would move into the NSG down their concentration and electrical gradients. The primary cation would be Na, although Ca and K could also enter. (*4*) As positive charges accumulated inside the NSG, anions, such as Cl, would also flow in through the NSG anion channel (*striped*). (*5*) This osmotic increase would force water (H_2O) to enter the NSG and cause them to swell. (*6*) This swelling would tend to drive the NSG and plasma membranes closer, especially with the channel anchoring them together, and promote adhesion. (*7*) If the area of adhesion was sufficient and enough of an osmotic gradient were formed, then the two membranes would eventually fuse and lead to subsequent lysis and expulsion (*large white arrow*) of the contents of the NSG: the neuropeptides oxytocin and vasopressin.

exocytosis such bimodal regulation by intracellular Ca^{2+} is well documented (Augustine et al., 1987). The hemi-channel in the nerve terminal plasma membrane could, furthermore, act as a "docking protein," thus ensuring that release only occurs at specific sites. Cations and other solutes would then be able to enter the vesicle, moving down their concentration and/or electrical gradients (Scherman et al., 1982) and increase the osmotic concentration inside the vesicle. Anions could enter as counter ions through the NSG anion channel, further elevating the NSG osmotic pressure. This osmotic gradient would cause water to move into the NSG with subsequent swelling. Such swelling after fusion pore formation has been observed in mast cell granules just prior to exocytosis (Breckenridge and Almers, 1987; Zimmerberg et al., 1987). Since the gap junction proteins anchor the vesicle to the plasma membrane, there might be considerable adhesion between the two membranes. Swelling would increase the NSG surface tension and this would promote fusion of the two bilayers (Finkelstein et al., 1986; Rand and Parsegian, 1986; Woodburry and Hall, 1988). Osmotic swelling may not be enough to cause fusion, however, and fusiogenic proteins or lipids might also be required. Finally, the osmotic pressure would cause expulsion of the NSG contents into the extracellular medium as has been shown to occur in the sea urchin egg cortical reaction (Whitaker and Zimmerberg, 1987).

In support of the possible role of NSG channels and osmolarity in the mechanism of release, are the following recent findings: (*a*) CCCP (carbonyl cyanide *m*-chlorophenylhydrazone), a H^+ ionophore, has been found to considerably reduce the transmembrane potential of isolated neurohypophysial granules. Using voltage-sensitive dyes Scherman et al. (1982) have shown that CCCP, by blocking electrogenic H^+ accumulation, dissipates the NSG transmembrane potential. CCCP inhibits Ca-dependent release from permeabilized nerve terminals (Cazalis et al., 1987*b*), suggesting that the NSG transmembrane potential is necessary for exocytosis to occur. The NSG potential could be important for the action of the NSG channels in increasing osmotic pressure. (*b*) Increasing osmotic pressure also inhibits neurohormone release from both permeabilized and intact neurohypophysial terminals, with increased sensitivity in the former (Cazalis et al., 1987*a*). These results would be expected if osmotic swelling of the NSG is a prerequisite for exocytosis, but osmotic stress could also be indirectly responsible for such inhibition (Finkelstein et al., 1986). (*c*) As previously mentioned, SITS not only blocks the NSG anion channel, it also inhibits neuropeptide release from this preparation. (*d*) Finally, the Ca^{2+}-dependent release of neuropeptides can occur in the absence of Na^+, K^+, and Mg^{2+}, but requires Cl^- and the presence of sucrose. In other words, it would seem that in the permeabilized nerve endings, calcium is the only cation necessary for inducing an increase in the amount of hormone secreted (Nordmann et al., 1987). Since sucrose can pass through gap junctions and contribute to increases in osmotic pressure, the model proposed here would be compatible with even this finding.

In conclusion, the characterization of two ionic channels in peptide-hormone–containing vesicles opens up new possibilities for molecular mechanisms underlying depolarization-secretion coupling in nerve terminals. Determination of any physiological role, however, for the NSG cation and anion channels will require further study.

Acknowledgments

We wish to acknowledge the excellent technical assistance of B. Muffly and S. Gatti-Favereau.

This work was supported by grants from the National Science Foundation and the National Institutes of Health to J. R. Lemos and from North Atlantic Treaty Organization to J. J. Nordmann and J. R. Lemos.

References

Augustine, G. J., M. P. Charlton, and S. J. Smith. 1987. Calcium action in synaptic transmitter release. *Annual Review of Neuroscience*. 10:633–693.

Breckenridge, L. J., and W. Almers. 1987b. Currents through the fusion pore that forms during exocytosis of a secretory vesicle. *Nature*. 328:814–817.

Breckenridge, L. J., and W. Almers. 1987*a*. Final steps in exocytosis observed in a cell with giant secretory granules. *Proceedings of the National Academy of Sciences*. 84:1945–1949.

Brethes, D., G. Dayanithi, L. Letellier, and J. J. Nordmann. 1987. Depolarization induced Ca^{2+} increase in isolated neurosecretory nerve terminals measured with fura-2. *Proceedings of the National Academy of Sciences*. 84:1439–1443.

Cazalis, M., G. Dayanithi, and J. J. Nordmann. 1987*a*. Hormone release from isolated nerve endings of the rat neurohypophysis. *Journal of Physiology*. 390:55–70.

Cazalis, M., G. Dayanithi, and J. J. Nordmann. 1987*b*. Requirements for hormone release from permeabilized nerve endings isolated from the rat neurohypophysis. *Journal of Physiology*. 390:71–91.

Cohen, F., M. Akabas, and A. Finkelstein. 1982. Osmotic swelling of phospholipid vesicles causes them to fuse with planar phospholipid bilayer membrane. *Science*. 217:458–460.

Coronado, R., and R. Latorre. 1983. Phospholipid bilayers made from monolayers in patch-clamp pipettes. *Biophysical Journal*. 13:231–236.

Dayanithi, G., and J. J. Nordmann. 1989. Release of neuropeptides does not only occur at nerve terminals. *Bioscience Reports*. 8:471–483.

DeRiemer, S. A., R. Martin, R. Rahamimoff, B. Sakmann, and H. Stadler. 1988. Use of fused synaptosomes or synaptic vesicles to study ion channels involved in neurotransmission. *In* Calcium and Ion Channel Modulation. Plenum Publishing Corp., New York. 407–414.

Douglas, W. W., and R. P. Rubin. 1963. The mechanism of catecholamine release from the adrenal medulla and the role of calcium in stimulus-secretion coupling. *Journal of Physiology*. 167:288–310.

Douglas, W. W., and A. M. Poisner. 1964*a*. Stimulus-secretion coupling in a neurosecretory organ: the role of calcium in the release of vasopressin from the neurohypophysis. *Journal of Physiology*. 172:1–18.

Douglas, W. W., and A. M. Poisner. 1964*b*. Calcium movement in the neurohypophysis of the rat and its relation to the release of vasopressin. *Journal of Physiology*. 172:19–30.

Finkelstein, A., J. Zimmerberg, and F. Cohen. 1986. Osmotic swelling of vesicles. *Annual Review of Physiology*. 48:163–174.

Hamill, O. P., A. Marty, E. Neher, B. Sakmann, and F. Sigworth. 1981. Improved patch-clamp techniques for high resolution current recordings from cells and cell-free membrane patches. *Pflügers Archiv*. 391:85–100.

Kley, N., J. P. Loeffler, and V. Hollt. 1987. Ca^{2+}-dependent histaminergic regulation of proenkephalin mRNA levels in cultured adrenal chromaffin cells. *Neuroendocrinology*. 46:89–92.

Latorre, R., O. Alvarez, X. Cecchi, and C. Vergara. 1985. Properties of reconstituted ion channels. *Annual Review of Biophysics and Biophysical Chemistry*. 14:79–111.

Lemos, J. R., and J. J. Nordmann. 1986. Ionic channels and hormone release from peptidergic nerve terminals. *Journal of Experimental Biology*. 124:53–72.

Lemos, J. R., J. J. Nordmann, I. M. Cooke, and E. L. Stuenkel. 1986. Single channels and ionic currents in peptidergic nerve terminals. *Nature*. 319:410–412.

Lemos, J. R., and M. Nowycky. 1987. One type of Ca channel in nerve terminals from the rat neurohypophysis responds to dihydropyridines. *Society for Neuroscience Abstracts*. 13:793. (Abstr.)

Lemos, J. R., and M. Nowycky. 1989. Two types of calcium channnel coexist in peptide-releasing vertebrate nerve terminals. *Neuron*. In press.

Lemos, J. R., K. A. Ocorr, and J. J. Nordmann. 1987. Ionic channels in neurosecretory granules from rat neural lobe. *Biophysical Journal*. 51:64*a*. (Abstr.)

Llinas, R., T. L. McGuinness, C. S. Leonard, M. Sugimori, and P. Greengard. 1985. Intraterminal injection of synapsin I or calcium/calmodulin–dependent protein kinase II alters neurotransmitter release at the squid giant synapse. *Proceedings of the National Academy of Sciences*. 82:3035–3039.

Lowenstein, W. R. 1987. The cell-to-cell channel of gap junctions. *Cell*. 48:725–726.

Ma, J., M. Fill, C. M. Knudson, K. P. Campbell, and R. Coronado. 1988. Ryanodine receptor of skeletal muscle is a gap junction type channel. *Science*. 242:99–102.

Nordmann, J. J. 1983. Stimulus-secretion coupling. *Progress in Brain Research*. 60:281–303.

Nordmann, J. J., and M. Cazalis. 1986. Characterization of newly formed and aged granules in the neurohypophysis. *Journal of Neurochemistry*. 47:1534–1543.

Nordmann, J. J., G. Dayanithi, and J. R. Lemos. 1987. Isolated neurosecretory nerve endings as a tool for studying the mechanism of stimulus-secretion coupling. *Bioscience Reports*. 7:411–425.

Nordmann, J. J., and R. E. J. Dyball. 1978. Effects of veratridine on Ca fluxes and the release of oxytocin and vasopressin from the isolated rat neurohypophysis. *Journal of General Physiology*. 72:297–304.

Nordmann, J. J., F. Louis, and S. J. Morris. 1979. Purification of two structurally and morphologically distinct populations of rat neurohypophysial secretory granules. *Neuroscience*. 4:1367–1375.

Nowycky, M., and J. R. Lemos. 1988. Calcium channels in nerve terminals isolated from the rat neurohypophysis. *Biophysical Journal*. 53:232*a*. (Abstr.)

Perrin, D., O. K. Langley, and D. Aunis. 1987. Anti–α-fodrin inhibits secretion from permeabilized chromaffin cells. *Nature*. 321:498–501.

Pollard, H. B., C. J. Pazoles, C. E. Creutz, and O. Zinder. 1979. The chromaffin granule and possible mechanisms of exocytosis. *International Review of Cytology*. 58:159–197.

Rahamimoff, R., S. A. DeRiemer, B. Sakmann, H. Stadler, and N. Yakir. 1988. Ion channels in synaptic vesicles from *Torpedo* electric organ. *Proceedings of the National Academy of Sciences*. 85:5310–5314.

Rand, R. P., and V. A. Parsegian. 1986. Mimicry and mechanism in phospholipid models of membrane fusion. *Annual Review of Physiology*. 48:201–212.

Salzberg, B. M., A. L. Obaid, D. M. Senseman, and H. Gainer. 1983. Optical recording of action potentials from vertebrate nerve terminals using potentiomemtric probes provides evidence for sodium and calcium components. *Nature*. 306:36–40.

Scherman, D., J. J. Nordmann, and J. P. Henry. 1982. Evidence for a an ATP-dependent proton translocase in neurosecretory granule membranes. *Biochemistry*. 21:687–694.

Spray, D. C., J. H. Stern, A. L. Harris, and M. V. L. Bennett. 1982. Gap junctional conductance: comparison of sensitivities to H ions and Ca ions. *Proceedings of the National Academy of Sciences.* 79:441–445.

Spray, D. C., J. C. Saez, D. Brosius, M. V. L. Bennett, and E. Hertzberg. 1986. Isolated liver gap junctions: gating of transjunctional currents is similar to that in intact pairs of rat hepatocytes. *Proceedings of the National Academy of Sciences.* 83:5494–5497.

Spray, D. C., R. L. White, F. Mazet, and M. V. L. Bennett. 1985. Regulation of gap junctional conductance. *American Journal of Physiology.* 248:H753–H764.

Stanley, E., and G. Ehrenstein. 1985. A model for exocytosis based on the opening of calcium-activated potassium channels in vesicles. *Life Sciences.* 37:1985–1995.

Stanley, E., G. Ehrenstein, and J. Russell. 1988. Evidence for anion channels in secretory vesicles. *Neuroscience.* 25:1035–1039.

Thieffry, M., J. F. Chich, D. Goldschmidt, and J. P. Henry. 1988. Incorporation in lipid bilayers of a large conductance cationic channel from mitochondrial membranes. *EMBO Journal.* 7:1449–1454.

Thorn, P., and J. R. Lemos. 1988. Ca-activated K channel in rat neurohypophysial nerve terminals. *Society for Neuroscience Abstracts.* 14:641. (Abstr.)

Whitaker, M., and J. Zimmerberg. 1987. Inhibition of secretory granule discharge during exocytosis in sea urchin eggs by polymer solutions. *Journal of Physiology.* 389:527–539.

Woodbury, D. J. 1989. Pure lipid vesicles can induce channel-like conductances in planar bilayers. *Journal of Membrane Biology.* In press.

Woodbury, D. J., and J. E. Hall. 1988. The role of channels in the fusion of vesicles with a planar bilayer. *Biophysical Journal.* 54:1053–1063.

Zimmerberg, J., M. Curran, F. S. Cohen, and M. Brodwick. 1987. Simultaneous electrical and optical measurements show that membrane fusion precedes secretory granule swelling during exocytosis of beige mouse mast cells. *Proceedings of the National Academy of Sciences.* 84:1585–1589.

List of Contributors

S. Akerman, Department of Neuroendocrinology, Agricultural and Food Research Council Institute of Animal Physiology and Genetics Research, Babraham, Cambridge, United Kingdom

Qais Al-Awqati, Department of Medicine, Columbia University, College of Physicians and Surgeons, New York, New York

Wolfhard Almers, Department of Physiology and Biophysics, School of Medicine, University of Washington, Seattle, Washington

George J. Augustine, Department of Biological Sciences, Section of Neurobiology, University of Southern California, Los Angeles, California

M. Berridge, Department of Zoology, Agricultural and Food Research Council Invertebrate Chemistry and Zoology Unit, University of Cambridge, Cambridge, United Kingdom

Lutz Birnbaumer, Departments of Physiology and Molecular Biophysics, Cell Biology, Baylor College of Medicine, Houston, Texas

Jonathan Brasch, Department of Medicine, Columbia University, College of Physicians and Surgeons, New York, New York

L. Breckenridge, Department of Physiology and Biophysics, School of Medicine, University of Washington, Seattle, Washington

Catherine Brion, Department of Physiology-Anatomy, University of California, Berkeley, California

Arthur M. Brown, Department of Physiology and Molecular Biophysics, Baylor College of Medicine, Houston, Texas

Edward Brown, Department of Medicine, Endocrine-Hypertension Unit, Brigham and Women's Hospital, Boston, Massachusetts

JoAnn Buchanan, Howard Hughes Medical Institute, Section of Molecular Neurobiology, Yale University Medical School, New Haven, Connecticut

Milton P. Charlton, Department of Physiology, University of Toronto, Toronto, Ontario, Canada

T. Cheek, Department of Zoology, Agricultural and Food Research Council Invertebrate Chemistry and Zoology Unit, University of Cambridge, Cambridge, United Kingdom

Chu Chen, Endocrinology Division, The Children's Hospital, Boston, Massachusetts

Koong-Nah Chung, Department of Physiology-Anatomy, University of California, Berkeley, California

David E. Clapham, Department of Pharmacology and Physiology, Mayo Foundation, Rochester, Minnesota

Gabriel Cota, Departments of Physiology, Biophysics, and Neurosciences, Centro de Investigacion y de Estudios Avanzados del IPN, Mexico City, DF, Mexico

Kathleen Dunlap, Department of Physiology, Tufts Medical School, Boston, Massachusetts

Leslie Einhorn, The Neurobiology Program, University of North Carolina–Chapel Hill, Chapel Hill, North Carolina

Chada El-Hajj, Department of Medicine, Endocrine-Hypertension Unit, Brigham and Women's Hospital, Boston, Massachusetts

Marvin C. Gershengorn, Department of Medicine, Division of Endocrinology and Metabolism, Cornell University Medical College and The New York Hospital, New York, New York

Benjamin S. Glick, Department of Molecular Biology, Princeton University, Princeton, New Jersey

Rolf Graf, Department of Cell Biology, College of Medicine, Houston, Texas

Karen A. Gregerson, Department of Pediatrics, University of Maryland, Baltimore, Maryland

P. B. Guthrie, Department of Anatomy and Neurobiology, Program in Neuronal Growth and Development, Colorado State University, Fort Collins, Colorado

Jurgen Hescheler, II. Physiologisches Institut, Universität des Saarlandes, Homburg, Federal Republic of Germany

Marcia Hiriart, Department of Neurosciences, Instituto de Fisiologia Celular, Mexico City, DF, Mexico

George G. Holz, Department of Physiology, Tufts Medical School, Boston, Massachusetts

R. Horn, Roche Institute for Molecular Biology, Nutley, New Jersey

Leonard K. Kaczmarek, Departments of Pharmacology and Physiology, Yale University School of Medicine, New Haven, Connecticut

S. B. Kater, Department of Anatomy and Neurobiology, Program in Neuronal Growth and Development, Colorado State University, Fort Collins, Colorado

Olga Kifor, Department of Medicine, Endocrine-Hypertension Unit, Brigham and Women's Hospital, Boston, Massachusetts

Donghee Kim, Department of Pharmacology, Mayo Foundation, Rochester, Minnesota

Donald Landry, Department of Medicine, Columbia University, College of Physicians and Surgeons, New York, New York

Meryl LeBoff, Department of Medicine, Endocrine-Hypertension Unit, Brigham and Women's Hospital, Boston, Massachusetts

Leo Lehmicke, Department of Physiology-Anatomy, University of California, Berkeley, California

Jose R. Lemos, Worcester Foundation for Experimental Biology, Shrewsbury, Massachusetts

Clark A. Lindgren, Department of Neurobiology, Duke University Medical Center, Durham, North Carolina

Michael Lisanti, Department of Cell Biology and Anatomy, Cornell University Medical Center, New York, New York

Karen J. Loechner, Department of Pharmacology, Yale University School of Medicine, New Haven, Connecticut

Vivek Malhotra, Department of Molecular Biology, Princeton University, Princeton, New Jersey

Alain Marty, Laboratoire de Neurophysiologie, Ecole Normale Supérieure, Paris, France

William T. Mason, Department of Neuroendocrinology, Agricultural and Food Research Council Institute of Animal Physiology and Genetics Research, Babraham, Cambridge, United Kingdom

Rafael Mattera, Department of Cell Biology and Medicine, Baylor College of Medicine, Houston, Texas

Juan Medina, Department of Cell Biology, Baylor College of Medicine, Houston, Texas

Paul Melançon, Department of Molecular Biology, Princeton University, Princeton, New Jersey

Hsiao-Ping H. Moore, Department of Physiology-Anatomy, University of California, Berkeley, California

John W. Moore, Department of Neurobiology, Duke University Medical Center, Durham, North Carolina

R. Moreton, Department of Zoology, Agricultural and Food Research Council Invertebrate Chemistry and Zoology Unit, University of Cambridge, Cambridge, United Kingdom

Erwin Neher, Max-Planck-Institut für biophysikalische Chemie, Göttingen, Federal Republic of Germany

Jean J. Nordmann, Centre de Neurochimie, Strasbourg, France

Ana Lia Obaid, Department of Physiology, School of Medicine, University of Pennsylvania, Philadelphia, Pennsylvania

Karen A. Ocorr, Department of Pharmacology, Stanford, University School of Medicine, Stanford, California

Marguerite H. Oetting, Department of Medicine, Endocrine-Hypertension Unit, Brigham and Women's Hospital, Boston, Massachusetts

Lelio Orci, Institute of Histology and Embryology, University of Geneva Medical School, Geneva, Switzerland

Luis R. Osses, Department of Biological Sciences, Section of Neurobiology, University of Southern California, Los Angeles, California

Gerry S. Oxford, Department of Physiology, University of North Carolina–Chapel Hill, Chapel Hill, North Carolina

Reinhold Penner, Max-Planck-Institut für biophysikalische Chemie, Göttingen, Federal Republic of Germany

David Quinn, Department of Physiology-Anatomy, University of California, Berkeley, California

S. Rawlings, Department of Zoology, Agricultural and Food Research Council Invertebrate Chemistry and Zoology Unit, University of Cambridge, Cambridge, United Kingdom

Rodolfo Rivas, Department of Physiology-Anatomy, University of California, Berkeley, California

E. J. Rodriguez-Boulan, Department of Cell Biology and Anatomy, Cornell University Medical Center, New York, New York

James E. Rothman, Department of Molecular Biology, Princeton University, Princeton, New Jersey

Brian Salzberg, Department of Physiology, School of Medicine, University of Pennsylvania, Philadelphia, Pennsylvania

Klaus Schröter, Abteilung Zellphysiologie, Max-Planck-Institut für biophysikalische Chemie, Göttingen, Federal Republic of Germany

Tito Serafini, Department of Molecular Biology, Princeton University, Princeton, New Jersey

S. K. Sikdar, Department of Neuroendocrinology, Agricultural and Food Research Council Institute of Animal Physiology and Genetics Research, Babraham, Cambridge, United Kingdom

Michele M. Smith, Department of Physiology, University of North Carolina, Chapel Hill, North Carolina

Stephen J. Smith, Section of Molecular Neurobiology, Howard Hughes Medical Institute, Yale University Medical School, New Haven, Connecticut

A. Spruce, Department of Physiology and Biophysics, School of Medicine, University of Washington, Seattle, Washington

Y. P. Tan, Laboratoire de Neurophysiologie, Ecole Normale Supérieure, Paris, France

Gerhard Trube, Pharmaceutical Research Department, F. Hoffmann-La Roche & Co., Ltd., Basle, Switzerland

Anthony M. J. VanDongen, Departments of Physiology and Molecular Biophysics, Baylor College of Medicine, Houston, Texas

Peggy J. Weidman, Department of Molecular Biology, Princeton University, Princeton, New Jersey

Atsuko Yatani, Department of Cell Biology, Baylor College of Medicine, Houston, Texas

J. Zimmerberg, Laboratoire de Neurophysiologie, Ecole Normale Supérieure, Paris, France

R. Zorec, Department of Neuroendocrinology, Agricultural and Food Research Council Institute of Animal Physiology and Genetics Research, Babraham, Cambridge, United Kingdom

Subject Index